Selected Numerical Constants

Fundamental Constants (n, k = positive integers)

Designation*	Symbol	Representation	Section
Archimedes number	π	$4\sum_{k=1}^{\infty}\left[\frac{(-1)^{k-1}}{2k-1}\left(\frac{4}{5^{2k-1}}-\frac{1}{239^{2k-1}}\right)\right]$	A.15
Base of nat. logarithms	e	$2\left(\frac{1}{1!}+\frac{2}{3!}+\frac{3}{5!}+\cdots\right)$	A.15
Bernoulli number	B_n	$2\frac{(2n)!}{(2\pi)^{2n}}\left(\frac{1}{1^{2n}}+\frac{1}{2^{2n}}+\frac{1}{2^{3n}}+\cdots\right)$	A.03
Binomial coefficient	$\binom{n}{k}$	$\dfrac{n(n-1)(n-2)(n-3)\cdots(n-k+1)}{k(k-1)(k-2)\cdots3\cdot2\cdot1}$	18.09
Catalan constant	G	$\frac{1}{1^2}-\frac{1}{3^2}+\frac{1}{5^2}-\cdots=\bar{Z}(2)$	A.06
Double factorial	$(2n)!!$	$2n(2n-2)(2n-4)\cdots6\cdot4\cdot2$	A.01
Double factorial	$(2n-1)!!$	$(2n-1)(2n-3)(2n-5)\cdots5\cdot3\cdot1$	A.01
Euler constant	C	$\lim_{n\to\infty}\left(\frac{1}{1}+\frac{1}{2}+\frac{1}{3}+\cdots+\frac{1}{n}-\ln n\right)$	A.15
Euler number	E_n	$2\left(\frac{2}{\pi}\right)^{2n+1}(2n)!\left(\frac{1}{1^{2n+1}}-\frac{1}{3^{2n+1}}+\frac{1}{5^{2n+1}}-\cdots\right)$	A.05
Factorial	$n!$	$n(n-1)(n-2)(n-3)\cdots3\cdot2\cdot1$	A.01
Zeta function	$Z(n)$	$\frac{1}{1^n}+\frac{1}{2^n}+\frac{1}{3^n}+\cdots$	A.06
Zeta function	$\bar{Z}(n)$	$\frac{1}{1^n}-\frac{1}{3^n}+\frac{1}{5^n}-\cdots$	A.06

Numerical Values

π	3.14159	26535	89793	23846	(+00)	e	2.71828	18284	59045	23536	(+00)
π^2	9.86960	44010	80358	61881	(+00)	e^2	7.38905	60989	30650	22723	(+00)
π^3	3.10062	76680	29982	01755	(+01)	e^3	2.00855	36923	18766	77409	(+01)
π^e	2.24591	57718	36104	54734	(+01)	e^π	2.31406	92632	77926	90057	(+01)
$\sqrt{\pi}$	1.77245	38509	05516	02730	(+00)	$\sqrt{e}$	1.64872	12707	00128	14883	(+00)
$\sqrt[3]{\pi}$	1.46459	18875	61523	26302	(+00)	$\sqrt[3]{e}$	1.39561	24250	86089	52862	(+00)
$1/\pi$	3.18309	88618	37906	71538	(−01)	$1/e$	3.67879	44117	14423	21596	(−01)
$1/\pi^e$	4.45252	67226	92290	61514	(−02)	$1/e^\pi$	4.32139	18263	77224	97742	(−02)
$\ln\pi$	1.14472	98858	49400	17414	(+00)	$\ln 10$	2.30258	50929	94045	68402	(+00)
$\log\pi$	4.97149	87269	41338	54351	(−01)	$\log e$	4.34294	48190	32518	27651	(−01)
C	5.77215	66490	15328	60607	(−01)	G	1.64493	69031	59594	28540	(+00)

*For additional information see Section identified in the last column.

Engineering Mathematics Handbook

ENGINEERING MATHEMATICS HANDBOOK

DEFINITIONS • THEOREMS • FORMULAS • TABLES

Second, Enlarged and Revised Edition

Jan J. Tuma, Ph.D.

Professor of Engineering
Arizona State University

McGraw-Hill Book Company

New York   St. Louis   San Francisco   Auckland   Bogotá
Düsseldorf   Johannesburg   London   Madrid
Mexico   Montreal   New Delhi   Panama
Paris   São Paulo   Singapore
Sydney   Tokyo   Toronto

Library of Congress Cataloging in Publication Data

Tuma, Jan J
    Engineering mathematics handbook.

    Includes index.
    1. Engineering mathematics — Handbooks, manuals, etc.
I. Title.
TA332.T85   1978   510'.21'2   77-17786
ISBN 0-07-065429-8

234567890   HDHD   7865432109

The editors for this book were Harold B. Crawford and Ruth L. Weine
and the production supervisor
was Thomas G. Kowalczyk. It was set in Baskerville
by The Clarinda Company.

Printed and bound by Halliday Lithograph.

To my mother Antonie,
and my wife Hana

Contents

Preface
to the Second Edition

The excellent reception, wide use, and continuous demand for additional reprintings of the first edition of this handbook resulted in the publisher's requesting the preparation of a revised and enlarged second edition.

In the design of this new edition, an effort was made to preserve *the telescopic form of presentation* which was found so useful by many users, to improve the material of the first edition, to include new material requested by the users, to make the book readily applicable to *the electronic pocket calculator operations*, and, yet, to keep the volume size in reasonable bounds for easy desk-top handling.

Also, it was assumed that the user of this second edition possesses a ten-digit display, four-register stack electronic pocket calculator with the standard complement of elementary functions, known as *the electronic slide rule*. In reference to this calculator the following abbreviations are used in this preface:

MNM = Micronumerical model	PAA = Phase angle application
NS = Nested sum	PCA = Pocket calculator application
SAS = Standard algebraic sum	PCS = Pocket calculator storage

The most common MNM is NS, illustrated by the following example:

$$\sum_{k=1}^{5} kx^{2k} = x^2 + 2x^4 + 3x^6 + 4x^8 + 5x^{10} = \text{SAS}$$

$$= S(1 + \tfrac{2}{1}S(1 + \tfrac{3}{2}S(1 + \tfrac{4}{3}S(1 + \tfrac{5}{4}S))))$$

$$= S \bigwedge_{k=1}^{4} \left[1 + \frac{k+1}{k}S \right] = S(\text{NS})$$

where $S = x^2$ is PCS and NS must be evaluated by proceeding from the right end of the nested series (Sec. 8.11).

The major changes and additions in this new edition appear in the following chapters:

Chapter 3, Trigonometry, includes revised and new tables of MNMs for plane and spherical triangles and quadrilaterals with PCS indicated.

Chapter 8, Sequences and Series, is completely reconstructed with new tables of series of constant terms, Bernoulli and Euler numbers and polynomials, series of powers of integers, definitions of and operations with NS, applications of NS to binomial series, and a comparative representation of elementary functions in SAS and NS forms.

Chapter 10, Vector Analysis, includes revised tables of basic operations and new tables of vectors in curvilinear, cylindrical, and spherical coordinates and their matrix transformations with PCS indicated.

Chapter 12, Fourier Series, displays expanded tables of rectangular, triangular, trapezoidal, curvilinear, and singular periodic functions for a direct PCA.

Chapter 14, Ordinary Differential Equations, contains enlarged tables of second-, third-, and fourth-order differential equations with an expanded catalog of particular solutions and their respective MNMs, with u and v indicated as PCS. Also included are NS forms of Legendre, Chebyshev, Laguerre, and Hermite functions and polynomials with their PCS indicated.

Chapter 15, Partial Differential Equations, includes new tables of first-order equations, wave equations, diffusion equations, and the solutions of vibration equations by orthogonal series.

Chapter 16, Laplace Transforms, displays new tables of shape functions and response functions of second- and fourth-order differential equations with PCS indicated and illustrated by graphs of their respective input functions.

Chapter 19, Tables of Indefinite Integrals, which initially displayed 720 cases, includes now 1,100 cases. Since the solutions of many indefinite integrals yield recurrent relations, these results are expressed in NS for a direct PCA. The equivalents given at the top of tables can be used efficiently as PCS. Finally, the solutions of many complicated trigonometric integrals are reduced by PAA to simple trigonometric formulas in which the phase angle is PCS.

Chapter 20, Tables of Definite Integrals, displays 400 cases expressed in terms of equivalents given in notation blocks at the top of each table. As in Chapter 19, these equivalents can be used efficiently as PCS. All definite integrals are tabulated according to the limits of integration and the principal function of the integrand, which greatly simplifies the location of the desired case.

Appendix A, Numerical Tables, is reduced in size since many numerical values can be directly computed by PCA. The new additions include tables of numerical values which cannot be computed directly by PCA, such as tables of factorials (PCA is restricted to $0 \leq n \leq 66$), double factorials, factorials of fractions, tables of Bernoulli, Euler, and Stirling numbers, and tables of zeta functions.

In closing, the author expresses his gratitude to the many users who helped to improve the handbook by suggesting corrections and additions, and earnestly solicits comments and recommendations for improvements and future additions. Finally, but not least, gratitude is expressed to Mrs. Aileene Sparling who again typed the final draft of the manuscript and to my wife Hana and son Peter for their assistance in reading the proofs.

Jan J. Tuma

Preface
to the First Edition

This Handbook presents in one volume a concise summary of the major tools of engineering mathematics and was prepared with the intent to serve as a desk-top reference book for engineers, scientists, and architects.

The subject matter is divided into twenty chapters arranged in a logical sequence and corresponding to the material required in the undergraduate and graduate courses of the same name in modern science and engineering curricula. Consequently, this book may also be used by college teachers and students as a desk-top reference book.

The material is grouped into five parts each related to a particular type of mathematics.

The first part (Chapters 1–6) covers *algebra*, *plane* and *solid geometry*, *trigonometry*, *analytic geometry*, and *elementary functions* (circular and hyperbolic) with major emphasis on topics frequently occurring in the solution of physical problems.

The second part (Chapters 7–13) presents the *differential calculus*, *infinite series*, *integral calculus*, *vectors*, *complex variables*, *Fourier Series*, and *special functions*, and their applications encountered in the applied sciences and engineering analysis.

The third part (Chapters 14–16) is an extensive summary of special cases of *ordinary* and *partial differential equations*, and *related topics* (Bessel functions, orthogonal polynomials, Laplace transforms).

The fourth part (Chapters 17–19) gives a summary of

the terminology and major formulas of *numerical methods, probability, and statistics* and presents *related tables of numerical coefficients*.

The fifth part (Chapter 20) consists of over 720 cases of *indefinite integrals* and their solutions indexed in 80 tables.

Finally, *Tables of Numerical Values of the most important functions* are assembled in the Appendixes.

The form of presentation has many special features facilitating an easy and rapid location of the desired information.

1. Each page presents the information in a graphical arrangement pertinent to the specific type of material, designated by a title and section number. Consequently each page is a table, designed by the author as an information unit.
2. Left and right pages of the book present related or similar material; all functions and formulas are arranged in logical sequences, and placed in blocks.
3. If odd, even functions are involved, they are presented on even, odd pages or in odd, even columns, respectively.

The graphical arrangement offers the possibility of using this Handbook as a pictorial dictionary of engineering mathematics, a review outline for examinations, and a manual of comparative study.

The organization and preparation of the material presented in this book spans a period of over thirty years, during which the author has been assisted by many individuals and has relied on an extensive wealth of reference material forming a body of general knowledge known as engineering mathematics. The space limitation prevents the inclusion of a complete list of references, yet a great effort was made to credit those sources which were directly used.

Although it is not possible to mention all collaborators by name, the assistance of the following individuals and institutions is acknowledged in the sequence of their participation.

For the assistance in the development of the indexed tables of indefinite integrals acknowledgment is made to the staff of the Institute of Programmed Learning Tukarny in Prague and particularly to my first assistant Ing. Miloslav Klimes, Associate Director of the Institute. Their enthusiastic cooperation during the years of 1942–1947 made the completion of the integral tables possible.

For the calculation of tables of numerical coefficients included in this volume credit is given to the computing centers of the following Institutions: Oklahoma State University, Stillwater, Oklahoma; Swiss Federal Institute of Technology, Zurich, Switzerland; and University of Colorado, Boulder, Colorado.

Mr. Fredrick W. Humburg, Graduate Assistant in the computing center of the University of Colorado provided the final computer check of the appendix tables, and Mrs. Aileene Sparling of Keyboard Secretarial Services in Boulder typed the final draft of the manuscript.

Finally, but not least, gratitude is expressed to my wife Hana and son Peter for their patience, understanding, and continued interest during the preparation of the manuscript and for their assistance in reading the final proofs.

Jan J. Tuma

ENGINEERING
MATHEMATICS
HANDBOOK

1

ALGEBRA

(1) General

Algebra is a systematic investigation of general numbers and their relationships. *General numbers* (real and complex) are letter symbols $(A, B, C, \ldots, a, b, c, \ldots, \alpha, \beta, \gamma, \ldots)$ representing *constant or variable quantities*. The study involves a finite number of *binary operations* (additions, multiplications), governed by algebraic laws and rules of signs.

(2) Algebraic Laws

Commutative law:	Associative law:
$a + b = b + a$	$a + (b + c) = (a + b) + c$
$ab = ba$	$a(bc) = (ab)c$
Distributive law:	**Division law:**
$a(b + c) = ab + ac$	If $ab = 0$, then $a = 0$ or $b = 0$
$(a + b)c = ac + bc$	

(3) Rules of Signs

Summation	Multiplication	Division
$a + (+b) = a + b$	$(+a)(+b) = +ab$	$\dfrac{+a}{+b} = +\dfrac{a}{b}$
$a + (-b) = a - b$	$(+a)(-b) = -ab$	$\dfrac{+a}{-b} = -\dfrac{a}{b}$
$a - (+b) = a - b$	$(-a)(+b) = -ab$	$\dfrac{-a}{+b} = -\dfrac{a}{b}$
$a - (-b) = a + b$	$(-a)(-b) = +ab$	$\dfrac{-a}{-b} = +\dfrac{a}{b}$

(4) Powers

$a^k = \underbrace{aa \cdots a}_{k \text{ times}}$	$a^0 = 1$	$a^1 = a$

$$a^{-k} = \frac{1}{a^k} \qquad \frac{1}{a^{-k}} = a^k$$

$$a^m a^n = a^{m+n} \qquad (ab)^k = a^k b^k$$

$$\frac{a^m}{a^n} = a^{m-n} \qquad \left(\frac{a}{b}\right)^k = \frac{a^k}{b^k}$$

$$(a^m)^n = +a^{mn} \qquad (a^m)^{-n} = \frac{1}{a^{mn}}$$

$$(\pm a)^{2k} = +a^{2k} \qquad (\pm a)^{2k+1} = \pm a^{2k+1}$$

(5) Roots

$\sqrt[k]{a} = a^{1/k}$	$\sqrt[0]{a} = \infty*$	$\sqrt[1]{a} = a$

$$\sqrt[-k]{a} = \frac{1}{\sqrt[k]{a}} \qquad \frac{1}{\sqrt[-k]{a}} = \sqrt[k]{a}$$

$$\sqrt[m]{a}\sqrt[n]{a} = a^{(m+n)/mn} \qquad \sqrt[k]{ab} = \sqrt[k]{a}\sqrt[k]{b}$$

$$\frac{\sqrt[m]{a}}{\sqrt[n]{a}} = a^{(n-m)/mn} \qquad \sqrt[k]{a/b} = \frac{\sqrt[k]{a}}{\sqrt[k]{b}}$$

$$\sqrt[n]{a^m} = a^{m/n} \qquad \sqrt[-n]{a^m} = \frac{1}{a^{m/n}}$$

$$\sqrt[2k]{+a^{2k}} = +a \qquad \sqrt[2k+1]{+a^{2k+1}} = +a$$

*If $a > 1$.

(1) Definitions

If $a^x = b$, then x is the logarithm of b to the base a. Logarithms to the *base 10* are called *common*, or *Brigg's*, logarithms and denoted as $\log x$. Logarithms to the *base* $e = 2.718\ 281\ 828$ are called *natural*, or *Napier's*, logarithms and denoted as $\ln x$.

(2) Basic Formulas

$10^{\infty} = \infty$	$\log \infty = \infty$	$\ln \infty = \infty$	$e^{\infty} = \infty$
$10^2 = 100$	$\log 100 = 2$	$\ln e^2 = 2$	$e^2 = e^2$
$10^1 = 10$	$\log 10 = 1$	$\ln e = 1$	$e^1 = e$
$10^0 = 1$	$\log 1 = 0$	$\ln 1 = 0$	$e^0 = 1$
$10^{-1} = \frac{1}{10}$	$\log \frac{1}{10} = -1$	$\ln \dfrac{1}{e} = -1$	$e^{-1} = \dfrac{1}{e}$
$10^{-2} = \frac{1}{100}$	$\log \frac{1}{100} = -2$	$\ln \dfrac{1}{e^2} = -2$	$e^{-2} = \dfrac{1}{e^2}$
$10^{-\infty} = 0$	$\log 0 = -\infty$	$\ln 0 = -\infty$	$e^{-\infty} = 0$

(3) Transformations

$$\log x = \frac{\ln x}{\ln 10} = 0.434\ 294\ 482 \ldots \ln x \qquad\qquad \ln x = \frac{\log x}{\log e} = 2.302\ 585\ 093 \ldots \log x$$

(4) Basic Operations

$\log xy = \log x + \log y$	$\ln xy = \ln x + \ln y$
$\log \dfrac{x}{y} = \log x - \log y$	$\ln \dfrac{x}{y} = \ln x - \ln y$
$\log x^k = k \log x$	$\ln x^k = k \ln x$
$\log \sqrt[k]{x} = \dfrac{1}{k} \log x$	$\ln \sqrt[k]{x} = \dfrac{1}{k} \ln x$
$\log 10^k = k$	$\ln e^k = k$
$\log e = 0.434\ 29 \ldots$	$\ln 10 = 2.302\ 59 \ldots$

(5) Characteristic of Common Logarithm

A common logarithm consists of an integer, called the *characteristic*, and a decimal, called the *mantissa*. The mantissa is a tabulated value. The characteristic is determined from the following:

$\log 100 = 2.00$	$\log 0.1 = 0.00 - 1 = 9.00 - 10$
$\log 10 = 1.00$	$\log 0.01 = 0.00 - 2 = 8.00 - 10$
$\log 1 = 0.00$	$\log 0.001 = 0.00 - 3 = 7.00 - 10$

$k, n, p = 0, 1, 2, \ldots$ $\Gamma(\ ) = $ gamma function (Sec. 13.03)

$\bar{B}_k = $ Bernoulli number (Sec. A.03) $\mathscr{S}_k^{(p)} = $ Stirling number (Sec. A.05)

(1) n Factorial   (Sec. A.01)

$$n! = n(n-1)(n-2) \cdots (3)(2)(1) = \Gamma(n+1)$$
$$= n(n-1)! \qquad\qquad\qquad = n\Gamma(n)$$
$$= n(n-1)(n-2)! \qquad\qquad = n(n-1)\Gamma(n-1)$$

....

$0! = 1$ (by definition)

$1! = 1, \quad 2! = 2 \cdot 1 = 2, \quad 3! = 3 \cdot 2 \cdot 1 = 6, \ldots$

(2) Coefficients of τ

$2k-1$	a_{2k-1}
1	+0.422 784 335
3	−0.067 352 301
5	−0.007 385 551
7	−0.001 192 754
9	−0.000 223 155
11	−0.000 044 926
13	−0.000 009 439

(3) u Factorial $(-0.5 \leq u \leq 0.5)$ *

$$u! = \Gamma(u+1) = e^\tau \sqrt{\frac{(1-u)\,u\pi}{(1+u)\,\sin u\pi}} + \epsilon \qquad |\epsilon| < 5 \times 10^{-10}$$

$$\tau = \sum_{k=1}^{7} a_{2k-1} u^{2k-1} \qquad [\text{for } a_{2k-1} \text{ see (2)}]$$

(4) v Factorial $(1 < v \leq 4)$

$$v! = (n+u)! = (n+u)(n-1+u) \cdots (1+u)\,u!$$
$[\text{for } u! \text{ use (3) or (5)}]$

(5) Special Values $(|u| < 1)$

x	$x!$
$-u$	$\dfrac{\pi u}{u! \sin \pi u}$
-0.5	$\sqrt{\pi}$
$+0.5$	$0.5\sqrt{\pi}$
$+u$	$\dfrac{\pi u(1-u)}{(1-u)! \sin \pi u}$

(6) N Factorial $(4 < N < \infty)$ †

$$N! = \Gamma(N+1) = e^\lambda \left(\frac{N}{e}\right)^N \sqrt{2\pi N} + \epsilon \qquad |\epsilon| < 5 \times 10^{-10}$$

$$\lambda = \frac{\bar{B}_2}{1 \cdot 2} N^{-1} + \frac{\bar{B}_4}{3 \cdot 4} N^{-3} + \frac{\bar{B}_6}{5 \cdot 6} N^{-5} + \frac{\bar{B}_8}{7 \cdot 8} N^{-7} = \frac{N^{-1}}{12} - \frac{N^{-3}}{360} + \frac{N^{-5}}{1,260} - \frac{N^{-7}}{1,680}$$

(7) Double Factorials   (Sec. A.01)

$$(2n)!! = 2n(2n-1)(2n-4) \cdots 6 \cdot 4 \cdot 2 = 2^n n! = 2^n \Gamma(n+1)$$

$$(2n-1)!! = (2n-1)(2n-3)(2n-5) \cdots 5 \cdot 3 \cdot 1 = 2^n \frac{\Gamma(n+\frac{1}{2})}{\sqrt{\pi}}$$

$0!! = 1$

$1!! = 1$

(8) Factorial Polynomials $(x, h = \text{real numbers}, p \neq 1)$

$$X_1^{(p)} = \prod_{k=0}^{p-1} (x-k) = x(x-1)(x-2) \cdots (x-p+1) = \frac{\Gamma(x+1)}{\Gamma(x-p+1)}$$

$$= \sum_{k=1}^{p} \mathscr{S}_k^{(p)} x^k = \mathscr{S}_1^{(p)} x + \mathscr{S}_2^{(p)} x^2 + \mathscr{S}_3^{(p)} x^3 + \cdots + \mathscr{S}_p^{(p)} x^p$$

$P_1^{(p)} = p!$

$$X_h^{(p)} = \prod_{k=0}^{p-1} (x-kh) = x(x-h)(x-2h) \cdots (x-ph+h) = \frac{h^p \Gamma\left(\frac{x}{h}+1\right)}{\Gamma\left(\frac{x}{h}-p+1\right)}$$

$$= \sum_{k=1}^{p} \mathscr{S}_k^{(p)} h^{p-k} x^k = \mathscr{S}_1^{(p)} h^{p-1} x + \mathscr{S}_2^{(p)} h^{p-2} x^2 + \mathscr{S}_3^{(p)} h^{p-3} x^3 + \cdots + \mathscr{S}_p^{(p)} x^p$$

$P_2^{(p)} = p!!$

*For $-1 < u < -\frac{1}{2}$ and $\frac{1}{2} < u < 1$ use (3) and (5); see also Sec. A.02.

†In nested form: $\lambda = \frac{NS}{12}\left(1 - \frac{S}{30}\left(1 - \frac{2S}{7}\left(1 - \frac{3S}{4}\right)\right)\right)$ with $S = N^{-2}$ used as pocket calculator storage.

$k, n, r = 0, 1, 2, \ldots$ $a, b =$ real numbers

$n! = n$ factorial (Sec. 1.03) $\Lambda[\ ] =$ nested sum (Sec. 8.11)

(1) Binomial Theorem $(b/a = S =$ pocket calculator storage)

$$(a+b)^n = a^n + \binom{n}{1}a^{n-1}b + \binom{n}{2}a^{n-2}b^2 + \binom{n}{3}a^{n-3}b^3 + \cdots + b^n = \sum_{k=0}^{n}\binom{n}{k}a^{n-k}b^k$$

$$= a^n\left(1 + \frac{n}{1}S\left(1 + \frac{n-1}{2}S\left(1 + \frac{n-2}{3}S\left(1 + \cdots + \frac{1}{n}S\right)\right)\right)\right) = a^n \overset{n-1}{\underset{k=0}{\Lambda}}\left[1 + \frac{n-k}{1+k}S\right]$$

(2) Binomial Coefficients   (Sec. 18.09)

$$\binom{n}{k} = \frac{n!}{(n-k)!\,k!} \qquad\qquad \binom{-n}{k} = (-1)^k\binom{n+k-1}{k}$$

$$= \frac{n(n-1)(n-2)\cdots(n-k+1)}{k!} \qquad\qquad = \frac{(-n)(-n-1)\cdots(-n-k+1)}{k!}$$

$\binom{n}{0} = 1$	$\binom{n}{1} = n$	$\binom{n}{k} = \binom{n}{n-k}$	$\binom{2n}{n} = \frac{(2n)!}{(n!)^2} = (-1)^n\binom{-n-1}{n}$
$\binom{n}{n} = 1$	$\binom{n}{n-1} = n$	$\binom{n}{k+1} = \frac{n-k}{k+1}\binom{n}{k}$	$\binom{2n-1}{n} = \frac{n(2n-1)!}{(n!)^2} = (-1)^n\binom{-n}{n}$
$\dfrac{\binom{n}{k}}{\binom{n}{k-1}} = \frac{n+1-k}{k}$	$\dfrac{\binom{n}{k+1}}{\binom{n}{k}} = \frac{n-k}{k+1}$	$\binom{n+1}{k} = \binom{n}{k} + \binom{n}{k-1}$	$\binom{n+1}{k+1} = \binom{n}{k} + \binom{n}{k+1}$

(3) Pascal's Triangle of $\binom{n}{k}$

$(a+b)^0 \rightarrow$ 1

$(a+b)^1 \rightarrow$ 1   1

$(a+b)^2 \rightarrow$ 1   2   1

$(a+b)^3 \rightarrow$ 1   3   3   1

$(a+b)^4 \rightarrow$ 1   4   6   4   1

(4) Binomial Sums   (Sec. 8.03)

$$\binom{n}{k} + \binom{n-1}{k} + \binom{n-2}{k} + \cdots + \binom{k}{k} = \binom{n+1}{k+1}$$

$$\binom{n}{0} + \binom{n}{1} + \binom{n}{2} + \cdots + \binom{n}{n} = 2^n$$

$$\binom{n}{0} - \binom{n}{1} + \binom{n}{2} - \cdots + (-1)^n\binom{n}{n} = 0$$

(5) Sums of Incomplete Binomial Sequence $(r \le n)$

$$\sum_{k=0}^{r}\binom{n}{k} = \overset{r-1}{\underset{k=0}{\Lambda}}\left[1 + \frac{n-k}{1+k}\right] = \left(1 + n\left(1 + \frac{n-1}{2}\left(1 + \frac{n-2}{3}\left(1 + \cdots + \frac{n+1-r}{r}\right)\right)\right)\right)$$

$$\sum_{k=0}^{r}(-1)^k\binom{n}{k} = \overset{r-1}{\underset{k=0}{\Lambda}}\left[1 - \frac{n-k}{1+k}\right] = \left(1 - n\left(1 - \frac{n-1}{2}\left(1 - \frac{n-2}{3}\left(1 - \cdots - \frac{n+1-r}{r}\right)\right)\right)\right)$$

(6) Special Cases   (Sec. 8.12)

$(1 \pm x)^n = 1 \pm \binom{n}{1}x + \binom{n}{2}x^2 \pm \cdots$ If $n =$ positive integer, the series is finite.

$\left(1 \pm \dfrac{1}{n}\right)^n = 1 \pm \binom{n}{1}n^{-1} + \binom{n}{2}n^{-2} \pm \cdots$ If $n =$ negative integer or fraction, the series is infinite.

If $n \rightarrow \infty$, then $\left(1 + \dfrac{1}{n}\right)^n = 1 + \dfrac{1}{1!} + \dfrac{1}{2!} + \dfrac{1}{3!} + \cdots = e = 2.718\ 281\ 828 =$ Euler's number.

(1) Integral Function

(a) Every integral rational algebraic function of independent variable x,

$$f(x) = a_0 + a_1 x + a_2 x^2 + \cdots + a_n x^n = \sum_{k=0}^{n} a_k x^k$$

can be expressed as a product of real linear factors of the form $c + dx$ and of real irreducible quadratic factors of the form $e + fx + gx^2$.

(b) If two of these functions of the same degree are equal for all values of x, the constant a's of like powers are equal.

$$\sum_{k=0}^{n} a_k x^k = \sum_{k=0}^{n} b_k x^k \qquad\qquad a_0 = b_0, a_1 = b_1, \ldots, a_n = b_n$$

(2) Fractional Function

(a) A rational algebraic function of independent variable x, where $g(x)$ and $h(x)$ are integral rational algebraic functions, is called a *rational algebraic fraction*.

$$f(x) = \frac{g(x)}{h(x)}$$

(b) If the degree of $g(x)$ is less than the degree of $h(x)$, $f(x)$ is called *proper*. If the opposite is true, $f(x)$ is called *improper*.

(c) Every proper, rational algebraic fraction can be resolved into a sum of simpler fractions, whose denominators are of the form $(c + dx)^k$ and $(e + fx + gx^2)^l$, k and l being positive integers.

$$\frac{g(x)}{h(x)} = \sum_{k} \left[\frac{b_{k1}}{(x - x_k)} + \frac{b_{k2}}{(x - x_k)^2} + \cdots + \frac{b_{km}}{(x - x_k)^m} \right]$$

(3) Coefficients

The coefficients b_{kj} are obtained by one of the following methods:

(a) If $m = 1$ ($x_k =$ distinct root), then

$$b_{kj} = \frac{g(x_k)}{f'(x_k)}$$

in which $f'(x_k)$ is the first derivative of $f(x)$ with respect to x evaluated for $x = x_k$.

(b) Multiply both sides of $h(x)$, and use the theorem of Sec. 1.04-1b.

(c) Multiply both sides by $h(x)$, and differentiate successively. Solve this set of equations for $b_{km}, b_{km-1}, \ldots, b_{k1}$.

The partial fractions corresponding to any pair of complex-conjugate roots $a_k + i\alpha_k$, $a_k - i\alpha_k$ of order m may be combined into

$$c_{kj} \frac{x + d_{kj}}{[(x - a_k)^2 + \alpha_k^2]^j}$$

(1) Quadratic and Biquadratic Equations[1]

$$ax^2 + bx + c = 0 \qquad a \neq 0$$

$$x_{1,2} = \frac{-b \pm \sqrt{b^2 - 4ac}}{2a}$$

If a, b, c are real and if

$b^2 - 4ac > 0$ the roots are real and unequal;

$b^2 - 4ac = 0$ the roots are real and equal;

$b^2 - 4ac < 0$ the roots are complex conjugate.

$$ax^4 + bx^2 + c = 0 \qquad a \neq 0$$

This reduces by the substitution

$$y = x^2 \qquad \text{to} \qquad y^2 + py + q = 0$$

in which

$$p = b/a \qquad q = c/a$$

$$x_{1,2,3,4} = \pm \sqrt{-\frac{p}{2} \pm \sqrt{\frac{p^2}{4} - q}}$$

(2) Binomial Equations[1]

$$x^n - a = 0 \qquad a \neq 0$$

$$x_{1,2,3,\ldots,n} = \sqrt[n]{a}\left(\cos\frac{2k\pi}{n} + i\sin\frac{2k\pi}{n}\right)$$

$$\sqrt[n]{i} = \cos\frac{(4k+1)\,\pi}{2n} + i\sin\frac{(4k+1)\,\pi}{2n}$$

$$k = 0, 1, 2, \ldots, n-1$$

$$x^n + a = 0 \qquad a \neq 0$$

$$x_{1,2,3,\ldots,n} = \sqrt[n]{a}\left[\cos\frac{(2k+1)\,\pi}{n} + i\sin\frac{(2k+1)\,\pi}{n}\right]$$

$$\sqrt[n]{-i} = \cos\frac{(4k-1)\,\pi}{2n} + i\sin\frac{(4k-1)\,\pi}{2n}$$

$$k = 0, 1, 2, \ldots, n-1$$

(3) Cubic Equations[1]

$$ax^3 + bx^2 + cx + d = 0 \qquad a \neq 0$$

This reduces by the substitution

$$x = y - \frac{b}{3a} \qquad \text{to} \qquad y^3 + py + q = 0$$

where

$$p = \frac{1}{3}\left[3\left(\frac{c}{a}\right) - \left(\frac{b}{a}\right)^2\right]$$

$$q = \frac{1}{27}\left[2\left(\frac{b}{a}\right)^3 - 9\left(\frac{b}{a}\right)\left(\frac{c}{a}\right) + 27\left(\frac{d}{a}\right)\right]$$

$$y_1 = u + v \qquad y_2 = -\frac{u+v}{2} + \frac{u-v}{2}i\sqrt{3}$$

$$y_3 = -\frac{u+v}{2} - \frac{u-v}{2}i\sqrt{3}$$

If $D < 0$, a trigonometric formulation is useful.

$$y_1 = 2\sqrt{\frac{|p|}{3}}\cos\frac{\phi}{3}$$

$$y_2 = -2\sqrt{\frac{|p|}{3}}\cos\frac{\phi+\pi}{3}$$

$$y_3 = -2\sqrt{\frac{|p|}{3}}\cos\frac{\phi-\pi}{3}$$

$$D = \left(\frac{p}{3}\right)^3 + \left(\frac{q}{2}\right)^2$$

$$u = \sqrt[3]{-\frac{q}{2} + \sqrt{D}} \qquad v = \sqrt[3]{-\frac{q}{2} - \sqrt{D}}$$

If a, b, c, d are real and if

$D > 0$ there are one real and two conjugate complex roots;

$D = 0$ there are three real roots of which at least two are equal;

$D < 0$ there are three real unequal roots.

The value of ϕ is calculated from the expression

$$\phi = \cos^{-1}\frac{-q/2}{\sqrt{|p|^3/27}}$$

---

[1]For $i = \sqrt{-1}$ refer to Sec. 11.01.

(1) Definition

(a) **A determinant** of the nth order contains $n \times n$ elements, arranged in n rows and n columns.

$$D = \begin{vmatrix} a_{11} & a_{12} & \cdots & a_{1n} \\ a_{21} & a_{22} & \cdots & a_{2n} \\ \cdots & \cdots & \cdots & \cdots \\ a_{n1} & a_{n2} & \cdots & a_{nn} \end{vmatrix}$$

(b) **A minor** D_{jk} of the element a_{jk} in the nth-order determinant D is the $(n-1)$st-order determinant obtained from D by deleting the jth row and the kth column.

(2) Evaluation

(a) **The cofactor**

$$A_{jk} = (-1)^{j+k} D_{jk}$$

(b) **The evaluation** of the determinant D is accomplished by summing the products of elements of any row or any column into their respective cofactors.

$$D = \sum_{i=1}^{n} a_{ij} A_{ij} = \sum_{j=1}^{n} a_{jk} A_{jk} = \sum_{k=1}^{n} a_{kl} A_{kl} = \cdots$$

(3) Special Cases

(a) **Second-order determinant**

$$\begin{vmatrix} a_{11} & a_{12} \\ a_{21} & a_{22} \end{vmatrix} = a_{11} a_{22} - a_{21} a_{12}$$

(b) **Third-order determinant**

$$\begin{vmatrix} a_{11} & a_{12} & a_{13} \\ a_{21} & a_{22} & a_{23} \\ a_{31} & a_{32} & a_{33} \end{vmatrix} = a_{11} A_{11} + a_{21} A_{21} + a_{31} A_{31}$$

where

$$A_{11} = \begin{vmatrix} a_{22} & a_{23} \\ a_{32} & a_{33} \end{vmatrix} \qquad A_{21} = -\begin{vmatrix} a_{12} & a_{13} \\ a_{32} & a_{33} \end{vmatrix} \qquad A_{31} = \begin{vmatrix} a_{12} & a_{13} \\ a_{22} & a_{23} \end{vmatrix}$$

(4) Basic Operations

(a) **Equal determinants**

Two determinants $|A|$ and $|B|$ are equal if they have the same dimensions and their corresponding elements are equal.

$$a_{jk} = b_{jk}$$

(b) **Transpose**

The value of a determinant is unchanged if the corresponding rows and columns are interchanged (transpose of determinant).

$$|D| = |D|^{T}$$

(1) Transformations Inducing no Change

(a) The *value* of a determinant is *unchanged* if the corresponding rows and columns are interchanged.

$$\begin{vmatrix} a_{11} & a_{12} & a_{13} \\ a_{21} & a_{22} & a_{23} \\ a_{31} & a_{32} & a_{33} \end{vmatrix} = \begin{vmatrix} a_{11} & a_{21} & a_{31} \\ a_{12} & a_{22} & a_{32} \\ a_{13} & a_{23} & a_{33} \end{vmatrix} = D$$

(b) If to each element of a row (or column) is *added m times* the corresponding element in another row (or column), the value of the determinant is unchanged.

$$\begin{vmatrix} a_{11} & a_{12} & a_{13} \\ a_{21} & a_{22} & a_{23} \\ a_{31} & a_{32} & a_{33} \end{vmatrix} = \begin{vmatrix} a_{11} & a_{12} & a_{13} \\ a_{21} + ma_{11} & a_{22} + ma_{12} & a_{23} + ma_{13} \\ a_{31} & a_{32} & a_{33} \end{vmatrix}$$

(c) If *two determinants differ* from each other only *in the elements of any one row* (or column), they may be added as follows:

$$\begin{vmatrix} a_{11} & a_{12} & a_{13} \\ a_{21} & a_{22} & a_{23} \\ a_{31} & a_{32} & a_{33} \end{vmatrix} + \begin{vmatrix} b_{11} & b_{12} & b_{13} \\ a_{21} & a_{22} & a_{23} \\ a_{31} & a_{32} & a_{33} \end{vmatrix} = \begin{vmatrix} a_{11} + b_{11} & a_{12} + b_{12} & a_{13} + b_{13} \\ a_{21} & a_{22} & a_{23} \\ a_{31} & a_{32} & a_{33} \end{vmatrix}$$

(2) Transformations Inducing Change

(a) The *sign* of a determinant is *changed* (unchanged) if an odd (even) number of interchanges of any two rows or of any two columns is introduced.

(b) If each element of a row (or column) is multiplied by m, the *new determinant is equal to mD.*

$$\begin{vmatrix} a_{11} & a_{12} & a_{13} \\ ma_{21} & ma_{22} & ma_{23} \\ a_{31} & a_{32} & a_{33} \end{vmatrix} = m\begin{vmatrix} a_{11} & a_{12} & a_{13} \\ a_{21} & a_{22} & a_{23} \\ a_{31} & a_{32} & a_{33} \end{vmatrix} = mD$$

(3) Zero Determinant

(a) If a determinant has *two identical rows* (or columns) or if all the elements of one row (or column) are zero, then the value of the determinant is zero.

(b) If the elements of any row (or column) are *linear combinations* of the corresponding elements of the other rows (or columns), the value of the determinant is zero.

$$\begin{vmatrix} ba_{21} + ca_{31} & ba_{22} + ca_{32} & ba_{23} + ca_{33} \\ a_{21} & a_{22} & a_{23} \\ a_{31} & a_{32} & a_{33} \end{vmatrix} = 0$$

(1) Definition

A *matrix* is a rectangular *array of elements* arranged in rows and columns.

m = number of rows	a_{jk} = any element
n = number of columns	$m \times n$ = dimension of matrix

$$[A] = \begin{bmatrix} a_{11} & a_{12} & \cdots & a_{1n} \\ a_{21} & a_{22} & \cdots & a_{2n} \\ \vdots & & & \vdots \\ a_{m1} & a_{m2} & \cdots & a_{mn} \end{bmatrix} = [a_{jk}]$$

(2) Shapes

Rectangular matrix	$m \neq n$	Column matrix	$n = 1$
Square matrix	$m = n$	Row matrix	$m = 1$

(3) Basic Types

(a) Unit matrix

$$\begin{bmatrix} 1 & 0 & 0 \\ 0 & 1 & 0 \\ 0 & 0 & 1 \end{bmatrix}$$

(b) Diagonal matrix

$$\begin{bmatrix} a & 0 & 0 \\ 0 & b & 0 \\ 0 & 0 & c \end{bmatrix}$$

(c) Zero matrix

$$\begin{bmatrix} 0 & 0 & 0 \\ 0 & 0 & 0 \\ 0 & 0 & 0 \end{bmatrix}$$

(d) Symmetrical matrix

$$\begin{bmatrix} a & d & l \\ d & b & f \\ l & f & c \end{bmatrix}$$

(e) Antisymmetrical matrix

$$\begin{bmatrix} 0 & +d & -l \\ -d & 0 & +f \\ +l & -f & 0 \end{bmatrix}$$

(f) Point symmetrical matrix

$$\begin{bmatrix} a & d & c \\ d & b & d \\ c & d & a \end{bmatrix}$$

(4) Basic Operations

(a) Equal matrices

Two matrices $[A]$ and $[B]$ are equal if they have the same dimensions and their corresponding elements are equal.

$$a_{ik} = b_{ik}$$

(b) Sum of matrices

The sum of two or several $m \times n$ matrices is an $m \times n$ matrix, each of whose elements is equal to the sum of the corresponding elements of the initial matrices.

$$a_{ik} + b_{ik} + c_{ik} + \cdots = s_{ik}$$

(c) Scalar-matrix multiplication

The product of a scalar k and an $m \times n$ matrix $[A]$ is an $m \times n$ matrix, each of whose elements is equal to the product of the scalar and the corresponding element of $[A]$.

$$k[A] = [kA]$$

(d) Matrix-matrix multiplication

A product of two rectangular conformable matrices of dimensions $m_1 \times n_1$ and $m_2 \times n_2$ is a rectangular (or square) matrix of dimensions $m_1 \times n_2$ whose elements are equal to the sum of products of the inner elements.

$$\begin{bmatrix} a_{11} & a_{12} & a_{13} \\ a_{21} & a_{22} & a_{23} \end{bmatrix} \begin{bmatrix} b_{11} & b_{12} \\ b_{21} & b_{22} \\ b_{31} & b_{32} \end{bmatrix} = \begin{bmatrix} a_{11}b_{11} + a_{12}b_{21} + a_{13}b_{31} & a_{11}b_{12} + a_{12}b_{22} + a_{13}b_{32} \\ a_{21}b_{11} + a_{22}b_{21} + a_{23}b_{31} & a_{21}b_{12} + a_{22}b_{22} + a_{23}b_{32} \end{bmatrix}$$

Two rectangular matrices are conformable if $n_1 = m_2$.

(1) Definition

The transpose $[A]^T$ of a matrix $[A]$ has each row identical with the corresponding column of $[A]$.

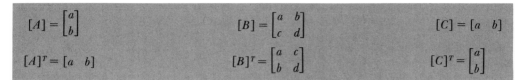

$$[A] = \begin{bmatrix} a \\ b \end{bmatrix} \qquad [B] = \begin{bmatrix} a & b \\ c & d \end{bmatrix} \qquad [C] = \begin{bmatrix} a & b \end{bmatrix}$$

$$[A]^T = \begin{bmatrix} a & b \end{bmatrix} \qquad [B]^T = \begin{bmatrix} a & c \\ b & d \end{bmatrix} \qquad [C]^T = \begin{bmatrix} a \\ b \end{bmatrix}$$

(2) Special Cases

Transpose of transpose

$$[[A]^T]^T = [A]$$

Transpose of unit matrix

$$[I]^T = [I]$$

Transpose of zero matrix

$$[0]^T = [0]$$

Transpose of diagonal matrix

$$\begin{bmatrix} a & 0 \\ 0 & d \end{bmatrix}^T = \begin{bmatrix} a & 0 \\ 0 & d \end{bmatrix}$$

$$[D]^T = [D]$$

Transpose of symmetrical matrix

$$\begin{bmatrix} a & b \\ b & c \end{bmatrix}^T = \begin{bmatrix} a & b \\ b & c \end{bmatrix}$$

$$[E]^T = [E]$$

Transpose of anti-symmetrical matrix

$$\begin{bmatrix} 0 & b \\ -b & 0 \end{bmatrix}^T = \begin{bmatrix} 0 & -b \\ b & 0 \end{bmatrix}$$

$$[F]^T = -[F]$$

(3) Basic Operations

(a) Transpose of product

The transpose of product of two or more matrices is equal to the product of their transposes in reverse order.

$$\left[[A][B][C][D] \right]^T = [D]^T[C]^T[B]^T[A]^T$$

(b) Matrix-transpose sum and difference

The sum of a square matrix and its transpose is a symmetrical matrix. The difference of a square matrix and its transpose is an antisymmetrical matrix.

$$\underbrace{\begin{bmatrix} a & b \\ c & d \end{bmatrix}}_{[B]} + \underbrace{\begin{bmatrix} a & c \\ b & d \end{bmatrix}}_{[B]^T} = \underbrace{\begin{bmatrix} 2a & b+c \\ c+b & 2d \end{bmatrix}}_{\text{Symmetrical}} \qquad \underbrace{\begin{bmatrix} a & b \\ c & d \end{bmatrix}}_{[B]} - \underbrace{\begin{bmatrix} a & c \\ b & d \end{bmatrix}}_{[B]^T} = \underbrace{\begin{bmatrix} 0 & b-c \\ c-b & 0 \end{bmatrix}}_{\text{Antisymmetrical}}$$

(c) Matrix-transpose product

The product of a square matrix and its transpose, or vice versa, is a symmetrical matrix.

$$\underbrace{\begin{bmatrix} a & b \\ c & d \end{bmatrix}}_{[B]} \underbrace{\begin{bmatrix} a & c \\ b & d \end{bmatrix}}_{[B]^T} = \underbrace{\begin{bmatrix} a^2+b^2 & ac+bd \\ ac+bd & c^2+d^2 \end{bmatrix}}_{\text{Symmetrical}} \qquad \underbrace{\begin{bmatrix} a & c \\ b & d \end{bmatrix}}_{[B]^T} \underbrace{\begin{bmatrix} a & b \\ c & d \end{bmatrix}}_{[B]} = \underbrace{\begin{bmatrix} a^2+c^2 & ab+cd \\ ab+cd & b^2+d^2 \end{bmatrix}}_{\text{Symmetrical}}$$

(d) Matrix resolution

Every unsymmetrical square matrix can be expressed as the sum of a symmetrical matrix and an antisymmetrical matrix.

(1) Definition

The inverse $[A]^{-1}$ of a square matrix $[A]$ is uniquely defined by the conditions

$$[A]^{-1}[A] = [I] = [A][A]^{-1} \quad \text{and} \quad |A| \neq 0$$

If $|A| = 0$, the inverse of $[A]$ does not exist, and $[A]$ is said to be a *singular matrix*. Only nonsingular square matrices have inverses.

$$[A]^{-1} = \begin{bmatrix} a_{11} & a_{12} & \cdots & a_{1n} \\ a_{21} & a_{22} & \cdots & a_{2n} \\ \cdots & \cdots & \cdots & \cdots \\ a_{n1} & a_{n2} & \cdots & a_{nn} \end{bmatrix}^{-1} = \frac{1}{|A|} \begin{bmatrix} A_{11} & A_{21} & \cdots & A_{n1} \\ A_{12} & A_{22} & \cdots & A_{n2} \\ \cdots & \cdots & \cdots & \cdots \\ A_{1n} & A_{2n} & \cdots & A_{nn} \end{bmatrix} = \frac{[A_{jk}]}{|A|}$$

in which $|A|$ = determinant of $[A]$, A_{jk} = cofactor jk of $|A|$, and $[A_{jk}]$ = adjoint matrix of cofactors of $|A|$.

(2) Special Cases

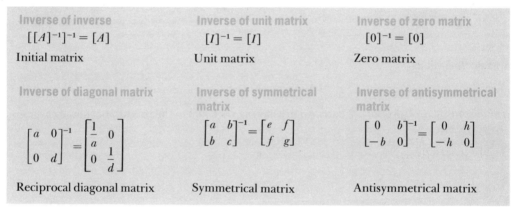

Inverse of inverse	Inverse of unit matrix	Inverse of zero matrix
$[[A]^{-1}]^{-1} = [A]$	$[I]^{-1} = [I]$	$[0]^{-1} = [0]$
Initial matrix	Unit matrix	Zero matrix

Inverse of diagonal matrix / Inverse of symmetrical matrix / Inverse of antisymmetrical matrix

$$\begin{bmatrix} a & 0 \\ 0 & d \end{bmatrix}^{-1} = \begin{bmatrix} \dfrac{1}{a} & 0 \\ 0 & \dfrac{1}{d} \end{bmatrix}$$

$$\begin{bmatrix} a & b \\ b & c \end{bmatrix}^{-1} = \begin{bmatrix} e & f \\ f & g \end{bmatrix}$$

$$\begin{bmatrix} 0 & b \\ -b & 0 \end{bmatrix}^{-1} = \begin{bmatrix} 0 & h \\ -h & 0 \end{bmatrix}$$

Reciprocal diagonal matrix   Symmetrical matrix   Antisymmetrical matrix

(3) Basic Operations

(a) Inverse of product

The inverse of the product of two or more matrices is equal to the product of their inverses in reverse order.

$$[[A][B][C][D]]^{-1} = [D]^{-1}[C]^{-1}[B]^{-1}[A]^{-1}$$

(b) Normal matrix

A square matrix is said to be normal if it is equal to its transpose. All symmetrical matrices are normal.

$$\text{If } [A] = [A]^T \quad \text{then} \quad [A][A]^T = [A]^T[A] = [A]^2$$

(c) Orthogonal matrix

A square matrix is said to be orthogonal if its transpose is equal to its inverse.

$$\text{If } [A]^T = [A]^{-1} \quad \text{then} \quad [A][A]^T = [A]^T[A] = [I]$$

(1) Basic Laws — Matrices

Commutative law:

$$A + B = B + A$$

$$[A + B]^T = [B + A]^T$$

$$[A + B]^{-1} = [B + A]^{-1}$$

$$AB \neq BA$$

$$[AB]^T \neq [BA]^T$$

$$[AB]^{-1} \neq [BA]^{-1}$$

Associative law:

$$A + (B + C) = (A + B) + C$$

$$[A + (B + C)]^T = [(A + B) + C]^T$$

$$[A + (B + C)]^{-1} = [(A + B) + C]^{-1}$$

$$A(BC) = (AB)C$$

$$[A(BC)]^T = [(AB)C]^T$$

$$[A(BC)]^{-1} = [(AB)C]^{-1}$$

Distributive law:

$$A[B + C] = AB + AC$$

$$[A + B]C = AC + BC$$

Division law:

If $AB = 0$, then A and/or B may or may not be zero.

(2) Basic Laws — Determinant Matrices $n \times n$

$$\text{Det}\,(A^T) = \text{Det}\,(A)$$

$$\text{Det}\,(A^{-1}) = \frac{1}{\text{Det}\,(A)}$$

$$\text{Det}\,(I) = 1$$

$$\text{Det}\,[\text{Adj}\,(A)] = [\text{Det}\,(A)]^{n-1}$$

$$\text{Det}\,(AB) = \text{Det}\,(A)\,\text{Det}\,(B)$$

$$\text{Det}\,(AB^{-1}) = \frac{\text{Det}\,(A)}{\text{Det}\,(B)}$$

$$\text{Det}\,(A^{-1}B^{-1}) = \frac{1}{\text{Det}\,(A)\,\text{Det}\,(B)}$$

$$\text{Adj}\,(AB) = \text{Adj}\,(B)\,\text{Adj}\,(A)$$

(3) Characteristics

(a) **The rank of a matrix** is the order of the largest nonzero determinant that can be obtained from the elements of the matrix. The matrix whose order exceeds its rank is singular.

(b) **The trace of a matrix** is the sum of its diagonal elements.

(4) Relationships of Two Matrices

(a) **Equivalence**

The square matrix A is equivalent to another square matrix B if there exist nonsingular matrices P and Q such that $A = PBQ$

(b) **Congruence**

The square matrix A is congruent to another square matrix B if there exists a nonsingular matrix Q such that $A = Q^T BQ$

(c) **Similarity**

The square matrix A is similar to another square matrix B if there exists a nonsingular matrix Q such that $A = Q^{-1}BQ$

(1) Methods of Solution

(a) Systemof n simultaneous nonhomogeneous linear equations

$$
\begin{aligned}
a_{11}x_1 + a_{12}x_2 + \cdots + a_{1n}x_n &= b_1 \\
a_{21}x_1 + a_{22}x_2 + \cdots + a_{2n}x_n &= b_2 \\
&\cdots\cdots\cdots\cdots\cdots\cdots\cdots \\
a_{n1}x_1 + a_{n2}x_2 + \cdots + a_{nn}x_n &= n
\end{aligned}
$$

has a *unique solution* for the unknowns $x_1, x_2, \ldots, x_n$ if

$$
D = \begin{vmatrix}
a_{11} & a_{12} & \cdots & a_{1n} \\
a_{21} & a_{22} & \cdots & a_{2n} \\
& \cdots\cdots\cdots\cdots & \\
a_{n1} & a_{n2} & \cdots & a_{nn}
\end{vmatrix} \neq 0
$$

and at least one of the terms $b_1, b_2, \ldots, b_n$ is different from zero.

(b) Determinant solutionfor the unknowns is

$$
x_1 = \frac{D_1}{D}, \qquad x_2 = \frac{D_2}{D}, \qquad \cdots, \qquad x_n = \frac{D_n}{D}
$$

where the *augmented determinants* are

$$
D_1 = \begin{vmatrix}
b_1 & a_{12} & \cdots & a_{1n} \\
b_2 & a_{22} & \cdots & a_{2n} \\
& \cdots\cdots\cdots\cdots & \\
b_n & a_{n2} & \cdots & a_{nn}
\end{vmatrix}
\qquad
D_2 = \begin{vmatrix}
a_{11} & b_1 & \cdots & a_{1n} \\
a_{21} & b_2 & \cdots & a_{2n} \\
& \cdots\cdots\cdots\cdots & \\
a_{n1} & b_n & \cdots & a_{nn}
\end{vmatrix}
\qquad
D_n = \begin{vmatrix}
a_{11} & a_{12} & \cdots & b_1 \\
a_{21} & a_{22} & \cdots & b_2 \\
& \cdots\cdots\cdots\cdots & \\
a_{n1} & a_{n2} & \cdots & b_n
\end{vmatrix}
$$

(c) Matrix solutionis represented symbolically as

$$
[x] = [A]^{-1}[b]
$$

where $[x]$ is the column matrix of the unknowns, $[A]^{-1}$ is the inverse of $[A]$ (Sec. 1.11), and $[b]$ is the column matrix of the terms b's.

(2) Classification of Solutions

(a) Unique solutionIf $D \neq 0$ and $[b] \neq 0$, the system has a unique solution in which some but not all x_j may be zero.

(b) Trivial solutionIf $D \neq 0$ and $[b] = 0$, the system has only one solution, $x_1 = x_2 = \cdots = x_n = 0$.

(c) Infinitely many solutionsIf $D = 0$ and $[b] = 0$, the system is called homogeneous, and it has infinitely many solutions, one of which is the trivial solution.

(d) No solutionIf $D = 0, D_1 = D_2 = \cdots = D_n = 0$, and $[b] \neq 0$, the system has no solution.

(1) Eigenvalues

(a) Initial system

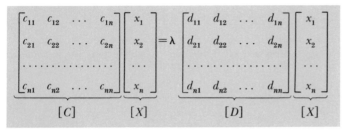

$$[C] \qquad [X] \qquad\qquad [D] \qquad [X]$$

or simply

$$
\begin{bmatrix}
m_{11}-\lambda & m_{12} & \cdots & m_{1n} \\
m_{21} & m_{22}-\lambda & \cdots & m_{2n} \\
\cdots & \cdots & \cdots & \cdots \\
m_{n1} & m_{n2} & \cdots & m_{nn}-\lambda
\end{bmatrix}
\begin{bmatrix}
x_1 \\ x_2 \\ \cdots \\ x_n
\end{bmatrix}
=
\begin{bmatrix}
0 \\ 0 \\ \cdots \\ 0
\end{bmatrix}
$$

$$[K] \qquad\qquad [X] \qquad [0]$$

where $c_{jk}, d_{jk} =$ given constants, λ, $x_j =$ unknowns, and $j = 1, 2, \ldots, n$, $k = 1, 2, \ldots, n$.

(b) Characteristic matrix

$$[K][X] = [0]$$

is called the characteristic matrix equation in which

$$[K] = [M] - \lambda[I] = [D]^{-1}[C] - \lambda[I]$$

is the *characteristic matrix.*

(c) Nontrivial solution of the characteristic matrix equation (Sec. 1.13−2c) exists if and only if det $(K) = 0$, which is a polynomial algebraic equation of nth degree in λ, called the *characteristic equation.* The roots of this equation $\lambda_1, \lambda_2, \ldots, \lambda_n$ are the *eigenvalues* of $[K]$.

(2) Eigenvectors

(a) Definition. Corresponding to each eigenvalue λ_j is a set of values $x_{j1}, x_{j2}, \ldots, x_{jn}$ forming a column matrix $[X_j]$ called the *eigenvector j.*

(b) Orthogonality. If $[M]$ is a symmetrical[1] matrix and $[X_j]$, $[X_k]$ are the eigenvectors corresponding to λ_j, λ_k, respectively, then

$$[X_j]^T[M][X_k] = \begin{cases} 0 & \text{if } j \neq k \\ \lambda_j & \text{if } j = k \end{cases}$$

and the eigenvectors are *orthogonal.*

(c) Normalization. If $[M]$ is the same as in (b) and

$$[Y_j] = [X_j/\sqrt{\lambda_j}] \qquad [Y_k] = [X_k/\sqrt{\lambda_k}]$$

are the *normalized eigenvectors* corresponding to λ_j, λ_k, respectively, then

$$[Y_j]^T[M][Y_k] = \begin{cases} 0 & \text{if } j \neq k \\ 1 & \text{if } j = k \end{cases}$$

and the normalized eisenvectors are also *orthogonal.*

[1]Or hermitian (Sec. 11.05).

(1) Permutations

(a) A permutation is an arrangement of n elements. The number of all possible permutations of n different elements is

$$_nP_n = n(n-1)(n-2) \cdots (3)(2)(1) = n!$$

(b) The number of all different permutations of n elements, among which there are a elements of equal value, is

$$_aP_n = \frac{n(n-1)(n-2) \cdots (3)(2)(1)}{a(a-1)(a-2) \cdots (3)(2)(1)} = \frac{n!}{a!}$$

(c) The number of all different permutations of n elements, among which there are a elements of one equal value and b elements of another equal value, is

$$_{a,b}P_n = \frac{n(n-1)(n-2) \cdots (3)(2)(1)}{a(a-1)(a-2) \cdots (3)(2)(1)b(b-1)(b-2) \cdots (3)(2)(1)} = \frac{n!}{a!b!}$$

(2) Variations

A variation is an arrangement of n elements into a sequence of k terms. The number of all possible variations is

$$_kV_n = \frac{n(n-1)(n-2) \cdots (3)(2)(1)}{(n-k)(n-k-1)(n-k-2) \cdots (3)(2)(1)} = \frac{n!}{(n-k)!} = \binom{n}{k}k!$$

(3) Combinations

A combination is an arrangement (without repetition) of n elements into a sequence of k terms. The number of all possible combinations is

$$_kC_n = \frac{n(n-1)(n-2) \cdots (n-k+2)(n-k+1)}{(n-k)(n-k-1)(n-k-2) \cdots (3)(2)(1)} = \frac{n!}{(n-k)!k!} = \binom{n}{k}$$

(4) Table – Example

Permutations	Elements A, B, C	$n = 3$
	$ABC \quad BCA \quad CAB$ $ACB \quad BAC \quad CBA$	$P_3 = (3)(2)(1) = 6$
Permutations	Elements A, A, C	$n = 3 \qquad a = 2$
	$AAC \quad ACA \quad CAA$	$_2P_3 = \dfrac{(3)(2)(1)}{(2)(1)} = 3$
Variations	Elements A, B, C	$n = 3 \qquad k = 2$
	$AB \quad BC \quad CA$ $BA \quad CB \quad AC$	$_2V_3 = \dfrac{(3)(2)(1)}{1} = 6$
Combinations	Elements A, B, C	$n = 3 \qquad k = 2$
	$AB \quad BC \quad CA$	$_2C_3 = \dfrac{(3)(2)(1)}{(2)(1)} = 3$

2
GEOMETRY

$a, b, c = $ sides	$A = $ area	$h = $ altitude
$\alpha, \beta, \gamma = $ angles	$R = $ circumradius	$m = $ median
$2p = a + b + c$	$r = $ inradius	$t = $ bisector

(a) Oblique triangle $\quad (\alpha + \beta + \gamma = 180°)$

$$A = \sqrt{p(p-a)(p-b)(p-c)} = \frac{abc}{4R} = pr$$

$$A = \frac{ah_a}{2} = \frac{bh_b}{2} = \frac{ch_c}{2} = 2R^2 \sin \alpha \sin \beta \sin \gamma$$

$$A = \frac{ab \sin \gamma}{2} = \frac{bc \sin \alpha}{2} = \frac{ac \sin \beta}{2} = p^2 \tan \frac{\alpha}{2} \tan \frac{\beta}{2} \tan \frac{\gamma}{2}$$

$$A = \frac{a^2 \sin \beta \sin \gamma}{2 \sin \alpha} = \frac{b^2 \sin \alpha \sin \gamma}{2 \sin \beta} = \frac{c^2 \sin \alpha \sin \beta}{2 \sin \gamma} = r^2 \cot \frac{\alpha}{2} \cot \frac{\beta}{2} \cot \frac{\gamma}{2}$$

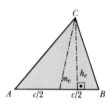

$$h_c = b \sin \alpha = a \sin \beta = \frac{2\sqrt{p(p-a)(p-b)(p-c)}}{c}$$

$$R = \frac{c}{2 \sin \gamma} = \frac{abc}{4A} = \frac{abc}{4\sqrt{p(p-a)(p-b)(p-c)}}$$

$$r = (p-c) \tan \frac{\gamma}{2} = \frac{2A}{a+b+c} = \frac{2\sqrt{p(p-a)(p-b)(p-c)}}{a+b+c}$$

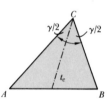

$$m_c = \sqrt{\frac{a^2}{2} + \frac{b^2}{2} - \frac{c^2}{4}} = \sqrt{b^2 + \left(\frac{c}{2}\right)^2 - bc \cos \alpha}$$

$$t_c = \sqrt{ab\left[1 - \left(\frac{c}{a+b}\right)^2\right]} = \frac{2ab}{a+b} \cos \frac{\gamma}{2}$$

$$a:b:c = \frac{1}{h_a}:\frac{1}{h_b}:\frac{1}{h_c} \qquad h_a:h_b:h_c = \frac{1}{a}:\frac{1}{b}:\frac{1}{c} \qquad \frac{1}{r} = \frac{1}{h_a} + \frac{1}{h_b} + \frac{1}{h_c}$$

(See also Sec. 3.03.)

(b) Right triangle $\quad (\alpha + \beta = 90°)$

$$A = \frac{ab}{2} = \frac{hc}{2}$$

$$h = \frac{ab}{c} \qquad R = \frac{c}{2}$$

$$r = \frac{a+b-c}{2}$$

$$a^2 + b^2 = c^2$$

$$p = \frac{b^2}{c} \qquad q = \frac{a^2}{c}$$

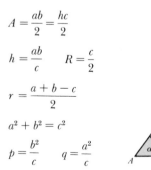

(See also Sec. 3.01.)

(c) Equilateral triangle $\quad (\alpha = \beta = \gamma = 60°)$

$$A = \frac{a^2}{4}\sqrt{3} = \frac{h^2}{3}\sqrt{3}$$

$$h = m = t = \frac{a}{2}\sqrt{3}$$

$$R = \frac{a}{3}\sqrt{3}$$

$$r = \frac{a}{6}\sqrt{3}$$

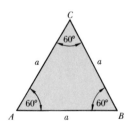

(See also Sec. 3.01.)

α = central angle	a = side	A = area
β = interior angle	n = number of sides	R = circumradius
γ = exterior angle		r = inradius

(a) General polygon

$$\sum_{1}^{n} \beta_j = (n-2)180°$$

$$\sum_{1}^{n} \gamma_j = 360°$$

$$\sum_{1}^{n-2} A_j = A$$

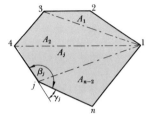

(b) Regular polygon   ($n\alpha = 360°$, $n\beta = (n-2)180°$, $n\gamma = 360°$)

$$A = \frac{na^2}{4} \cot \frac{\pi}{n} = \frac{nar}{2} = \frac{nR^2}{2} \sin \frac{2\pi}{n}$$

$$R = \frac{a}{2} \csc \frac{\pi}{n} \qquad r = \frac{a}{2} \cot \frac{\pi}{n}$$

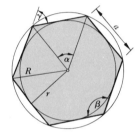

(c) Regular polygon — Table of coefficients

n	$180°/n$	A/a^2	A/R^2	A/r^2	R/a	r/a	R/r
3	60.000	0.433013	1.299038	5.196152	0.577350	0.288675	2.000000
4	45.000	1.000000	2.000000	4.000000	0.707107	0.500000	1.414214
5	36.000	1.720477	2.377642	3.632713	0.850651	0.688191	1.236068
6	30.000	2.598076	2.598076	3.464102	1.000000	0.866025	1.154701
7	25.714* …	3.633914	2.736408	3.371021	1.152383	1.038261	1.109916
8	22.500	4.828427	2.828427	3.313710	1.306563	1.207107	1.082392
9	20.000	6.181825	2.892544	3.275732	1.461902	1.373739	1.064177
10	18.000	7.694208	2.938926	3.249197	1.618034	1.538842	1.051462
12	15.000	11.196154	3.000000	3.215389	1.931852	1.866025	1.035277
15	12.000	17.642362	3.050524	3.188348	2.404867	2.352314	1.022341
16	11.250	20.109363	3.061464	3.182596	2.562917	2.513670	1.019592
20	9.000	31.568769	3.090168	3.167687	3.196228	3.156877	1.012465
24	7.500	45.574519	3.105827	3.159659	3.830649	3.797877	1.008629
32	5.625	81.225378	3.121442	3.151724	5.101151	5.076586	1.004839
48	3.750	183.084812	3.132619	3.146082	7.644910	7.628533	1.002147
64	2.8125*	325.687826	3.136541	3.144114	10.190024	10.177744	1.001206

*$180°/7 = 25.714\,285\,714 \cdots°$ (periodic), $180°/64 = 2.8125°$ (finite)

a, b, c, d = sides	$2s = a + b + c + d$	A = area
e, f = diagonals	h = altitude	R = circumradius
$\alpha, \beta, \gamma, \delta$ = angles		r = inradius

(a) Square $(\alpha = \beta = \gamma = \delta = 90°)$

$$e = a\sqrt{2} = 1.4142a \qquad R = \frac{a}{2}\sqrt{2} = 0.7071a$$

$$A = a^2 \qquad r = \frac{a}{2}$$

(b) Rectangle $(\alpha = \beta = \gamma = \delta = 90°)$

$$e = f = \sqrt{a^2 + b^2} \qquad R = \frac{e}{2} = \frac{\sqrt{a^2 + b^2}}{2}$$

$$A = ab$$

(c) Rhombus $(\alpha + \beta = \gamma + \delta = 180°)$ $(\alpha = \gamma, \beta = \delta)$

$$e = 2a\cos\frac{\alpha}{2} \qquad f = 2a\sin\frac{\alpha}{2}$$

$$e^2 + f^2 = 4a^2$$

$$h = a\sin\alpha \qquad r = \frac{a}{2}\sin\alpha$$

$$A = ah = a^2\sin\alpha = \frac{ef}{2}$$

(d) Rhomboid $(\alpha + \beta = \gamma + \delta = 180°)$ $(\alpha = \gamma, \beta = \delta)$

$$e = \sqrt{a^2 + b^2 - 2ab\cos\beta} \qquad f = \sqrt{a^2 + b^2 - 2ab\cos\alpha}$$

$$e^2 + f^2 = 2(a^2 + b^2)$$

$$h_a = b\sin\alpha \qquad h_b = a\sin\alpha$$

$$A = ah_a = ab\sin\alpha = bh_b$$

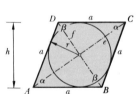

$a, b, c, d =$ sides	$2s = a + b + c + d$	$A =$ area
$e, f =$ diagonals	$h =$ altitude	$R =$ circumradius
$\alpha, \beta, \gamma, \delta =$ angles	$2t = a + b - c + d$	$r =$ inradius

(a) Trapezoid $(\alpha + \delta = \beta + \gamma = 180°)$

$$e = \sqrt{a^2 + b^2 - 2ab\cos\beta}$$

$$f = \sqrt{a^2 + d^2 - 2ad\cos\alpha}$$

$$h = \frac{2}{a - c}\sqrt{t(t - a + c)(t - b)(t - d)}$$

$$A = \frac{(a + c)h}{2}$$

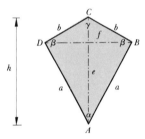

(b) Deltoid $\left(\dfrac{\alpha}{2} + \beta + \dfrac{\gamma}{2} = 180°\right)$

$$e = a\cos\frac{\alpha}{2} + b\cos\frac{\gamma}{2}$$

$$f = 2b\sin\frac{\gamma}{2} = 2a\sin\frac{\alpha}{2}$$

$$A = \frac{ef}{2} = \frac{a^2\sin\alpha + b^2\sin\gamma}{2}$$

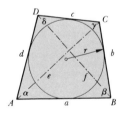

(c) Tangent – Quadrilateral $(a + c = b + d)$

$$A = sr$$

If $\alpha + \gamma = \beta + \delta = 180°$,

$$A = \sqrt{abcd} \qquad r = \frac{\sqrt{abcd}}{s}$$

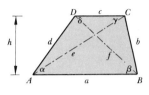

(d) Secant – Quadrilateral $(\alpha + \gamma = \beta + \delta = 180°)$

$$ef = ac + bd = g$$

$$e = \sqrt{\frac{(ad + bc)g}{ab + cd}}$$

$$f = \sqrt{\frac{(ab + cd)g}{ad + bc}} \qquad \sin\omega = \frac{2A}{g}$$

$$A = \sqrt{(s - a)(s - b)(s - c)(s - d)}$$

$$R = \frac{\sqrt{(ab + cd)(ad + bc)g}}{4A}$$

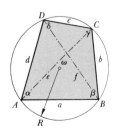

C = circumference S = length of arc $2l$ = chord

A = area α = angle h = altitude

R = radius D = diameter v = rise

Arc $1° = \dfrac{\pi}{180}$ Arc 1 minute $= \dfrac{\pi}{10{,}800}$ Arc 1 second $= \dfrac{\pi}{648{,}000}$

$= 0.017453293$ $= 0.000290888$ $= 0.000004848$

radian   radian   radian

(a) Circle   **(b) Sector**   **(c) Segment**

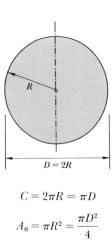

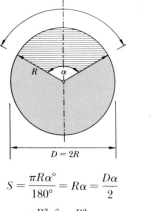

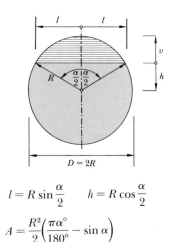

$C = 2\pi R = \pi D$

$A_0 = \pi R^2 = \dfrac{\pi D^2}{4}$

$S = \dfrac{\pi R \alpha°}{180°} = R\alpha = \dfrac{D\alpha}{2}$

$A = \dfrac{\pi R^2 \alpha°}{360°} = \dfrac{R^2 \alpha}{2}$

$l = R \sin \dfrac{\alpha}{2}$ $h = R \cos \dfrac{\alpha}{2}$

$A = \dfrac{R^2}{2}\left(\dfrac{\pi \alpha°}{180°} - \sin \alpha\right)$

(d) π constants

n	$\nu\pi$	$\dfrac{1}{n\pi}$	$\dfrac{\pi}{n}$	$\dfrac{n}{\pi}$
1	3.141 592 653 6	0.318 309 886 2	3.141 592 653 6	0.318 309 886 2
2	6.283 185 307 2	0.159 154 943 1	1.570 796 326 8	0.636 619 772 4
3	9.424 777 960 8	0.106 103 295 4	1.047 197 551 2	0.954 929 658 6
4	12.566 370 614 4	0.079 577 471 5	0.785 398 163 4	1.273 239 544 7
5	15.707 963 267 9	0.063 661 977 2	0.628 318 530 7	1.591 549 430 9
6	18.849 555 921 5	0.053 051 647 7	0.523 598 775 6	1.909 859 317 1
7	21.991 148 575 1	0.045 472 840 9	0.448 798 950 5	2.228 169 203 3
8	25.132 741 228 7	0.039 788 735 8	0.392 699 081 7	2.546 479 089 5
9	28.274 338 882 3	0.035 367 765 1	0.349 065 850 4	2.864 788 975 7

a, b = semiaxis	S = length of curve	A = area
$\dfrac{1}{r_a}$ = curvature at A	$\dfrac{1}{r_b}$ = curvature at B	C = circumference

(a) Ellipse

$$r_b = \frac{a^2}{b} \qquad r_a = \frac{b^2}{a} \qquad m = \frac{a-b}{a+b}$$

$$C = \pi(a+b)\left(1 + \frac{m^2}{4} + \frac{m^4}{64} + \frac{m^6}{256} + \cdots\right)$$

$$C_{\text{approx}} = \pi[1.5(a+b) - \sqrt{ab}]$$

$$= \pi(a+b)\frac{64 - 3m^4}{64 - 16m^2}$$

$$A_{\bigcirc} = \pi ab \qquad\qquad A_{\square} = \frac{\pi ab}{4}$$

$$A_{\triangle} = \frac{ab}{2}\cos^{-1}\frac{x}{a} \qquad\qquad A_{\triangleright} = \frac{ab}{2}\sin^{-1}\frac{x}{a}$$

$$A_{\square} = -xy + ab\cos^{-1}\frac{x}{a}$$

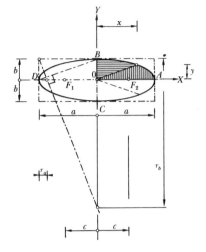

(b) Hyperbola

$$r_a = r_b = \frac{b^2}{a}$$

$$\alpha = \tan^{-1}\frac{b}{a}$$

$$A_{\triangleleft} = ab\ln\left(\frac{x}{a} + \frac{y}{b}\right) = ab\cosh^{-1}\frac{x}{a}$$

$$A_{\triangleleft} = xy - ab\ln\left(\frac{x}{a} + \frac{y}{b}\right)$$

$$= xy - ab\cosh^{-1}\frac{x}{a}$$

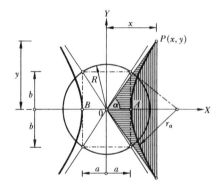

(c) Parabola

$$r_a = 2a \qquad\qquad b, h = \text{segments}$$

$$A_{\triangleleft} = \tfrac{2}{3}bh \qquad\qquad A_{\triangleleft} = \tfrac{4}{3}xy$$

$$S = a\left[\sqrt{\frac{x}{a}\left(1 + \frac{x}{a}\right)} + \ln\left(\sqrt{\frac{x}{a}} + \sqrt{1 + \frac{x}{a}}\right)\right]$$

$$= \sqrt{x(x+a)} + a\sin h^{-1}\sqrt{\frac{x}{a}}$$

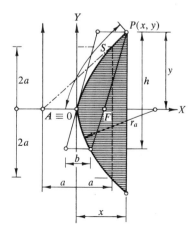

a = edge	R = circumradius	S = surface
e = diagonal	r = inradius	V = volume
v = altitude	ω = dihedral angle	

(a) Tetrahedron   **(4 triangles, 6 edges, 4 vertices)** $(\omega = 70° \, 31' \, 44'')$

$$R = \frac{a}{4}\sqrt{6} \qquad r = \frac{a}{12}\sqrt{6} \qquad v = \frac{a}{3}\sqrt{6}$$

$$S = a^2\sqrt{3} \approx 1.7321a^2$$

$$V = \frac{a^3\sqrt{2}}{12} \approx 0.1179a^3$$

(b) Cube   **(6 squares, 12 edges, 8 vertices)** $(\omega = 90°)$

$$R = \frac{a}{2}\sqrt{3} \qquad r = \frac{a}{2} \qquad e = a\sqrt{3}$$

$$S = 6a^2$$

$$V = a^3$$

(c) Octahedron   **(8 triangles, 12 edges, 6 vertices)** $(\omega = 109° \, 28' \, 16'')$

$$R = \frac{a}{2}\sqrt{2} \qquad r = \frac{a}{6}\sqrt{6}$$

$$S = 2a^2\sqrt{3} \approx 3.4641a^2$$

$$V = \frac{a^3}{3}\sqrt{2} \approx 0.4714a^3$$

(d) Dodecahedron   **(12 pentagons, 30 edges, 20 vertices)** $(\omega = 116° \, 33' \, 54'')$

$$R = \frac{a(1 + \sqrt{5})\sqrt{3}}{4} \qquad r = \frac{a}{4}\sqrt{\frac{50 + 22\sqrt{5}}{5}}$$

$$S = 3a^2\sqrt{5(5 + 2\sqrt{5})} \approx 20.6457a^2$$

$$V = \frac{a^3}{4}(15 + 7\sqrt{5}) \approx 7.6631a^3$$

(e) Icosahedron   **(20 triangles, 30 edges, 12 vertices)** $(\omega = 138° \, 11' \, 23'')$

$$R = \frac{a}{4}\sqrt{2(5 + \sqrt{5})} \qquad r = \frac{a}{2}\sqrt{\frac{7 + 3\sqrt{5}}{6}}$$

$$S = 5a^2\sqrt{3} \approx 8.6603a^2$$

$$V = \frac{5a^3}{12}(3 + \sqrt{5}) \approx 2.1817a^3$$

a, b, c = edges	R = circumradius	A = lateral area
e = diagonal	B = area of base	S = surface
h = lateral edge	v = altitude	V = volume

(a) Rectangular parallelepiped

$$e = \sqrt{a^2 + b^2 + c^2} \qquad R = \frac{\sqrt{a^2 + b^2 + c^2}}{2}$$

$$S = 2(ab + bc + ca)$$

$$V = abc$$

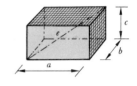

(b) Prism

$2p$ = perimeter of right section

$$A = 2ph$$

$$V = Bv$$

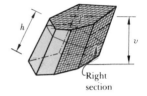

Right section

(c) Right pyramid

$$A = a\sqrt{v^2 + \left(\frac{b}{2}\right)^2} + b\sqrt{v^2 + \left(\frac{a}{2}\right)^2}$$

$$B = ab \qquad S = A + B$$

$$V = \frac{Bv}{3} \qquad \text{(valid for any pyramid)}$$

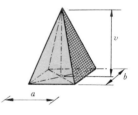

(d) Frustum of right pyramid   (subscript b = bottom, t = top)

$$A = (a_b + a_t)\sqrt{v^2 + \left(\frac{b_b - b_t}{2}\right)^2} + (b_b + b_t)\sqrt{v^2 + \left(\frac{a_b - a_t}{2}\right)^2}$$

$$S = A + a_b b_b + a_t b_t$$

$$V = \frac{v}{3}(B_b + B_t + \sqrt{B_b B_t}) \qquad \text{(valid for any frustum)}$$

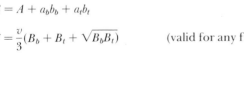

(e) Right wedge

$$A = 2(a + c)\sqrt{v^2 + b^2} + 2b\sqrt{v^2 + (a - c)^2}$$

$$S = A + 4ab$$

$$V = \frac{2bv}{3}(2a + c)$$

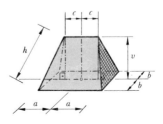

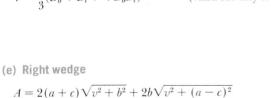

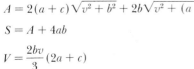

$\pi = 3.141\ 59 \cdots$	$R =$ radius	$A =$ lateral area
$r =$ radius	$B =$ area of base	$S =$ surface
$h =$ height	$v =$ altitude	$V =$ volume

(a) Right circular cylinder

$A = 2\pi R v$

$B = \pi R^2 \qquad S = 2\pi R(R + v)$

$V = \pi R^2 v$

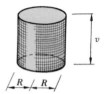

(b) Truncated frustum of right circular cylinder

$A = \pi R(h_1 + h_2)$

$S = \pi R \left[h_1 + h_2 + R + \sqrt{R^2 + \left(\dfrac{h_2 - h_1}{2}\right)^2} \right]$

$V = \pi R^2 \dfrac{h_1 + h_2}{2}$

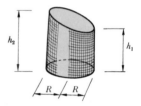

(c) Ungula of right circular cylinder $\qquad (2\omega =$ central angle)

$A = \dfrac{2Rh}{b}[(b - R)\omega + a]$

$V = \dfrac{h}{3b}[a(3R^2 - a^2) + 3R^2(b - R)\omega]$

If $a = b = R$, then

$A = 2Rh \qquad\qquad V = \tfrac{2}{3}R^2 h$

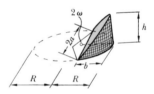

(d) Hollow right circular cylinder

$t = R - r \qquad\qquad \rho = \dfrac{R + r}{2}$

$A = 2\pi R v$

$B = \pi(R^2 - r^2)$

$V = \pi v(R^2 - r^2) = 2\pi v t \rho$

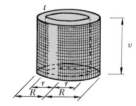

(e) General circular cylinder

$C =$ circumference of right section

$A = Ch \qquad\qquad B = \pi R^2$

$V = Bv$

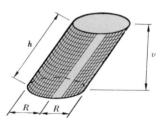

$\pi = 3.141\ 59 \cdots$	$R = $ radius	$A = $ lateral area
$r = $ radius	$B = $ area of base	$S = $ surface
$h = $ slant height	$v = $ altitude	$V = $ volume

(a) Circular right cone

$$A = \pi R \sqrt{v^2 + R^2} = \pi R h$$

$$B = \pi R^2$$

$$S = \pi R(R + h)$$

$$V = \frac{\pi R^2 v}{3}$$

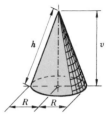

(b) Frustum of right cone　　　(subscript b = bottom, t = top)

$$A = \pi(R_1 + R_2)\sqrt{v^2 + (R_1 - R_2)^2} = \pi(R_1 + R_2)h$$

$$B_b = \pi R_1^2 \qquad B_t = \pi R_2^2$$

$$S = \pi[R_1^2 + (R_1 + R_2)h + R_2^2]$$

$$V = \frac{\pi v}{3}(R_1^2 + R_1 R_2 + R_2^2)$$

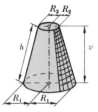

(c) General cone

$$V = \frac{Bv}{3}$$

For the frustum,

$$V = \frac{v_1}{3}(B_b + B_t + \sqrt{B_b B_t})$$

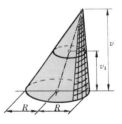

(d) Torus

$$S = 4\pi^2 Rr \approx 39.4784 Rr$$

$$V = 2\pi^2 Rr^2 \approx 19.7392 Rr^2$$

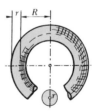

(e) Circular barrel

For circular curvature,

$$V = \tfrac{1}{3}\pi v(2R^2 + r^2)$$

For parabolic curvature,

$$V = \tfrac{1}{15}\pi v(8R^2 + 4Rr + 3r^2)$$

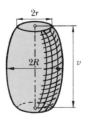

$$\pi = 3.141\ 59 \cdots \qquad R = \text{radius} \qquad A = \text{lateral area}$$
$$a, b = \text{radii} \qquad D = \text{diameter} \qquad S = \text{surface}$$
$$v = \text{altitude} \qquad V = \text{volume}$$

(a) Sphere

$$S = 4\pi R^2$$
$$= \pi D^2 = \sqrt[3]{36\pi V^2}$$
$$V = \tfrac{4}{3}\pi R^3$$
$$= \frac{\pi D^3}{6} = \frac{1}{6}\sqrt{\frac{S^3}{\pi}}$$

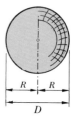

(b) Spherical sector

$$S = \pi R(2v + a)$$
$$V = \frac{2\pi}{3}R^2 v$$

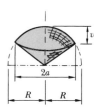

(c) Spherical sector   (one base)

$$a = \sqrt{v(2R - v)}$$
$$A = 2\pi R v$$
$$S = \pi v(4R - v)$$
$$V = \frac{\pi}{3}v^2(3R - v)$$

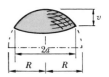

(d) Spherical sector   (two bases)

$$R^2 = a^2 + \left(\frac{a^2 - b^2 - v^2}{2v}\right)^2$$
$$A = 2\pi R v$$
$$S = \pi(2Rv + a^2 + b^2)$$
$$V = \frac{\pi v}{6}(3a^2 + 3b^2 + v^2)$$

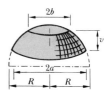

(e) Conical ring

$$S = 2\pi R\left(v + \sqrt{R^2 - \frac{v^2}{4}}\right)$$
$$V = \frac{2\pi}{3}R^2 v$$

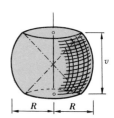

3

TRIGONOMETRY

a, b = legs	c = hypotenuse	A = area
A, B, C = vertices	p, q = segment of c	R = circumradius
α, β, γ = angles	h = height	r = inradius

(1) Relationships

$$a^2 + b^2 = c^2 \qquad \alpha + \beta = 90°$$

$$\sin \alpha = \frac{a}{c} \qquad\qquad \sin \beta = \frac{b}{c}$$

$$\cos \alpha = \frac{b}{c} \qquad\qquad \cos \beta = \frac{a}{c}$$

$$\tan \alpha = \frac{a}{b} \qquad\qquad \tan \beta = \frac{b}{a}$$

$$\cot \alpha = \frac{b}{a} \qquad\qquad \cot \beta = \frac{a}{b}$$

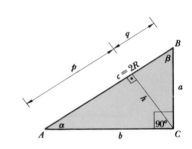

$$\alpha = \sin^{-1}\frac{a}{c} = \cos^{-1}\frac{b}{c} = \tan^{-1}\frac{a}{b} \qquad\qquad \beta = \sin^{-1}\frac{b}{c} = \cos^{-1}\frac{a}{c} = \tan^{-1}\frac{b}{a}$$

(2) General Formulas

$$h = a \sin \beta \qquad\qquad p = b \cos \alpha \qquad\qquad q = a \cos \beta$$
$$= b \sin \alpha \qquad\qquad = h \cot \alpha \qquad\qquad = h \cot \beta$$
$$h = \sqrt{ab \cos \alpha \cos \beta} \qquad\qquad c = b \cos \alpha + a \cos \beta$$

$$A = \frac{ab}{2} = \frac{c^2}{4} \sin 2\alpha = \frac{a^2}{2} \cot \alpha = \frac{b^2}{2} \cot \beta \qquad r = \frac{a + b - c}{2} = \frac{c(\sin \alpha + \sin \beta - 1)}{2} \qquad R = \frac{c}{2}$$

(3) Solutions

Known	Solution					
	a	b	c	α	β	A
a, b			$\sqrt{a^2 + b^2}$	$\tan^{-1}\frac{a}{b}$	$\tan^{-1}\frac{b}{a}$	$\frac{ab}{2}$
a, c		$\sqrt{c^2 - a^2}$		$\sin^{-1}\frac{a}{c}$	$\cos^{-1}\frac{a}{c}$	$\frac{a\sqrt{c^2 - a^2}}{2}$
a, α		$a \cot \alpha$	$\frac{a}{\sin \alpha}$		$90° - \alpha$	$\frac{a^2 \cot \alpha}{2}$
b, α	$b \tan \alpha$		$\frac{b}{\cos \alpha}$		$90° - \alpha$	$\frac{b^2 \tan \alpha}{2}$
c, α	$c \sin \alpha$	$c \cos \alpha$			$90° - \alpha$	$\frac{c^2 \sin 2\alpha}{4}$

Note: For definition of $\sin^{-1}(\ )$, $\cos^{-1}(\ )$, $\tan^{-1}(\ )$ see Sec. 6.08.

$a, b =$ legs	$c =$ hypotenuse	$\bar{R} =$ radius of sphere
$A, B, C =$ vertices	$h =$ height	$R =$ circumradius
$\alpha, \beta =$ angles	$A =$ area	$r =$ inradius

(1) Relationships

In the circle diagram of elements $\alpha, c, \beta, \bar{a}, \bar{b}$, the cosine of any element equals the product of the cotangents of the adjacent elements, and the cosine of any element equals the product of the sines of the opposite elements.

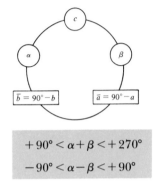

$$
\begin{aligned}
\cos \bar{a} &= \cot \bar{b} \cot \beta & \cos \bar{a} &= \sin \alpha \sin c \\
\cos \bar{b} &= \cot \bar{a} \cot \alpha & \cos \bar{b} &= \sin \beta \sin c \\
\cos c &= \cot \alpha \cot \beta & \cos c &= \sin \bar{a} \sin \bar{b} \\
\cos \alpha &= \cot \bar{b} \cot c & \cos \alpha &= \sin \bar{a} \sin \beta \\
\cos \beta &= \cot \bar{a} \cot c & \cos \beta &= \sin \bar{b} \sin \alpha
\end{aligned}
$$

$$+90° < \alpha + \beta < +270°$$

$$-90° < \alpha - \beta < +90°$$

(2) General Formulas $(2p = a + b + c, \ 2\sigma = \alpha + \beta + 90°)$

$$h = \sin^{-1}(\sin \alpha \sin b) = \sin^{-1}(\sin \beta \sin a)$$

$$A = \pi \bar{R}^2 (\alpha + \beta - 90°)/180°$$

$$R = \cot^{-1} \sqrt{\frac{\cos(\sigma - \alpha) \cos(\sigma - \beta) \cos(\sigma - 90°)}{-\cos \sigma}}$$

$$r = \tan^{-1} \sqrt{\frac{\sin(p - a) \sin(p - b) \sin(p - c)}{\sin p}}$$

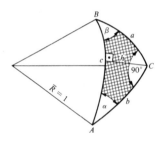

(3) Solutions

Known	Solution				
	a	b	c	α	β
a, b			$\cos^{-1}(\cos a \cos b)$	$\tan^{-1} \dfrac{\tan a}{\sin b}$	$\tan^{-1} \dfrac{\tan b}{\sin a}$
a, c		$\cos^{-1} \dfrac{\cos c}{\cos a}$		$\sin^{-1} \dfrac{\sin a}{\sin c}$	$\cos^{-1} \dfrac{\tan a}{\tan c}$
a, α		$\sin^{-1} \dfrac{\tan a}{\tan \alpha}$	$\sin^{-1} \dfrac{\sin a}{\sin \alpha}$		$\sin^{-1} \dfrac{\cos \alpha}{\cos a}$
b, α	$\tan^{-1}(\sin b \tan \alpha)$		$\tan^{-1} \dfrac{\tan b}{\cos \alpha}$		$\cos^{-1}(\cos b \sin \alpha)$
c, α	$\sin^{-1}(\sin c \sin \alpha)$	$\tan^{-1}(\tan c \cos \alpha)$			$\cot^{-1}(\cos c \tan \alpha)$

Note: For definition of $\sin^{-1}(\ )$, $\cos^{-1}(\ )$, $\tan^{-1}(\ )$, $\cot^{-1}(\ )$ see Sec. 6.08.

a, b, c = sides	h = altitude	A = area
A, B, C = vertices	m = median	R = circumradius
α, β, γ = angles	t = bisector	r = inradius

(1) Basic Laws $(\alpha + \beta + \gamma = 180°)$

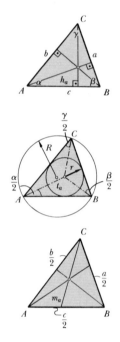

(a) Law of sines

$$a : b : c = \sin\alpha : \sin\beta : \sin\gamma$$

(b) Law of cosines

$$a^2 = b^2 + c^2 - 2bc\cos\alpha$$

(c) Law of tangents

$$(a+b) : (a-b) = \tan\frac{\alpha+\beta}{2} : \tan\frac{\alpha-\beta}{2}$$

(d) Law of projection

$$a = b\cos\gamma + c\cos\beta$$

(e) Law of angles

$$\sin(\alpha+\beta) = \sin\gamma$$

$$\cos(\alpha+\beta) = -\cos\gamma$$

$$\tan(\alpha+\beta) = -\tan\gamma$$

$$\cot(\alpha+\beta) = -\cot\gamma$$

(2) General Formulas $(2p = a + b + c)$

(a) Angles

$$\alpha = 2\sin^{-1}\sqrt{\frac{(p-b)(p-c)}{bc}}$$

$$= 2\cos^{-1}\sqrt{\frac{p(p-a)}{bc}}$$

$$= 2\tan^{-1}\sqrt{\frac{(p-b)(p-c)}{p(p-a)}}$$

(b) Radii

$$R = \frac{a}{2\sin\alpha}$$

$$r = (p-a)\tan\frac{\alpha}{2}$$

$$r = 4R\sin\frac{\alpha}{2}\sin\frac{\beta}{2}\sin\frac{\gamma}{2}$$

(c) Segments

$$h_a = b\sin\gamma = c\sin\beta$$

$$m_a = \tfrac{1}{2}\sqrt{b^2 + c^2 + 2bc\cos\alpha}$$

$$t_a = \frac{2bc\cos(\alpha/2)}{b+c}$$

(d) Area

$$A = \frac{ah_a}{2} = \frac{ab\sin\gamma}{2}$$

$$= \frac{abc}{4R} = \frac{a^2\sin\beta\sin\gamma}{2\sin\alpha}$$

$$= \sqrt{p(p-a)(p-b)(p-c)}$$

Note: Additional formulas are obtained by simultaneous cyclic substitution. For definition of $\sin^{-1}(\ )$, $\cos^{-1}(\ )$, $\tan^{-1}(\ )$ see Sec. 6.08.

Known	Solution* $(2p = a + b + c)$
a, b, c	$\alpha = \cos^{-1}\dfrac{b^2 + c^2 - a^2}{2bc}$ or $\alpha = 2\cos^{-1}\sqrt{\dfrac{p(p-a)}{bc}}$ $\beta = \cos^{-1}\dfrac{a^2 + c^2 - b^2}{2ac}$ or $\beta = 2\cos^{-1}\sqrt{\dfrac{p(p-b)}{ac}}$ $\gamma = 180° - (\alpha + \beta)$ $A = \sqrt{p(p-a)(p-b)(p-c)}$
a, b, α	$\beta = \sin^{-1}\dfrac{b\sin\alpha}{a}$ $\gamma = 180° - (\alpha + \beta)$ $c = \dfrac{a\sin\gamma}{\sin\alpha}$ $A = \dfrac{ab}{2}\sin\gamma$
a, b, γ	$X = 90° - \dfrac{\gamma}{2}$ $Y = \tan^{-1}\left(\dfrac{a-b}{a+b}\cot\dfrac{\gamma}{2}\right)$ $\alpha = X + Y$ $\beta = X - Y$ $c = \sqrt{a^2 + b^2 - 2ab\cos\gamma}$ $A = \dfrac{ab}{2}\sin\gamma$
a, α, β	$\gamma = 180° - (\alpha + \beta)$ $b = \dfrac{a\sin\beta}{\sin\alpha}$ $c = \dfrac{a\sin\gamma}{\sin\alpha} = \dfrac{a\sin(\alpha+\beta)}{\sin\alpha}$ $A = \dfrac{ab}{2}\sin\gamma$
a, β, γ	$\alpha = 180° - (\beta + \gamma)$ $b = \dfrac{a\sin\beta}{\sin\alpha} = \dfrac{a\sin\beta}{\sin(\beta+\gamma)}$ $c = \dfrac{a\sin\gamma}{\sin\alpha} = \dfrac{a\sin\gamma}{\sin(\beta+\gamma)}$ $A = \dfrac{a^2\sin\beta\sin\gamma}{2\sin(\beta+\gamma)}$

*p, X, Y = pocket calculator storage.

$a, b, c = $ sides	$h = $ height	$\bar{R} = $ radius of sphere
$A, B, C = $ vertices	$\epsilon = $ spherical excess	$R = $ circumradius
$\alpha, \beta, \gamma = $ angles	$d = $ spherical defect	$r = $ inradius
$2p = a + b + c$	$2\sigma = \alpha + \beta + \gamma$	$A = $ area

(1) Basic Laws

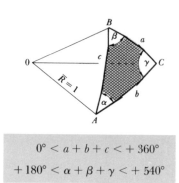

(a) Law of sines

$$\sin a : \sin b : \sin c = \sin \alpha : \sin \beta : \sin \gamma$$

(b) Law of cosines I

$$\cos a = \cos b \cos c + \sin b \sin c \cos \alpha$$

(c) Law of cosines II

$$\cos \alpha = -\cos \beta \cos \gamma + \sin \beta \sin \gamma \cos a$$

$$0° < a + b + c < +360°$$
$$+180° < \alpha + \beta + \gamma < +540°$$

(2) General Formulas

(a) Delambre's equations

$$\sin \frac{\alpha + \beta}{2} \cos \frac{c}{2} = \cos \frac{a - b}{2} \cos \frac{\gamma}{2}$$

$$\sin \frac{\alpha - \beta}{2} \sin \frac{c}{2} = \sin \frac{a - b}{2} \cos \frac{\gamma}{2}$$

$$\cos \frac{\alpha + \beta}{2} \cos \frac{c}{2} = \cos \frac{a + b}{2} \sin \frac{\gamma}{2}$$

$$\cos \frac{\alpha - \beta}{2} \sin \frac{c}{2} = \sin \frac{a + b}{2} \sin \frac{\gamma}{2}$$

(b) Napier's equations

$$\tan \frac{\alpha + \beta}{2} \cos \frac{a + b}{2} = \cos \frac{a - b}{2} \cot \frac{\gamma}{2}$$

$$\tan \frac{\alpha - \beta}{2} \sin \frac{a + b}{2} = \sin \frac{a - b}{2} \cot \frac{\gamma}{2}$$

$$\tan \frac{a + b}{2} \cos \frac{\alpha + \beta}{2} = \cos \frac{\alpha - \beta}{2} \tan \frac{c}{2}$$

$$\tan \frac{a - b}{2} \sin \frac{\alpha + \beta}{2} = \sin \frac{\alpha - \beta}{2} \tan \frac{c}{2}$$

(c) Circumradius

$$R = \cot^{-1} \sqrt{\frac{\cos (\sigma - \alpha) \cos (\sigma - \beta) \cos (\sigma - \gamma)}{-\cos \sigma}}$$

(d) Inradius

$$r = \tan^{-1} \sqrt{\frac{\sin (p - a) \sin (p - b) \sin (p - c)}{\sin p}}$$

(e) Spherical angle

$$A = \frac{\pi \bar{R}^2 \alpha°}{90°}$$

(f) Spherical triangle

$$A = \frac{\pi \bar{R}^2 \epsilon°}{180°}$$

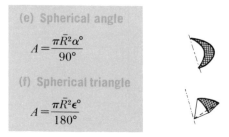

(g) Angles

$$\alpha = 2 \tan^{-1} \sqrt{\frac{\sin (p - b) \sin (p - c)}{\sin p \sin (p - a)}} \qquad a = 2 \tan^{-1} \sqrt{\frac{-\cos \sigma \cos (\sigma - \alpha)}{\cos (\sigma - \beta) \cos (\sigma - \gamma)}}$$

(h) Spherical excess and defect

$$\epsilon = \alpha + \beta + \gamma - 180° = 4 \tan^{-1} \sqrt{\tan \frac{p}{2} \tan \frac{p - a}{2} \tan \frac{p - b}{2} \tan \frac{p - c}{2}}$$

$$d = 360° - (a + b + c) = 2a - 4 \tan^{-1} \sqrt{\cot \frac{\sigma}{2} \cot \frac{\sigma - \alpha}{2} \tan \frac{\sigma - \beta}{2} \tan \frac{\sigma - \gamma}{2}}$$

Note: Additional formulas are obtained by simultaneous cyclic substitution. For definition of $\sin^{-1}( )$, $\cos^{-1}( )$, $\tan^{-1}( )$, $\cot^{-1}( )$ see Sec. 6.08.

Known	Solution* ($\alpha + \beta + \gamma > 180°$)
a, b, c	$p = \dfrac{a+b+c}{2}$ $X = \sqrt{\dfrac{\sin (p-a) \sin (p-b) \sin (p-c)}{\sin p}}$ $\alpha = 2 \tan^{-1} \dfrac{X}{\sin (p-a)} \qquad \beta = 2 \tan^{-1} \dfrac{X}{\sin (p-b)} \qquad \gamma = 2 \tan^{-1} \dfrac{X}{\sin (p-c)}$
α, β, γ	$\sigma = \dfrac{\alpha + \beta + \gamma}{2}$ $Y = \sqrt{\dfrac{\cos (\sigma - \alpha) \cos (\sigma - \beta) \cos (\sigma - \gamma)}{-\cos \sigma}}$ $a = 2 \tan^{-1} \dfrac{\cos (\sigma - \alpha)}{Y} \qquad b = 2 \tan^{-1} \dfrac{\cos (\sigma - \beta)}{Y} \qquad c = 2 \tan^{-1} \dfrac{\cos (\sigma - \gamma)}{Y}$
a, b, α	$X = \dfrac{\sin b \sin \alpha}{\sin a}$ $Y = \tan \dfrac{a+b}{2} \dfrac{\cos [(\alpha + \beta)/2]}{\cos [(\alpha - \beta)/2]}$ $Z = \tan \dfrac{\alpha + \beta}{2} \dfrac{\cos [(a+b)/2]}{\cos [(a-b)/2]}$ $\beta = \sin^{-1} X \qquad c = 2 \tan^{-1} Y \qquad \gamma = 2 \cot^{-1} Z$
a, b, γ	$X = \tan^{-1} \dfrac{\cos (\gamma/2) \cos [(a-b)/2]}{\sin (\gamma/2) \cos [(a+b)/2]}$ $Y = \tan^{-1} \dfrac{\cos (\gamma/2) \sin [(a-b)/2]}{\sin (\gamma/2) \sin [(a+b)/2]}$ $Z = \dfrac{\cos (\gamma/2) \cos [(a-b)/2]}{\sin X}$ $\alpha = X + Y \qquad \beta = X - Y \qquad c = 2 \cos^{-1} Z$
a, β, γ	$X = \tan^{-1} \dfrac{\sin (a/2) \cos [(\beta - \gamma)/2]}{\cos (a/2) \cos [(\beta + \gamma)/2]}$ $Y = \tan^{-1} \dfrac{\sin (a/2) \sin [(\beta - \gamma)/2]}{\cos (a/2) \sin [(\beta + \gamma)/2]}$ $Z = \dfrac{\cos (a/2) \cos [(\beta + \gamma)/2]}{\cos X}$ $b = X + Y \qquad c = X - Y \qquad \alpha = 2 \sin^{-1} Z$

*p, σ, X, Y, Z = pocket calculator storage.

$a, b, c, d = $ sides	$e, f = $ diagonals	$A = $ area
$A, B, C, D = $ vertices	$\alpha, \beta, \gamma, \delta = $ angles	$\omega = $ central angle $\leq \dfrac{\pi}{2}$

(1) General Formulas $(2s = a + b + c + d)$

$$\alpha = \alpha_1 + \alpha_2 \qquad \beta = \beta_1 + \beta_2$$
$$\gamma = \gamma_1 + \gamma_2 \qquad \delta = \delta_1 + \delta_2$$

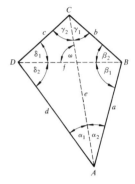

$$\alpha + \beta + \gamma + \delta = 360°$$

$$\omega = \cos^{-1} \frac{b^2 + d^2 - a^2 - c^2}{2ef} = \sin^{-1} \frac{2A}{ef}$$

$$A = \frac{ef}{2} \sin \omega = (b^2 + d^2 - a^2 - c^2) \frac{\tan \omega}{4}$$

$$= \sqrt{(s-a)(s-b)(s-c)(s-d) - abcd \cos^2 \frac{\alpha + \gamma}{2}}$$

(2) Solutions

Known	Solution*
a, d α, β, δ	$X = 90° - \dfrac{\alpha}{2} \qquad\qquad Y = \tan^{-1}\left(\dfrac{a-d}{a+d} \cot \dfrac{\alpha}{2}\right) \qquad \gamma = 360° - \alpha - \beta - \delta$ $\beta_1 = X - Y \qquad\qquad\qquad \delta_2 = X + Y \qquad\qquad\qquad f = a\cos(X-Y) + d\cos(X+Y)$ $b = \dfrac{f \sin(\delta - X - Y)}{\sin \gamma} \qquad\qquad\qquad\qquad\qquad c = \dfrac{f \sin(\beta - X + Y)}{\sin \gamma}$
a, d $\alpha, \gamma_1, \gamma_2$	$X = \tan^{-1} \dfrac{d \sin \gamma_1}{a \sin(\alpha + \gamma) \sin \gamma_2} \qquad\qquad\qquad Y = \tan^{-1} \dfrac{a \sin \gamma_2}{d \sin(\alpha + \gamma) \sin \gamma_1}$ $\beta = \cot^{-1}\left[-\dfrac{\cos(\alpha + \gamma - Y)}{\sin(\alpha + \gamma) \cos Y}\right] \qquad \delta = \cot^{-1}\left[-\dfrac{\cos(\alpha + \gamma - X)}{\sin(\alpha + \gamma) \cos X}\right]$ $\alpha_1 = 180° - \delta - \gamma_2 \qquad\qquad\qquad \alpha_2 = 180° - \beta - \gamma_1$ $b = \dfrac{a \sin \alpha_2}{\sin \gamma_1} \qquad\qquad\qquad\qquad c = \dfrac{d \sin \alpha_1}{\sin \gamma_2}$
a α_1, α_2 β_1, β_2	$b = \dfrac{a \sin \alpha_2}{\sin(\beta + \alpha_2)} \qquad\qquad\qquad\qquad d = \dfrac{a \sin \beta_1}{\sin(\alpha + \beta_1)}$ $e = \dfrac{a \sin \beta}{\sin(\beta + \alpha_2)} \qquad\qquad\qquad\qquad f = \dfrac{a \sin \alpha}{\sin(\alpha + \beta_1)}$ $c = \sqrt{d^2 + e^2 - 2de \cos \alpha_1} = \sqrt{b^2 + f^2 - 2bf \cos \beta_2}$

*First four quantities = pocket calculator storage.

4
PLANE ANALYTIC GEOMETRY

(1) Systems of Coordinates

(a) Cartesian coordinates

A point P is given by two mutually perpendicular distances x,y (coordinates) measured from two mutually perpendicular axes X,Y (coordinate axes) intersecting at the origin 0.

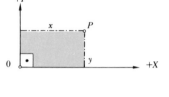

(b) Skew coordinates

A point P is given by the coordinates u,v parallel to two skew axes U,V intersecting at the origin 0.

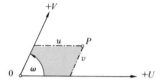

(c) Polar coordinates

A point P is given by two polar coordinates associated with a fixed axis X (polar axis) and a fixed point 0 on this axis (pole). The first coordinate is the radius r, the distance from 0 to P, and the second coordinate is the position angle θ, measured from $+X$ to r.

(2) Relationships

	Cartesian coordinates	Skew coordinates	Polar coordinates
Cartesian coordinates	$x = x$ $y = y$	$x = u + v\cos\omega$ $y = v\sin\omega$	$x = r\cos\theta$ $y = r\sin\theta$
Skew coordinates	$u = x - y\cot\omega$ $v = y\csc\omega$	$u = u$ $v = v$	$u = r\dfrac{\sin(\omega - \theta)}{\sin\omega}$ $v = r\dfrac{\sin\theta}{\sin\omega}$
Polar coordinates	$r = \sqrt{x^2 + y^2}$ $\theta = \tan^{-1}\dfrac{y}{x}$	$r = \sqrt{u^2 + v^2 + 2uv\cos\omega}$ $\theta = \tan^{-1}\dfrac{v\sin\omega}{u + v\cos\omega}$	$r = r$ $\theta = \theta$

(1) **Distance of Two Points, Segment in Plane** $\overline{P_1P_2}$

(a) **Cartesian coordinates** $P_1(x_1, y_1); P_2(x_2, y_2)$

$$d = \sqrt{(x_2 - x_1)^2 + (y_2 - y_1)^2}$$

$$\tan \alpha = \frac{y_2 - y_1}{x_2 - x_1}$$

$$\cos \alpha = \frac{x_2 - x_1}{d}$$

$$\cos \beta = \frac{y_2 - y_1}{d}$$

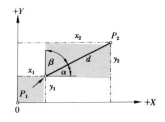

(b) **Skew coordinates** $P_1(u_1, v_1); P_2(u_2, v_2)$

$$d = \sqrt{(u_2 - u_1)^2 + (v_2 - v_1)^2 + 2(u_2 - u_1)(v_2 - v_1)\cos \omega}$$

$$\tan \alpha = \frac{(v_2 - v_1)\sin \omega}{(u_2 - u_1) + (v_2 - v_1)\cos \omega}$$

$$\cos \alpha = \frac{(u_2 - u_1) + (v_2 - v_1)\cos \omega}{d}$$

$$\cos \gamma = \frac{(v_2 - v_1) + (u_2 - u_1)\cos \omega}{d}$$

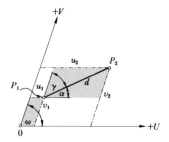

(c) **Polar coordinates** $P_1(r_1, \theta_1); P_2(r_2, \theta_2)$

$$d = \sqrt{r_1^2 + r_2^2 - 2r_1r_2\cos(\theta_1 - \theta_2)}$$

$$\tan \alpha = \frac{r_2 \sin \theta_2 - r_1 \sin \theta_1}{r_2 \cos \theta_2 - r_1 \cos \theta_1}$$

$$\cos \alpha = \frac{r_2 \cos \theta_2 - r_1 \cos \theta_1}{d}$$

$$\cos \beta = \frac{r_2 \sin \theta_2 - r_1 \sin \theta_1}{d}$$

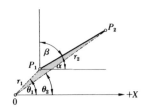

(2) **Three Points** $P_1(x_1, y_1); P_2(x_2, y_2); P_3(x_3, y_3)$

(a) **Area of triangle** $(P_1P_2P_3)$

$$A = \tfrac{1}{2}\begin{vmatrix} x_1 & y_1 & 1 \\ x_2 & y_2 & 1 \\ x_3 & y_3 & 1 \end{vmatrix}$$

(b) $P_1P_2P_3$ **on a straight line**

$$0 = \begin{vmatrix} x_1 & y_1 & 1 \\ x_2 & y_2 & 1 \\ x_3 & y_3 & 1 \end{vmatrix}$$

(1) Algebraic Transformations

(a) Translation   **(b) Rotation**   **(c) Translation and rotation**

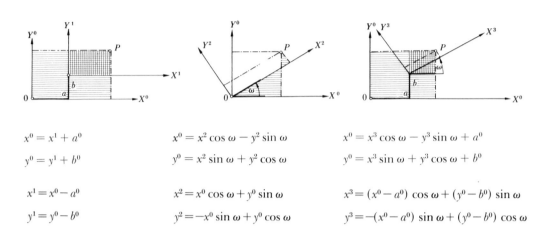

$$x^0 = x^1 + a^0$$
$$y^0 = y^1 + b^0$$

$$x^0 = x^2 \cos \omega - y^2 \sin \omega$$
$$y^0 = x^2 \sin \omega + y^2 \cos \omega$$

$$x^0 = x^3 \cos \omega - y^3 \sin \omega + a^0$$
$$y^0 = x^3 \sin \omega + y^3 \cos \omega + b^0$$

$$x^1 = x^0 - a^0$$
$$y^1 = y^0 - b^0$$

$$x^2 = x^0 \cos \omega + y^0 \sin \omega$$
$$y^2 = -x^0 \sin \omega + y^0 \cos \omega$$

$$x^3 = (x^0 - a^0) \cos \omega + (y^0 - b^0) \sin \omega$$
$$y^3 = -(x^0 - a^0) \sin \omega + (y^0 - b^0) \cos \omega$$

Note: 0, 1, 2, 3 are superscripts and not exponents.

(2) Matrix Transformations

(a) Translation   **(b) Rotation**   **(c) Translation and rotation**

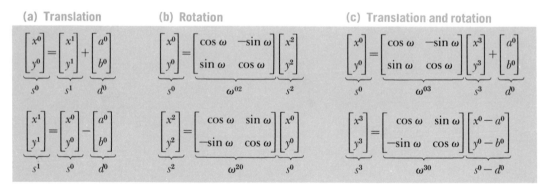

Note: ω matrices are orthogonal, det $(\omega) = 1$, $\omega^T = \omega^{-1}$.

(3) Transformations in Complex Plane $(i = \sqrt{-1}, \text{Sec. } 11.01)$

(a) Translation

$$(x^0 + iy^0) = (x^1 + iy^1) + (a^0 + ib^0)$$

$$(x^1 + iy^1) = (x^0 + iy^0) - (a^0 + ib^0)$$

(b) Rotation

$$(x^0 + iy^0) = (x^2 + iy^2) e^{i\omega}$$

$$(x^2 + iy^2) = (x^0 + iy^0) e^{-i\omega}$$

(c) Translation and rotation

$$(x^0 + iy^0) = (x^3 + iy^3) e^{i\omega} + (a^0 + ib^0)$$

$$(x^3 + iy^3) = [(x^0 + iy^0) - (a^0 + ib^0)] e^{-i\omega}$$

(1) Basic Forms

Direction form

$$y = kx + l$$

Intercept form

$$\frac{x}{a} + \frac{y}{b} = 1$$

Normal form

$$x \cos \beta + y \cos \alpha = n$$

General form

$$Ax + By + C = 0$$

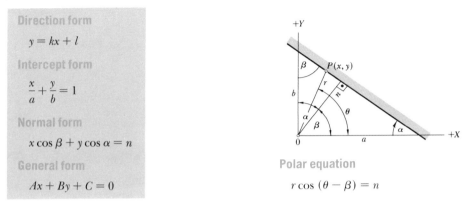

Polar equation

$$r \cos(\theta - \beta) = n$$

(2) Parameters

$A = b$	$\cos \alpha = \pm \dfrac{B}{\sqrt{A^2 + B^2}}$	$\cos \alpha = \pm \dfrac{a}{\sqrt{a^2 + b^2}}$
$B = a$	$\cos \beta = \pm \dfrac{A}{\sqrt{A^2 + B^2}}$	$\cos \beta = \pm \dfrac{b}{\sqrt{a^2 + b^2}}$
$C = -ab$	$n = \pm \dfrac{C}{\sqrt{A^2 + B^2}}$	$n = \pm \dfrac{ab}{\sqrt{a^2 + b^2}}$
$k = \tan \alpha$	$k = -\dfrac{A}{B} \qquad l = -\dfrac{C}{B}$	$k = -\dfrac{b}{a} \qquad l = b$

(3) Two Straight Lines in Plane　　$(A_1 x + B_1 y + C_1 = 0;\ A_2 x + B_2 y + C_2 = 0)$

$\dfrac{A_1}{B_1} \neq \dfrac{A_2}{B_2}$　　Lines intersect at a point　　　　$\dfrac{A_1}{A_2} = \dfrac{B_1}{B_2} = \dfrac{C_1}{C_2}$　　Lines coincide

$\dfrac{A_1}{B_1} = -\dfrac{B_2}{A_2}$　　Lines are normal　　　　$\dfrac{A_1}{B_1} = \dfrac{A_2}{B_2}$　　Lines are parallel

The angle ω between two lines is
the clockwise rotation required
to transform line 2 into line 1.　　　　$\tan \omega = \dfrac{A_1 B_2 - B_1 A_2}{A_1 A_2 + B_1 B_2}$

(4) Distances　　$P(x_0, y_0);\ (A_1 x + B_1 y + C_1 = 0;\ A_2 x + B_2 y + C_2 = 0)$

(a) From a line to a point

$$d = \frac{A_1 x_0 + B_1 y_0 + C_1}{\pm \sqrt{A_1^2 + B_1^2}}$$

(b) Between two parallel lines

$$d = \frac{C_2}{\pm \sqrt{A_2^2 + B_2^2}} - \frac{C_1}{\pm \sqrt{A_1^2 + B_1^2}}$$

Note:　The sign of the denominator is opposite to the sign of C.

(a) General equation

$$Ax^2 + Ay^2 + 2Dx + 2Ey + F = 0$$

Center: $\qquad x_M = -\dfrac{D}{A} \qquad y_M = -\dfrac{E}{A}$

Radius: $\qquad R = \dfrac{\sqrt{D^2 + E^2 - AF}}{A}$

$$(x - x_M)^2 + (y - y_M)^2 = R^2$$

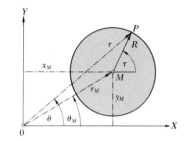

(b) Parametric equation

$$x = x_M + R \cos \tau \qquad y = y_M + R \sin \tau$$

(c) Polar equation

$$r^2 - 2r_M r \cos(\theta - \theta_M) + r_M{}^2 = R^2$$

(d) Special positions

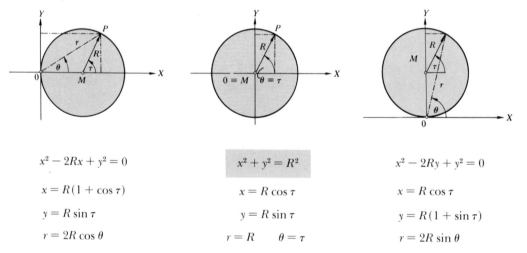

$$x^2 - 2Rx + y^2 = 0$$

$$x = R(1 + \cos \tau)$$

$$y = R \sin \tau$$

$$r = 2R \cos \theta$$

$$x^2 + y^2 = R^2$$

$$x = R \cos \tau$$

$$y = R \sin \tau$$

$$r = R \qquad \theta = \tau$$

$$x^2 - 2Ry + y^2 = 0$$

$$x = R \cos \tau$$

$$y = R(1 + \sin \tau)$$

$$r = 2R \sin \theta$$

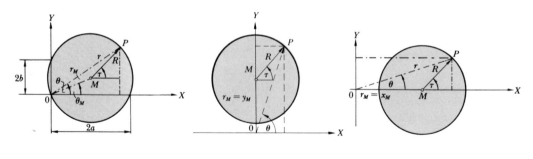

$$x(x - 2a) + y(y - 2b) = 0$$

$$x = a + R \cos \tau$$

$$y = b + R \sin \tau$$

$$r = 2a \cos \theta + 2b \sin \theta$$

$$x^2 + (y - y_M)^2 = R^2$$

$$x = R \cos \tau$$

$$y = y_M + R \sin \tau$$

$$r^2 - 2rr_M \sin \theta + r_M{}^2 = R^2$$

$$(x - x_M)^2 + y^2 = R^2$$

$$x = x_M + R \cos \tau$$

$$y = R \sin \tau$$

$$r^2 - 2rr_M \cos \theta + r_M{}^2 = R^2$$

(a) Notation

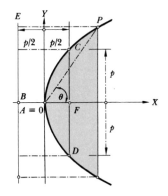

$F = \text{focus}$ $PF = PE$ $0 = \text{vertex}$

$\overline{AF} = \dfrac{p}{2}$ $\overline{BF} = \overline{FC} = p$ $\overline{BA} = \dfrac{p}{2}$

$2p = \text{parameter} = \text{latus rectum}$

$\epsilon = 1$ (numerical eccentricity)

(b) Cartesian equation $(A \equiv 0, X \text{ axis})$

$y^2 = 2px$ $\begin{cases} p > 0 & \text{Open right} \\ p < 0 & \text{Open left} \end{cases}$

(c) Parametric equation $(A \equiv 0, X \text{ axis})$

$x = \dfrac{p}{2}\tau^2$ $y = p\tau$

(d) Polar equations $(X \text{ axis})$

Pole at 0 $r = 2p \cos\theta \,(1 + \cot^2\theta)$ Pole at F $r = \dfrac{p}{1 - \cos\theta}$

(e) Normal position $(X \text{ axis})$

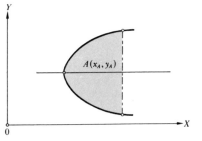

$Cy^2 + 2Dx + 2Ey + F = 0$

Vertex: $x_A = \dfrac{E^2 - CF}{2CD}$ $y_A = -\dfrac{E}{C}$

$p = -\dfrac{D}{C}$

$(y - y_A)^2 = 2p(x - x_A)$

(f) Normal position $(Y \text{ axis})$

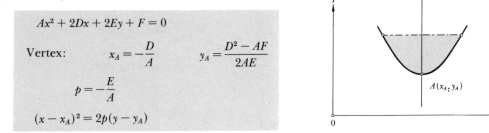

$Ax^2 + 2Dx + 2Ey + F = 0$

Vertex: $x_A = -\dfrac{D}{A}$ $y_A = \dfrac{D^2 - AF}{2AE}$

$p = -\dfrac{E}{A}$

$(x - x_A)^2 = 2p(y - y_A)$

(g) Special cases   (Sec. 4.12)

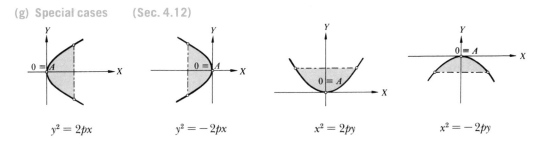

$y^2 = 2px$ $y^2 = -2px$ $x^2 = 2py$ $x^2 = -2py$

(a) Notation

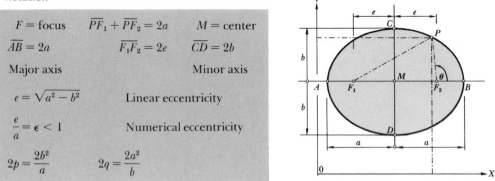

$$F = \text{focus} \qquad \overline{PF_1} + \overline{PF_2} = 2a \qquad M = \text{center}$$

$$\overline{AB} = 2a \qquad \overline{F_1F_2} = 2e \qquad \overline{CD} = 2b$$

Major axis $\qquad\qquad$ Minor axis

$$e = \sqrt{a^2 - b^2} \qquad \text{Linear eccentricity}$$

$$\frac{e}{a} = \epsilon < 1 \qquad \text{Numerical eccentricity}$$

$$2p = \frac{2b^2}{a} \qquad 2q = \frac{2a^2}{b}$$

(b) Cartesian equation $\quad (M \equiv 0)$

$$\frac{x^2}{a^2} + \frac{y^2}{b^2} = 1$$

(c) Parametric equation $\quad (M \equiv 0)$

$$x = a \cos \tau \qquad y = b \sin \tau$$

(d) Polar equations

Pole at $0 \quad r^2 = \dfrac{b^2}{1 - \epsilon^2 \cos^2 \theta}$
$\qquad$
Pole at $F_2 \quad r = \dfrac{p}{1 + \epsilon \cos \theta}$

(e) Normal position

$$Ax^2 + Cy^2 + 2Dx + 2Ey + F = 0$$

Center: $\qquad x_M = -\dfrac{D}{A} \qquad y_M = -\dfrac{E}{C}$

$$a = \sqrt{\frac{CD^2 + AE^2 - ACF}{A^2 C}}$$

$$b = \sqrt{\frac{CD^2 + AE^2 - ACF}{AC^2}}$$

$$\frac{(x - x_M)^2}{a^2} + \frac{(y - y_M)^2}{b^2} = 1$$

(f) Special positions

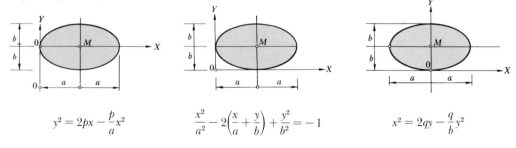

$$y^2 = 2px - \frac{p}{a}x^2$$

$$\frac{x^2}{a^2} - 2\left(\frac{x}{a} + \frac{y}{b}\right) + \frac{y^2}{b^2} = -1$$

$$x^2 = 2qy - \frac{q}{b}y^2$$

(a) Notation

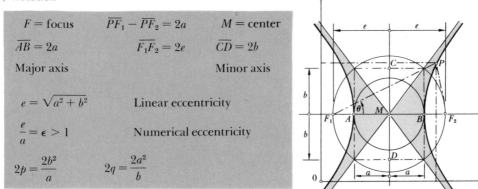

$$F = \text{focus} \qquad \overline{PF_1} - \overline{PF_2} = 2a \qquad M = \text{center}$$

$$\overline{AB} = 2a \qquad \overline{F_1 F_2} = 2e \qquad \overline{CD} = 2b$$

Major axis   Minor axis

$$e = \sqrt{a^2 + b^2} \qquad \text{Linear eccentricity}$$

$$\frac{e}{a} = \epsilon > 1 \qquad \text{Numerical eccentricity}$$

$$2p = \frac{2b^2}{a} \qquad 2q = \frac{2a^2}{b}$$

(b) Cartesian equation $(M \equiv 0)$

$$\frac{x^2}{a^2} - \frac{y^2}{b^2} = 1$$

(c) Parametric equation $(M \equiv 0)$

$$x = \frac{a}{\cos \tau} \qquad y = \pm b \tan \tau$$

(d) Polar equations

Pole at 0 $r^2 = \dfrac{b^2}{\epsilon^2 \cos^2 \theta - 1}$ Pole at F $r = \dfrac{p}{1 + \epsilon \cos \theta}$

(e) Normal position

$$Ax^2 - Cy^2 + 2Dx + 2Ey + F = 0$$

Center: $x_M = -\dfrac{D}{A}$ $y_M = \dfrac{E}{C}$

$$a = \sqrt{\frac{CD^2 - AE^2 - ACF}{A^2 C}}$$

$$b = \sqrt{\frac{CD^2 - AE^2 - ACF}{AC^2}}$$

$$\frac{(x - x_M)^2}{a^2} - \frac{(y - y_M)^2}{b^2} = 1$$

(f) Special positions (Sec. 4.12)

$$y^2 = -2px + \frac{p}{a}x^2$$

$$\frac{x^2}{a^2} - 2\left(\frac{x}{a} - \frac{y}{b}\right) - \frac{y^2}{b^2} = 1$$

$$x^2 = -2qy + \frac{2}{b}y^2 + 2a^2$$

(1) General Equation

The general equation of the second degree,

$$a_{11}x^2 + 2a_{12}xy + a_{22}y^2 + 2a_{13}x + 2a_{23}y + a_{33} = 0$$

defines the following conic sections: a circle, an ellipse, a hyperbola, a parabola, a pair of straight lines, a straight line, or a point (the last three cases are degenerate curves of the second degree).

(2) Invariants

The quantities (note that $a_{ik} = a_{ki}$)

$$I_3 = \begin{vmatrix} a_{11} & a_{12} & a_{13} \\ a_{21} & a_{22} & a_{23} \\ a_{31} & a_{32} & a_{33} \end{vmatrix} \qquad I_2 = \begin{vmatrix} a_{11} & a_{12} \\ a_{21} & a_{22} \end{vmatrix} \qquad I_1 = a_{11} + a_{22}$$

and the sign of the quantity

$$A = \begin{vmatrix} a_{22} & a_{23} \\ a_{32} & a_{33} \end{vmatrix} + \begin{vmatrix} a_{11} & a_{13} \\ a_{31} & a_{33} \end{vmatrix}$$

are invariants of transformation (they are not affected by the translation or the rotation of coordinate axes) and define properties of conics.

(3) Principal Axis

The direction angle of the principal axis of a real ellipse, hyperbola, and parabola is

$$\omega = \frac{1}{2}\tan^{-1}\frac{2a_{12}}{a_{11} - a_{22}} \qquad \leq \frac{\pi}{2}$$

(4) Classification   (CS = conic section)

Type	$I_2 \neq 0$		$I_2 = 0$		
	Central CS		Noncentral CS		
	$I_2 > 0$	$I_2 < 0$			
Proper $I_3 \neq 0$	Real ellipse $I_1 I_3 < 0$	Hyperbola	Parabola		
	Imaginary ellipse $I_1 I_3 > 0$				
Improper $I_3 = 0$	Two nonparallel straight lines		Two parallel straight lines		One straight line
	Imaginary	Real	Real $A < 0$	Imaginary $A > 0$	Real $A = 0$

a_{ij} = constants in X, Y axes (Sec. 4.09)	b_{ij} = constants in U, V axes (Sec. 4.10)
$i, j = 1, 2, 3$	ω = position angle of $\bar{X}$ axis
I_1, I_2, I_3 = invariants (Sec. 4.09)	a, b = semiaxes in U, V axes

(1) Central Conic Section $(I_2 \neq 0)$

(a) Coordinates of center M

$$x_M = -\frac{a_{13}a_{22} - a_{12}a_{23}}{I_2} \qquad y_M = -\frac{a_{11}a_{23} - a_{12}a_{13}}{I_2}$$

(b) Equation of CS in U, V axes

$$b_{11}u^2 + b_{22}v^2 + b_{33} = 0$$

$$b_{11}, b_{22} = \frac{I_1}{2} \pm \sqrt{\left(\frac{I_1}{2}\right)^2 - I_2} \qquad b_{33} = \frac{I_3}{I_2}$$

(c) Principal semiaxes

$$a^2 = \left| -\frac{b_{33}}{b_{11}} \right| \qquad b^2 = \left| -\frac{b_{33}}{b_{22}} \right|$$

For ellipse $\dfrac{b_{33}}{b_{11}} < 0 \qquad \dfrac{b_{33}}{b_{22}} < 0$. For hyperbola $\dfrac{b_{33}}{b_{11}} < 0 \qquad \dfrac{b_{33}}{b_{22}} > 0$

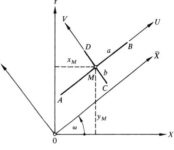

(2) Noncentral Conic Section $(I_2 = 0)$

(a) Equation of principal axis U or V

$$a_{11}x + a_{12}y + \frac{a_{11}a_{13} + a_{12}a_{23}}{I_1} = 0$$

which is tangent to the conic section at the vertex A.

(b) Equation of CS in U, V axes

$$b_{11}u^2 + 2b_{23}v = 0 \qquad b_{22}v^2 + 2b_{13}u = 0$$

$$b_{11} = I_1 \qquad b_{13} = a_{13}\sqrt{\frac{a_{11}}{I_1}} + a_{23}\sqrt{\frac{a_{22}}{I_1}}$$

$$b_{22} = I_1 \qquad b_{23} = a_{23}\sqrt{\frac{a_{11}}{I_1}} - a_{13}\sqrt{\frac{a_{22}}{I_1}}$$

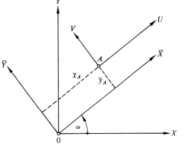

(c) Coordinates of vertex A in $\bar{X}$, $\bar{Y}$ axes

	Axis of symmetry	
	$\bar{X}$ axis	**$\bar{Y}$ axis**
$\bar{x}_A$	$\dfrac{b_{23}}{b_{22}}$	$\dfrac{b_{13}}{b_{11}}$
$\bar{y}_A$	$\dfrac{a_{33}}{2b_{23}} - \dfrac{b_{23}^2}{2b_{22}b_{13}}$	$\dfrac{a_{33}}{2b_{23}} - \dfrac{b_{13}^2}{2b_{11}b_{23}}$

Exponential curve

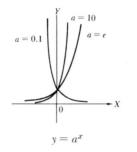

$$y = a^x$$

Logarithmic curve

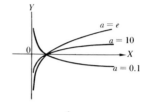

$$y = \log_a x$$

Catenary

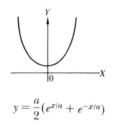

$$y = \frac{a}{2}\left(e^{x/a} + e^{-x/a}\right)$$

Probability curve

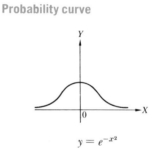

$$y = e^{-x^2}$$

Exponential curve with end point

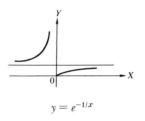

$$y = e^{-1/x}$$

Logarithmic curve with end point

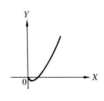

$$y = x \ln x$$

Cubic parabola

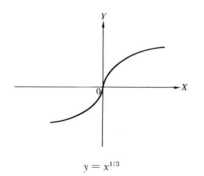

$$y = x^{1/3}$$

Semicubic parabola

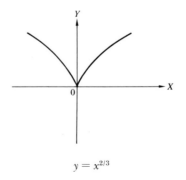

$$y = x^{2/3}$$

Parabola

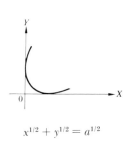

$$x^{1/2} + y^{1/2} = a^{1/2}$$

Equilateral hyperbola

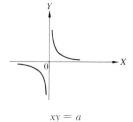

$$xy = a$$

Cusp of first kind

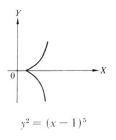

$$y^2 = (x-1)^5$$

Cusp of second kind

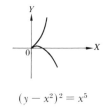

$$(y - x^2)^2 = x^5$$

Power curve

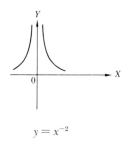

$$y = x^{-2}$$

Cissoid

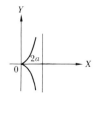

$$y^2(2a - x) = x^3$$

Strophoid

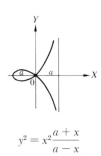

$$y^2 = x^2 \frac{a + x}{a - x}$$

Folium of Descartes

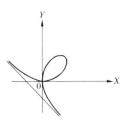

$$x^3 + y^3 - 3axy = 0$$

Parabolic spiral

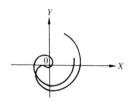

$$(r - a)^2 = 4ak\theta$$

Logarithmic spiral

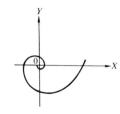

$$\ln r = a\theta$$

Hyperbolic spiral

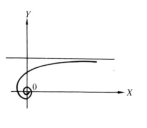

$$r\theta = a$$

Lituus

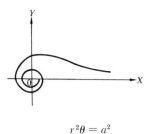

$$r^2\theta = a^2$$

Limacon $(a > b > 0)$

$$r = a + b \cos\theta$$

Cardioid $(a = b)$

$$r = a(1 + \cos\theta)$$

Limacon $(b > a > 0)$

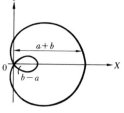

$$r = a + b \cos\theta$$

Ordinary Cycloids

Cusp at origin

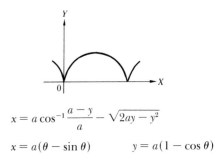

$$x = a \cos^{-1} \frac{a - y}{a} - \sqrt{2ay - y^2}$$

$$x = a(\theta - \sin \theta) \qquad\qquad y = a(1 - \cos \theta)$$

Vertex at origin

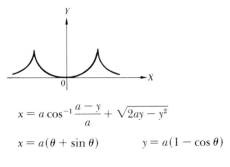

$$x = a \cos^{-1} \frac{a - y}{a} + \sqrt{2ay - y^2}$$

$$x = a(\theta + \sin \theta) \qquad\qquad y = a(1 - \cos \theta)$$

Prolate cycloid   (a < b)

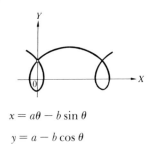

$$x = a\theta - b \sin \theta$$

$$y = a - b \cos \theta$$

Curtate cycloid   (a > b)

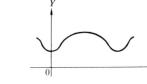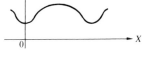

$$x = a\theta - b \sin \theta$$

$$y = a - b \cos \theta$$

Hypocycloid

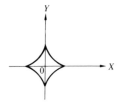

$$x^{2/3} + y^{2/3} = a^{2/3}$$

Evolute of ellipse

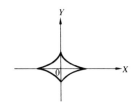

$$(ax)^{2/3} + (by)^{2/3} = (a^2 - b^2)^{2/3}$$

Conchoids

a < b

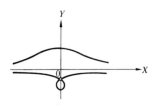

a > b

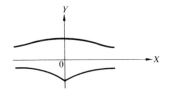

$$(y - a)^2 (x^2 + y^2) = b^2 y^2$$

$$r = a \csc \theta \pm b$$

Lemniscates of Bernoulli

Two-leaved rose

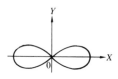

$$(x^2 + y^2)^2 = a^2(x^2 - y^2)$$

$$r^2 = a^2 \cos 2\theta$$

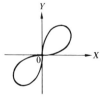

$$(x^2 + y^2)^2 = 2a^2xy$$

$$r^2 = a^2 \sin 2\theta$$

Three-leaved rose

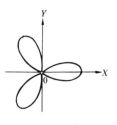

$$r = a \cos 3\theta$$

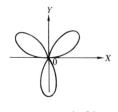

$$r = a \sin 3\theta$$

Four-leaved rose

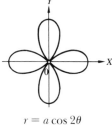

$$r = a \cos 2\theta$$

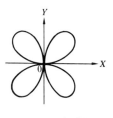

$$r = a \sin 2\theta$$

n-leaved rose

The roses given by $r = a \sin n\theta$ and $r = a \cos n\theta$ have $2n$ leaves if n is even and n leaves if n is odd. The roses given by $r^2 = a^2 \sin n\theta$ and $r^2 = a^2 \cos n\theta$ have n leaves if n is even and $2n$ leaves if n is odd.

5

SPACE ANALYTIC GEOMETRY

(1) Systems of Coordinates

(a) Cartesian coordinates

A point P is given by three mutually perpendicular distances x, y, z (coordinates) measured from three mutually perpendicular YZ, ZX, XY planes, respectively. The lines of intersection of these planes are the coordinate axes X, Y, Z, and the point of their intersection is the origin 0. The right-hand system is shown.

(b) Cylindrical coordinates

A point P is given by its polar coordinates r and θ, in the XY plane, and the cartesian coordinate z.

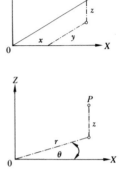

(c) Spherical coordinates

A point P is given by a position angle θ measured from a fixed axis X, a position angle ϕ measured from another fixed axis Z, normal to X, and the position radius ρ, measured from the point of intersection of X and Y, designated as the pole 0.

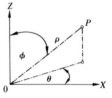

(2) Relationships

	Cartesian	Cylindrical	Spherical
Cartesian	$x = x$ $y = y$ $z = z$	$x = r\cos\theta$ $y = r\sin\theta$ $z = z$	$x = \rho\cos\theta\sin\phi$ $y = \rho\sin\theta\sin\phi$ $z = \rho\cos\phi$
Cylindrical	$r = \sqrt{x^2 + y^2}$ $\theta = \tan^{-1}\dfrac{y}{x}$ $z = z$	$r = r$ $\theta = \theta$ $z = z$	$r = \rho\sin\phi$ $\theta = \theta$ $z = \rho\cos\phi$
Spherical	$\rho = \sqrt{x^2 + y^2 + z^2}$ $\theta = \tan^{-1}\dfrac{y}{x}$ $\phi = \cos^{-1}\dfrac{z}{\sqrt{x^2 + y^2 + z^2}}$	$\rho = \sqrt{r^2 + z^2}$ $\theta = \theta$ $\phi = \cos^{-1}\dfrac{z}{\sqrt{r^2 + z^2}}$	$\rho = \rho$ $\theta = \theta$ $\phi = \phi$

(1) Distance of Two Points, Segment $P_1P_2 = d_{12}$

Segment

$$d_{12} = \sqrt{(x_2 - x_1)^2 + (y_2 - y_1)^2 + (z_2 - z_1)^2}$$

Direction cosines

$$\alpha_{12} = \cos X d_{12} = \frac{x_2 - x_1}{d_{12}} = \frac{d_{12x}}{d_{12}}$$

$$\beta_{12} = \cos Y d_{12} = \frac{y_2 - y_1}{d_{12}} = \frac{d_{12y}}{d_{12}}$$

$$\gamma_{12} = \cos Z d_{12} = \frac{z_2 - z_1}{d_{12}} = \frac{d_{12z}}{d_{12}}$$

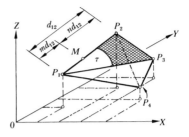

Relationship

$$\alpha^2 + \beta^2 + \gamma^2 = +1$$

(2) Components of Segment d_{12}

$$d_{12x} = d_{12}\alpha_{12} \qquad\qquad d_{12y} = d_{12}\beta_{12} \qquad\qquad d_{12z} = d_{12}\gamma_{12}$$

$$d_{12} = d_{12x}\alpha_{12} + d_{12y}\beta_{12} + d_{12z}\gamma_{12}$$

(3) Coordinates of M Dividing P_1P_2 in Ratio $m:n$

$$x_M = \frac{nx_1 + mx_2}{m + n} \qquad\qquad y_M = \frac{ny_1 + my_2}{m + n} \qquad\qquad z_M = \frac{nz_1 + mz_2}{m + n}$$

(4) Area and Centroid of Triangle with Vertices $P_1P_2P_3$

$$A = \sqrt{A_1^2 + A_2^2 + A_3^2}$$

$$A_1 = \tfrac{1}{2}\begin{vmatrix} y_1 & z_1 & 1 \\ y_2 & z_2 & 1 \\ y_3 & z_3 & 1 \end{vmatrix} \qquad A_2 = \tfrac{1}{2}\begin{vmatrix} z_1 & x_1 & 1 \\ z_2 & x_2 & 1 \\ z_3 & x_3 & 1 \end{vmatrix} \qquad A_3 = \tfrac{1}{2}\begin{vmatrix} x_1 & y_1 & 1 \\ x_2 & y_2 & 1 \\ x_3 & y_3 & 1 \end{vmatrix}$$

$$x_c = \frac{x_1 + x_2 + x_3}{3} \qquad\qquad y_c = \frac{y_1 + y_2 + y_3}{3} \qquad\qquad z_c = \frac{z_1 + z_2 + z_3}{3}$$

(5) Volume of Tetrahedron with Vertices $P_1 P_2 P_3 P_4$

$$V = \tfrac{1}{6}\begin{vmatrix} x_1 & y_1 & z_1 & 1 \\ x_2 & y_2 & z_2 & 1 \\ x_3 & y_3 & z_3 & 1 \\ x_4 & y_4 & z_4 & 1 \end{vmatrix}$$

(6) Angle τ between d_{12} and d_{13}

$$\cos \tau = \alpha_{12}\alpha_{13} + \beta_{12}\beta_{13} + \gamma_{12}\gamma_{13}$$

$$= \frac{d_{12x}d_{13x} + d_{12y}d_{13y} + d_{12z}d_{13z}}{d_{12}d_{13}}$$

(1) Basic Equations of a Plane

Direction form

$$z = k_1 x + k_2 y + l$$

Intercept form

$$\frac{x}{a} + \frac{y}{b} + \frac{z}{c} = 1$$

Normal form

$$\alpha x + \beta y + \gamma z = n$$

General form

$$Ax + By + Cz + D = 0$$

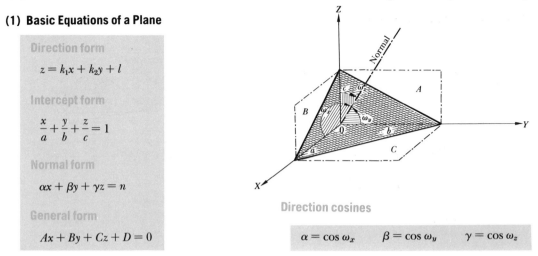

Direction cosines

$$\alpha = \cos \omega_x \qquad \beta = \cos \omega_y \qquad \gamma = \cos \omega_z$$

(2) Relationships

$A = bc$	$\alpha = \dfrac{A}{\pm \sqrt{A^2 + B^2 + C^2}}$	$\alpha = \dfrac{bc}{\pm \sqrt{(ab)^2 + (bc)^2 + (ca)^2}}$
$B = ca$	$\beta = \dfrac{B}{\pm \sqrt{A^2 + B^2 + C^2}}$	$\beta = \dfrac{ca}{\pm \sqrt{(ab)^2 + (bc)^2 + (ca)^2}}$
$C = ab$	$\gamma = \dfrac{C}{\pm \sqrt{A^2 + B^2 + C^2}}$	$\gamma = \dfrac{ab}{\pm \sqrt{(ab)^2 + (bc)^2 + (ca)^2}}$
$D = -abc$	$n = \dfrac{-D}{\pm \sqrt{A^2 + B^2 + C^2}}$	$n = \dfrac{abc}{\pm \sqrt{(ab)^2 + (bc)^2 + (ca)^2}}$
$a = -\dfrac{D}{A}$	$k_1 = -\dfrac{A}{C}$	$k_1 = -\dfrac{c}{a}$
$b = -\dfrac{D}{B}$	$k_2 = -\dfrac{B}{C}$	$k_2 = -\dfrac{c}{b}$
$c = -\dfrac{D}{C}$	$l = -\dfrac{D}{C}$	$l = c$

(3) Plane Passing through a Point P_i in a Given Direction

$$A(x - x_i) + B(y - y_i) + C(z - z_i) = 0$$

(4) Plane Passing through Three Points P_i, P_j, P_k

$$\begin{vmatrix} y_i & z_i & 1 \\ y_j & z_j & 1 \\ y_k & z_k & 1 \end{vmatrix} x + \begin{vmatrix} z_i & x_i & 1 \\ z_j & x_j & 1 \\ z_k & x_k & 1 \end{vmatrix} y + \begin{vmatrix} x_i & y_i & 1 \\ x_j & y_j & 1 \\ x_k & y_k & 1 \end{vmatrix} z = \begin{vmatrix} x_i & y_i & z_i \\ x_j & y_j & z_j \\ x_k & y_k & z_k \end{vmatrix}$$

(5) Distance between the Point P_i and the Plane $Ax + By + Cz + D = 0$

$$d = \frac{Ax_i + By_i + Cz_i + D}{\pm \sqrt{A^2 + B^2 + C^2}}$$

Note: The sign of the denominator is opposite to the sign of D.

(1) Relationships of Two Planes

If two planes are given by their equations as

$$A_1x + B_1y + C_1z + D_1 = 0 \qquad\qquad A_2x + B_2y + C_2z + D_2 = 0$$

then the following are their relationships:

$A_1 : B_1 : C_1 \neq A_2 : B_2 : C_2$	Planes intersect
$A_1 : B_1 : C_1 = A_2 : B_2 : C_2$	Planes are parallel
$A_1A_2 + B_1B_2 + C_1C_2 = 0$	Planes are normal
$A_1 : A_2 = B_1 : B_2 = C_1 : C_2 = D_1 : D_2$	Planes coincide

(2) Angle between the Normals of Two Planes

$$\cos \tau = \frac{A_1A_2 + B_1B_2 + C_1C_2}{\sqrt{A_1^2 + B_1^2 + C_1^2}\,\sqrt{A_2^2 + B_2^2 + C_2^2}}$$

(3) Distance between Two Parallel Planes

$$d = \frac{D_1 - D_2}{\pm\sqrt{A^2 + B^2 + C^2}}$$

$$A_1 = A_2 = A$$
$$B_1 = B_2 = B$$
$$C_1 = C_2 = C$$

(4) Intersection of Two Planes

$$\begin{vmatrix} C_1 & C_2 \\ A_1 & A_2 \end{vmatrix} x + \begin{vmatrix} C_1 & C_2 \\ B_1 & B_2 \end{vmatrix} y + \begin{vmatrix} C_1 & C_2 \\ D_1 & D_2 \end{vmatrix} = 0$$

$$\begin{vmatrix} A_1 & A_2 \\ B_1 & B_2 \end{vmatrix} y + \begin{vmatrix} A_1 & A_2 \\ C_1 & C_2 \end{vmatrix} z + \begin{vmatrix} A_1 & A_2 \\ D_1 & D_2 \end{vmatrix} = 0$$

$$\begin{vmatrix} B_1 & B_2 \\ C_1 & C_2 \end{vmatrix} z + \begin{vmatrix} B_1 & B_2 \\ A_1 & A_2 \end{vmatrix} x + \begin{vmatrix} B_1 & B_2 \\ D_1 & D_2 \end{vmatrix} = 0$$

Each one of these equations represents the projected line of intersection in the respective coordinate plane.

(5) Point of Intersection of Three Planes

$$x = -\frac{\begin{vmatrix} D_1 & B_1 & C_1 \\ D_2 & B_2 & C_2 \\ D_3 & B_3 & C_3 \end{vmatrix}}{\begin{vmatrix} A_1 & B_1 & C_1 \\ A_2 & B_2 & C_2 \\ A_3 & B_3 & C_3 \end{vmatrix}} \qquad y = -\frac{\begin{vmatrix} A_1 & D_1 & C_1 \\ A_2 & D_2 & C_2 \\ A_3 & D_3 & C_3 \end{vmatrix}}{\begin{vmatrix} A_1 & B_1 & C_1 \\ A_2 & B_2 & C_2 \\ A_3 & B_3 & C_3 \end{vmatrix}} \qquad z = -\frac{\begin{vmatrix} A_1 & B_1 & D_1 \\ A_2 & B_2 & D_2 \\ A_3 & B_3 & D_3 \end{vmatrix}}{\begin{vmatrix} A_1 & B_1 & C_1 \\ A_2 & B_2 & C_2 \\ A_3 & B_3 & C_3 \end{vmatrix}}$$

(6) Plane of Symmetry of Two Planes

$$\frac{A_1x + B_1y + C_1z + D_1}{\pm\sqrt{A_1^2 + B_1^2 + C_1^2}} \pm \frac{A_2x + B_2y + C_2z + D_2}{\pm\sqrt{A_2^2 + B_2^2 + C_2^2}} = 0$$

Note: The sign of the denominator is opposite to the sign of the respective D.

(1) General Form

Two linearly independent equations,

$$A_1 x + B_1 y + C_1 z + D_1 = 0 \qquad\qquad A_2 x + B_2 y + C_2 z + D_2 = 0$$

represent a straight line in space. The projections of this line in the coordinate planes and their constants are as follows:

$$\bar{B}x + \bar{A}y + \bar{D}_{xy} = 0 \qquad\qquad \bar{A}z + \bar{C}x + \bar{D}_{zx} = 0 \qquad\qquad \bar{C}y + \bar{B}z + \bar{D}_{yz} = 0$$

$$\bar{A} = \begin{vmatrix} B_1 & C_1 \\ B_2 & C_2 \end{vmatrix} \qquad\qquad \bar{D}_{xy} = \begin{vmatrix} C_1 & D_1 \\ C_2 & D_2 \end{vmatrix} = \bar{D}_{yx} \qquad\qquad \bar{\alpha} = \frac{\bar{A}}{\sqrt{\bar{A}^2 + \bar{B}^2 + \bar{C}^2}}$$

$$\bar{B} = \begin{vmatrix} C_1 & A_1 \\ C_2 & A_2 \end{vmatrix} \qquad\qquad \bar{D}_{zx} = \begin{vmatrix} A_1 & D_1 \\ A_2 & D_2 \end{vmatrix} = \bar{D}_{xz} \qquad\qquad \bar{\beta} = \frac{\bar{B}}{\sqrt{\bar{A}^2 + \bar{B}^2 + \bar{C}^2}}$$

$$\bar{C} = \begin{vmatrix} A_1 & B_1 \\ A_2 & B_2 \end{vmatrix} \qquad\qquad \bar{D}_{yz} = \begin{vmatrix} B_1 & D_1 \\ B_2 & D_2 \end{vmatrix} = \bar{D}_{zy} \qquad\qquad \bar{\gamma} = \frac{\bar{C}}{\sqrt{\bar{A}^2 + \bar{B}^2 + \bar{C}^2}}$$

(2) Direction Form

$y = k_{yx}x + l_{yx}$	$x = k_{xy}y + l_{xy}$	$x = k_{xz}z + l_{xz}$
$z = k_{zx}x + l_{zx}$	$z = k_{zy}y + l_{zy}$	$y = k_{yz}z + l_{yz}$
$k_{yx} = -\dfrac{\bar{B}}{\bar{A}}$	$k_{xy} = -\dfrac{\bar{A}}{\bar{B}}$	$k_{xz} = -\dfrac{\bar{A}}{\bar{C}}$
$k_{zx} = -\dfrac{\bar{C}}{\bar{A}}$	$k_{zy} = -\dfrac{\bar{C}}{\bar{B}}$	$k_{yz} = -\dfrac{\bar{B}}{\bar{C}}$
$l_{yx} = -\dfrac{\bar{D}_{xy}}{\bar{A}}$	$l_{xy} = -\dfrac{\bar{D}_{yx}}{\bar{B}}$	$l_{xz} = -\dfrac{\bar{D}_{xz}}{\bar{C}}$
$l_{zx} = -\dfrac{\bar{D}_{zx}}{\bar{A}}$	$l_{zy} = -\dfrac{\bar{D}_{zy}}{\bar{B}}$	$l_{yz} = -\dfrac{\bar{D}_{yz}}{\bar{C}}$

(3) Straight Line Passing through a Point P_i in a Given Direction

$$\frac{x - x_i}{\bar{\alpha}} = \frac{y - y_i}{\bar{\beta}} = \frac{z - z_i}{\bar{\gamma}}$$

(4) Straight Line Passing through Points P_i and P_j

$$\frac{x - x_i}{x_j - x_i} = \frac{y - y_i}{y_j - y_i} = \frac{z - z_i}{z_j - z_i}$$

(5) Parametric Equation of a Straight Line through a Point P_i

$$x = x_i + \bar{\alpha}t \qquad\qquad y = y_i + \bar{\beta}t \qquad\qquad z = z_i + \bar{\gamma}t$$

(6) Distance of the Point P_k to a Straight Line through a Point P_i

$$d = \sqrt{\left|\begin{matrix} \dfrac{x_k - x_i}{\bar{\alpha}} & \dfrac{y_k - y_i}{\bar{\beta}} \end{matrix}\right|^2 + \left|\begin{matrix} \dfrac{y_k - y_i}{\bar{\beta}} & \dfrac{z_k - z_i}{\bar{\gamma}} \end{matrix}\right|^2 + \left|\begin{matrix} \dfrac{z_k - z_i}{\bar{\gamma}} & \dfrac{x_k - x_i}{\bar{\alpha}} \end{matrix}\right|^2}$$

(1) Straight Line and Plane

If a plane and a straight line are given, respectively, as

$$Ax + By + Cz + D = 0 \qquad \bar{B}x + \bar{A}y + \bar{D}_{xy} = 0 \qquad \bar{A}z + \bar{C}x + \bar{D}_{zx} = 0 \qquad \bar{C}y + \bar{B}z + \bar{D}_{yz} = 0$$

Then the following are their relationships:

$A : B : C \neq \bar{A} : \bar{B} : \bar{C}$	Line and plane intersect
$A : \bar{A} = B : \bar{B} = C : \bar{C}$	Line normal to the plane
$A\bar{A} + B\bar{B} + C\bar{C} = 0$	Line parallel to the plane

Note: The line lies in the plane if they have a common point and $A\bar{A} + B\bar{B} + C\bar{C} = 0$.

(2) Angle between a Straight Line and a Plane

α, β, γ = direction cosines, plane

$\bar{\alpha}, \bar{\beta}, \bar{\gamma}$ = direction cosines, line

$$\sin \tau = \alpha\bar{\alpha} + \beta\bar{\beta} + \gamma\bar{\gamma}$$

(3) Two Straight Lines   (Sec. 5.05 – 3)

$$\frac{x - x_i}{\bar{\alpha}_1} = \frac{y - y_i}{\bar{\beta}_1} = \frac{z - z_i}{\bar{\gamma}_1} \qquad\qquad \frac{x - x_j}{\bar{\alpha}_2} = \frac{y - y_j}{\bar{\beta}_2} = \frac{z - z_j}{\bar{\gamma}_2}$$

(a) **The lines are parallel** if

$$\bar{\alpha}_1 = \bar{\alpha}_2 \qquad \bar{\beta}_1 = \bar{\beta}_2 \qquad \bar{\gamma}_1 = \bar{\gamma}_2$$

(b) **The lines are normal** if

$$\bar{\alpha}_1\bar{\alpha}_2 + \bar{\beta}_1\bar{\beta}_2 + \bar{\gamma}_1\bar{\gamma}_2 = 0$$

(c) **The lines are coplanar** if

$$\begin{vmatrix} x_j - x_i & y_j - y_i & z_j - z_i \\ \bar{\alpha}_1 & \bar{\beta}_1 & \bar{\gamma}_1 \\ \bar{\alpha}_2 & \bar{\beta}_2 & \bar{\gamma}_2 \end{vmatrix} = \Delta = 0$$

(d) **The angle** between these lines is given by

$$\cos \tau = \bar{\alpha}_1\bar{\alpha}_2 + \bar{\beta}_1\bar{\beta}_2 + \bar{\gamma}_1\bar{\gamma}_2$$

(e) **The distance** between these lines if $\Delta \neq 0$ is

$$d = \frac{\begin{vmatrix} x_j - x_i & y_j - y_i & z_j - z_i \\ \bar{\alpha}_1 & \bar{\beta}_1 & \bar{\gamma}_1 \\ \bar{\alpha}_2 & \bar{\beta}_2 & \bar{\gamma}_2 \end{vmatrix}}{\sqrt{\begin{vmatrix} \bar{\alpha}_1 & \bar{\alpha}_2 \\ \bar{\beta}_1 & \bar{\beta}_2 \end{vmatrix}^2 + \begin{vmatrix} \bar{\beta}_1 & \bar{\beta}_2 \\ \bar{\gamma}_1 & \bar{\gamma}_2 \end{vmatrix}^2 + \begin{vmatrix} \bar{\gamma}_1 & \bar{\gamma}_2 \\ \bar{\alpha}_1 & \bar{\alpha}_2 \end{vmatrix}^2}}$$

(1) Transformation Matrices

Translation   Rotation

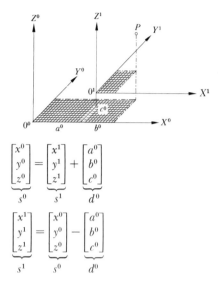

$$\begin{bmatrix} x^0 \\ y^0 \\ z^0 \end{bmatrix}_{s^0} = \begin{bmatrix} x^1 \\ y^1 \\ z^1 \end{bmatrix}_{s^1} + \begin{bmatrix} a^0 \\ b^0 \\ c^0 \end{bmatrix}_{d^0}$$

$$\begin{bmatrix} x^1 \\ y^1 \\ z^1 \end{bmatrix}_{s^1} = \begin{bmatrix} x^0 \\ y^0 \\ z^0 \end{bmatrix}_{s^0} - \begin{bmatrix} a^0 \\ b^0 \\ c^0 \end{bmatrix}_{d^0}$$

$$\begin{bmatrix} x^0 \\ y^0 \\ z^0 \end{bmatrix}_{s^0} = \begin{bmatrix} \alpha_x & \alpha_y & \alpha_z \\ \beta_x & \beta_y & \beta_z \\ \gamma_x & \gamma_y & \gamma_z \end{bmatrix}_{\omega^{0l}} \begin{bmatrix} x^l \\ y^l \\ z^l \end{bmatrix}_{s^l}$$

$$\begin{bmatrix} x^l \\ y^l \\ z^l \end{bmatrix}_{s^l} = \begin{bmatrix} \alpha_x & \beta_x & \gamma_x \\ \alpha_y & \beta_y & \gamma_y \\ \alpha_z & \beta_z & \gamma_z \end{bmatrix}_{\omega^{l0}} \begin{bmatrix} x^0 \\ y^0 \\ z^0 \end{bmatrix}_{s^0}$$

Note: 0, 1, and *l* are superscripts designating the system.

(2) Direction Cosines   (for derivation see Sec. 5.08)

$$\alpha_x = \cos(x^0 x^l) = \cos(x^l x^0) \qquad \alpha_y = \cos(x^0 y^l) = \cos(y^l x^0) \qquad \alpha_z = \cos(x^0 z^l) = \cos(z^l x^0)$$

$$\beta_x = \cos(y^0 x^l) = \cos(x^l y^0) \qquad \beta_y = \cos(y^0 y^l) = \cos(y^l y^0) \qquad \beta_z = \cos(y^0 z^l) = \cos(z^l y^0)$$

$$\gamma_x = \cos(z^0 x^l) = \cos(x^l z^0) \qquad \gamma_y = \cos(z^0 y^l) = \cos(y^l z^0) \qquad \gamma_z = \cos(z^0 z^l) = \cos(z^l z^0)$$

(3) Properties of ω Matrices

$$s^0 = \omega^{0l} s^l$$

$$s^l = \omega^{l0} s^0$$

$$\omega^{0l} \omega^{l0} = I$$

$$\omega^{l0} \omega^{0l} = I$$

$$\omega^{0l} = (\omega^{l0})^T = (\omega^{l0})^{-1}$$

$$\omega^{l0} = (\omega^{0l})^T = (\omega^{0l})^{-1}$$

Note: ω matrices are orthogonal.

(4) Properties of Direction Cosines

$$\alpha_x^2 + \alpha_y^2 + \alpha_z^2 = 1$$

$$\beta_x^2 + \beta_y^2 + \beta_z^2 = 1$$

$$\gamma_x^2 + \gamma_y^2 + \gamma_z^2 = 1$$

Diagonal terms of matrix product $\omega^{0l} \omega^{l0}$

$$\alpha_x \alpha_y + \beta_x \beta_y + \gamma_x \gamma_y = 0$$

$$\alpha_y \alpha_z + \beta_y \beta_z + \gamma_y \gamma_z = 0$$

$$\alpha_z \alpha_x + \beta_z \beta_x + \gamma_z \gamma_x = 0$$

Off-diagonal terms of matrix product $\omega^{l0} \omega^{0l}$

$$\alpha_x \beta_x + \alpha_y \beta_y + \alpha_z \beta_z = 0$$

$$\beta_x \gamma_x + \beta_y \gamma_y + \beta_z \gamma_z = 0$$

$$\gamma_x \alpha_x + \gamma_y \alpha_y + \gamma_z \alpha_z = 0$$

Off-diagonal terms of matrix product $\omega^{0l} \omega^{l0}$

$$\alpha_x^2 + \beta_x^2 + \gamma_x^2 = 1$$

$$\alpha_y^2 + \beta_y^2 + \gamma_y^2 = 1$$

$$\alpha_z^2 + \beta_z^2 + \gamma_z^2 = 1$$

Diagonal terms of matrix product $\omega^{l0} \omega^{0l}$

(1) Successive Rotation

Every space rotation can be resolved into three components, and every component rotation can be computed independently.

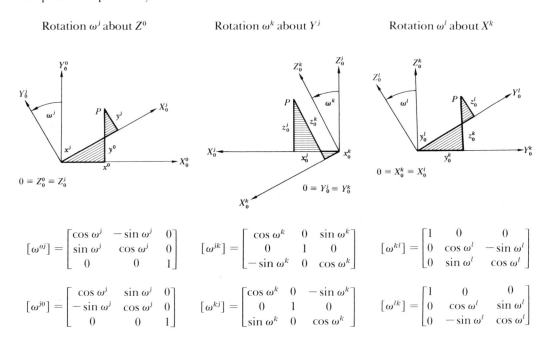

Rotation ω^j about Z^0 Rotation ω^k about Y^j Rotation ω^l about X^k

$$[\omega^{0j}] = \begin{bmatrix} \cos \omega^j & -\sin \omega^j & 0 \\ \sin \omega^j & \cos \omega^j & 0 \\ 0 & 0 & 1 \end{bmatrix}$$

$$[\omega^{jk}] = \begin{bmatrix} \cos \omega^k & 0 & \sin \omega^k \\ 0 & 1 & 0 \\ -\sin \omega^k & 0 & \cos \omega^k \end{bmatrix}$$

$$[\omega^{kl}] = \begin{bmatrix} 1 & 0 & 0 \\ 0 & \cos \omega^l & -\sin \omega^l \\ 0 & \sin \omega^l & \cos \omega^l \end{bmatrix}$$

$$[\omega^{j0}] = \begin{bmatrix} \cos \omega^j & \sin \omega^j & 0 \\ -\sin \omega^j & \cos \omega^j & 0 \\ 0 & 0 & 1 \end{bmatrix}$$

$$[\omega^{kj}] = \begin{bmatrix} \cos \omega^k & 0 & -\sin \omega^k \\ 0 & 1 & 0 \\ \sin \omega^k & 0 & \cos \omega^k \end{bmatrix}$$

$$[\omega^{lk}] = \begin{bmatrix} 1 & 0 & 0 \\ 0 & \cos \omega^l & \sin \omega^l \\ 0 & -\sin \omega^l & \cos \omega^l \end{bmatrix}$$

(2) Successive Matrix Multiplication

The resulting rotational transformation matrix (ω^{0l} or ω^{l0}) is equal to the chain-matrix product of the component matrices, executed in the order of rotation.

$$\omega^{0l} = \omega^{0j}\omega^{jk}\omega^{kl}$$

$$\begin{bmatrix} \alpha_x & \alpha_y & \alpha_z \\ \beta_x & \beta_y & \beta_z \\ \gamma_x & \gamma_y & \gamma_z \end{bmatrix} = \begin{bmatrix} +\cos \omega^j \cos \omega^k & \begin{array}{c} -\sin \omega^j \cos \omega^l \\ +\cos \omega^j \sin \omega^k \sin \omega^l \end{array} & \begin{array}{c} +\sin \omega^j \sin \omega^l \\ +\cos \omega^j \sin \omega^k \cos \omega^l \end{array} \\ +\sin \omega^j \cos \omega^k & \begin{array}{c} +\cos \omega^j \cos \omega^l \\ +\sin \omega^j \sin \omega^k \sin \omega^l \end{array} & \begin{array}{c} -\cos \omega^j \sin \omega^l \\ +\sin \omega^j \sin \omega^k \cos \omega^l \end{array} \\ -\sin \omega^k & +\cos \omega^k \sin \omega^l & +\cos \omega^k \cos \omega^l \end{bmatrix}$$

$$\omega^{l0} = \omega^{lk}\omega^{kj}\omega^{j0}$$

$$\begin{bmatrix} \alpha_x & \beta_x & \gamma_x \\ \alpha_y & \beta_y & \gamma_y \\ \alpha_z & \beta_z & \gamma_z \end{bmatrix} = \begin{bmatrix} +\cos \omega^j \cos \omega^k & +\sin \omega^j \cos \omega^k & -\sin \omega^k \\ \begin{array}{c} -\sin \omega^j \cos \omega^l \\ +\cos \omega^j \sin \omega^k \sin \omega^l \end{array} & \begin{array}{c} +\cos \omega^j \cos \omega^l \\ +\sin \omega^j \sin \omega^k \sin \omega^l \end{array} & +\cos \omega^k \sin \omega^l \\ \begin{array}{c} +\sin \omega^j \sin \omega^l \\ +\cos \omega^j \sin \omega^k \cos \omega^l \end{array} & \begin{array}{c} -\cos \omega^j \sin \omega^l \\ +\sin \omega^j \sin \omega^k \cos \omega^l \end{array} & +\cos \omega^k \cos \omega^l \end{bmatrix}$$

$$\eta_x = \cos(x, N)$$
$$\eta_y = \cos(y, N)$$
$$\eta_z = \cos(z, N)$$

$$\Delta_1 = 1 - \cos\theta$$
$$\Delta_2 = \sin\theta$$
$$\Delta_3 = \cos\theta$$

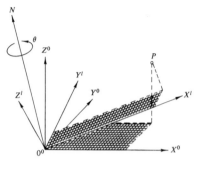

(1) Pure Rotation

The coordinate axes X^0, Y^0, Z^0, are rotated about the fixed axis N of given direction cosines η_x, η_y, η_z through the right-handed angle θ. The relationship of the initial coordinates x^0, y^0, z^0 of a point P to the new coordinates x^l, y^l, z^l of the same point are

$$\underbrace{\begin{bmatrix} x^0 \\ y^0 \\ z^0 \end{bmatrix}}_{s^0} = \underbrace{\begin{bmatrix} \eta_x^2\Delta_1 + \Delta_3 & \eta_x\eta_y\Delta_1 + \eta_z\Delta_2 & \eta_x\eta_z\Delta_1 - \eta_y\Delta_2 \\ \eta_y\eta_x\Delta_1 - \eta_z\Delta_2 & \eta_y^2\Delta_1 + \Delta_3 & \eta_y\eta_z\Delta_1 + \eta_x\Delta_2 \\ \eta_z\eta_x\Delta_1 + \eta_y\Delta_2 & \eta_z\eta_y\Delta_1 - \eta_x\Delta_2 & \eta_z^2\Delta_1 + \Delta_3 \end{bmatrix}}_{\theta^{0l}} \underbrace{\begin{bmatrix} x^l \\ y^l \\ z^l \end{bmatrix}}_{s^l}$$

Inversely, $s^l = \theta^{l0}s^0$
where $\theta^{l0} = (\theta^{0l})^T = (\theta^{0l})^{-1}$

(2) Properties of θ Matrices

The elements of the θ^{0l} matrix are numerically equal to the respective elements of the ω^{0l} matrix in Sec. 5.08–2 such that

$$\theta^{0l} = \omega^{0l} \qquad \theta^{l0} = \omega^{l0}$$

In addition,

$$\theta^{0l} = A\Delta_1 + B\Delta_2 + I\Delta_3 = \omega^{0l}$$

where $I = 3 \times 3$ unit matrix and

$$A = \begin{bmatrix} \eta_x & 0 & 0 \\ 0 & \eta_y & 0 \\ 0 & 0 & \eta_z \end{bmatrix} \begin{bmatrix} 1 & 1 & 1 \\ 1 & 1 & 1 \\ 1 & 1 & 1 \end{bmatrix} \begin{bmatrix} \eta_x & 0 & 0 \\ 0 & \eta_y & 0 \\ 0 & 0 & \eta_z \end{bmatrix} \qquad B = \begin{bmatrix} 0 & \eta_z & -\eta_y \\ -\eta_z & 0 & \eta_x \\ \eta_y & -\eta_x & 0 \end{bmatrix}$$

(3) Inverse Relationship

If the elements of ω^{0l} are known (Sec. 5.08–2), then

$$\cos\theta = \frac{\alpha_x + \beta_y + \gamma_z - 1}{2}$$

$$\eta_x = \sqrt{\frac{\alpha_x - \cos\theta}{1 - \cos\theta}} \qquad \eta_y = \sqrt{\frac{\beta_y - \cos\theta}{1 - \cos\theta}} \qquad \eta_z = \sqrt{\frac{\gamma_z - \cos\theta}{1 - \cos\theta}}$$

(1) General Equation

The general equation of the second degree.

$$a_{11}x^2 + a_{22}y^2 + a_{33}z^2 + 2a_{12}xy + 2a_{13}xz + 2a_{23}yz + 2a_{14}x + 2a_{24}y + 2a_{34}z + a_{44} = 0$$

defines the following quadratic surfaces: a sphere, an ellipsoid, a hyperboloid, a paraboloid, a cone, a cylinder, two planes, a line, and a point (the last three are the degenerated surfaces of the second degree).

(2) Invariants

The quantities $(a_{ik} = a_{ki})$

$$J_4 = \begin{vmatrix} a_{11} & a_{12} & a_{13} & a_{14} \\ a_{21} & a_{22} & a_{23} & a_{24} \\ a_{31} & a_{32} & a_{33} & a_{34} \\ a_{41} & a_{42} & a_{43} & a_{44} \end{vmatrix} \qquad\qquad J_3 = \begin{vmatrix} a_{11} & a_{12} & a_{13} \\ a_{21} & a_{22} & a_{23} \\ a_{31} & a_{32} & a_{33} \end{vmatrix}$$

$$J_2 = \begin{vmatrix} a_{11} & a_{12} \\ a_{21} & a_{22} \end{vmatrix} + \begin{vmatrix} a_{11} & a_{13} \\ a_{31} & a_{33} \end{vmatrix} + \begin{vmatrix} a_{22} & a_{23} \\ a_{32} & a_{33} \end{vmatrix} \qquad\qquad J_1 = a_{11} + a_{22} + a_{33}$$

are invariants of the general equation; they remain unchanged under transformation of coordinates.

(3) Center

The center of a central quadratic surface ($J_3 \neq 0$) is the point of intersection of the three planes

$$a_{11}x + a_{12}y + a_{13}z + a_{14} = 0$$
$$a_{21}x + a_{22}y + a_{23}z + a_{24} = 0$$
$$a_{31}x + a_{32}y + a_{33}z + a_{34} = 0$$

(4) Classification

Type		$J_3 \neq 0$			$J_3 = 0$
		Central surface			Noncentral surface
		$J_3 J_1 > 0$	$J_2 > 0$	$J_3 J_1$ or $J_2 > 0$	
$J_4 \neq 0$	$J_4 < 0$	Real ellipsoid		Hyperboloid of two sheets	Elliptic paraboloid
	$J_4 > 0$	Imaginary ellipsoid		Hyperboloid of one sheet	Hyperbolic paraboloid
$J_4 = 0$		Imaginary cone		Real cone	Cylinder or two planes

Sphere

$$\frac{x^2}{a^2} + \frac{y^2}{a^2} + \frac{z^2}{a^2} = 1$$

Ellipsoid

$$\frac{x^2}{a^2} + \frac{y^2}{b^2} + \frac{z^2}{c^2} = 1$$

Elliptic paraboloid

$$\frac{x^2}{a^2} + \frac{y^2}{b^2} = z$$

Elliptic hyperboloid of one sheet

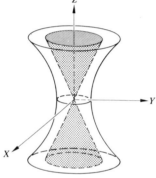

$$\frac{x^2}{a^2} + \frac{y^2}{b^2} - \frac{z^2}{c^2} = 1$$

Hyperbolic cylinder

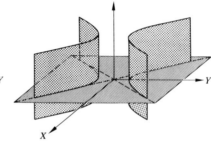

$$-\frac{x^2}{a^2} + \frac{y^2}{b^2} = 1$$

Elliptic hyperboloid of two sheets

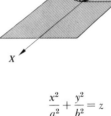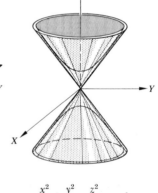

$$\frac{x^2}{a^2} + \frac{y^2}{b^2} - \frac{z^2}{c^2} = -1$$

Circular cylinder

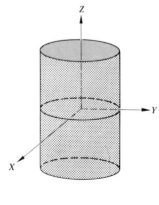

$$\frac{x^2}{a^2} + \frac{y^2}{a^2} = 1$$

Elliptic cone

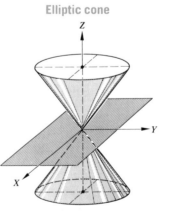

$$\frac{x^2}{a^2} + \frac{y^2}{b^2} - \frac{z^2}{c^2} = 0$$

Hyperbolic paraboloid

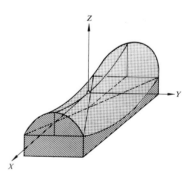

$$\frac{x^2}{a^2} - \frac{y^2}{b^2} = z$$

6
ELEMENTARY FUNCTIONS

(1) Definitions

Trigonometric functions of an angle ω are defined as follows:

$$\text{sine } \omega = \sin \omega = \frac{y}{R} \qquad\qquad \text{cosecant } \omega = \csc \omega = \frac{R}{y}$$

$$\text{cosine } \omega = \cos \omega = \frac{x}{R} \qquad\qquad \text{secant } \omega = \sec \omega = \frac{R}{x}$$

$$\text{tangent } \omega = \tan \omega = \frac{y}{x} \qquad\qquad \text{cotangent } \omega = \cot \omega = \frac{x}{y}$$

$$\text{versine } \omega = \text{vers } \omega = \frac{R - x}{R} \qquad\qquad \text{coversine } \omega = \text{covers } \omega = \frac{R - y}{R}$$

(2) Angle

The independent variable ω is measured in radians.

$$180° = \pi \qquad\qquad 1° = \frac{\pi}{180} \qquad\qquad 1 \text{ radian} = \frac{180}{\pi}$$

$$= 3.141\ 592\ 653\ 5 \text{ radians} \qquad = 0.017\ 453\ 292\ 5 \text{ radian} \qquad = 57.295\ 779\ 513\ 0°$$

(3) Relationships

$$\sin^2 \omega + \cos^2 \omega = 1$$

$$\tan^2 \omega + 1 = \sec^2 \omega$$

$$\cot^2 \omega + 1 = \csc^2 \omega$$

$$\sin \omega \csc \omega = 1$$

$$\cos \omega \sec \omega = 1$$

$$\tan \omega \cot \omega = 1$$

$$\sin \omega + \text{covers } \omega = +1$$

$$\cos \omega + \text{vers } \omega = +1$$

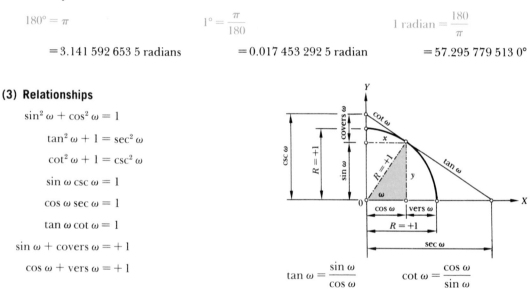

$$\tan \omega = \frac{\sin \omega}{\cos \omega} \qquad\qquad \cot \omega = \frac{\cos \omega}{\sin \omega}$$

(4) Reductions $(k = 0, 1, 2, \ldots)$

	$\pm \alpha$	$\frac{\pi}{2} \pm \alpha$	$2k\pi \pm \alpha$	$(4k+1)\frac{\pi}{2} \pm \alpha$	$(4k+2)\frac{\pi}{2} \pm \alpha$	$(4k+3)\frac{\pi}{2} \pm \alpha$
sin	$\pm \sin \alpha$	$+ \cos \alpha$	$\pm \sin \alpha$	$+ \cos \alpha$	$\mp \sin \alpha$	$- \cos \alpha$
cos	$+ \cos \alpha$	$\mp \sin \alpha$	$+ \cos \alpha$	$\mp \sin \alpha$	$- \cos \alpha$	$\pm \sin \alpha$
tan	$\pm \tan \alpha$	$\mp \cot \alpha$	$\pm \tan \alpha$	$\mp \cot \alpha$	$\pm \tan \alpha$	$\mp \cot \alpha$
cot	$\pm \cot \alpha$	$\mp \tan \alpha$	$\pm \cot \alpha$	$\mp \tan \alpha$	$\pm \cot \alpha$	$\mp \tan \alpha$
sec	$+ \sec \alpha$	$\mp \csc \alpha$	$+ \sec \alpha$	$\mp \csc \alpha$	$- \sec \alpha$	$+ \csc \alpha$
csc	$\pm \csc \alpha$	$+ \sec \alpha$	$\pm \csc \alpha$	$+ \sec \alpha$	$\mp \csc \alpha$	$- \sec \alpha$

(1) Signs in Quadrants

Quadrant	sin	cos	tan	cot	sec	csc	vers	covers
I	+	+	+	+	+	+	+	+
II	+	−	−	−	−	+	+	+
III	−	−	+	+	−	−	+	+
IV	−	+	−	−	+	−	+	+

(2) Graphs

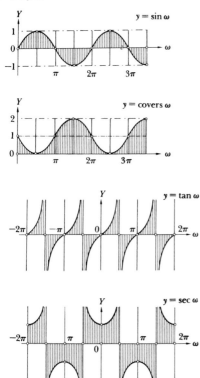

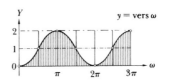

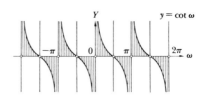

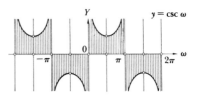

(3) Limit Values

Degrees	Radians	sin ω	cos ω	tan ω	cot ω	sec ω	csc ω	vers ω	covers ω
0°	0	0	+1	0	$\mp\infty$	+1	$\mp\infty$	0	+1
90°	$\dfrac{\pi}{2}$	+1	0	$\pm\infty$	0	$\pm\infty$	+1	+1	0
180°	π	0	−1	0	$\mp\infty$	−1	$\pm\infty$	+2	+1
270°	$\dfrac{3\pi}{2}$	−1	0	$\pm\infty$	0	$\mp\infty$	−1	+1	+2
360°	2π	0	+1	0	$\mp\infty$	+1	$\mp\infty$	0	+1

(1) Functions of Selected Angles

Degrees	Radians	$\sin\alpha$	$\cos\alpha$	$\tan\alpha$	$\cot\alpha$	$\sec\alpha$	$\csc\alpha$	vers α	covers α
0	0	0	$+1$	0	$\mp\infty$	$+1$	$\mp\infty$	0	$+1$
30	$\dfrac{\pi}{6}$	$+\dfrac{1}{2}$	$+\dfrac{\sqrt{3}}{2}$	$+\dfrac{\sqrt{3}}{3}$	$+\sqrt{3}$	$+\dfrac{2\sqrt{3}}{3}$	$+2$	$+1-\dfrac{\sqrt{3}}{2}$	$+\dfrac{1}{2}$
45	$\dfrac{\pi}{4}$	$+\dfrac{\sqrt{2}}{2}$	$+\dfrac{\sqrt{2}}{2}$	$+1$	$+1$	$+\sqrt{2}$	$+\sqrt{2}$	$+1-\dfrac{\sqrt{2}}{2}$	$+1-\dfrac{\sqrt{2}}{2}$
60	$\dfrac{\pi}{3}$	$+\dfrac{\sqrt{3}}{2}$	$+\dfrac{1}{2}$	$+\sqrt{3}$	$+\dfrac{\sqrt{3}}{3}$	$+2$	$+\dfrac{2\sqrt{3}}{3}$	$+\dfrac{1}{2}$	$+1-\dfrac{\sqrt{3}}{2}$
90	$\dfrac{\pi}{2}$	$+1$	0	$\pm\infty$	0	$\pm\infty$	$+1$	$+1$	0
120	$\dfrac{2\pi}{3}$	$+\dfrac{\sqrt{3}}{2}$	$-\dfrac{1}{2}$	$-\sqrt{3}$	$-\dfrac{\sqrt{3}}{3}$	-2	$+\dfrac{2\sqrt{3}}{3}$	$+\dfrac{3}{2}$	$+1-\dfrac{\sqrt{3}}{2}$
135	$\dfrac{3\pi}{4}$	$+\dfrac{\sqrt{2}}{2}$	$-\dfrac{\sqrt{2}}{2}$	-1	-1	$-\sqrt{2}$	$+\sqrt{2}$	$+1+\dfrac{\sqrt{2}}{2}$	$+1-\dfrac{\sqrt{2}}{2}$
150	$\dfrac{5\pi}{6}$	$+\dfrac{1}{2}$	$-\dfrac{\sqrt{3}}{2}$	$-\dfrac{\sqrt{3}}{3}$	$-\sqrt{3}$	$-\dfrac{2\sqrt{3}}{3}$	$+2$	$+1+\dfrac{\sqrt{3}}{2}$	$+\dfrac{1}{2}$
180	π	0	-1	0	$\mp\infty$	-1	$\mp\infty$	$+2$	$+1$
210	$\dfrac{7\pi}{6}$	$-\dfrac{1}{2}$	$-\dfrac{\sqrt{3}}{2}$	$+\dfrac{\sqrt{3}}{3}$	$+\sqrt{3}$	$-\dfrac{2\sqrt{3}}{3}$	-2	$+1+\dfrac{\sqrt{3}}{2}$	$+\dfrac{3}{2}$
225	$\dfrac{5\pi}{4}$	$-\dfrac{\sqrt{2}}{2}$	$-\dfrac{\sqrt{2}}{2}$	$+1$	$+1$	$-\sqrt{2}$	$-\sqrt{2}$	$+1+\dfrac{\sqrt{2}}{2}$	$+1+\dfrac{\sqrt{2}}{2}$
240	$\dfrac{4\pi}{3}$	$-\dfrac{\sqrt{3}}{2}$	$-\dfrac{1}{2}$	$+\sqrt{3}$	$+\dfrac{\sqrt{3}}{3}$	-2	$-\dfrac{2\sqrt{3}}{3}$	$+\dfrac{3}{2}$	$+1+\dfrac{\sqrt{3}}{2}$
270	$\dfrac{3\pi}{2}$	-1	0	$\pm\infty$	0	$\mp\infty$	-1	$+1$	$+2$
300	$\dfrac{5\pi}{3}$	$-\dfrac{\sqrt{3}}{2}$	$+\dfrac{1}{2}$	$-\sqrt{3}$	$-\dfrac{\sqrt{3}}{3}$	$+2$	$-\dfrac{2\sqrt{3}}{3}$	$+\dfrac{1}{2}$	$+1+\dfrac{\sqrt{3}}{2}$
315	$\dfrac{7\pi}{4}$	$-\dfrac{\sqrt{2}}{2}$	$+\dfrac{\sqrt{2}}{2}$	-1	-1	$+\sqrt{2}$	$-\sqrt{2}$	$+1-\dfrac{\sqrt{2}}{2}$	$+1+\dfrac{\sqrt{2}}{2}$
330	$\dfrac{11\pi}{6}$	$-\dfrac{1}{2}$	$+\dfrac{\sqrt{3}}{2}$	$-\dfrac{\sqrt{3}}{3}$	$-\sqrt{3}$	$+\dfrac{2\sqrt{3}}{3}$	-2	$+1-\dfrac{\sqrt{3}}{2}$	$+\dfrac{3}{2}$
360	2π	0	$+1$	0	$\mp\infty$	$+1$	$\mp\infty$	0	$+1$

(2) Functions of Very Small Angles

$$\Delta\omega \le 0.05 \text{ radian} \qquad \omega > 0.05 \text{ radian} \qquad |\epsilon| \le 1.25 \times 10^{-3}$$
$$\sin\Delta\omega \cong \Delta\omega \qquad\qquad \cos\Delta\omega \cong 1 \qquad\qquad \tan\Delta\omega \cong \Delta\omega$$
$$\sin(\omega \pm \Delta\omega) \cong \sin\omega \pm \Delta\omega\cos\omega \qquad \cos(\omega \pm \Delta\omega) \cong \cos\omega \mp \Delta\omega\sin\omega$$

(1) Transformations

	$\sin\omega$	$\cos\omega$	$\tan\omega$	$\cot\omega$	$\sec\omega$	$\csc\omega$
$\sin\omega$	$\sin\omega$	$\pm\sqrt{1-\cos^2\omega}$	$\dfrac{\tan\omega}{\pm\sqrt{1+\tan^2\omega}}$	$\dfrac{1}{\pm\sqrt{1+\cot^2\omega}}$	$\dfrac{\pm\sqrt{\sec^2\omega-1}}{\sec\omega}$	$\dfrac{1}{\csc\omega}$
$\cos\omega$	$\pm\sqrt{1-\sin^2\omega}$	$\cos\omega$	$\dfrac{1}{\pm\sqrt{1+\tan^2\omega}}$	$\dfrac{\cot\omega}{\pm\sqrt{1+\cot^2\omega}}$	$\dfrac{1}{\sec\omega}$	$\dfrac{\pm\sqrt{\csc^2\omega-1}}{\csc\omega}$
$\tan\omega$	$\dfrac{\sin\omega}{\pm\sqrt{1-\sin^2\omega}}$	$\dfrac{\pm\sqrt{1-\cos^2\omega}}{\cos\omega}$	$\tan\omega$	$\dfrac{1}{\cot\omega}$	$\pm\sqrt{\sec^2\omega-1}$	$\dfrac{1}{\pm\sqrt{\csc^2\omega-1}}$
$\cot\omega$	$\dfrac{\pm\sqrt{1-\sin^2\omega}}{\sin\omega}$	$\dfrac{\cos\omega}{\pm\sqrt{1-\cos^2\omega}}$	$\dfrac{1}{\tan\omega}$	$\cot\omega$	$\dfrac{1}{\pm\sqrt{\sec^2\omega-1}}$	$\pm\sqrt{\csc^2\omega-1}$
$\sec\omega$	$\dfrac{1}{\pm\sqrt{1-\sin^2\omega}}$	$\dfrac{1}{\cos\omega}$	$\pm\sqrt{1+\tan^2\omega}$	$\dfrac{\pm\sqrt{1+\cot^2\omega}}{\cot\omega}$	$\sec\omega$	$\dfrac{\csc\omega}{\pm\sqrt{\csc^2\omega-1}}$
$\csc\omega$	$\dfrac{1}{\sin\omega}$	$\dfrac{1}{\pm\sqrt{1-\cos^2\omega}}$	$\dfrac{\pm\sqrt{1+\tan^2\omega}}{\tan\omega}$	$\pm\sqrt{1+\cot^2\omega}$	$\dfrac{\sec\omega}{\pm\sqrt{\sec^2\omega-1}}$	$\csc\omega$

Note: The sign is governed by the quadrant in which the argument terminates.

(2) Expansions

$$\sin n\omega = n\sin\omega\cos^{n-1}\omega - \binom{n}{3}\sin^3\omega\cos^{n-3}\omega + \binom{n}{5}\sin^5\omega\cos^{n-5}\omega - \cdots$$

$$\cos n\omega = \cos^n\omega - \binom{n}{2}\sin^2\omega\cos^{n-2}\omega + \binom{n}{4}\sin^4\omega\cos^{n-4}\omega - \cdots$$

$$\sin^{2n}\omega = \frac{(-1)^n}{2^{2n-1}}\left[\cos 2n\omega - \binom{2n}{1}\cos(2n-2)\omega + \cdots (-1)^{n-1}\binom{2n}{n-1}\cos 2\omega\right] + \binom{2n}{n}\frac{1}{2^{2n}}$$

$$\sin^{2n-1}\omega = \frac{(-1)^{n-1}}{2^{2n-2}}\left[\sin(2n-1)\omega - \binom{2n-1}{1}\sin(2n-3)\omega + \cdots (-1)^{n-1}\binom{2n-1}{n-1}\sin\omega\right]$$

$$\cos^{2n}\omega = \frac{1}{2^{2n-1}}\left[\cos 2n\omega + \binom{2n}{1}\cos(2n-2)\omega + \cdots + \binom{2n}{n-1}\cos 2\omega\right] + \binom{2n}{n}\frac{1}{2^{2n}}$$

$$\cos^{2n-1}\omega = \frac{1}{2^{2n-2}}\left[\cos(2n-1)\omega + \binom{2n-1}{1}\cos(2n-3)\omega + \cdots + \binom{2n-1}{n-1}\cos\omega\right]$$

(1) Sums of Angles

$$\sin(\alpha \pm \beta) = \sin\alpha\cos\beta \pm \cos\alpha\sin\beta \qquad \cos(\alpha \pm \beta) = \cos\alpha\cos\beta \mp \sin\alpha\sin\beta$$

$$\tan(\alpha \pm \beta) = \frac{\tan\alpha \pm \tan\beta}{1 \mp \tan\alpha\tan\beta} \qquad \cot(\alpha \pm \beta) = \frac{\cot\alpha\cot\beta \mp 1}{\cot\beta \pm \cot\alpha}$$

(2) Sums of Functions

$$\sin\alpha + \sin\beta = 2\sin\frac{\alpha+\beta}{2}\cos\frac{\alpha-\beta}{2} \qquad \cos\alpha + \cos\beta = 2\cos\frac{\alpha+\beta}{2}\cos\frac{\alpha-\beta}{2}$$

$$\sin\alpha - \sin\beta = 2\sin\frac{\alpha-\beta}{2}\cos\frac{\alpha+\beta}{2} \qquad \cos\alpha - \cos\beta = -2\sin\frac{\alpha-\beta}{2}\sin\frac{\alpha+\beta}{2}$$

$$\tan\alpha \pm \tan\beta = \frac{\sin(\alpha \pm \beta)}{\cos\alpha\cos\beta} \qquad \cot\alpha \pm \cot\beta = \frac{\sin(\beta \pm \alpha)}{\sin\alpha\sin\beta}$$

$$\sin\alpha + \cos\alpha = \sqrt{2}\sin\left(\frac{\pi}{4}+\alpha\right) \qquad \sin\alpha - \cos\alpha = -\sqrt{2}\cos\left(\frac{\pi}{4}+\alpha\right)$$

$$\tan\alpha + \cot\alpha = 2\csc 2\alpha \qquad \tan\alpha - \cot\alpha = -2\cot 2\alpha$$

$$\frac{1+\tan\alpha}{1-\tan\alpha} = \tan\left(\frac{\pi}{4}+\alpha\right) \qquad \frac{1+\cot\alpha}{1-\cot\alpha} = -\cot\left(\frac{\pi}{4}-\alpha\right)$$

(3) Sums of Sums

$$\sin(\alpha+\beta) + \sin(\alpha-\beta) = 2\sin\alpha\cos\beta \qquad \cos(\alpha+\beta) + \cos(\alpha-\beta) = 2\cos\alpha\cos\beta$$

$$\sin(\alpha+\beta) - \sin(\alpha-\beta) = 2\sin\beta\cos\alpha \qquad \cos(\alpha+\beta) - \cos(\alpha-\beta) = -2\sin\alpha\sin\beta$$

$$\tan(\alpha+\beta) + \tan(\alpha-\beta) \qquad\qquad \cot(\alpha+\beta) + \cot(\alpha-\beta)$$

$$= \frac{\sin 2\alpha}{\cos(\alpha+\beta)\cos(\alpha-\beta)} \qquad\qquad = \frac{\sin 2\alpha}{\sin(\alpha+\beta)\sin(\alpha-\beta)}$$

$$\tan(\alpha+\beta) - \tan(\alpha-\beta) \qquad\qquad \cot(\alpha+\beta) - \cot(\alpha-\beta)$$

$$= \frac{\sin 2\beta}{\cos(\alpha+\beta)\cos(\alpha-\beta)} \qquad\qquad = -\frac{\sin 2\beta}{\sin(\alpha+\beta)\sin(\alpha-\beta)}$$

(1) Multiple Angles

$$\sin 2\alpha = 2 \sin \alpha \cos \alpha$$

$$\sin 3\alpha = (\sin \alpha)(3 - 4 \sin^2 \alpha)$$

$$\sin 4\alpha = (4 \sin \alpha \cos \alpha)(1 - 2 \sin^2 \alpha)$$

$$\sin 5\alpha = (\sin \alpha)(5 - 20 \sin^2 \alpha + 16 \sin^4 \alpha)$$

$$\cos 2\alpha = 1 - 2 \sin^2 \alpha$$

$$\cos 3\alpha = 4 \cos^3 \alpha - 3 \cos \alpha$$

$$\cos 4\alpha = 8 \cos^4 \alpha - 8 \cos^2 \alpha + 1$$

$$\cos 5\alpha = 16 \cos^5 \alpha - 20 \cos^3 \alpha + 5 \cos \alpha$$

$$\tan 2\alpha = \frac{2 \tan \alpha}{1 - \tan^2 \alpha}$$

$$\tan 3\alpha = \frac{3 \tan \alpha - \tan^3 \alpha}{1 - 3 \tan^2 \alpha}$$

$$\tan 4\alpha = \frac{4 \tan \alpha - 4 \tan^3 \alpha}{1 - 6 \tan^2 \alpha + \tan^4 \alpha}$$

$$\tan 5\alpha = \frac{5 \tan \alpha - 10 \tan^3 \alpha + \tan^5 \alpha}{1 - 10 \tan^2 \alpha + 5 \tan^4 \alpha}$$

$$\cot 2\alpha = \frac{\cot^2 \alpha - 1}{2 \cot \alpha}$$

$$\cot 3\alpha = \frac{3 \cot \alpha - \cot^3 \alpha}{1 - 3 \cot^2 \alpha}$$

$$\cot 4\alpha = \frac{\cot^4 \alpha - 6 \cot^2 \alpha + 1}{4 \cot^3 \alpha - 4 \cot \alpha}$$

$$\cot 5\alpha = \frac{5 \cot \alpha - 10 \cot^3 \alpha + \cot^5 \alpha}{1 - 10 \cot^2 \alpha + 5 \cot^4 \alpha}$$

(2) Half Angles $(\alpha < \pi)$

$$\sin \frac{\alpha}{2} = \sqrt{\frac{1 - \cos \alpha}{2}}$$

$$\cos \frac{\alpha}{2} = \sqrt{\frac{1 + \cos \alpha}{2}}$$

$$\tan \frac{\alpha}{2} = \sqrt{\frac{1 - \cos \alpha}{1 + \cos \alpha}} = \frac{\sin \alpha}{1 + \cos \alpha} = \frac{1 - \cos \alpha}{\sin \alpha}$$

$$\cot \frac{\alpha}{2} = \sqrt{\frac{1 + \cos \alpha}{1 - \cos \alpha}} = \frac{1 + \cos \alpha}{\sin \alpha} = \frac{\sin \alpha}{1 - \cos \alpha}$$

(3) Relations $(\alpha < \pi/2)$

$$\sin \alpha = 2 \sin \frac{\alpha}{2} \cos \frac{\alpha}{2}$$

$$\tan \alpha = \frac{2 \tan (\alpha/2)}{1 - \tan^2 (\alpha/2)}$$

$$= \frac{2 \sin (\alpha/2) \cos (\alpha/2)}{\cos^2 (\alpha/2) - \sin^2 (\alpha/2)}$$

$$\cos \alpha = \cos^2 \frac{\alpha}{2} - \sin^2 \frac{\alpha}{2}$$

$$\cot \alpha = \frac{\cot^2 (\alpha/2) - 1}{2 \cot (\alpha/2)}$$

$$= \frac{\cos^2 (\alpha/2) - \sin^2 (\alpha/2)}{2 \sin (\alpha/2) \cos (\alpha/2)}$$

$$\sin \alpha = \sqrt{\frac{1 - \cos 2\alpha}{2}}$$

$$\tan \alpha = \sqrt{\frac{1 - \cos 2\alpha}{1 + \cos 2\alpha}}$$

$$= \frac{\sin 2\alpha}{1 + \cos 2\alpha} = \frac{1 - \cos 2\alpha}{\sin 2\alpha}$$

$$\cos \alpha = \sqrt{\frac{1 + \cos 2\alpha}{2}}$$

$$\cot \alpha = \sqrt{\frac{1 + \cos 2\alpha}{1 - \cos 2\alpha}}$$

$$= \frac{1 + \cos 2\alpha}{\sin 2\alpha} = \frac{\sin 2\alpha}{1 - \cos 2\alpha}$$

(1) Powers

$$\sin^2 \alpha = \tfrac{1}{2}(-\cos 2\alpha + 1)$$

$$\cos^2 \alpha = \tfrac{1}{2}(\cos 2\alpha + 1)$$

$$\sin^3 \alpha = \tfrac{1}{4}(-\sin 3\alpha + 3 \sin \alpha)$$

$$\cos^3 \alpha = \tfrac{1}{4}(\cos 3\alpha + 3 \cos \alpha)$$

$$\sin^4 \alpha = \tfrac{1}{8}(\cos 4\alpha - 4 \cos 2\alpha + 3)$$

$$\cos^4 \alpha = \tfrac{1}{8}(\cos 4\alpha + 4 \cos 2\alpha + 3)$$

$$\sin^5 \alpha = \tfrac{1}{16}(\sin 5\alpha - 5 \sin 3\alpha + 10 \sin \alpha)$$

$$\cos^5 \alpha = \tfrac{1}{16}(\cos 5\alpha + 5 \cos 3\alpha + 10 \cos \alpha)$$

$$\tan^2 \alpha = \frac{1 - \cos 2\alpha}{1 + \cos 2\alpha}$$

$$\cot^2 \alpha = \frac{1 + \cos 2\alpha}{1 - \cos 2\alpha}$$

$$\tan^3 \alpha = \frac{-\sin 3\alpha + 3 \sin \alpha}{\cos 3\alpha + 3 \cos \alpha}$$

$$\cot^3 \alpha = \frac{\cos 3\alpha + 3 \cos \alpha}{-\sin 3\alpha + 3 \sin \alpha}$$

$$\tan^4 \alpha = \frac{\cos 4\alpha - 4 \cos 2\alpha + 3}{\cos 4\alpha + 4 \cos 2\alpha + 3}$$

$$\cot^4 \alpha = \frac{\cos 4\alpha + 4 \cos 2\alpha + 3}{\cos 4\alpha - 4 \cos 2\alpha + 3}$$

$$\tan^5 \alpha = \frac{\sin 5\alpha - 5 \sin 3\alpha + 10 \sin \alpha}{\cos 5\alpha + 5 \cos 3\alpha + 10 \cos \alpha}$$

$$\cot^5 \alpha = \frac{\cos 5\alpha + 5 \cos 3\alpha + 10 \cos \alpha}{\sin 5\alpha - 5 \sin 3\alpha + 10 \sin \alpha}$$

(2) Products

$$\sin \alpha \sin \beta = \tfrac{1}{2} \cos (\alpha - \beta) - \tfrac{1}{2} \cos (\alpha + \beta)$$

$$\tan \alpha \tan \beta = \frac{\cos (\alpha - \beta) - \cos (\alpha + \beta)}{\cos (\alpha - \beta) + \cos (\alpha + \beta)}$$

$$\sin \alpha \cos \beta = \tfrac{1}{2} \sin (\alpha - \beta) + \tfrac{1}{2} \sin (\alpha + \beta)$$

$$\tan \alpha \cot \beta = \frac{\sin (\alpha - \beta) + \sin (\alpha + \beta)}{-\sin (\alpha - \beta) + \sin (\alpha + \beta)}$$

$$\cos \alpha \cos \beta = \tfrac{1}{2} \cos (\alpha - \beta) + \tfrac{1}{2} \cos (\alpha + \beta)$$

$$\cot \alpha \cot \beta = \frac{\cos (\alpha - \beta) + \cos (\alpha + \beta)}{\cos (\alpha - \beta) - \cos (\alpha + \beta)}$$

If $\alpha + \beta + \gamma = \pi$,

$$4 \sin \alpha \sin \beta \sin \gamma = \sin (\alpha + \beta - \gamma) - \sin (\beta + \gamma - \alpha) + \sin (\gamma + \alpha - \beta) - \sin (\alpha + \beta + \gamma)$$

$$4 \sin \alpha \sin \beta \cos \gamma = -\cos (\alpha + \beta - \gamma) + \cos (\beta + \gamma - \alpha) + \cos (\gamma + \alpha - \beta) - \cos (\alpha + \beta + \gamma)$$

$$4 \sin \alpha \cos \beta \cos \gamma = \sin (\alpha + \beta - \gamma) - \sin (\beta + \gamma - \alpha) + \sin (\gamma + \alpha - \beta) - \sin (\alpha + \beta + \gamma)$$

$$4 \cos \alpha \cos \beta \cos \gamma = \cos (\alpha + \beta - \gamma) + \cos (\beta + \gamma - \alpha) + \cos (\gamma + \alpha - \beta) + \cos (\alpha + \beta + \gamma)$$

(1) Definitions

Inverse trigonometric functions are defined as follows:

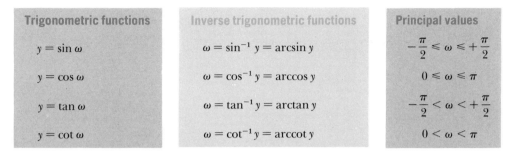

Trigonometric functions	Inverse trigonometric functions	Principal values
$y = \sin \omega$	$\omega = \sin^{-1} y = \arcsin y$	$-\dfrac{\pi}{2} \leqslant \omega \leqslant +\dfrac{\pi}{2}$
$y = \cos \omega$	$\omega = \cos^{-1} y = \arccos y$	$0 \leqslant \omega \leqslant \pi$
$y = \tan \omega$	$\omega = \tan^{-1} y = \arctan y$	$-\dfrac{\pi}{2} < \omega < +\dfrac{\pi}{2}$
$y = \cot \omega$	$\omega = \cot^{-1} y = \text{arccot } y$	$0 < \omega < \pi$

which means that ω is the arc of an angle of which the trigonometric function is y.

(2) Relationships

$$\sin^{-1} y + \cos^{-1} y = \frac{\pi}{2}$$

$$\tan^{-1} y + \cot^{-1} y = \frac{\pi}{2}$$

$$\sin^{-1}(-y) = -\sin^{-1} y$$

$$\tan^{-1}(-y) = -\tan^{-1} y$$

(3) Graphs

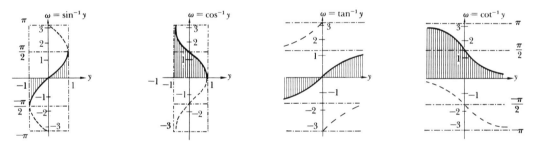

(4) Transformation Table

$y \geq 0$	$\sin^{-1} y$	$\cos^{-1} y$	$\tan^{-1} y$	$\cot^{-1} y$
$\sin^{-1} y$	$\sin^{-1} y$	$\cos^{-1} \sqrt{1 - y^2}$	$\tan^{-1} \dfrac{y}{\sqrt{1 - y^2}}$	$\cot^{-1} \dfrac{\sqrt{1 - y^2}}{y}$
$\cos^{-1} y$	$\sin^{-1} \sqrt{1 - y^2}$	$\cos^{-1} y$	$\tan^{-1} \dfrac{\sqrt{1 - y^2}}{y}$	$\cot^{-1} \dfrac{y}{\sqrt{1 - y^2}}$
$\tan^{-1} y$	$\sin^{-1} \dfrac{y}{\sqrt{1 + y^2}}$	$\cos^{-1} \dfrac{1}{\sqrt{1 + y^2}}$	$\tan^{-1} y$	$\cot^{-1} \dfrac{1}{y}$
$\cot^{-1} y$	$\sin^{-1} \dfrac{1}{\sqrt{1 + y^2}}$	$\cos^{-1} \dfrac{y}{\sqrt{1 + y^2}}$	$\tan^{-1} \dfrac{1}{y}$	$\cot^{-1} y$

(1) Definitions

A hyperbolic function is a combination of e^x and e^{-x} and is introduced as follows:

$$\text{Hyperbolic sine of } x = \sinh x = \frac{e^x - e^{-x}}{2}$$

$$\text{Hyperbolic cosine of } x = \cosh x = \frac{e^x + e^{-x}}{2}$$

$$\text{Hyperbolic tangent of } x = \tanh x = \frac{e^x - e^{-x}}{e^x + e^{-x}}$$

$$\text{Hyperbolic cotangent of } x = \coth x = \frac{e^x + e^{-x}}{e^x - e^{-x}}$$

$$\text{Hyperbolic secant of } x = \operatorname{sech} x = \frac{2}{e^x + e^{-x}}$$

$$\text{Hyperbolic cosecant of } x = \operatorname{csch} x = \frac{2}{e^x - e^{-x}}$$

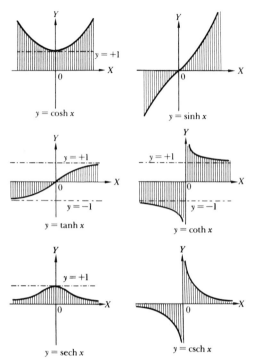

$y = \cosh x$

$y = \sinh x$

$y = \tanh x$

$y = \coth x$

$y = \operatorname{sech} x$

$y = \operatorname{csch} x$

(2) Relationships

$$\cosh^2 x - \sinh^2 x = 1$$

$$\tanh^2 x + \operatorname{sech}^2 x = 1$$

$$\coth^2 x - \operatorname{csch}^2 x = 1$$

$$\tanh x = \frac{\sinh x}{\cosh x}$$

$$\coth x = \frac{\cosh x}{\sinh x}$$

$$\operatorname{sech} x \cosh x = 1$$

$$\operatorname{csch} x \sinh x = 1$$

$$\tanh x \coth x = 1$$

$$\sinh (-x) = -\sinh x$$

$$\operatorname{sech} (-x) = \operatorname{sech} x$$

$$\tanh (-x) = -\tanh x$$

$$\coth (-x) = -\coth x$$

$$\cosh (-x) = \cosh x$$

$$\operatorname{csch} (-x) = -\operatorname{csch} x$$

(3) Limit Values

x	$\sinh x$	$\cosh x$	$\tanh x$	$\coth x$	$\operatorname{sech} x$	$\operatorname{csch} x$
$-\infty$	$-\infty$	$+\infty$	-1	-1	0	0
-1	-1.1752	$+1.5431$	-0.7616	-1.3130	$+0.6480$	-0.8509
0	0	$+1$	0	$\mp\infty$	$+1$	$\mp\infty$
$+1$	$+1.1752$	$+1.5431$	$+0.7616$	$+1.3130$	$+0.6480$	$+0.8509$
$+\infty$	$+\infty$	$+\infty$	$+1$	$+1$	0	0

(1) Transformations

$x > 0$	$\sinh x$	$\cosh x$	$\tanh x$	$\coth x$	$\operatorname{sech} x$	$\operatorname{csch} x$
$\sinh x$	$\sinh x$	$\sqrt{\cosh^2 x - 1}$	$\dfrac{\tanh x}{\sqrt{1 - \tanh^2 x}}$	$\dfrac{1}{\sqrt{\coth^2 x - 1}}$	$\dfrac{\sqrt{1 - \operatorname{sech}^2 x}}{\operatorname{sech} x}$	$\dfrac{1}{\operatorname{csch} x}$
$\cosh x$	$\sqrt{1 + \sinh^2 x}$	$\cosh x$	$\dfrac{1}{\sqrt{1 - \tanh^2 x}}$	$\dfrac{\coth x}{\sqrt{\coth^2 x - 1}}$	$\dfrac{1}{\operatorname{sech} x}$	$\dfrac{\sqrt{1 + \operatorname{csch}^2 x}}{\operatorname{csch} x}$
$\tanh x$	$\dfrac{\sinh x}{\sqrt{1 + \sinh^2 x}}$	$\dfrac{\sqrt{\cosh^2 x - 1}}{\cosh x}$	$\tanh x$	$\dfrac{1}{\coth x}$	$\sqrt{1 - \operatorname{sech}^2 x}$	$\dfrac{1}{\sqrt{1 + \operatorname{csch}^2 x}}$
$\coth x$	$\dfrac{\sqrt{1 + \sinh^2 x}}{\sinh x}$	$\dfrac{\cosh x}{\sqrt{\cosh^2 x - 1}}$	$\dfrac{1}{\tanh x}$	$\coth x$	$\dfrac{1}{\sqrt{1 - \operatorname{sech}^2 x}}$	$\sqrt{1 + \operatorname{csch}^2 x}$
$\operatorname{sech} x$	$\dfrac{1}{\sqrt{1 + \sinh^2 x}}$	$\dfrac{1}{\cosh x}$	$\sqrt{1 - \tanh^2 x}$	$\dfrac{\sqrt{\coth^2 x - 1}}{\coth x}$	$\operatorname{sech} x$	$\dfrac{\operatorname{csch} x}{\sqrt{1 + \operatorname{csch}^2 x}}$
$\operatorname{csch} x$	$\dfrac{1}{\sinh x}$	$\dfrac{1}{\sqrt{\cosh^2 x - 1}}$	$\dfrac{\sqrt{1 - \tanh^2 x}}{\tanh x}$	$\sqrt{\coth^2 x - 1}$	$\dfrac{\operatorname{sech} x}{\sqrt{1 - \operatorname{sech}^2 x}}$	$\operatorname{csch} x$

(2) Expansions

$$\sinh nx = n \sinh x \cosh^{n-1} x + \binom{n}{3} \sinh^3 x \cosh^{n-3} x + \binom{n}{5} \sinh^5 x \cosh^{n-5} x + \cdots$$

$$\cosh nx = \cosh^n x + \binom{n}{2} \sinh^2 x \cosh^{n-2} x + \binom{n}{4} \sinh^4 x \cosh^{n-4} x + \cdots$$

$$\sinh^{2n} x = \frac{1}{2^{2n-1}}\left[\cosh 2nx - \binom{2n}{1} \cosh (2n-2)x + \cdots (-1)^{n-1}\binom{2n}{n-1} \cosh 2x \right] + \binom{2n}{n}\frac{1}{2^{2n}}$$

$$\sinh^{2n-1} x = \frac{1}{2^{2n-2}}\left[\sinh (2n-1)x - \binom{2n-1}{1} \sinh (2n-3)x + \cdots (-1)^{n-1}\binom{2n-1}{n-1} \sinh x \right]$$

$$\cosh^{2n} x = \frac{1}{2^{2n-1}}\left[\cosh 2nx + \binom{2n}{1} \cosh (2n-2)x + \cdots + \binom{2n}{n-1} \cosh 2x \right] - \binom{2n}{n}\frac{1}{2^{2n}}$$

$$\cosh^{2n-1} x = \frac{1}{2^{2n-2}}\left[\cosh (2n-1)x + \binom{2n-1}{1} \cosh (2n-3)x + \cdots + \binom{2n-1}{n-1} \cosh x \right]$$

(1) Sums of Angles

$$\sinh (a \pm b) = \sinh a \cosh b \pm \cosh a \sinh b$$

$$\tanh (a \pm b) = \frac{\tanh a \pm \tanh b}{1 \pm \tanh a \tanh b}$$

$$\cosh (a \pm b) = \cosh a \cosh b \pm \sinh a \sinh b$$

$$\coth (a \pm b) = \frac{\coth a \coth b \pm 1}{\coth b \pm \coth a}$$

(2) Sums of Functions

$$\sinh a + \sinh b = 2 \sinh \frac{a+b}{2} \cosh \frac{a-b}{2}$$

$$\tanh a + \tanh b = \frac{\sinh (a+b)}{\cosh a \cosh b}$$

$$\sinh a - \sinh b = 2 \cosh \frac{a+b}{2} \sinh \frac{a-b}{2}$$

$$\tanh a - \tanh b = \frac{\sinh (a-b)}{\cosh a \cosh b}$$

$$\cosh a + \cosh b = 2 \cosh \frac{a+b}{2} \cosh \frac{a-b}{2}$$

$$\coth a + \coth b = \frac{\sinh (a+b)}{\sinh a \sinh b}$$

$$\cosh a - \cosh b = 2 \sinh \frac{a+b}{2} \sinh \frac{a-b}{2}$$

$$\coth a - \coth b = \frac{\sinh (b-a)}{\sinh a \sinh b}$$

$$\sinh a + \cosh a = e^a$$

$$\tanh a + \coth a = 2 \coth 2a$$

$$\sinh a - \cosh a = -e^{-a}$$

$$\tanh a - \coth a = -2 \operatorname{csch} 2a$$

(3) Sums of Sums

$$A = a + b \qquad B = a - b$$

$$\sinh A + \sinh B = 2 \sinh a \cosh b$$

$$\cosh A + \cosh B = 2 \cosh a \cosh b$$

$$\sinh A - \sinh B = 2 \cosh a \sinh b$$

$$\cosh A - \cosh B = 2 \sinh a \sinh b$$

$$\tanh A + \tanh B = \frac{\sinh 2a}{\cosh A \cosh B}$$

$$\coth A + \coth B = \frac{\sinh 2a}{\sinh A \sinh B}$$

$$\tanh A - \tanh B = \frac{\sinh 2b}{\cosh A \cosh B}$$

$$\coth A - \coth B = \frac{-\sinh 2b}{\sinh A \sinh B}$$

(4) Products

$$\sinh a \sinh b = \tfrac{1}{2} \cosh A - \tfrac{1}{2} \cosh B$$

$$\cosh a \cosh b = \tfrac{1}{2} \cosh A + \tfrac{1}{2} \cosh B$$

$$\sinh a \cosh b = \tfrac{1}{2} \sinh A + \tfrac{1}{2} \sinh B$$

$$\tanh a \tanh b = \frac{\tanh a + \tanh b}{\coth a + \coth b}$$

$$\coth a \coth b = \frac{\coth a + \coth b}{\tanh a + \tanh b}$$

(1) Half Angles

$$\sinh\frac{a}{2} = \sqrt{\tfrac{1}{2}(\cosh a - 1)} = \tfrac{1}{2}\sqrt{\cosh a + \sinh a} - \tfrac{1}{2}\sqrt{\cosh a - \sinh a}$$

$$\cosh\frac{a}{2} = \sqrt{\tfrac{1}{2}(\cosh a + 1)} = \tfrac{1}{2}\sqrt{\cosh a + \sinh a} + \tfrac{1}{2}\sqrt{\cosh a - \sinh a}$$

$$\tanh\frac{a}{2} = \frac{\sinh a}{\cosh a + 1} = \frac{\cosh a - 1}{\sinh a} = \sqrt{\frac{\cosh a - 1}{\cosh a + 1}}$$

$$\coth\frac{a}{2} = \frac{\sinh a}{\cosh a - 1} = \frac{\cosh a + 1}{\sinh a} = \sqrt{\frac{\cosh a + 1}{\cosh a - 1}}$$

(2) Multiple Angles

$$\sinh 2a = 2\sinh a \cosh a \qquad\qquad \cosh 2a = \sinh^2 a + \cosh^2 a$$

$$\tanh 2a = \frac{2\tanh a}{1 + \tanh^2 a} \qquad\qquad \coth 2a = \frac{1 + \coth^2 a}{2\coth a}$$

$$\sinh 3a = (\sinh a)(4\cosh^2 a - 1) \qquad \cosh 3a = (\cosh a)(4\sinh^2 a - 1)$$

$$\tanh 3a = \frac{\tanh^3 a + 3\tanh a}{3\tanh^2 a + 1} \qquad \coth 3a = \frac{\coth^3 a + 3\coth a}{3\coth^2 a + 1}$$

(3) Powers

$$\sinh^2 a = \tfrac{1}{2}(\cosh 2a - 1) \qquad\qquad \cosh^2 a = \tfrac{1}{2}(\cosh 2a + 1)$$

$$\tanh^2 a = \frac{\cosh 2a - 1}{\cosh 2a + 1} \qquad\qquad \coth^2 a = \frac{\cosh 2a + 1}{\cosh 2a - 1}$$

$$\sinh^3 a = \tfrac{1}{4}(\sinh 3a - 3\sinh a) \qquad \cosh^3 a = \tfrac{1}{4}(\cosh 3a + 3\cosh a)$$

$$\sinh^4 a = \tfrac{1}{8}(\cosh 4a - 4\cosh 2a + 3) \qquad \cosh^4 a = \tfrac{1}{8}(\cosh 4a + 4\cosh 2a + 3)$$

$$\sinh^2 a + \cosh^2 a = \cosh 2a \qquad\qquad (\sinh a \pm \cosh a)^2 = \cosh 2a \pm \sinh 2a$$

$$\sinh^2 a - \cosh^2 a = -1$$

$$\tanh^2 a + \coth^2 a = -8\frac{\cosh 2a}{\cosh 4a - 1} \qquad (\tanh a + \coth a)^2 = 4\frac{\cosh 4a + 1}{\cosh 4a - 1}$$

$$\tanh^2 a - \coth^2 a = 2\frac{\cosh 4a + 3}{\cosh 4a - 1} \qquad (\tanh a - \coth a)^2 = \frac{8}{\cosh 4a - 1}$$

$$(\cosh a \pm \sinh a)^n = \cosh na \pm \sinh na$$

(1) Definitions

Inverse hyperbolic functions (area functions) are defined as follows:

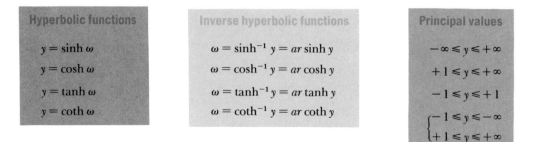

Hyperbolic functions	Inverse hyperbolic functions	Principal values
$y = \sinh \omega$	$\omega = \sinh^{-1} y = ar\sinh y$	$-\infty \leqslant y \leqslant +\infty$
$y = \cosh \omega$	$\omega = \cosh^{-1} y = ar\cosh y$	$+1 \leqslant y \leqslant +\infty$
$y = \tanh \omega$	$\omega = \tanh^{-1} y = ar\tanh y$	$-1 \leqslant y \leqslant +1$
$y = \coth \omega$	$\omega = \coth^{-1} y = ar\coth y$	$\begin{cases} -1 \leqslant y \leqslant -\infty \\ +1 \leqslant y \leqslant +\infty \end{cases}$

This means that ω is the area of the hyperbolic segment of which the hyperbolic function is y.

(2) Relationships

$$\sinh^{-1} y = \ln (y + \sqrt{y^2 + 1})$$

$$\tanh^{-1} y = \tfrac{1}{2} \ln \frac{1 + y}{1 - y} \qquad |y| < 1$$

$$\cosh^{-1} y = \pm \ln (y + \sqrt{y^2 - 1}) \qquad y \geqslant 1$$

$$\coth^{-1} y = \tfrac{1}{2} \ln \frac{y + 1}{y - 1} \qquad |y| > 1$$

For other relationships use the logarithmic transformations, for example,

$$\sinh^{-1} y + \cosh^{-1} y = \ln (y + \sqrt{y^2 + 1}) \pm \ln (y + \sqrt{y^2 - 1})$$

(3) Graphs

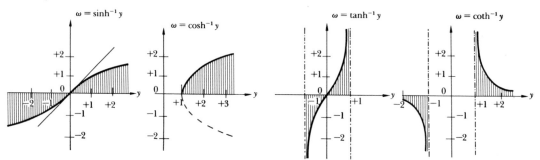

(4) Transformation Table $(k = +1$ if $y > 0$, or $k = -1$ if $y < 0)$

	$\sinh^{-1} y$	$\cosh^{-1} y$	$\tanh^{-1} y$	$\coth^{-1} y$
$\sinh^{-1} y$	$\sinh^{-1} y$	$k\cosh^{-1} \sqrt{y^2 + 1}$	$\tanh^{-1} \dfrac{y}{\sqrt{y^2 + 1}}$	$\coth^{-1} \dfrac{\sqrt{y^2 + 1}}{y}$
$\cosh^{-1} y$	$k\sinh^{-1} \sqrt{y^2 - 1}$	$\cosh^{-1} y$	$k\tanh^{-1} \dfrac{\sqrt{y^2 - 1}}{y}$	$k\coth^{-1} \dfrac{y}{\sqrt{y^2 - 1}}$
$\tanh^{-1} y$	$\sinh^{-1} \dfrac{y}{\sqrt{1 - y^2}}$	$k\cosh^{-1} \dfrac{1}{\sqrt{1 - y^2}}$	$\tanh^{-1} y$	$\coth^{-1} \dfrac{1}{y}$
$\coth^{-1} y$	$\sinh^{-1} \dfrac{1}{\sqrt{y^2 - 1}}$	$k\cosh^{-1} \dfrac{y}{\sqrt{y^2 - 1}}$	$\tanh^{-1} \dfrac{1}{y}$	$\coth^{-1} y$

7

DIFFERENTIAL CALCULUS

(1) Basic Terms

(a) The quality which changes its value is called a *variable*, and the range of its variation is known as the *interval*.

(b) The quality which remains unchanged is called a *constant*, and its range is a *single number*.

(c) A *function* is a relationship between two or more *variables* and *constants*.

(2) Definitions

(a) A variable y is a *function of another variable x*, $$y = f(x)$$

if for each value in the range of x there are one or more values in the range of y.

$x =$ independent variable or argument $y =$ dependent variable

(b) A variable y is a *function of several variables* $x_1, x_2, \ldots, x_n$, $$y = f(x_1, x_2, \ldots, x_n)$$

if for each set of values in the range of x's there is one or more values in the range of y.

(3) Forms

(a) The three most common forms of representation of a function are *tabular*, *graphical*, and *analytical*.

(b) Analytical representation

Explicit: $y = f(x_1, x_2, \ldots, x_n)$
Implicit: $0 = g(x_1, x_2, \ldots, x_n, y)$
Parametric: $x_1 = h_1(\tau), x_2 = h_2(\tau), \ldots, y = h(\tau)$

(4) Characteristic Properties

(a) Single-valued

A function is single-valued if it has a single value y for any given value of x.

(b) Multivalued

A function is multivalued if it has more than one value of y for any given value of x.

(c) Even and odd

A function is even if $f(-x) = f(x)$ and odd if $f(-x) = -f(x)$.

(d) Periodic

A function is periodic with the period T if $f(x + T) = f(x)$.

(e) Inverse

An inverse of a function $y = f(x)$ is another function $x = g(y)$.

(5) Interval

$a < x < b$	Bounded open interval	(a, b)
$a < x$	Unbounded open interval	(a, ∞)
$x < b$	Unbounded open interval	$(-\infty, b)$
$a \leq x \leq b$	Bounded closed interval	$[a, b]$
$a \leq x$	Unbounded closed interval	$[a, \infty]$
$x \leq b$	Unbounded closed interval	$[-\infty, b]$

(1) Definition of Limit

A variable x is said to have a limit a ($\lim x = a$, or $x \to a$) as x takes on consecutively the values $x_1, x_2, x_3, \ldots, x_n$ if, for every positive number ϵ, however small, the numerical value of

$$|x - a| < \epsilon$$

A function $f(x)$ is said to have a limit b $[\lim_{x \to a} f(x) = b]$ as x takes on consecutively the values $x_1, x_2, x_3, \ldots, x_n$ and approaches a, without assuming the value a, if, for every positive number δ, however small, the numerical value of

$$|f(x) - b| < \epsilon$$

(2) Operations with Limits

$$\lim_{x \to a} [f(x) + g(x)] = \lim_{x \to a} f(x) + \lim_{x \to a} g(x)$$

$$\lim_{x \to a} [f(x)g(x)] = \lim_{x \to a} f(x) \lim_{x \to a} g(x)$$

$$\lim_{x \to a} \frac{f(x)}{g(x)} = \frac{\lim_{x \to a} f(x)}{\lim_{x \to a} g(x)} \qquad \lim_{x \to a} g(x) \neq 0$$

(3) Special Cases

$$\lim_{x \to 0} a^x = 1 \qquad a > 0$$

$$\lim_{x \to \infty} \sqrt[x]{x} = 1$$

$$\lim_{m \to \infty} \frac{a^m}{m!} = 0$$

$$\lim_{x \to \infty} \frac{(\ln x)^m}{m} = 0$$

$$\lim_{x \to 0} \frac{\sin x}{x} = 1$$

$$\lim_{x \to 0} (1 + x)^{1/x} = e$$

$$\lim_{x \to \infty} \left(1 + \frac{y}{x}\right)^x = e^y$$

$$\lim_{x \to \infty} \left(1 + \frac{1}{x}\right)^x = e$$

$$\lim_{x \to 1} \frac{x - 1}{\ln x} = 1$$

$$\lim_{x \to 0} \frac{1 - \cos x}{x} = 0$$

$$\lim_{x \to 0} \frac{e^x - 1}{x} = 1$$

$$\lim_{x \to \infty} \frac{a^x - 1}{x} = \ln a \qquad a > 0$$

$$\lim_{x \to \infty} \frac{x^m}{e^x} = 0$$

$$\lim_{x \to \infty} \frac{\ln (x + 1)}{x} = 1$$

$$\lim_{x \to 0} \frac{\tan x}{x} = 1$$

$$\lim_{m \to \infty} \left(1 + \frac{1}{2} + \frac{1}{3} + \cdots + \frac{1}{m} - \ln m\right) = 0.577\ 215\ 665 \qquad \text{(Euler's constant, Sec. 13.01)}$$

$$\lim_{m \to \infty} \frac{m!}{m^m e^{-m} \sqrt{m}} = \sqrt{2\pi} \qquad \text{(Stirling's formula, Sec. 1.03)}$$

(4) Continuity of Function

(a) A single-valued function is *continuous throughout the neighborhood* of $x = a$ if and only if $\lim_{x \to a} f(x)$ exists and is equal to $f(a)$.

(b) A single-valued function is *continuous in an interval* (a, b) or $[a, b]$ if and only if it is continuous at each point of this interval.

(c) A single-valued function has a *discontinuity of the first kind* at the point $x = a$ if $f(a + 0) \neq f(a - 0)$. The difference of these two values is known as the *jump* (status) of $f(x)$.

(d) A single-valued function is *piecewise-continuous* on a given interval (a, b) or $[a, b]$ if and only if $f(x)$ is continuous throughout this interval except for a finite number of discontinuities of the first kind.

(1) Definitions

(a) **First derivative** of $y = f(x)$ with respect to x is defined as

$$\tan \phi = \frac{dy}{dx} = \lim_{\Delta x \to 0} \frac{\Delta y}{\Delta x} = \lim_{\Delta x \to 0} \frac{f(x + \Delta x) - f(x)}{\Delta x}$$

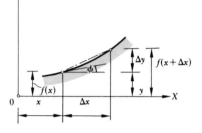

Alternative notations are $f'(x)$, $\dfrac{df(x)}{dx}$, and y'. If $y = f(t)$,

$$\frac{dy}{dt} = \frac{df(t)}{dt} = \dot{y}$$

(b) **Second and higher derivatives** of the same function are

$$\frac{d^2 y}{dx^2} = \frac{d}{dx}\left(\frac{dy}{dx}\right) = \frac{d}{dx}[f'(x)] = y''$$

$$\frac{d^n y}{dx^n} = \frac{d}{dx}\left(\frac{d^{n-1} y}{dx}\right) = \frac{d}{dx}[f^{(n-1)}(x)] = f^{(n)}(x)$$

(c) **First partial derivatives** of $y = f(x_1, x_2, \ldots, x_i, x_j, \ldots)$ with respect to one of the independent variables x_i or x_j are

$$\frac{\partial y}{\partial x_i} = \frac{\partial f(\ )}{\partial x_i} = F_i$$

$$\frac{\partial y}{\partial x_j} = \frac{\partial f(\ )}{\partial x_j} = F_j$$

Thus there are as many possible first partial derivatives as there are independent variables.

(d) **Second and higher derivatives** are

$$F_{jj} = \frac{\partial F_j}{\partial x_j} \qquad F_{ij} = \frac{\partial F_j}{\partial x_i} \qquad F_{ji} = \frac{\partial F_i}{\partial x_j} \qquad F_{ii} = \frac{\partial F_i}{\partial x_i}$$

The same process defines derivatives of any order. When the highest derivatives involved are continuous, the result is independent of the order in which the differentiation is performed.

$$F_{ijj} = F_{jij} = F_{jji} \qquad\qquad F_{ij} = F_{ji}$$

The number of differentiations performed is the *order of the partial derivatives*.

(2) Rules

$y = f(x_1)$	$x_1 = g(x_2)$	$x_2 = h(x_3)$

$$\frac{dy}{dx_3} = \frac{dy}{dx_1}\frac{dx_1}{dx_2}\frac{dx_2}{dx_3}$$

$y = f(x)$	$x = g(y)$	$\dfrac{dx}{dy} \neq 0$

$$\frac{dy}{dx} = \frac{1}{dx/dy} \qquad\qquad \frac{d^2 y}{dx^2} = -\frac{d^2 x/dy^2}{(dx/dy)^3}$$

$F(x, y) = 0$	$F_y \neq 0$

$$\frac{dy}{dx} = -F_x : F_y$$

$$\frac{d^2 y}{dx^2} = -(F_{xx}F_y{}^2 - 2F_x F_{xy}F_y + F_x{}^2 F_{yy}) : F_y{}^3$$

$x = x(t)$	$y = y(t)$	$\dot{x}(t) = \dfrac{dx}{dt} \neq 0$	$\dot{y}(t) = \dfrac{dy}{dt} \neq 0$

$$\frac{dy}{dx} = \frac{\dot{y}(t)}{\dot{x}(t)}$$

$$\frac{d^2 y}{dx^2} = \frac{\dot{x}(t)\ddot{y}(t) - \ddot{x}(t)\dot{y}(t)}{[\dot{x}(t)]^3}$$

(1) Rolle's Theorem

If a function $f(x)$ is continuous in the closed interval $[a, b]$ and is differentiable in the open interval (a, b) and if $f(a) = f(b)$, then there is at least one point $(x = c)$ in (a, b) in which

$$f'(c) = 0$$

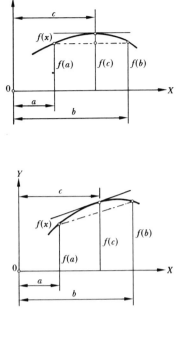

(2) Lagrange's Theorem (first mean-value theorem)

If a function $f(x)$ is continuous in the closed interval $[a, b]$ and is differentiable in the open interval (a, b), then there is at least one point $(x = c)$ in (a, b) in which

$$\frac{f(b) - f(a)}{b - a} = f'(c)$$

(3) Cauchy's Theorem (second mean-value theorem)

If the functions $f(x)$ and $g(x)$ are continuous in the closed interval $[a, b]$ and are differentiable in the open interval (a, b), if $f(x)$ and $g(x)$ are not simultaneously equal to zero at any point of this open interval, and if $g(a) \neq g(b)$, then there is at least one point $(x = c)$ in (a, b) in which

$$\frac{f(b) - f(a)}{g(b) - g(a)} = \frac{f'(c)}{g'(c)}$$

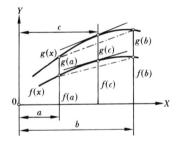

(4) L'Hôpital's Rules

$f(x)$ and $g(x)$ are two continuous functions of x having continuous derivatives at x.

If $f(x)/g(x)$ for $x = a$ is $0/0$ or ∞/∞, then

$$\lim_{x \to a} \frac{f(x)}{g(x)} = \lim_{x \to a} \frac{f'(x)}{g'(x)}$$

If $f(x) - g(x)$ for $x = a$ is $\infty - \infty$, then

$$\lim_{x \to a} [f(x) - g(x)] = \lim_{x \to a} \frac{[1/f(x) - 1/g(x)]'}{[1/f(x)]'[1/g(x)]'}$$

If $f(x)g(x)$ for $x = a$ is $(0)(\infty)$ or $(\infty)(0)$, then

$$\lim_{x \to a} [f(x)g(x)] = \lim_{x \to a} \frac{f'(x)}{[1/g(x)]'}$$

If $f(x)^{g(x)}$ for $x = a$ is 0^0 or ∞^0 or $1^{-\infty}$, then

$$\lim_{x \to a} f(x)^{g(x)} = \lim_{x \to a} e^{g(x)\ln f(x)}$$

$u, v, w = $ differentiable functions of x; $u', v', w' = $ first derivatives of these functions with respect to x; $a, b, c, m = $ constants.

(1) General Formulas

$$(a)' = 0$$

$$(au)' = au'$$

$$(u + v + w + \cdots)' = u' + v' + w' + \cdots$$

$$\left(\frac{uv}{w}\right)' = \left(\frac{uv}{w}\right)\left(\frac{u'}{u} + \frac{v'}{v} - \frac{w'}{w}\right)$$

$$(uv)' = u'v + uv'$$

$$\left(\frac{u}{v}\right)' = \frac{u'v - uv'}{v^2}$$

$$(uvw \cdots)' = (uvw \cdots)\left(\frac{u'}{u} + \frac{v'}{v} + \frac{w'}{w} + \cdots\right)$$

$$(u^v)' = vu^{v-1}u' + u^v v' \ln u$$

(2) Algebraic Functions

$$(au^m)' = amu^{m-1}u'$$

$$(a\sqrt{u})' = \frac{a}{2\sqrt{u}}u'$$

$$(a\sqrt[m]{u})' = \frac{a\sqrt[m]{u}}{mu}u'$$

$$\left(\frac{a}{u^m}\right)' = -\frac{am}{u^{m+1}}u'$$

$$\left(\frac{a}{\sqrt{u}}\right)' = -\frac{a}{2\sqrt{u^3}}u'$$

$$\left(\frac{a}{\sqrt[m]{u}}\right)' = -\frac{a}{mu\sqrt[m]{u}}u'$$

(3) Exponential Functions $\quad (e = 2.718\ 28 \cdots)$

$$(e^{mu})' = me^{mu}u'$$

$$(a^{mu})' = (ma^{mu}\ln a)u'$$

$$(u^{mu})' = mu^{mu}(1 + \ln u)u'$$

$$(e^{-mu})' = -me^{-mu}u'$$

$$(a^{-mu})' = -(m \ln a)a^{-mu}u'$$

$$(u^{-mu})' = -\frac{m(1 + \ln u)}{u^{mu}}u'$$

(4) Logarithmic Functions

$$(\ln au)' = \frac{u'}{u}$$

$$(\log u)' = \frac{u'}{u}\log e$$

$$(\log au)' = \frac{u'}{u}\log e$$

$$\left(\ln \frac{a}{u}\right)' = -\frac{u'}{u}$$

$$\left(\log \frac{1}{u}\right)' = -\frac{u'}{u}\log e$$

$$\left(\log \frac{a}{u}\right)' = -\frac{u'}{u}\log e$$

$u = $ differentiable function of x	$u' = $ first derivative of u with respect to x

(1) Trigonometric Functions

$$(\sin u)' = u' \cos u \qquad\qquad \left(\frac{1}{\sin u}\right)' = -\frac{u' \cos u}{\sin^2 u}$$

$$(\cos u)' = -u' \sin u \qquad\qquad \left(\frac{1}{\cos u}\right)' = \frac{u' \sin u}{\cos^2 u}$$

$$(\tan u)' = \frac{u'}{\cos^2 u} \qquad\qquad \left(\frac{1}{\tan u}\right)' = -\frac{u'}{\sin^2 u}$$

$$(\cot u)' = -\frac{u'}{\sin^2 u} \qquad\qquad \left(\frac{1}{\cot u}\right)' = \frac{u'}{\cos^2 u}$$

$$(\sec u)' = \frac{u' \sin u}{\cos^2 u} \qquad\qquad \left(\frac{1}{\sec u}\right)' = -u' \sin u$$

$$(\csc u)' = -\frac{u' \cos u}{\sin^2 u} \qquad\qquad \left(\frac{1}{\csc u}\right)' = u' \cos u$$

(2) Inverse Trigonometric Functions   ($u = $ principal value)

$$(\sin^{-1} u)' = \frac{u'}{\sqrt{1 - u^2}} \qquad\qquad (\cos^{-1} u)' = -\frac{u'}{\sqrt{1 - u^2}}$$

$$(\tan^{-1} u)' = \frac{u'}{1 + u^2} \qquad\qquad (\cot^{-1} u)' = -\frac{u'}{1 + u^2}$$

(3) Hyperbolic Functions

$$(\sinh u)' = u' \cosh u \qquad\qquad (\cosh u)' = u' \sinh u$$

$$(\tanh u)' = \frac{u'}{\cosh^2 u} \qquad\qquad (\coth u)' = -\frac{u'}{\sinh^2 u}$$

$$(\operatorname{sech} u)' = -\frac{u' \sinh u}{\cosh^2 u} \qquad\qquad (\operatorname{csch} u)' = -\frac{u' \cosh u}{\sinh^2 u}$$

(4) Inverse Hyperbolic Functions   ($u = $ principal value)

$$(\sinh^{-1} u)' = \frac{u'}{\sqrt{u^2 + 1}} \qquad\qquad (\cosh^{-1} u)' = \frac{u'}{\sqrt{u^2 - 1}}$$

$$(\tanh^{-1} u)' = \frac{u'}{1 - u^2} \qquad\qquad (\coth^{-1} u)' = \frac{u'}{1 - u^2}$$

a, k = constants	m = integer	n = order of derivative

(1) Algebraic Functions

y	$y^{(n)} = \dfrac{d^n y}{dx^n}$	n
$(ax)^m$	$\begin{cases} m(m-1)(m-2)\cdots(m-n+1)\dfrac{y}{x^n} \\ m!\,a^m \\ 0 \end{cases}$	$\begin{array}{l} n < m \\ n = m \\ n > m \end{array}$
$\sqrt[m]{ax}$	$(-1)^{n-1}(m-1)(2m-1)\cdots[(n-1)m-1]\dfrac{y}{(mx)^n}$	$n \leqq m$
$\left(\dfrac{1}{ax}\right)^m$	$(-1)^n m(m+1)(m+2)\cdots(m+n-1)\dfrac{y}{x^n}$	$n \leqq m$
$\sqrt[m]{\dfrac{1}{ax}}$	$(-1)^n(m+1)(2m+1)\cdots[(n-1)m+1]\dfrac{y}{(mx)^n}$	$n \leqq m$

(2) Exponential and Logarithmic Functions

y	$y^{(n)} = \dfrac{d^n y}{dx^n}$	y	$y^{(n)} = \dfrac{d^n y}{dx^n}$
e^x	e^x	a^x	$(\ln a)^n a^x$
e^{kx}	$k^n e^{kx}$	a^{kx}	$(k \ln a)^n a^{kx}$
$\ln x$	$\dfrac{(-1)^{n-1}(n-1)!}{x^n}$	$\log x$	$\dfrac{(-1)^{n-1}(n-1)!\log e}{x^n}$
$\ln kx$	$\dfrac{(-1)^{n-1}(n-1)!}{x^n}$	$\log kx$	$\dfrac{(-1)^{n-1}(n-1)!\log e}{x^n}$

(3) Trigonometric and Hyperbolic Functions

y	$y^{(n)} = \dfrac{d^n y}{dx^n}$	y	$y^{(n)} = \dfrac{d^n y}{dx^n}$
$\sin x$	$\sin\left(x + \dfrac{n\pi}{2}\right)$	$\cos x$	$\cos\left(x + \dfrac{n\pi}{2}\right)$
$\sin kx$	$k^n \sin\left(kx + \dfrac{n\pi}{2}\right)$	$\cos kx$	$k^n \cos\left(kx + \dfrac{n\pi}{2}\right)$
$\sinh x$	$\begin{cases} \sinh x & (n \text{ even}) \\ \cosh x & (n \text{ odd}) \end{cases}$	$\cosh x$	$\begin{cases} \cosh x & (n \text{ even}) \\ \sinh x & (n \text{ odd}) \end{cases}$
$\sinh kx$	$\begin{cases} k^n \sinh kx & (n \text{ even}) \\ k^n \cosh kx & (n \text{ odd}) \end{cases}$	$\cosh kx$	$\begin{cases} k^n \cosh kx & (n \text{ even}) \\ k^n \sinh kx & (n \text{ odd}) \end{cases}$

k, m, p, r = positive integers
$\bar{B}_r(x)$ = Bernoulli polynomial (Sec. 8.05)
$X_p^{(r)}$ = factorial polynomial (Sec. 1.03)

n = order of derivative
$\bar{E}_r(x)$ = Euler polynomial (Sec. 8.06)
$\mathscr{S}_k^{(r)}$ = Stirling number (Sec. A.05)

(1) Factorial Functions

y	$y^{(n)} = \dfrac{d^n y}{dx^n}$	n
$\binom{x}{r}$	$\begin{cases} \dfrac{n!}{r!x^n}\left[\binom{n}{n}\mathscr{S}_n^{(r)}x^n + \binom{n+1}{n}\mathscr{S}_{n+1}^{(r)}x^{n+1} + \cdots + \binom{r}{r}\mathscr{S}_r^{(r)}x^r\right] \\ 1 \\ 0 \end{cases}$	$n < r$ $n = r$ $n > r$
$X_p^{(r)}$	$\begin{cases} \dfrac{n!\,p^r}{x^n}\left[\binom{n}{n}\mathscr{S}_n^{(r)}\left(\dfrac{x}{p}\right)^n + \binom{n+1}{n}\mathscr{S}_{n+1}^{(r)}\left(\dfrac{x}{p}\right)^{n+1} + \cdots + \binom{r}{n}\mathscr{S}_r^{(r)}\left(\dfrac{x}{p}\right)^r\right] \\ n! \\ 0 \end{cases}$	$n < r$ $n = r$ $n > r$
$x^m X_p^{(r)}$	$\begin{cases} \dfrac{n!\,p^r x^m}{x^n}\left[\binom{m+n}{n}\mathscr{S}_n^{(r)}\left(\dfrac{x}{p}\right)^n + \binom{m+n+1}{n}\mathscr{S}_{n+1}^{(r)}\left(\dfrac{x}{p}\right)^{n+1} + \cdots + \binom{m+r}{n}\mathscr{S}_r^{(r)}\left(\dfrac{x}{p}\right)^r\right] \\ \dfrac{(r+m)!}{m!}x^m \\ \dfrac{(r+m)!}{(r+m-n)!}x^{r+m-n} \\ n! \\ 0 \end{cases}$	$n < r$ $n = r$ $r < n < r+m$ $n = r+m$ $n > r+m$

(2) Bernoulli and Euler Polynomials

y	$y^{(n)} = \dfrac{d^n y}{dx^n}$	y	$y^{(n)} = \dfrac{d^n y}{dx^n}$	n
$\bar{B}_r(x)$	$\begin{cases} n!\binom{r}{n}\bar{B}_{r-n}(x) \\ n! \\ 0 \end{cases}$	$\bar{E}_r(x)$	$\begin{cases} n!\binom{r}{n}\bar{E}_{r-n}(x) \\ n! \\ 0 \end{cases}$	$n < r$ $n = r$ $n > r$

(3) Product of Two Functions

If u, v are differentiable functions of x and

$$\frac{d^n u}{dx^n} = u^{(n)} \qquad \frac{d^n v}{dx^n} = v^{(n)}$$

are the nth derivatives of u, v, respectively, then

$$\frac{d^n(uv)}{dx^n} = uv^{(n)} + \binom{n}{1}u^{(1)}v^{(n-1)} + \binom{n}{2}u^{(2)}v^{(n-2)} + \cdots + u^{(n)}v$$

(1) Differential

The first differential of $y = f(x)$ is

$$dy = \frac{dy}{dx} dx = df(x) = f'(x)\, dx = y'\, dx$$

The higher differentials are obtained by successive differentiations of the first differential.

$$d^2y = f''(x)\, dx^2 \qquad\qquad d^ny = f^{(n)}(x)\, dx^n$$

If $x = x(t)$ and $y = y(t)$, then

$$dx = \frac{\partial[x(t)]}{\partial t}\, dt = \dot{x}\, dt \qquad\qquad\qquad dy = \frac{\partial[y(t)]}{\partial t}\, dt = \dot{y}\, dt$$

(2) Total Differential of $z = f(x, y),\, x = x(t),\, y = y(t)$

$$dz = \frac{\partial z}{\partial x} dx + \frac{\partial z}{\partial y} dy \qquad\qquad\qquad dz = \frac{\partial z}{\partial x}\frac{\partial x}{\partial t} dt + \frac{\partial z}{\partial y}\frac{\partial y}{\partial t} dt$$

(3) Exact Differential

In order that

$$dF[x, y] = A(x, y)\, dx + B(x, y)\, dy$$

$$dF[x, y, z] = A(x, y, z)\, dx + B(x, y, z)\, dy + C(x, y, z)\, dz$$

be the exact differentials, the following (necessary and sufficient) conditions must be satisfied, respectively:

$$\frac{\partial A}{\partial y} = \frac{\partial B}{\partial x} \qquad\qquad \frac{\partial B}{\partial z} = \frac{\partial C}{\partial y} \qquad\qquad \frac{\partial C}{\partial x} = \frac{\partial A}{\partial z}$$

(4) Jacobian Determinant

If $x = x(t),\, y = y(t),$ and $z = z(t),$

$$A(x, y, z, t) = 0 \qquad\qquad B(x, y, z, t) = 0 \qquad\qquad C(x, y, z, t) = 0$$

and if the jacobian determinant

$$J\!\left[\frac{A, B, C}{x, y, z}\right] = \begin{vmatrix} \dfrac{\partial A}{\partial x} & \dfrac{\partial A}{\partial y} & \dfrac{\partial A}{\partial z} \\[2mm] \dfrac{\partial B}{\partial x} & \dfrac{\partial B}{\partial y} & \dfrac{\partial B}{\partial y} \\[2mm] \dfrac{\partial C}{\partial x} & \dfrac{\partial C}{\partial y} & \dfrac{\partial C}{\partial z} \end{vmatrix} = \frac{\partial(A, B, C)}{\partial(x, y, z)} \neq 0$$

then the derivatives and the differentials of A, B, C are given by

$$\begin{bmatrix} \dfrac{\partial A}{\partial t} \\[2mm] \dfrac{\partial B}{\partial t} \\[2mm] \dfrac{\partial C}{\partial t} \end{bmatrix} = \begin{bmatrix} \dfrac{\partial A}{\partial x} & \dfrac{\partial A}{\partial y} & \dfrac{\partial A}{\partial z} \\[2mm] \dfrac{\partial B}{\partial x} & \dfrac{\partial B}{\partial y} & \dfrac{\partial B}{\partial z} \\[2mm] \dfrac{\partial C}{\partial x} & \dfrac{\partial C}{\partial y} & \dfrac{\partial C}{\partial z} \end{bmatrix} \begin{bmatrix} \dfrac{\partial x}{\partial t} \\[2mm] \dfrac{\partial y}{\partial t} \\[2mm] \dfrac{\partial z}{\partial t} \end{bmatrix}$$

The higher derivatives and differentials can be found in the same manner.

(a) The rates of change of $f(x)$ at $x = x$ are

$f'(x) > 0$; $f(x)$ is increasing $f'(x) < 0$; $f(x)$ is decreasing

$f'(x) = 0$; $f(x)$ has a tangent parallel to the X axis

(b) The curvature of $f(x)$ at $x = x$ is as follows:

$f''(x) > 0$; $f(x)$ is convex $f''(x) < 0$; $f(x)$ is concave

$f''(x) = 0$ and changes in sign; $f(x)$ has an inflection point

$f''(x) = 0$ and does not change in sign; $f(x)$ has a flat point

(c) The necessary condition for a maximum or a minimum of $f(x)$ at $x = a$ is

$$f'(a) = 0$$

(d) Sufficient conditions for extrema of $f(x)$ at $x = a$ are

$$f'(a) = 0 \qquad f''(a) < 0 \qquad \text{(maximum)}$$
$$f'(a) = 0 \qquad f''(a) > 0 \qquad \text{(minimum)}$$

(e) Necessary and sufficient conditions for extrema of $f(x)$ at $x = a$ are if $f'(a) = f''(a) = \cdots = f^{(n-1)}(a) = 0$ and $f^{(n)}(a) \neq 0$, then

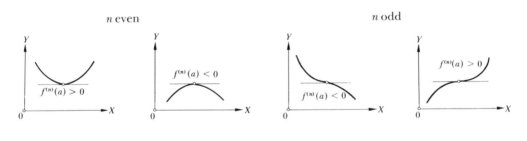

n even n odd

$f^{(n)}(a) > 0$ $f^{(n)}(a) < 0$ $f^{(n)}(a) < 0$ $f^{(n)}(a) > 0$

7.11   DIRECTION DERIVATIVES

(a) If $f = f(x, y)$, the derivative of f in the s direction is

$$\frac{\partial f}{\partial s} = \frac{\partial f}{\partial x} \cos \alpha + \frac{\partial f}{\partial y} \cos \beta$$

in which α and β are the angles between the s direction and the positive directions of the X axis and Y axis, respectively.

(b) If $f = f(x, y, z)$, the direction derivative is, analogically,

$$\frac{\partial f}{\partial s} = \frac{\partial f}{\partial x} \cos \alpha + \frac{\partial f}{\partial y} \cos \beta + \frac{\partial f}{\partial z} \cos \gamma$$

(1) Notation

S = point of contact	ds = element arc
C = center of curvature	ρ = radius of curvature
x, y = coordinates of contact point	X, Y = coordinates of running point
$\overline{AS}$ = tangent	$\overline{DS}$ = normal

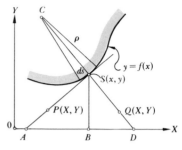

(2) Basic Equations

Curve	$y = f(x)$	$F(x, y) = 0$	$x = x(t)$ $\quad$ $y = y(t)$
Derivatives	$y' = \dfrac{dy}{dx}$	$y' = -\dfrac{F_x}{F_y}$	$\dot{x} = \dfrac{dx}{dt} \qquad \dot{y} = \dfrac{dy}{dt}$
Arc length ds	$\sqrt{1 + (y')^2}\, dx$	$\sqrt{1 + \left(\dfrac{F_x}{F_y}\right)^2}\, dx$	$\sqrt{(\dot{x})^2 + (\dot{y})^2}\, dt$
Tangent at S	$Y - y = y'(X - x)$	$F_x(X - x) = -F_y(Y - y)$	$(Y - y)\dot{x} = (X - x)\dot{y}$
Normal at S	$X - x = -y'(Y - y)$	$F_x(Y - y) = F_y(X - x)$	$(X - x)\dot{x} = -(Y - y)\dot{y}$
Length of $\overline{AS}$	$\dfrac{y}{y'}\sqrt{1 + (y')^2}$	$\dfrac{y}{F_x}\sqrt{F_x{}^2 + F_y{}^2}$	$\dfrac{y\sqrt{(\dot{x})^2 + (\dot{y})^2}}{\dot{y}}$
Length of $\overline{DS}$	$y\sqrt{1 + (y')^2}$	$\dfrac{y}{F_y}\sqrt{F_x{}^2 + F_y{}^2}$	$\dfrac{y\sqrt{(\dot{x})^2 + (\dot{y})^2}}{\dot{x}}$
Coordinates of center of curvature $\quad X_C$	$x - \dfrac{y'[1 + (y')^2]}{y''}$	$x + F_x\dfrac{F_x{}^2 + F_y{}^2}{R}$ †	$x - \dot{y}\dfrac{(\dot{x})^2 + (\dot{y})^2}{Q}$ ‡
$\quad Y_C$	$y + \dfrac{1 + (y')^2}{y''}$	$y + F_y\dfrac{F_x{}^2 + F_y{}^2}{R}$ †	$y + \dot{x}\dfrac{(\dot{x})^2 + (\dot{y})^2}{Q}$ ‡
Radius of curvature ρ	$\dfrac{[\sqrt{1 + (y')^2}]^3}{y''}$	$\dfrac{[\sqrt{F_x{}^2 + F_y{}^2}]^3}{R}$ †	$\dfrac{[\sqrt{(\dot{x})^2 + (\dot{y})^2}]^3}{Q}$ ‡

$$†R = F_x\begin{vmatrix} F_{xy} & F_x \\ F_{yy} & F_y \end{vmatrix} - F_y\begin{vmatrix} F_{xx} & F_x \\ F_{yx} & F_y \end{vmatrix} \qquad ‡Q = \begin{vmatrix} \dot{x} & \dot{y} \\ \ddot{x} & \ddot{y} \end{vmatrix}$$

(3) Shape of a Curve

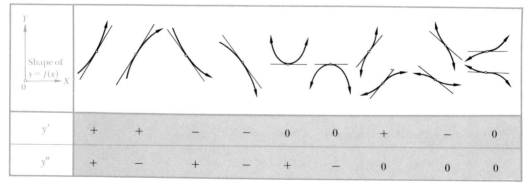

Shape of $y = f(x)$									
y'	+	+	−	−	0	0	+	−	0
y''	+	−	+	−	+	−	0	0	0

8

SEQUENCES AND SERIES

(1) Series of Constant Terms

(a) **A sequence** is a set of numbers $u_1, u_2, u_3, \ldots$ arranged in a prescribed order and formed according to a definite rule. Each member of the sequence is called a *term*, and the sequence is defined by the number of terms as *finite* or *infinite*.

(b) **An infinite series**

$$\sum_{i=1}^{\infty} u_i = u_1 + u_2 + u_3 + \cdots$$

is the sum of an infinite sequence.
If

$$s_n = \sum_{i=1}^{n} u_i \qquad \text{and} \qquad \lim_{n \to \infty} s_n = S$$

exist, the series is called *convergent*, and S is the sum. The infinite series which does not converge is *divergent*.

(c) **An absolutely convergent series** is a series whose absolute terms form a convergent series.

$$\sum_{i=1}^{\infty} |u_i| = |S|$$

(d) **A double series**

$$\sum_{i=0}^{\infty} \sum_{x=0}^{\infty} a_{ix} = \sum_{i=0}^{\infty} \left(\sum_{x=0}^{\infty} a_{ix} \right) = \sum_{x=0}^{\infty} \left(\sum_{i=0}^{\infty} a_{ix} \right)$$

converges to the limit D if

$$\lim_{\substack{m \to \infty \\ n \to \infty}} \sum_{i=0}^{m} \left(\sum_{x=0}^{n} a_{ix} \right) = D$$

(e) **A product series**

$$\sum_{i=0}^{\infty} \sum_{x=0}^{\infty} a_i b_x = \left(\sum_{i=0}^{\infty} a_i \right) \left(\sum_{x=0}^{\infty} b_x \right)$$

is an *absolutely convergent series* if

$$\sum_{i=0}^{\infty} a_i \qquad \text{and} \qquad \sum_{x=0}^{\infty} b_x$$

are *absolutely convergent series*.

(2) Series of Functions

(a) **A power series** in $x - a$ is of the form

$$S(x) = \sum_{i=0}^{\infty} c_i (x - a)^i = c_0 + c_1 (x - a) + c_2 (x - a)^2 + c_3 (x - a)^3 \cdots$$

where a and $c_0, c_1, c_2, c_3, \ldots$ are *constants*. For every power series there exists a value $b \geqslant 0$, such that $S(x)$ is absolutely convergent for all $|x| < b$ and divergent for all $|x| > b$. Then b is the *radius of convergence*, and the totality $-b < x < b$ is the *interval of convergence*.

(b) **A function series** in x is of the form

$$F(x) = \sum_{i=0}^{\infty} a_i f_i(x) = a_0 f_0(x) + a_1 f_1(x) + a_2 f_2(x) + \cdots$$

where $a_0, a_1, a_2, \ldots$ are constants. The series converges to $F(x)$ if the sequence of partial sums converges to $F(x)$.

(1) Tests of Convergence

The following tests are available for the analysis of convergence and divergence of the series

$$\sum_{n=1}^{\infty} u_n = u_1 + u_2 + u_3 + \cdots$$

in which *each term is a constant.*

(a) Comparison test

$$\text{If} \quad |u_n| < a|v_n|$$

where a is a constant independent of n and v_n is the nth term of another series which is known to be absolutely convergent, the series consisting of u terms is also absolutely convergent.

$$\text{If} \quad |u_n| > a|v_n|$$

and the series consisting of v terms is known to be absolutely divergent, then the series consisting of u terms is also divergent.

(b) Cauchy's test (nth root test)

$$\text{If} \quad \lim_{n \to \infty} \sqrt[n]{|u_n|} = L$$

then the series is absolutely convergent for $L < 1$ and divergent for $L > 1$, and the test fails for $L = 1$.

(c) D'Alambert's test (ratio test)

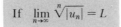

$$\text{If} \quad \lim_{n \to \infty} \left| \frac{u_{n+1}}{u_n} \right| = L$$

then the series is absolutely convergent for $L < 1$ and divergent for $L > 1$, and the test fails for $L = 1$.

(d) Raabe's test

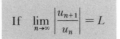

$$\text{If} \quad \lim_{n \to \infty} n \left(1 - \left| \frac{u_{n+1}}{u_n} \right| \right) = L$$

then the series is absolutely convergent for $L > 1$ and divergent for $L < 1$, and the test fails for $L = 1$.

(e) Integral test

If each term of the u series is a function of its suffix n and if the function $f(x)$ which represents $f(n)$ when $x = n$ is continuous and monotonic, the given series is convergent if

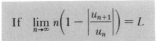

$$\int_n^{\infty} f(x)\, dx = \lim_{M \to \infty} \int_n^{M} f(x)\, dx$$

is convergent, and it is divergent if this integral is divergent.

(2) Operations with Absolutely Convergent Series

(a) The terms of an absolutely convergent series can be rearranged in any order, and the new series will converge to the same sum.

(b) The sum, difference, and product of two or more absolutely convergent series is absolutely convergent.

a, b, c = signed numbers $\qquad$ n = number of terms $\qquad$ $\alpha = (-1)^{n+1}$

A_k, B_k, C_k = kth term of series $\qquad$ k, m = positive integers $\qquad$ $\beta = (-1)^{k+1}$

(1) Arithmetic Series $\qquad\qquad$ $\boxed{A_k = a + (k-1)b}$ $\qquad$ $\boxed{B_k = \beta A_k}$

$$\sum_{k=1}^{n} A_k = a + (a+b) + (a+2b) + \cdots + [a+(n-1)b] = \frac{n}{2}(A_1 + A_n) = \frac{n}{2}[2a + (n-1)b] \qquad n < \infty$$

$$\sum_{k=1}^{n} B_k = a - (a+b) + (a+2b) - \cdots + \alpha[a+(n-1)b] = \frac{1+\alpha}{2}[a + b(n-\tfrac{1}{2})] - \frac{nb}{2} \qquad n < \infty$$

(2) Geometric Series $\qquad\qquad$ $\boxed{A_k = ab^{k-1}}$ $\qquad$ $\boxed{C_k = a^{m+1-k}b^{k-1}}$

$$\sum_{k=1}^{n} A_k = a + ab + ab^2 + \cdots + ab^{n-1} = \frac{bA_n - A_1}{b-1} = \frac{a(b^n-1)}{b-1} \qquad b \neq 1 \qquad n < \infty$$

$$\sum_{k=1}^{\infty} A_k = a + ab + ab^2 + \cdots = \frac{a}{1-b} \qquad -1 < b < 1 \qquad n = \infty$$

$$\sum_{k=1}^{m+1} C_k = a^m + a^{m-1}b + a^{m-2}b^2 + \cdots + b^m = \frac{a^n - b^n}{a-b} \qquad a \neq b \qquad n = m+1 < \infty$$

$$\sum_{k=1}^{\infty} C_k = a^m + a^{m-1}b + a^{m-2}b^2 + \cdots = \frac{a^{m+1}}{a-b} \qquad -1 < b < 1 \qquad a > b \qquad n = \infty$$

(3) Arithmogeometric Series $\qquad$ $\boxed{|c| \neq 1}$ $\quad$ $\boxed{A_k = [a+(k-1)b]c^{k-1}}$ $\quad$ $\boxed{B_k = \beta A_k}$

$$\sum_{k=1}^{n} A_k = a + (a+b)c + (a+2b)c^2 + \cdots + [a+(n-1)b]c^{n-1} = \frac{nbc^n}{c-1} - \frac{[a-c(a-b)](c^n-1)}{(c-1)^2} \qquad n < \infty$$

$$\sum_{k=1}^{\infty} A_k = a + (a+b)c + (a+2b)c^2 + \cdots = \frac{a+(b-a)c}{(1-c)^2} \qquad -1 < c < 1 \qquad n = \infty$$

$$\sum_{k=1}^{n} B_k = a - (a+b)c + (a+2b)c^2 - \cdots \alpha[a+(n-1)n]c^{n-1} = \frac{\alpha nbc^n}{1+c} + \frac{[a+c(a-b)](\alpha c^n + 1)}{(1+c)^2} \qquad n < \infty$$

$$\sum_{k=1}^{\infty} B_k = a - (a+b)c + (a+2b)c^2 - \cdots = \frac{a-(b-a)c}{(1+c)^2} \qquad -1 < c < 1 \qquad n = \infty$$

(4) Series of Binomial Coefficients $\quad$ **(Secs. 1.04 and 8.11)** $\quad$ $n = 1, 2, 3, \ldots$ $\quad$ $n < \infty$

$$\binom{2n}{2} + \binom{2n}{4} + \binom{2n}{6} + \cdots + \binom{2n}{2n} = 2^{2n-1} \qquad \binom{2n-1}{1} + \binom{2n-1}{3} + \binom{2n-1}{5} + \cdots + \binom{2n-1}{2n-1} = 2^{2n-2}$$

$$\binom{n}{1} + 2\binom{n}{2} + 3\binom{n}{3} + \cdots + n\binom{n}{n} = n2^{n-1} \qquad \binom{n}{1} + 3\binom{n}{2} + 5\binom{n}{3} + \cdots + (2n-1)\binom{n}{n} = (n-1)2^n + 1$$

$$\binom{n}{1}^2 + \binom{n}{2}^2 + \binom{n}{3}^2 + \cdots + \binom{n}{n}^2 = \binom{2n}{n} - 1 \qquad \binom{n}{1}^2 + 2\binom{n}{2}^2 + 3\binom{n}{3}^2 + \cdots + n\binom{n}{n}^2 = \frac{n}{2}\binom{2n}{n}$$

(5) Harmonic Series $\qquad\qquad$ $\psi(n) = $ Digamma function (Sec. 13.04) $\qquad$ $n <' \infty$

$$\sum_{k=1}^{n} \frac{1}{k} = C + \psi(n) \qquad \sum_{k=1}^{n} \frac{1}{2k-1} = \tfrac{1}{2}[C + \psi(n-\tfrac{1}{2})] + \ln 2 \qquad C = 0.577\,215\,665$$

$a, b =$ signed numbers	$n =$ number of terms	$\alpha = (-1)^{n+1}$
	$k, m =$ positive integers	$\beta = (-1)^{k+1}$

(1) Series of Algebraic Terms

$n < \infty$	$a > 0$	$n = \infty$	$0 < a < 1$
$\displaystyle\sum_{k=1}^{n} ka^k = \frac{na^{n+1}}{a-1} - \frac{a(a^n-1)}{(a-1)^2}$		$\displaystyle\sum_{k=1}^{\infty} ka^k = \frac{a}{(1-a)^2}$	
$\displaystyle\sum_{k=1}^{n} \beta k a^k = \frac{\alpha n a^{n+1}}{1+a} + \frac{a(1+\alpha a^n)}{(1+a)^2}$		$\displaystyle\sum_{k=1}^{\infty} \beta k a^k = \frac{a}{(1+a)^2}$	
$\displaystyle\sum_{k=1}^{n} (2k-1)a^k = \frac{2na^{n+1}}{a-1} - \frac{a(a+1)(a^n-1)}{(a-1)^2}$		$\displaystyle\sum_{k=1}^{\infty} (2k-1)a^k = \frac{a(1+a)}{(1-a)^2}$	
$\displaystyle\sum_{k=1}^{n} \beta(2k-1)a^k = \frac{2\alpha n a^{n+1}}{1+a} + \frac{a(a-1)(1+\alpha a^n)}{(1+a)^2}$		$\displaystyle\sum_{k=1}^{\infty} \beta(2k-1)a^k = \frac{a(1-a)}{(1+a)^2}$	
$\displaystyle\sum_{k=1}^{n} \frac{k}{b^k} = \frac{b(b^n-1) - n(b-1)}{b^n(b-1)^2}$	$b > 0$	$\displaystyle\sum_{k=1}^{\infty} \frac{k}{b^k} = \frac{b}{(b-1)^2}$	$b > 1$
$\displaystyle\sum_{k=1}^{n} \frac{\beta k}{b^k} = \frac{b(1+\alpha b^n) + \alpha n(1+b)}{b^n(1+b)^2}$		$\displaystyle\sum_{k=1}^{\infty} \frac{\beta k}{b^k} = \frac{b}{(b+1)^2}$	
$\displaystyle\sum_{k=1}^{n} \frac{2k-1}{b^k} = \frac{(b+1)(b^n-1) - 2n(b-1)}{b^n(b-1)^2}$		$\displaystyle\sum_{k=1}^{\infty} \frac{2k-1}{b^k} = \frac{b+1}{(b-1)^2}$	
$\displaystyle\sum_{k=1}^{n} \frac{\beta(2k-1)}{b^k} = \frac{(b-1)(1+\alpha b^n) + 2n(1+\alpha b)}{b^n(1+b)^2}$		$\displaystyle\sum_{k=1}^{\infty} \frac{\beta(2k-1)}{b^k} = \frac{b+1}{(b+1)^2}$	

(2) Series of Trigonometric Terms

$n < \infty$	$a > 0$	$n < \infty$	$a > 0$
$\displaystyle\sum_{k=1}^{n} \sin ka = \frac{\sin(na/2)\sin[(n+1)a/2]}{\sin(a/2)}$		$\displaystyle\sum_{k=1}^{n} \sin(2k-1)a = \frac{\sin^2 na}{\sin a}$	
$\displaystyle\sum_{k=1}^{n} \beta \sin ka = \frac{\sin a + \alpha \sin na + \alpha \sin(n+1)a}{2(1+\cos a)}$		$\displaystyle\sum_{k=1}^{n} \beta \sin(2k-1)a = \frac{\alpha \sin 2na}{2\cos a}$	
$\displaystyle\sum_{k=1}^{n} \cos ka = \frac{\sin(na/2)\cos[(n+1)a/2]}{\sin(a/2)}$		$\displaystyle\sum_{k=1}^{n} \cos(2k-1)a = \frac{\sin na \cos na}{\sin a}$	
$\displaystyle\sum_{k=1}^{n} \beta \cos ka = \frac{1 + \cos a + \alpha \cos na + \alpha \cos(n+1)a}{2(1+\cos a)}$		$\displaystyle\sum_{k=1}^{n} \beta \cos(2k-1)a = \frac{\alpha \cos 2na + 1}{2\cos a}$	
$\displaystyle\sum_{k=1}^{n} k \sin ka = \frac{\sin(n+1)a}{4\sin^2(a/2)} - \frac{(n+1)\cos[(2n+1)a/2]}{2\sin(a/2)}$		$\displaystyle\sum_{k=1}^{n} \sin^2 ka = \frac{n}{2} - \frac{\sin na \cos(n+1)a}{2\sin a}$	
$\displaystyle\sum_{k=1}^{n} k \cos ka = \frac{\cos(n+1)a - 1}{4\sin^2(a/2)} + \frac{(n+1)\sin[(2n+1)a/2]}{2\sin(a/2)}$		$\displaystyle\sum_{k=1}^{n} \cos^2 ka = \frac{n}{2} + \frac{\sin na \cos(n+1)a}{2\sin a}$	

$\bar{B}_m$ = Bernoulli number
B_m = auxiliary Bernoulli number
$\bar{B}_m(x)$ = Bernoulli polynomial

Z( ) = zeta function (Sec. A.06)
m = positive integer
$\alpha = (-1)^{m+1}$

(1) Bernoulli Numbers   (Sec. A.03−2)

(a) Generating function

$$\frac{x}{e^x - 1} = \sum_{m=0}^{\infty} \bar{B}_m \frac{x^m}{m!} = \frac{\bar{B}_0}{0!} + \frac{\bar{B}_1 x}{1!} + \frac{\bar{B}_2 x^2}{2!} + \frac{\bar{B}_3 x^3}{3!} + \cdots \qquad |x| < 2\pi$$

where

$\bar{B}_0 = 1$	$\bar{B}_2 = \frac{1}{6}$	$\bar{B}_4 = -\frac{1}{30}$	$\bar{B}_6 = \frac{1}{42}$	$\bar{B}_8 = -\frac{1}{30}$	$\bar{B}_{10} = \frac{5}{66}$	$\cdots$
$\bar{B}_1 = -\frac{1}{2}$	$\bar{B}_3 = 0$	$\bar{B}_5 = 0$	$\bar{B}_7 = 0$	$\bar{B}_9 = 0$	$\bar{B}_{11} = 0$	$\cdots$

are *Bernoulli numbers* of order $m = 0, 1, 2, 3, \ldots$.

(b) Auxiliary generating function

$$2 - \frac{x}{2} \cot \frac{x}{2} = \sum_{m=0}^{\infty} B_m \frac{x^{2m}}{(2m)!} = \frac{B_0}{0!} + \frac{B_1 x^2}{2!} + \frac{B_2 x^4}{4!} + \frac{B_3 x^6}{6!} + \cdots \qquad |x| < \pi$$

where

$B_0 = 1$	$B_1 = \frac{1}{6}$	$B_2 = \frac{1}{30}$	$B_3 = \frac{1}{42}$	$B_4 = \frac{1}{30}$	$B_5 = \frac{5}{66}$ $\cdots$

are *auxiliary Bernoulli numbers* of order $m = 0, 1, 2, 3, \ldots$.

(c) Series representation. For $m = 1, 2, 3, \ldots$,

$$B_m = \alpha \bar{B}_{2m} = 2 \frac{(2m)!}{(2\pi)^{2m}} \left(\frac{1}{1^{2m}} + \frac{1}{2^{2m}} + \frac{1}{3^{2m}} + \frac{1}{4^{2m}} + \cdots \right) = 2 \frac{(2m)!}{(2\pi)^{2m}} Z(2m)$$

(2) Bernoulli Polynomials   (Sec. A.03−1)

(a) Definition. The Bernoulli polynomial $\bar{B}_m(x)$ of order $m = 0, 1, 2, 3, \ldots$ is defined as

$$\bar{B}_m(x) = x^m \bar{B}_0 + \binom{m}{1} x^{m-1} \bar{B}_1 + \binom{m}{2} x^{m-2} \bar{B}_2 + \cdots + \binom{m}{m} \bar{B}_m$$

where $\bar{B}_0, \bar{B}_1, \bar{B}_2, \ldots, \bar{B}_m$ are Bernoulli numbers defined in (1a) above.

(b) First six polynomials

$\bar{B}_0(x) = 1$	$\bar{B}_3(x) = x^3 - \frac{3}{2}x^2 + \frac{1}{2}x$
$\bar{B}_1(x) = x - \frac{1}{2}$	$\bar{B}_4(x) = x^4 - 2x^3 + x^2 - \frac{1}{30}$
$\bar{B}_2(x) = x^2 - x + \frac{1}{6}$	$\bar{B}_5(x) = x^5 - \frac{5}{2}x^4 + \frac{5}{3}x^3 - \frac{1}{6}x$

(c) Properties   ($m > 0$)

$$\bar{B}_{2m}(0) = \bar{B}_{2m} = \alpha B_m \qquad \bar{B}_{2m+1}(0) = \bar{B}_{2m+1} = 0 \qquad \frac{d\bar{B}_m(x)}{dx} = m\bar{B}_{m-1}(x)$$

$\bar{E}_m$ = Euler number
E_m = auxiliary Euler number
$\bar{E}_m(x)$ = Euler polynomial

$\bar{Z}(\ )$ = complementary zeta function (Sec. A.06)
m = positive integer
$\alpha = (-1)^{m+1}$

(1) Euler Numbers   (Sec. A.04–2)

(a) Generating function

$$\frac{2\sqrt{e^x}}{e^x+1} = \sum_{m=0}^{\infty} \frac{\bar{E}_m x^m}{2^m\, m!} = \frac{\bar{E}_0}{0!} + \frac{\bar{E}_1 x}{2(1!)} + \frac{\bar{E}_2 x^2}{4(2!)} + \frac{\bar{E}_3 x^3}{8(3!)} + \cdots \qquad |x| < \pi$$

where

$\bar{E}_0 = 1$	$\bar{E}_2 = -1$	$\bar{E}_4 = 5$	$\bar{E}_6 = -61$	$\bar{E}_8 = 1{,}385$	$\bar{E}_{10} = -50{,}521$	$\cdots$
$\bar{E}_1 = 0$	$\bar{E}_3 = 0$	$\bar{E}_5 = 0$	$\bar{E}_7 = 0$	$\bar{E}_9 = 0$	$\bar{E}_{11} = 0$	$\cdots$

are *Euler numbers* of order $m = 0, 1, 2, 3, \ldots$.

(b) Auxiliary generating function

$$\sec x = \sum_{m=0}^{\infty} E_m \frac{x^{2m}}{(2m)!} = \frac{E_0}{0!} + \frac{E_1 x^2}{2!} + \frac{E_2 x^4}{4!} + \frac{E_3 x^6}{6!} + \cdots \qquad |x| < \pi$$

where

$E_0 = 1$	$E_1 = 1$	$E_2 = 5$	$E_3 = 61$	$E_4 = 1{,}385$	$E_5 = 50{,}521$ $\cdots$

are *auxiliary Euler numbers* of order $m = 0, 1, 2, 3, \ldots$.

(c) Series representation. For $m = 1, 2, 3, \ldots$,

$$E_m = -\alpha \bar{E}_{2m} = 2\left(\frac{2}{\pi}\right)^{2m+1}(2m)!\left(\frac{1}{1^{2m+1}} - \frac{1}{3^{2m+1}} + \frac{1}{5^{2m+1}} - \cdots\right) = 2\left(\frac{2}{\pi}\right)^{2m+1}(2m)!\bar{Z}(2m+1)$$

(2) Euler Polynomials   (Sec. A.04–1)

(a) Definition. The Euler polynomial $\bar{E}_m(x)$ of order $m = 0, 1, 2, 3, \ldots$ is defined as

$$\bar{E}_m(x) = \frac{2}{m+1}\left[(1-2)\binom{m+1}{1}x^m \bar{B}_1 + (1-2^2)\binom{m+1}{2}x^{m-1}\bar{B}_2 + \cdots + (1-2^{m+1})\binom{m+1}{m+1}\bar{B}_{m+1}\right]$$

where $\bar{B}_1, \bar{B}_2, \bar{B}_3, \ldots, \bar{B}_{m+1}$ are Bernoulli numbers defined in Sec. 8.05 – 1a.

(b) First six polynomials

$\bar{E}_0(x) = 1$	$\bar{E}_3(x) = x^3 - \frac{3}{2}x^2 + \frac{1}{4}$
$\bar{E}_1(x) = x - \frac{1}{2}$	$\bar{E}_4(x) = x^4 - 2x^3 + x$
$\bar{E}_2(x) = x^2 - x$	$\bar{E}_5(x) = x^5 - \frac{5}{2}x^4 + \frac{5}{2}x^2 - \frac{1}{2}$

(c) Properties $(m > 0)$

$$\bar{E}_{2m}(\tfrac{1}{2}) = 2^{-2m}\bar{E}_{2m} = -2^{-2m}\alpha E_m \qquad \bar{E}_{2m+1}(\tfrac{1}{2}) = \bar{E}_{2m+1} = 0 \qquad \frac{d\bar{E}_m(x)}{dx} = m\bar{E}_{m-1}(x)$$

$\bar{B}_{m+1}(\ ) =$ Bernoulli polynomial (Sec. 8.05)

$\bar{E}_m(\ ) =$ Euler polynomial (Sec. 8.06)

$\mathscr{P}_{m,n}, \mathscr{P}^*_{m,n}, \overline{\mathscr{P}}_{m,n}, \overline{\mathscr{P}}^*_{m,n} =$ sums

$n =$ number of terms

$k, m =$ positive integers

$\alpha = (-1)^{n+1} \qquad \beta = (-1)^{k+1}$

(1) Monotonic Series $\quad (n < \infty)$

(a) General cases

$$\sum_{k=1}^{n} k^m = 1^m + 2^m + 3^m + \cdots + n^m = \frac{\bar{B}_{m+1}(n+1) - \bar{B}_{m+1}(0)}{m+1} = \mathscr{P}_{m,n}$$

$$\sum_{k=1}^{n} (2k)^m = 2^m + 4^m + 6^m + \cdots + (2n)^m = 2^m \mathscr{P}_{m,n}$$

$$\sum_{k=1}^{n} (2k-1)^m = 1^m + 3^m + 5^m + \cdots + (2n-1)^m = 2^m \sum_{k=0}^{m} \left[\binom{m}{k} \frac{\mathscr{P}_{m-k,n}}{2^k} \right] = \mathscr{P}^*_{m,n}$$

(b) Particular cases

m	$\mathscr{P}_{m,n}$	$\mathscr{P}^*_{m,n}$
1	$\dfrac{n(n+1)}{2}$	n^2
2	$\dfrac{n(n+1)(2n+1)}{6}$	$\dfrac{n(4n^2-1)}{3}$
3	$\left[\dfrac{n(n+1)}{2}\right]^2$	$n^2(2n^2-1)$
4	$\dfrac{n^5}{5}+\dfrac{n^4}{2}+\dfrac{n^3}{3}-\dfrac{n}{30}$	$\dfrac{(2n)^5}{10}-\dfrac{(2n)^3}{3}+\dfrac{7(2n)}{30}$

(2) Alternating Series $\quad (n < \infty)$

(a) General cases

$$\sum_{k=1}^{n} \beta k^m = 1^m - 2^m + 3^m - \cdots \alpha n^m = \frac{\bar{E}_m(n+1) - \alpha \bar{E}_m(0)}{2} = \overline{\mathscr{P}}_{m,n}$$

$$\sum_{k=1}^{n} \beta(2k)^m = 2^m - 4^m + 6^m - \cdots \alpha(2n)^m = 2^m \overline{\mathscr{P}}_{m,n}$$

$$\sum_{k=1}^{n} \beta(2k-1)^m = 1^m - 3^m + 5^m - \cdots \alpha(2n-1)^m = 2^m \sum_{k=0}^{m} \left[\beta\binom{m}{k} \frac{\overline{\mathscr{P}}_{m-k,n}}{2^k} \right] = \overline{\mathscr{P}}^*_{m,n}$$

(b) Particular cases

m	$\overline{\mathscr{P}}_{m,n}$	$\overline{\mathscr{P}}^*_{m,n}$
1	$\dfrac{\alpha n + (1+\alpha)/2}{2}$	αn
2	$\dfrac{\alpha n(n+1)}{2}$	$\dfrac{4\alpha n^2 - (\alpha+1)}{2}$
3	$\dfrac{\alpha n^2(2n+3) - (\alpha+1)/2}{4}$	$\alpha n(4n^2-3)$
4	$\dfrac{\alpha n(n^3+2n^2-1)}{2}$	$\dfrac{8\alpha n^2(2n^2-3)+5(\alpha+1)}{2}$

$Z(\ ) = $ zeta function (Sec. A.06) $n = $ number of terms

$\overline{Z}(\ ) = $ complementary zeta function (Sec. A.06) $k, m = $ positive integers

$\mathcal{D}_{m,n}, \mathcal{D}^*_{m,n}, \overline{\mathcal{D}}_{m,n}, \overline{\mathcal{D}}^*_{m,n} = $ sums $\beta = (-1)^{k+1}$

(1) Monotonic Series $(n = \infty)$

(a) General cases

$$\sum_{k=1}^{\infty} \frac{1}{k^m} = \frac{1}{1^m} + \frac{1}{2^m} + \frac{1}{3^m} + \cdots = Z(m) = \mathcal{D}_{m,n}$$

$$\sum_{k=1}^{\infty} \frac{1}{(2k)^m} = \frac{1}{2^m} + \frac{1}{4^m} + \frac{1}{6^m} + \cdots = \frac{1}{2^m} Z(m) = \frac{1}{2^m} \mathcal{D}_{m,n}$$

$$\sum_{k=1}^{\infty} \frac{1}{(2k-1)^m} = \frac{1}{1^m} + \frac{1}{3^m} + \frac{1}{5^m} + \cdots = \left(1 - \frac{1}{2^m}\right) Z(m) = \mathcal{D}^*_{m,n}$$

(b) Particular cases   (Sec. A.06)

m	$\mathcal{D}_{m,n}$	$\mathcal{D}^*_{m,n}$
1	∞	∞
2	$\frac{\pi^2}{6} = 1.644\ 934\ 066\ 848\ 226$	$\frac{\pi^2}{8} = 1.233\ 700\ 550\ 136\ 170$
3	$1.202\ 056\ 903\ 159\ 594$	$1.051\ 799\ 790\ 264\ 644$
4	$\frac{\pi^4}{90} = 1.082\ 323\ 233\ 711\ 114$	$\frac{\pi^4}{96} = 1.014\ 678\ 031\ 604\ 192$

(2) Alternating Series $(n = \infty)$

(a) General cases

$$\sum_{k=1}^{\infty} \frac{\beta}{k^m} = \frac{1}{1^m} - \frac{1}{2^m} + \frac{1}{3^m} - \cdots = \left(1 - \frac{2}{2^m}\right) Z(m) = \overline{\mathcal{D}}_{m,n}$$

$$\sum_{k=1}^{\infty} \frac{\beta}{(2k)^m} = \frac{1}{2^m} - \frac{1}{4^m} + \frac{1}{6^m} - \cdots = \frac{1}{2^m}\left(1 - \frac{2}{2^m}\right) Z(m) = \frac{1}{2^m} \overline{\mathcal{D}}_{m,n}$$

$$\sum_{k=1}^{\infty} \frac{\beta}{(2k-1)^m} = \frac{1}{1^m} - \frac{1}{3^m} + \frac{1}{5^m} - \cdots = \overline{Z}(m) = \overline{\mathcal{D}}^*_{m,n}$$

(b) Particular cases   (Sec. A.06)

m	$\overline{\mathcal{D}}_{m,n}$	$\overline{\mathcal{D}}^*_{m,n}$
1	$\ln 2 = 0.693\ 147\ 180\ 559\ 945$	$\frac{\pi}{4} = 0.785\ 398\ 163\ 397\ 448$
2	$\frac{\pi^2}{12} = 0.822\ 467\ 033\ 424\ 113$	$0.915\ 965\ 594\ 177\ 219$
3	$0.901\ 542\ 677\ 369\ 696$	$\frac{\pi^3}{32} = 0.968\ 946\ 146\ 259\ 369$
4	$\frac{7\pi^4}{720} = 0.947\ 032\ 829\ 497\ 246$	$0.988\ 944\ 551\ 741\ 105$

(1) Tests of Convergence

The following tests are available for the analysis of the convergence of the series

$$F(x) = \sum_{n=1}^{\infty} f_n(x) = f_1(x) + f_2(x) + f_3(x) + \cdots$$

in which each term is a function.

(a) Cauchy's test for uniform convergence

A series of real (or complex) functions converges uniformly on $F(x)$ in $[a, b]$ if for every real number $\epsilon > 0$ there exists a real number $N > 0$, independent of x in $[a, b]$, such that

$$|F(x) - f_n(x)| < \epsilon \qquad \text{for all } n > N$$

This is a necessary and sufficient condition for uniform convergence for a function series.

(b) Weierstrass's test for uniform and absolute convergence

A series of real (or complex) functions converges uniformly and absolutely on every $F(x)$ in $[a, b]$ if

$$|f_n(x)| \leq M_n \qquad \text{for all } n$$

and $M_1 + M_2 + M_3 + \cdots$ is a convergent comparison series of real positive terms. Since this test establishes the absolute (as well as the uniform) convergence, it is applicable only to series which converge absolutely. It must be noted that a function series may converge uniformly but not absolutely, and vice versa.

(c) Dirichlet's test for uniform convergence

If $\displaystyle\sum_{n=1}^{\infty} a_n = a_1 + a_2 + a_3 + \cdots$ is a monotonic decreasing sequence of real numbers

then the infinite function series $\displaystyle\sum_{n=1}^{\infty} a_n f_n(x) = a_1 f_1(x) + a_2 f_2(x) + a_3 f_3(x) + \cdots$

converges uniformly on a set $G(x)$ of values of x if the infinite series

$$\sum_{n=1}^{\infty} f_n(x) = f_1(x) + f_2(x) + f_3(x) + \cdots$$

converges uniformly on the same set $G(x)$ of values of x.

(2) Properties of Uniformly Convergent Function Series

(a) Theorem of continuity

If any term of a uniformly convergent function series is a continuous function of x in $[a, b]$, then the sum of the series is also a continuous function of x in $[a, b]$.

(b) Theorem of differentiability of a function series

A uniformly convergent series in (a, b) can be differentiated term by term in (a, b). If each term of the differentiated series is continuous and the differentiated series is uniformly convergent in (a, b), then it will converge to the derivative of the function it represents in (a, b).

(c) Theorem of integrability of a function series

A uniformly convergent series in $[a, b]$ can be integrated term by term in $[a, b]$, and the integrated series will converge uniformly to the integral of the function it represents in $[a, b]$.

(1) Interval of Convergence (ratio test)

The power series in the real (or complex) variable x

$$S(x) = \sum_{n=0}^{\infty} a_n x^n = a_0 + a_1 x + a_2 x^2 + \cdots$$

where the coefficients $a_0, a_1, a_2, \ldots$ are real or complex numbers, independent of x,

is *convergent* if
$$\lim_{n \to \infty} \left| \frac{a_{n+1} x^{n+1}}{a_n x^n} \right| = \lim_{n \to \infty} \left| \frac{a_{n+1}}{a_n} \right| |x| = r|x| < 1$$

and is *divergent* if
$$\lim_{n \to \infty} \left| \frac{a_{n+1} x^{n+1}}{a_n x^n} \right| = \lim_{n \to \infty} \left| \frac{a_{n+1}}{a_n} \right| |x| = r|x| > 1$$

The interval of convergence is then

$$r|x| < 1 \qquad \text{or} \qquad -\frac{1}{r} < x < \frac{1}{r}$$

and it is symmetrical about the origin of x. The series is convergent in this interval and diverges outside this interval. It may or may not converge at the end points of the interval.

(2) Uniform and Absolute Convergence

The power series which converges in the interval

$$\alpha < x < \beta$$

converges absolutely and uniformly for every value of x within this interval. Since a uniformly convergent series represents a continuous function, a *uniformly convergent series defines a continuous function within the interval of convergence*.

(3) Operations with Power Series

(a) Uniqueness theorem

If two power series

$$S(x) = \sum_{n=0}^{\infty} a_n x^n \qquad \text{and} \qquad S(x) = \sum_{n=0}^{\infty} b_n x^n$$

converge to the same sum $S(x)$ for all real values of x, then

$$a_0 = b_0, a_1 = b_1, a_2 = b_2, \ldots$$

(b) Summation theorem

Two power series can be added or subtracted term by term for each value of x common to their interval of convergence.

(c) Product theorem

Two power series can be multiplied term by term for each value of x common to their interval of convergence. Thus

$$\left(\sum_{m=0}^{\infty} a_m x^m \right) \left(\sum_{n=0}^{\infty} b_n x^n \right) = \sum_{m=0}^{\infty} \sum_{n=0}^{\infty} a_m b_n x^{m+n}$$

(d) Theorem of differentiability and integrability

A power series can be differentiated and integrated term by term in any closed interval if and only if this interval lies entirely within the interval of uniform convergence of the power series.

a, b = signed numbers $\qquad$ k, n = positive integers

$\Lambda[\,]$ = nested sum $\qquad$ $n + 1$ = number of terms

(1) Series of Constant Terms

(a) Geometric series (Sec. 8.03)

$$\sum_{k=0}^{n} \bar{\beta} a b^k = a \pm ab + ab^2 \pm ab^3 + \cdots + \bar{\alpha} ab^n$$

$$= a(1 \pm b(1 \pm b(1 \pm b(1 \pm \cdots \pm b)))) = a \bigwedge_{k=1}^{n}\left[1 \pm \frac{kb}{k}\right]$$

$\bar{\alpha} = (\pm 1)^n$

$\bar{\beta} = (\pm 1)^k$

(b) Series of factorials (Sec. 1.03)

$$\sum_{k=0}^{n} \bar{\beta} k! = 0! \pm 1! + 2! \pm 3! + \cdots + \bar{\alpha} n!$$

$$= (1 \pm 1(1 \pm 2(1 \pm 3(1 \pm \cdots \pm n)))) = \bigwedge_{k=1}^{n}[1 \pm k]$$

$$\sum_{k=0}^{n} \bar{\beta}(2k+1)! = 1! \pm 3! + 5! \pm 7! + \cdots + \bar{\alpha}(2n+1)!$$

$$= (1 \pm 2 \cdot 3(1 \pm 4 \cdot 5(1 \pm 6 \cdot 7(1 \pm \cdots \pm 2n(2n+1))))) = \bigwedge_{k=1}^{n}[1 \pm 2k(2k+1)]$$

(c) Series of double factorials (Sec. 1.03)

$$\sum_{k=0}^{n} \bar{\beta}(2k)!! = 0!! \pm 2!! + 4!! \pm 6!! + \cdots + \bar{\alpha}(2n)!!$$

$$= (1 \pm 2(1 \pm 4(1 \pm 6(1 \pm \cdots \pm 2n)))) = \bigwedge_{k=1}^{n}[1 \pm 2k]$$

$$\sum_{k=0}^{n} \bar{\beta}(2k+1)!! = 1!! \pm 3!! + 5!! \pm 7!! + \cdots + \bar{\alpha}(2n+1)!!$$

$$= (1 \pm 3(1 \pm 5(1 \pm 7(1 \pm \cdots \pm (2n+1))))) = \bigwedge_{k=1}^{n}[1 \pm (2k+1)]$$

(d) Series of binomial coefficients (Sec. 1.04)

$$\sum_{k=0}^{n} \bar{\beta}\binom{n}{k} = \binom{n}{0} \pm \binom{n}{1} + \binom{n}{2} \pm \binom{n}{3} + \cdots + \bar{\alpha}\binom{n}{n}$$

$$= \left(1 \pm \frac{n}{1}\left(1 \pm \frac{n-1}{2}\left(1 \pm \frac{n-2}{3}\left(1 \pm \cdots \pm \frac{1}{n}\right)\right)\right)\right) = \bigwedge_{k=1}^{n}\left[1 \pm \frac{n+1-k}{k}\right]$$

(2) Power Series

(a) Basic form

$b_k = \dfrac{a_k}{a_{k-1}}$

$$\sum_{k=0}^{n} \bar{\beta} a_k x^k = a_0 \pm a_1 x + a_2 x^2 \pm a_3 x^3 + \cdots + \bar{\alpha} a_n x^n$$

$$= a_0(1 \pm b_1 x(1 \pm b_2 x(1 \pm b_3 x(1 \pm \cdots \pm b_n x)))) = a_0 \bigwedge_{k=1}^{n}[1 \pm b_k x]$$

(b) First derivative

$$\frac{d}{dx}\left[\sum_{k=0}^{n} \bar{\beta} a_k x^k\right] = \sum_{k=1}^{n} \bar{\beta} k a_k x^{k-1} = \pm a_1 + 2a_2 x \pm 3a_3 x^2 + \cdots + \bar{\alpha} n a_n x^{n-1}$$

$$= \pm a_1\left(1 \pm \frac{2b_2 x}{1}\left(1 \pm \frac{3b_3}{2} x\left(1 \pm \cdots \pm \frac{nb_n}{n-1} x\right)\right)\right) = \pm a_1 \bigwedge_{k=2}^{n}\left[1 \pm \frac{kb_k}{(k-1)} x\right]$$

(c) Indefinite integral (C = constant of integration)

$$\int\left[\sum_{k=0}^{n} \bar{\beta} a_k x^k\right] dx = \sum_{k=0}^{n} \bar{\beta}\frac{a_k x^{k+1}}{k+1} = a_0 x + \frac{a_1}{2} x^2 + \frac{a_2}{3} x^3 + \frac{a_3}{4} x^4 + \cdots + \bar{\alpha}\frac{a_n}{n+1} x^{n+1} + C$$

$$= a_0 x\left(1 \pm \frac{b_1}{2} x\left(1 \pm \frac{2b_2}{3} x\left(1 \pm \frac{3b_3}{4} x\left(1 \pm \cdots \pm \frac{nb_n}{n+1} x\right)\right)\right)\right) + C = a_0 x \bigwedge_{k=1}^{n}\left[1 \pm \frac{kb_k}{k+1} x\right] + C$$

n = signed number	k, p, q = positive integers
N = nested	$\Lambda[\ ]$ = N−sum (Sec. 8.11)

(1) Basic Cases

(a) Symbolic form (Sec. 1.04)

$$(1 \pm x)^n = 1 \pm \binom{n}{1}x + \binom{n}{2}x^2 \pm \binom{n}{3}x^3 + \cdots \begin{cases} n = 0, 1, 2, \ldots, x \gtreqless 0 & \text{Finite series} \\[4pt] n \neq 0, 1, 2, \ldots, |x| < 1 & \text{Infinite convergent series} \\[4pt] n \neq 0, 1, 2, \ldots, |x| > 1 & \text{Infinite divergent series} \end{cases}$$

(b) Standard form $(x^2 < 1, n \neq 0, 1, 2, \ldots)$

$$(1 \pm x)^n = 1 \pm \frac{n}{1!}x + \frac{n(n-1)}{2!}x^2 \pm \frac{n(n-1)(n-2)}{3!}x^3 + \cdots = \sum_{k=0}^{\infty}(\pm 1)^k\binom{n}{k}x^k$$

(c) Nested form $(x^2 < 1, n \neq 0, 1, 2, \ldots)$

$$(1 \pm x)^n = \left(1 \pm \frac{n}{1}x\left(1 \pm \frac{n-1}{2}x\left(1 \pm \frac{n-2}{3}x(1 \pm \cdots)\right)\right)\right) = \bigwedge_{k=1}^{\infty}\left[1 \pm \frac{n+k-1}{k}x\right]$$

$$A = \frac{x^*}{q}$$

(2) Special Cases in Nested Form $(x^2 < 1)$

n	N-series	N-sum
-1	$(1 \mp x(1 \mp x(1 \mp x(1 \mp x(1 \mp \cdots)))))$	$\displaystyle\bigwedge_{k=1}^{\infty}\left[1 \mp \frac{k}{k}x\right]$
-2	$\left(1 \mp \frac{2}{1}x\left(1 \mp \frac{3}{2}x\left(1 \mp \frac{4}{3}x\left(1 \mp \frac{5}{4}x(1 \mp \cdots)\right)\right)\right)\right)$	$\displaystyle\bigwedge_{k=1}^{\infty}\left[1 \mp \frac{k+1}{k}x\right]$
$-p$	$\left(1 \mp \frac{p}{1}x\left(1 \mp \frac{p+1}{2}x\left(1 \mp \frac{p+2}{3}x\left(1 \mp \frac{p+3}{4}x(1 \mp \cdots)\right)\right)\right)\right)$	$\displaystyle\bigwedge_{k=1}^{\infty}\left[1 \mp \frac{p+k-1}{k}x\right]$
$\dfrac{1}{2}$	$1 \pm \frac{x}{2}\left(1 \mp \frac{1}{2}\left(\frac{x}{2}\right)\left(1 \mp \frac{3}{3}\left(\frac{x}{2}\right)\left(1 \mp \frac{5}{4}\left(\frac{x}{2}\right)(1 \mp \cdots)\right)\right)\right)$	$1 \pm \frac{x}{2}\displaystyle\bigwedge_{k=1}^{\infty}\left[1 \mp \frac{2k-1}{k+1}\frac{x}{2}\right]$
$\dfrac{1}{3}$	$1 \pm \frac{x}{3}\left(1 \mp \frac{2}{2}\left(\frac{x}{3}\right)\left(1 \mp \frac{5}{3}\left(\frac{x}{3}\right)\left(1 \mp \frac{8}{4}\left(\frac{x}{3}\right)(1 \mp \cdots)\right)\right)\right)$	$1 \pm \frac{x}{3}\displaystyle\bigwedge_{k=1}^{\infty}\left[1 \mp \frac{3k-1}{k+1}\frac{x}{3}\right]$
$\dfrac{1}{q}$	$1 \pm A\left(1 \mp \frac{q-1}{2}A\left(1 \mp \frac{2q-1}{3}A\left(1 \mp \frac{3q-1}{4}A(1 \mp \cdots)\right)\right)\right)$	$1 \pm A\displaystyle\bigwedge_{k=1}^{\infty}\left[1 \mp \frac{kq-1}{k+1}A\right]$
$-\dfrac{1}{2}$	$1 \mp \frac{x}{2}\left(1 \mp \frac{3}{2}\left(\frac{x}{2}\right)\left(1 \mp \frac{5}{3}\left(\frac{x}{2}\right)\left(1 \mp \frac{7}{4}\left(\frac{x}{2}\right)(1 \mp \cdots)\right)\right)\right)$	$1 \mp \frac{x}{2}\displaystyle\bigwedge_{k=1}^{\infty}\left[1 \mp \frac{2k+1}{k+1}\frac{x}{2}\right]$
$-\dfrac{1}{3}$	$1 \mp \frac{x}{3}\left(1 \mp \frac{4}{2}\left(\frac{x}{3}\right)\left(1 \mp \frac{7}{3}\left(\frac{x}{3}\right)\left(1 \mp \frac{10}{4}\left(\frac{x}{3}\right)(1 \mp \cdots)\right)\right)\right)$	$1 \mp \frac{x}{3}\displaystyle\bigwedge_{k=1}^{\infty}\left[1 \mp \frac{3k+1}{k+1}\frac{x}{3}\right]$
$-\dfrac{1}{q}$	$1 \mp A\left(1 \mp \frac{q+1}{2}A\left(1 \mp \frac{2q+1}{3}A\left(1 \mp \frac{3q+1}{4}A(1 \mp \cdots)\right)\right)\right)$	$1 \mp A\displaystyle\bigwedge_{k=1}^{\infty}\left[1 \mp \frac{kq+1}{k+1}A\right]$
$\dfrac{p}{q}$	$1 \pm pA\left(1 \mp \frac{q-p}{2}A\left(1 \mp \frac{2q-p}{3}A\left(1 \mp \frac{3q-p}{4}A(1 \mp \cdots)\right)\right)\right)$	$1 \pm pA\displaystyle\bigwedge_{k=1}^{\infty}\left[1 \mp \frac{kq-p}{k+1}A\right]$
$-\dfrac{p}{q}$	$1 \mp pA\left(1 \mp \frac{q+p}{2}A\left(1 \mp \frac{2q+p}{3}A\left(1 \mp \frac{3q+p}{4}A(1 \mp \cdots)\right)\right)\right)$	$1 \mp pA\displaystyle\bigwedge_{k=1}^{\infty}\left[1 \mp \frac{kq+p}{k+1}A\right]$

*A = pocket calculator storage.

(1) Single Variable

(a) MacLaurin's series at $x = 0$

If a function $f(x)$ is continuous and single-valued and has all derivatives on an interval including $x = 0$, then

$$f(x) = f(0) + \frac{f'(0)}{1!}x + \frac{f''(0)}{2!}x^2 + \cdots + \frac{f^{(n)}(0)}{n!}x^n + R_n$$

in which $\qquad R_n = \frac{f^{(n+1)}(\theta x)}{(n+1)!}x^{n+1} \qquad\qquad 0 < \theta < 1$

This series represents $f(x)$ for those values of x for which $R_n \to 0$ as $n \to \infty$.

(b) Taylor's series at $x = a$

If a function $f(x)$ is continuous and single-valued and has all derivatives on an interval including $x = a$, then

$$f(x) = f(a) + \frac{f'(a)}{1!}(x-a) + \frac{f''(a)}{2!}(x-a)^2 + \cdots + \frac{f^{(n)}(a)}{n!}(x-a)^n + R_n$$

in which $\qquad R_n = \frac{f^{(n+1)}(\theta x)}{(n+1)!}(x-a)^{n+1} \qquad\qquad a < \theta x < x$

The series represents $f(x)$ for those values of x for which $R_n \to 0$ as $n \to \infty$.

(c) Modified Taylor's series at $x = a + h$

If a function $f(x)$ is continuous and single-valued and has all derivatives on an interval including $x = a + h$, then

$$f(a+h) = f(a) + \frac{f'(a)}{1!}h + \frac{f''(a)}{2!}h^2 + \cdots + \frac{f^{(n)}(a)}{n!}h^n + R_n$$

in which $\qquad R_n = \frac{f^{(n+1)}(a+\theta h)}{(n+1)!}h^{n+1} \qquad\qquad a < a + \theta h < a + h$

(2) Two Variables

Taylor's series for a function of two variables is

$$f(x+a, y+b) = f(x, y) + \frac{1}{1!}D_1[f(x, y)] + \frac{1}{2!}D_2[f(x, y)] + \cdots + \frac{1}{n!}D_n[f(x, y)] + R_n$$

in which $\quad D_n = \left(a\dfrac{\partial}{\partial x} + b\dfrac{\partial}{\partial y}\right)^n \quad$ and $\quad R_n = \dfrac{1}{(n+1)!}D_{n+1}[f(x + \theta_1 a, y + \theta_2 b)]$

or at $x = 0, y = 0$, $\qquad\qquad\qquad\qquad\qquad\qquad\qquad\qquad 0 < \theta_1 < 1, 0 < \theta_2 < 1$

$$f(x, y) = f(0, 0) + \frac{1}{1!}D_1[f(0, 0)] + \frac{1}{2!}D_2[f(0, 0)] + \cdots + \frac{1}{n!}D_n[f(0, 0)] + R_n$$

in which $\quad D_n = \left(x\dfrac{\partial}{\partial x} + y\dfrac{\partial}{\partial y}\right)^n \quad$ and $\quad R_n = \dfrac{1}{(n+1)!}D_{n+1}[f(\theta_1 x, \theta_2 y)]$

$$0 < \theta_1 < 1, 0 < \theta_2 < 1$$

m = real signed number j, k, r = positive integers

$$a_0 = 1, b_0 = 1$$

(1) Basic Operations

(a) Sum of two series

$$\sum_{k=0}^{\infty} a_k x^k \pm \sum_{k=0}^{\infty} b_k x^k = \sum_{k=0}^{\infty} (a_k + b_k) x^k$$

(b) Product and quotient of two series

$$\left(\sum_{k=0}^{\infty} a_k x^k\right)\left(\sum_{k=0}^{\infty} b_k x^k\right) = \sum_{k=0}^{\infty} A_k x^k \qquad A_0 = 1, A_1 = a_1 + b_1, \cdots, A_k = b_k + \sum_{j=1}^{k} a_j b_{k-j}$$

$$\left(\sum_{k=0}^{\infty} a_k x^k\right) : \left(\sum_{k=0}^{\infty} b_k x^k\right) = \sum_{k=0}^{\infty} B_k x^k \qquad B_0 = 1, B_1 = a_1 - b_1, \cdots, B_k = a_k - \sum_{j=1}^{k} b_j B_{k-j}$$

(2) Powers and Roots of a Series

$$a_0 = 1, \omega_j = jm - k + j$$

(a) General case

$$\left(\sum_{k=0}^{\infty} a_k x^k\right)^m = \sum_{k=0}^{\infty} C_{m,k} x^k \qquad C_{m,0} = 1, C_{m,1} = ma_1 C_{m,0}, \cdots, C_{m,k} = \frac{1}{k}\left[\sum_{j=1}^{k} \omega_j a_j C_{m,k-j}\right]$$

(b) Particular cases

m	ω_j	m	ω_j	m	ω_j	m	ω_j
2	$3j - k$	-1	$-k$	$\frac{1}{2}$	$3j/2 - k$	$-\frac{1}{2}$	$j/2 - k$
3	$4j - k$	-2	$-k - j$	$\frac{1}{3}$	$4j/3 - k$	$-\frac{1}{3}$	$2j/3 - k$
4	$5j - k$	-3	$-k - 2j$	$\frac{1}{4}$	$5j/4 - k$	$-\frac{1}{4}$	$3j/4 - k$

(3) Special Operations

$$a_0 = 0, b_0 = 0$$

(a) Substitution. If $y = a_1 x + a_2 x^2 + a_3 x^3 + \cdots$, then

$$\sum_{r=0}^{\infty} b_r y^r = D_1 x + D_2 x^2 + D_3 x^3 + \cdots = \sum_{r=1}^{\infty} D_r x^r$$

$$D_1 = b_1 C_{1,1}, D_2 = b_1 C_{1,2} + b_2 C_{2,2}, D_3 = b_1 C_{1,3} + b_2 C_{2,3} + b_3 C_{3,3} \cdots$$

where $C_{r,k}$ is $C_{m,k}$ given in (2a) above, $b_1, b_2, \ldots, b_r$ are known values, and $C_{r,r} = a_1^r$.

(b) Reversion. If $y = x - a_2 x^2 - a_3 x^3 - a_4 x^4 - a_5 x^5 - a_6 x^6 - a_7 x^7 - \cdots$, then

$$x = y + R_2 y^2 + R_3 y^3 + R_4 y^4 + R_5 y^5 + R_6 y^6 + R_7 y^7 + \cdots = y + \sum_{k=2}^{\infty} R_k y^k$$

$$R_2 = a_2 \qquad\qquad\qquad R_3 = 2a_2^2 + a_3$$

$$R_4 = 5a_2^3 + 5a_2 a_3 + a_4 \qquad R_5 = 14a_2^4 + 21a_2^2 a_3 + 3a_3^2 + a_5$$

$$R_6 = 42a_2^5 + 84a_2^3 a_3 + 28(a_2^2 a_4 + a_2 a_3^2) + 7(a_2 a_3 + a_3 a_4) + a_6$$

$$R_7 = 132a_2^6 + 330a_2^4 a_3 + 60(3a_2^2 a_3^2 + 2a_2^3 a_4) + 12(6a_2 a_3 a_4 + 3a_2^2 a_5 + a_3^3) + 4(2a_2 a_6 + 2a_3 a_5 + a_4^2) + a_7 \cdots$$

B_k = auxiliary Bernoulli number (Sec. 8.05)	a, b = constants
E_k = auxiliary Euler number (Sec. 8.06)	$\lambda = \sqrt{a^2 + b^2}$ $\omega = \tan^{-1}\dfrac{b}{a}$

(1) Trigonometric Functions[1]

$f(x)$	Standard series	Nested series	Interval
$\sin x$	$x \sum\limits_{k=0}^{\infty} (-1)^k \dfrac{x^{2k}}{(2k+1)!}$	$x \bigwedge\limits_{k=1}^{\infty} \left[1 - \dfrac{x^2}{2k(2k+1)} \right]$	
$\sin^2 x$	$2x^2 \sum\limits_{k=0}^{\infty} (-1)^k \dfrac{(2x)^{2k}}{(2k+2)!}$	$x^2 \bigwedge\limits_{k\le 1}^{\infty} \left[1 - \dfrac{(2x)^2}{(2k+1)(2k+2)} \right]$	$-\infty < x < \infty$
$\cos x$	$\sum\limits_{k=0}^{\infty} (-1)^k \dfrac{x^{2k}}{(2k)!}$	$\bigwedge\limits_{k=1}^{\infty} \left[1 - \dfrac{x^2}{(2k-1)2k} \right]$	
$\cos^2 x$	$1 + \dfrac{1}{2} \sum\limits_{k=1}^{\infty} (-1)^k \dfrac{(2x)^{2k}}{(2k)!}$	$1 - x^2 \bigwedge\limits_{k=1}^{\infty} \left[1 - \dfrac{(2x)^2}{(2k+1)(2k+2)} \right]$	
$e^{ax} \sin bx$	$\sum\limits_{k=1}^{\infty} \dfrac{(\lambda x)^k}{k!} \sin k\omega$	$\lambda x \sin \omega \bigwedge\limits_{k=1}^{\infty} \left[1 + \dfrac{\lambda x \sin(k+1)\omega}{(k+1)\sin k\omega} \right]$	$-\infty < x < \infty$
$e^{ax} \cos bx$	$\sum\limits_{k=0}^{\infty} \dfrac{(\lambda x)^k}{k!} \cos k\omega$	$\bigwedge\limits_{k=1}^{\infty} \left[1 + \dfrac{\lambda x \cos k\omega}{k \cos(k-1)\omega} \right]$	
$\tan x$	$\dfrac{1}{x} \sum\limits_{k=1}^{\infty} a_k (2x)^{2k}$	$x \bigwedge\limits_{k=1}^{\infty} \left[1 + \dfrac{a_{k+1}}{a_k}(2x)^2 \right]$	$\|x\| < \dfrac{\pi}{2}$
$\cot x$	$\dfrac{1}{x} - \dfrac{1}{x} \sum\limits_{k=1}^{\infty} b_k (2x)^{2k}$	$\dfrac{1}{x} - \dfrac{x}{3} \bigwedge\limits_{k=1}^{\infty} \left[1 + \dfrac{b_{k+1}}{b_k}(2x)^2 \right]$	$0 < \|x\| < \pi$
$\sec x$	$1 + \sum\limits_{k=1}^{\infty} c_k x^{2k}$	$1 + \dfrac{x^2}{2} \bigwedge\limits_{k=1}^{\infty} \left[1 + \dfrac{c_{k+1}}{c_k}x^2 \right]$	$\|x\| < \dfrac{\pi}{2}$
$\csc x$	$\dfrac{1}{x} + \dfrac{1}{x} \sum\limits_{k=1}^{\infty} d_k x^{2k}$	$\dfrac{1}{x} + \dfrac{x}{6} \bigwedge\limits_{k=1}^{\infty} \left[1 + \dfrac{d_{k+1}}{d_k}x^2 \right]$	$0 < \|x\| < \pi$

(2) Factors a_k, b_k, c_k, d_k for $k = 1, 2, \ldots, 5$*

k	$a_k = \dfrac{4^k - 1}{(2k)!}B_k$	$b_k = \dfrac{1}{(2k)!}B_k$	$c_k = \dfrac{1}{(2k)!}E_k$	$d_k = \dfrac{4^k - 2}{(2k)!}B_k$
1	2.500 000 000 (−01)	8.333 333 333 (−02)	5.000 000 000 (−01)	1.666 666 667 (−01)
2	2.083 333 333 (−02)	1.388 888 889 (−03)	2.083 333 333 (−01)	1.944 444 444 (−02)
3	2.083 333 333 (−03)	3.306 878 307 (−05)	8.472 222 222 (−02)	2.050 264 550 (−03)
4	2.108 134 921 (−04)	8.267 195 767 (−07)	3.435 019 841 (−02)	2.099 867 725 (−04)
5	2.135 692 240 (−05)	2.087 675 699 (−08)	1.392 223 325 (−02)	2.133 604 564 (−05)

[1]For $\Sigma(\ )$ and $\Lambda[\ ]$ refer to Secs. 8.01 and 8.11, respectively.
*For $k = 6, 7, \ldots, 10$ see opposite page.

B_k = auxiliary Bernoulli number (Sec. 8.05)	a, b = constants
E_k = auxiliary Euler number (Sec. 8.06)	$\lambda = a + b \qquad \omega = a - b$

(1) Hyperbolic Functions[1]

$f(x)$	Standard series	Nested series	Interval
$\sinh x$	$x \sum_{k=0}^{\infty} \dfrac{x^{2k}}{(2k+1)!}$	$x \bigwedge_{k=1}^{\infty} \left[1 + \dfrac{x^2}{2k(2k+1)} \right]$	
$\sinh^2 x$	$2x^2 \sum_{k=0}^{\infty} \dfrac{(2x)^{2k}}{(2k+2)!}$	$x^2 \bigwedge_{k=1}^{\infty} \left[1 + \dfrac{(2x)^2}{(2k+1)(2k+2)} \right]$	$-\infty < x < \infty$
$\cosh x$	$\sum_{k=0}^{\infty} \dfrac{x^{2k}}{(2k)!}$	$\bigwedge_{k=1}^{\infty} \left[1 + \dfrac{x^2}{(2k-1)2k} \right]$	
$\cosh^2 x$	$1 + \dfrac{1}{2} \sum_{k=1}^{\infty} \dfrac{(2x)^{2k}}{(2k)!}$	$1 + x^2 \bigwedge_{k=1}^{\infty} \left[1 + \dfrac{(2x)^2}{(2k+1)(2k+2)} \right]$	
$e^{ax} \sinh bx$	$\dfrac{1}{2} \sum_{k=1}^{\infty} \dfrac{(\lambda^k - \omega^k) x^k}{k!}$	$bx \bigwedge_{k=2}^{\infty} \left[1 + \dfrac{\lambda^k - \omega^k}{\lambda^{k-1} + \omega^{k-1}} \dfrac{x}{k} \right]$	$-\infty < x < \infty$
$e^{ax} \cosh bx$	$\dfrac{1}{2} \sum_{k=0}^{\infty} \dfrac{(\lambda^k + \omega^k) x^k}{k!}$	$\bigwedge_{k=1}^{\infty} \left[1 + \dfrac{\lambda^k + \omega^k}{\lambda^{k-1} + \omega^{k-1}} \dfrac{x}{k} \right]$	
$\tanh x$	$\dfrac{1}{x} \sum_{k=1}^{\infty} (-1)^{k+1} a_k (2x)^{2k}$	$x \bigwedge_{k=1}^{\infty} \left[1 - \dfrac{a_{k+1}}{a_k} (2x)^2 \right]$	$\|x\| < \dfrac{\pi}{2}$
$\coth x$	$\dfrac{1}{x} + \dfrac{1}{x} \sum_{k=1}^{\infty} (-1)^{k+1} b_k (2x)^{2k}$	$\dfrac{1}{x} + \dfrac{x}{3} \bigwedge_{k=1}^{\infty} \left[1 - \dfrac{b_{k+1}}{b_k} (2x)^2 \right]$	$0 < \|x\| < \pi$
$\operatorname{sech} x$	$1 + \sum_{k=1}^{\infty} (-1)^k c_k x^{2k}$	$1 - \dfrac{x^2}{2} \bigwedge_{k=1}^{\infty} \left[1 - \dfrac{c_{k+1}}{c_k} x^2 \right]$	$\|x\| < \dfrac{\pi}{2}$
$\operatorname{csch} x$	$\dfrac{1}{x} - \dfrac{1}{x} \sum_{k=1}^{\infty} (-1)^{k+1} d_k x^{2k}$	$\dfrac{1}{x} - \dfrac{x}{6} \bigwedge_{k=1}^{\infty} \left[1 - \dfrac{d_{k+1}}{d_k} x^2 \right]$	$0 < \|x\| < \pi$

(2) Factors a_k, b_k, c_k, d_k for $k = 6, 7, \ldots, 10$*

k	$a_k = \dfrac{4^k - 1}{(2k)!} B_k$	$b_k = \dfrac{1}{(2k)!} B_k$	$c_k = \dfrac{1}{(2k)!} E_k$	$d_k = \dfrac{4^k - 2}{(2k)!} B_k$
6	2.163 875 862 (−06)	5.284 190 139 (−10)	5.642 496 810 (−03)	2.163 347 443 (−06)
7	2.192 460 960 (−07)	1.338 253 653 (−11)	2.286 819 095 (−03)	2.192 327 134 (−07)
8	2.221 426 982 (−08)	3.389 680 296 (−13)	5.268 129 274 (−04)	2.221 393 085 (−08)
9	2.250 776 066 (−09)	8.586 062 056 (−15)	2.756 231 338 (−04)	2.250 767 480 (−09)
10	2.280 512 946 (−10)	2.174 868 699 (−16)	1.522 343 222 (−04)	2.280 510 771 (−10)

∞For $\Sigma(\ )$ and $\Lambda[\ ]$ refer to Secs. 8.01 and 8.11, respectively.
*For $k = 1, 2, \ldots, 5$ see opposite page.

a = positive constant $e = 2.718\,281\,828 \cdots$ (Sec. 1.04)

(1) Exponential Functions[1]

$f(x)$	Standard series	Nested series	Interval
e	$\displaystyle\sum_{k=0}^{\infty}\frac{1}{k!}$	$\displaystyle\bigwedge_{k=1}^{\infty}\left[1+\frac{1}{k}\right]$	
e^{x}	$\displaystyle\sum_{k=0}^{\infty}\frac{x^{k}}{k!}$	$\displaystyle\bigwedge_{k=1}^{\infty}\left[1+\frac{x}{k}\right]$	$-\infty < x < \infty$
e^{-x}	$\displaystyle\sum_{k=0}^{\infty}\frac{(-x)^{k}}{k!}$	$\displaystyle\bigwedge_{k=1}^{\infty}\left[1-\frac{x}{k}\right]$	
a	$\displaystyle\sum_{k=0}^{\infty}\frac{(\ln a)^{k}}{k!}$	$\displaystyle\bigwedge_{k=1}^{\infty}\left[1+\frac{\ln a}{k}\right]$	
a^{x}	$\displaystyle\sum_{k=0}^{\infty}\frac{(x\ln a)^{k}}{k!}$	$\displaystyle\bigwedge_{k=1}^{\infty}\left[1+\frac{x\ln a}{k}\right]$	$-\infty < x < \infty$
a^{-x}	$\displaystyle\sum_{k=0}^{\infty}\frac{(-x\ln a)^{k}}{k!}$	$\displaystyle\bigwedge_{k=1}^{\infty}\left[1-\frac{x\ln a}{k}\right]$	

(2) Logarithmic Functions[1]

$f(x)$	Standard series	Nested series	Interval
$\ln x$	$2\left(\dfrac{x-1}{x+1}\right)\displaystyle\sum_{k=0}^{\infty}\frac{1}{2k+1}\left(\frac{x-1}{x+1}\right)^{2k}$	$2\left(\dfrac{x-1}{x+1}\right)\displaystyle\bigwedge_{k=1}^{\infty}\left[1+\frac{2k-1}{2k+1}\left(\frac{x-1}{x+1}\right)^{2}\right]$	$0 < x < \infty$
	$(x-1)\displaystyle\sum_{k=0}^{\infty}\frac{(1-x)^{k}}{k+1}$	$(x-1)\displaystyle\bigwedge_{k=1}^{\infty}\left[1-\frac{k(x-1)}{k+1}\right]$	$0 < x < 2$
	$\dfrac{x-1}{x}\displaystyle\sum_{k=0}^{\infty}\frac{1}{k+1}\left(\frac{x-1}{x}\right)^{k}$	$\dfrac{x-1}{x}\displaystyle\bigwedge_{k=1}^{\infty}\left[1+\frac{k(x-1)}{(k+1)x}\right]$	$\dfrac{1}{2} \le x < \infty$
$\ln(x+1)$	$x\displaystyle\sum_{k=0}^{\infty}\frac{(-x)^{k}}{k+1}$	$x\displaystyle\bigwedge_{k=1}^{\infty}\left[1-\frac{kx}{k+1}\right]$	$-1 < x \le 1$
$\ln(x-1)$	$-x\displaystyle\sum_{k=0}^{\infty}\frac{x^{k}}{k+1}$	$-x\displaystyle\bigwedge_{k=1}^{\infty}\left[1+\frac{kx}{k+1}\right]$	$-1 \le x < 1$
$\ln\dfrac{x+1}{x-1}$	$\dfrac{2}{x}\displaystyle\sum_{k=0}^{\infty}\frac{1}{2k+1}\left(\frac{1}{x}\right)^{2k}=2\coth^{-1}x$	$\dfrac{2}{x}\displaystyle\bigwedge_{k=1}^{\infty}\left[1+\frac{2k-1}{2k+1}\left(\frac{1}{x}\right)^{2}\right]$	$|x| > 1$
$\ln\dfrac{1+x}{1-x}$	$2x\displaystyle\sum_{k=0}^{\infty}\frac{x^{2k}}{2k+1}=2\tanh^{-1}x$	$2x\displaystyle\bigwedge_{k=1}^{\infty}\left[1+\frac{(2k-1)x^{2}}{2k+1}\right]$	$|x| < 1$

[1]For $\Sigma(\ )$ and $\Lambda[\ ]$ refer to Secs. 8.01 and 8.11, respectively.

$(\ )!! = $ double factorial (Sec. 1.03) $\dbinom{2k}{k} = \dfrac{(2k)!}{(k!)^2} = 2^k\dfrac{(2k-1)!!}{k!}$ (Sec. 1.04)

(1) Inverse Trigonometric Functions[1]

$f(x)$	Standard series	Nested series	Interval
$\sin^{-1}x$	$x\displaystyle\sum_{k=0}^{\infty}\frac{1}{2k+1}\binom{2k}{k}\left(\frac{x}{2}\right)^{2k}$	$x\displaystyle\bigwedge_{k=1}^{\infty}\left[1+\frac{(2k-1)^2x^2}{2k(2k+1)}\right]$	$|x|<1$
$\cos^{-1}x$	$\dfrac{\pi}{2}-\sin^{-1}x$ (for $\sin^{-1}x$ use the series above)		
$\tan^{-1}x$	$\begin{cases} x\displaystyle\sum_{k=0}^{\infty}(-1)^k\dfrac{x^{2k}}{2k+1} \\[2mm] \dfrac{\pi}{2}-\dfrac{1}{x}\displaystyle\sum_{k=0}^{\infty}\dfrac{(-1)^k}{(2k+1)x^{2k}} \end{cases}$	$x\displaystyle\bigwedge_{k=1}^{\infty}\left[1-\dfrac{2k-1}{2k+1}x^2\right]$ $\dfrac{\pi}{2}-\dfrac{1}{x}\displaystyle\bigwedge_{k=1}^{\infty}\left[1-\dfrac{2k-1}{(2k+1)x^2}\right]$	$|x|<1$ $|x|\geq 1$
$\cot^{-1}x$	$\dfrac{\pi}{2}-\tan^{-1}x$ (for $\tan^{-1}x$ use the respective series above)		
$\sec^{-1}x$	$\dfrac{\pi}{2}-\dfrac{1}{x}\displaystyle\sum_{k=0}^{\infty}\dfrac{1}{2k+1}\binom{2k}{k}\left(\dfrac{1}{2x}\right)^{2k}$	$\dfrac{\pi}{2}-\dfrac{1}{x}\displaystyle\bigwedge_{k=1}^{\infty}\left[1+\dfrac{(2k-1)^2}{2k(2k+1)x^2}\right]$	$|x|>1$
$\csc^{-1}x$	$\dfrac{1}{x}\displaystyle\sum_{k=0}^{\infty}\dfrac{1}{2k+1}\binom{2k}{k}\left(\dfrac{1}{2x}\right)^{2k}$	$\dfrac{1}{x}\displaystyle\bigwedge_{k=1}^{\infty}\left[1+\dfrac{(2k-1)^2}{2k(2k+1)x^2}\right]$	

(2) Inverse Hyperbolic Functions[1]

$f(x)$	Standard series	Nested series	Interval
$\sinh^{-1}x$	$\begin{cases} x\displaystyle\sum_{k=0}^{\infty}\dfrac{(-1)^k}{2k+1}\binom{2k}{k}\left(\dfrac{x}{2}\right)^{2k} \\[3mm] \ln 2x-\displaystyle\sum_{k=1}^{\infty}\dfrac{(-1)^k}{2k}\binom{2k}{k}\left(\dfrac{1}{2x}\right)^{2k} \end{cases}$	$x\displaystyle\bigwedge_{k=1}^{\infty}\left[1-\dfrac{(2k-1)^2x^2}{2k(2k+1)}\right]$ $\ln 2x+\dfrac{1}{4x^2}\displaystyle\bigwedge_{k=1}^{\infty}\left[1-\dfrac{k(2k+1)}{2(k+1)^2x^2}\right]$	$|x|<1$ $x\geq 1$
$\cosh^{-1}x$	$\ln 2x-\displaystyle\sum_{k=1}^{\infty}\dfrac{1}{2k}\binom{2k}{k}\left(\dfrac{1}{2x}\right)^{2k}$	$\ln 2x-\dfrac{1}{4x^2}\displaystyle\bigwedge_{k=1}^{\infty}\left[1+\dfrac{k(2k+1)}{2(k+1)^2x^2}\right]$	$x\geq 1$
$\tanh^{-1}x$	$x\displaystyle\sum_{k=0}^{\infty}\dfrac{x^{2k}}{2k+1}$	$x\displaystyle\bigwedge_{k=1}^{\infty}\left[1+\dfrac{2k-1}{2k+1}x^2\right]$	$|x|<1$
$\coth^{-1}x$	$\dfrac{1}{x}\displaystyle\sum_{k=0}^{\infty}\dfrac{1}{(2k+1)x^{2k}}$	$\dfrac{1}{x}\displaystyle\bigwedge_{k=1}^{\infty}\left[1+\dfrac{2k-1}{(2k+1)x^2}\right]$	$|x|>1$
$\operatorname{sech}^{-1}x$	$\ln\dfrac{2}{x}-\displaystyle\sum_{k=1}^{\infty}\dfrac{1}{2k}\binom{2k}{k}\left(\dfrac{x}{2}\right)^{2k}$	$\ln\dfrac{2}{x}-\dfrac{x^2}{4}\displaystyle\bigwedge_{k=1}^{\infty}\left[1+\dfrac{k(2k+1)x^2}{2(k+1)^2}\right]$	$0<x<1$
$\operatorname{csch}^{-1}x$	$\ln\dfrac{2}{x}-\displaystyle\sum_{k=1}^{\infty}\dfrac{(-1)^k}{2k}\binom{2k}{k}\left(\dfrac{x}{2}\right)^{2k}$	$\ln\dfrac{2}{x}+\dfrac{x^2}{4}\displaystyle\bigwedge_{k=1}^{\infty}\left[1-\dfrac{k(2k+1)x^2}{2(k+1)^2}\right]$	$0<x<1$

[1]For $\Sigma(\ )$ and $\Lambda[\ ]$ refer to Secs. 8.01 and 8.11, respectively.

(1) Finite Products

$$1 + x^{2n} = \left(x^2 + 2x \cos \frac{\pi}{2n} + 1\right)\left(x^2 + 2x \cos \frac{3\pi}{2n} + 1\right) \cdots \left(x^2 + 2x \cos \frac{(2n-1)\pi}{2n} + 1\right)$$

$$1 + x^{2n+1} = (1 + x)\left[\left(x^2 - 2x \cos \frac{\pi}{2n+1} + 1\right)\left(x^2 - 2x \cos \frac{3\pi}{2n+1} + 1\right) \cdots \left(x^2 - 2x \cos \frac{(2n-1)\pi}{2n+1} + 1\right)\right]$$

$$\sin 2nx = n \sin 2x \left[(1 - a_1 \sin^2 x)(1 - a_2 \sin^2 x)(1 - a_3 \sin^2 x) \cdots (1 - a_{n-1} \sin^2 x)\right]$$

$$\sin (2n+1)x = n \sin x[(1 - b_1 \sin^2 x)(1 - b_2 \sin^2 x)(1 - b_3 \sin^2 x) \cdots (1 - b_n \sin^2 x)]$$

$$a_r = \frac{1}{\sin^2 (r\pi/2n)} \qquad b_r = \frac{1}{\sin^2 [r\pi/(2n+1)]} \qquad r = 1, 2, 3, \ldots$$

$$\cos 2nx = (1 - c_1 \sin^2 x)(1 - c_3 \sin^2 x)(1 - c_5 \sin^2 x) \cdots (1 - c_{2n-1} \sin^2 x)$$

$$\cos (2n+1)x = \cos x[(1 - d_1 \sin^2 x)(1 - d_3 \sin^2 x)(1 - d_5 \sin^2 x) \cdots (1 - d_{2n-1} \sin^2 x)]$$

$$c_r = \frac{1}{\sin^2 (r\pi/4n)} \qquad d_r = \frac{1}{\sin^2 [r\pi/(4n+2)]} \qquad r = 1, 3, 5, \ldots$$

(2) Infinite Products

$$\sin nx = nx\left[1 - \left(\frac{nx}{\pi}\right)^2\right]\left[1 - \left(\frac{nx}{2\pi}\right)^2\right] \times \left[1 - \left(\frac{nx}{3\pi}\right)^2\right] \cdots$$

$$\sinh nx = nx\left[1 + \left(\frac{nx}{\pi}\right)^2\right]\left[1 + \left(\frac{nx}{2\pi}\right)^2\right] \times \left[1 + \left(\frac{nx}{3\pi}\right)^2\right] \cdots$$

$$\cos nx = \left[1 - \left(\frac{2nx}{\pi}\right)^2\right]\left[1 - \left(\frac{2nx}{3\pi}\right)^2\right] \times \left[1 - \left(\frac{2nx}{5\pi}\right)^2\right] \cdots$$

$$\cosh nx = \left[1 + \left(\frac{2nx}{\pi}\right)^2\right]\left[1 + \left(\frac{2nx}{3\pi}\right)^2\right] \times \left[1 + \left(\frac{2nx}{5\pi}\right)^2\right] \cdots$$

$$\sin (x + y) = \left[\left(1 + \frac{y}{x}\right)\left(1 + \frac{y}{\pi + x}\right)\left(1 - \frac{y}{\pi - x}\right)\left(1 + \frac{y}{2\pi + x}\right)\left(1 - \frac{y}{2\pi - x}\right) \cdots\right] \sin x$$

$$\sin (x - y) = \left[\left(1 - \frac{y}{x}\right)\left(1 + \frac{y}{\pi - x}\right)\left(1 - \frac{y}{\pi + x}\right)\left(1 + \frac{y}{2\pi - x}\right)\left(1 - \frac{y}{2\pi + x}\right) \cdots\right] \sin x$$

$$\cos (x + y) = \left[\left(1 + \frac{2y}{\pi + 2x}\right)\left(1 - \frac{2y}{\pi - 2x}\right)\left(1 + \frac{2y}{3\pi + 2x}\right)\left(1 - \frac{2y}{3\pi - 2x}\right) \cdots\right] \cos x$$

$$\cos (x - y) = \left[\left(1 + \frac{2y}{\pi - 2x}\right)\left(1 - \frac{2y}{\pi + 2x}\right)\left(1 + \frac{2y}{3\pi - 2x}\right)\left(1 - \frac{2y}{3\pi + 2x}\right) \cdots\right] \cos x$$

$$\frac{\pi}{2} = \left(\frac{2}{1}\right)\left(\frac{2}{3}\right)\left(\frac{4}{3}\right)\left(\frac{4}{5}\right)\left(\frac{6}{5}\right)\left(\frac{6}{7}\right) \cdots$$

$$\frac{\sin x}{x} = \cos \frac{x}{2} \cos \frac{x}{4} \cos \frac{x}{8} \cos \frac{x}{16} \cdots$$

9
INTEGRAL CALCULUS

(1) Definitions

$F(x)$ is an *indefinite integral* (antiderivative) of $f(x)$ if

$$\frac{dF(x)}{dx} = f(x)$$

Since the derivative of $F(x) + C$ is also equal to $f(x)$, all integrals of $f(x)$ are included in the expression

$$\int f(x)\, dx = F(x) + C$$

in which $f(x)$ is called the *integrand* and C is an *arbitrary constant*. Because of the indeterminacy of C, there is an infinite number of $F(x) + C$, differing by their relative position to X axis only. The adjacent graph illustrates the meaning of C for a given function.

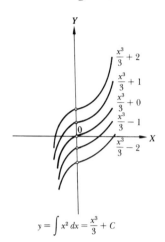

$$y = \int x^2\, dx = \frac{x^3}{3} + C$$

(2) Integration Methods

(a) Antiderivative method

If $f(x)$ is a derivative of a known function, the integral is this function plus the constant of integration.

$$\int \frac{dF(x)}{dx}\, dx = F(x) + C$$

(b) Integration by parts

If the integrand can be expressed as a product of two functions

$$f(x) = u(x)v'(x)$$

then

$$\int u(x)v'(x)\, dx = u(x)v(x) - \int u'(x)v(x)\, dx$$

in which the integral of the right side may be known or can be calculated by one or more repetitions of the same.

(c) Substitution method

The introduction of a new variable

$$x = \phi(t) \qquad\qquad dx = \phi'(t)\, dt$$

yields

$$\int f(x)\, dx = \int f(\phi(t))\phi'(t)\, dt + C$$

The integral of this transformed function may be known, or can be calculated by other methods. The most common substitutions are listed in Secs. 9.07 and 9.08. For particular cases refer to Chap. 19.

(d) Integration by series

If the integrand can be expressed as a uniformly convergent series of powers of x (within its interval of convergence) and if the result of integration of this series term by term is also a uniformly convergent series, the sum of this series is also the value of the integral.

(1) Definitions

If $f(x)$ is continuous in the closed interval $[a, b]$ and this interval is divided into n equal parts by the points $a, x_1, x_2, \ldots, x_{n-1}, b$ such that $\Delta x = (b - a)/n$, then the *definite integral* of $f(x)$ with respect to x, between the limits $x = a$ to $x = b$, is

$$\int_a^b f(x)\, dx = \lim_{n \to \infty} \sum_1^n f(X_i)\, \Delta x = \left[\int f(x)\, dx\right]_a^b = \left[F(x)\right]_a^b = F(b) - F(a)$$

where $F(x)$ is a function, the derivative of which with respect to x is $f(x)$.

The numbers a and b are called, respectively, the *lower* and *upper limits of integration*, and $[a, b]$ is called the *range of integration*.

Geometrically, the definite integral of $f(x)$ with respect to x, between limits $x = a$ to $x = b$, is the *area* bounded by $f(x)$, the X axis, and the verticals through the end points of a and b.

(2) Rules of Limits

$$\int_a^b = -\int_b^a \qquad \int_a^b + \int_b^c = \int_a^c \qquad \int_a^c - \int_b^c = \int_a^b \qquad \int_a^a = 0$$

(3) Fundamental Theorems

$$\int_a^b dx = b - a \qquad\qquad \int_a^b \lambda f(x)\, dx = \lambda \int_a^b f(x)\, dx$$

$$\int_a^b (f(x) + g(x))\, dx = \int_a^b f(x)\, dx + \int_a^b g(x)\, dx$$

$$\int_a^b f(x)\frac{dg(x)}{dx}\, dx = \left[f(x)g(x)\right]_a^b - \int_a^b \frac{df(x)}{dx}g(x)\, dx$$

$$\int_a^x f(t)\, dt = F(x) - F(a) \qquad\qquad \int_a^{\phi(x)} f(t)\, dt = F[\phi(x)] - F(a)$$

$$\frac{d}{dx}\int_a^x f(t)\, dt = f(x) \qquad\qquad \frac{d}{dx}\int_a^{\phi(x)} f(t)\, dt = F[\phi(x)]\frac{d\phi(x)}{dx}$$

$$\frac{\partial}{\partial \alpha}\int_{\phi_1(\alpha)}^{\phi_2(\alpha)} f(x, \alpha)\, dx = \int_{\phi_1(\alpha)}^{\phi_2(\alpha)} \frac{\partial f(x, \alpha)}{\partial \alpha}\, dx + f(\phi_2(\alpha), \alpha)\frac{\partial \phi_2(\alpha)}{\partial \alpha} - f(\phi_1(\alpha), \alpha)\frac{\partial \phi_1(\alpha)}{\partial \alpha}$$

For additional relationships and particular cases refer to Chap. 20.

(4) Mean Values

$$\text{Arithmetic mean value} = \frac{\int_a^b f(x)\, dx}{b - a} \qquad\qquad \text{Quadratic mean value} = \sqrt{\frac{\int_a^b f(x)^2\, dx}{b - a}}$$

(1) Notation

u = differentiable function of x $u', u'', \ldots, u^{(n)}$ = successive derivatives of u

$U_1, U_2, \ldots, U_n$ = successive integrals of u a, b, m = constants

The constant of integration, C, is omitted.

(2) Relationships

$$\int (0)\, dx = \text{constant}$$

$$\int (a)\, dx = ax$$

$$\int f'(x)\, dx = f(x)$$

$$\int f(a) f'(x)\, dx = f(a) f(x)$$

$$\int f(x)\, dx = x f(0) + \frac{x^2}{2!} f'(0) + \frac{x^3}{3!} f''(0) + \cdots$$

$$\int f(x)\, dx = x f(x) - \frac{x^2}{2!} f'(x) + \frac{x^3}{3!} f''(x) - \cdots$$

$$\int f(0)\, dx = x f(0) = f(0) e^{\ln x}$$

$$\int f(x)\, dx = x f(x) - \int x f'(x)\, dx$$

$$\int u u'\, dx = \frac{u^2}{2}$$

$$\int \frac{u'}{u}\, dx = \ln u$$

$$\int u^m u'\, dx = \frac{u^{m+1}}{m+1}$$

$$\int \frac{u'}{u^m}\, dx = -\frac{1}{(m-1) u^{m-1}} \qquad m \neq 1$$

$$\int (a + bu)^m u'\, dx = \frac{(a + bu)^{m+1}}{b(m+1)} \qquad m \neq -1, b \neq 0$$

$$\int \frac{u'\, dx}{a + bu} = \frac{\ln (a + bu)}{b} \qquad b \neq 0$$

$$\int \frac{u'\, dx}{(a + bu)^m} = -\frac{1}{b(m-1)(a + bu)^{m-1}} \qquad m \neq 1, b \neq 0$$

$$\int \frac{u'\, dx}{\sqrt{a + bu}} = \frac{2\sqrt{a + bu}}{b} \qquad b \neq 0$$

$$\int \frac{u'\, dx}{\sqrt{(a + u)(b + u)}} = 2 \ln (\sqrt{a + u} + \sqrt{b + u})$$

$$\int u x\, dx = x U_1 - U_2$$

$$\int u x^2\, dx = x^2 U_1 - 2x U_2 + 2 U_3$$

$$\int u x^m\, dx = x^m U_1 - m x^{m-1} U_2 + m(m-1) x^{m-2} U_3 - m(m-1)(m-2) x^{m-3} U_4 + \cdots$$

(1) Notation

> $u, v =$ differentiable functions of x
>
> $\left. \begin{array}{l} u', u'', \ldots, u^{(n)} \\ v', v'', \ldots, v^{(n)} \end{array} \right\} =$ successive derivatives of u and v, respectively
>
> $\left. \begin{array}{l} U_1, U_2, \ldots, U_n \\ V_1, V_2, \ldots, V_n \end{array} \right\} =$ successive integrals of u and v, respectively
>
> $a, b, m =$ constants
>
> The constant of integration, C, is omitted.

(2) Relationships

$$\int (u + v)\, dx = \int u\, dx + \int v\, dx$$

$$\int (u + v)^m\, dx = x(u + v) - \frac{x^2}{2!}(u' + v') + \frac{x^3}{3!}(u'' + v'') - \cdots$$

$$= \int u(u + v)^{m-1}\, dx + \int v(u + v)^{m-1}\, dx$$

$$\int uv\, dx = U_1 v - \underbrace{U_2 v' + U_3 v'' - U_4 v''' + \cdots}$$

$$= U_1 v - \int U_1 v'\, dx$$

$$= uV_1 - \underbrace{u'V_2 + u''V_3 - u'''V_4 + \cdots}$$

$$= uV_1 - \int u'V_1\, dx$$

$$\int uv'\, dx = uv - \int u'v\, dx$$

$$\int \frac{uv'}{v^2}\, dx = -\frac{u}{v} + \int \frac{u'}{v}\, dx$$

$$\int u'v\, dx = uv - \int uv'\, dx$$

$$\int \frac{u'v - uv'}{v^2}\, dx = \frac{u}{v}$$

$$\int (u'v + uv')\, dx = uv$$

$$\int \frac{u'v - uv'}{uv}\, dx = \ln\frac{u}{v}$$

$$\int \frac{u'v - uv'}{(u + v)^2}\, dx = -\frac{v}{u + v}$$

$$\int \frac{u'v - uv'}{u^2 + v^2}\, dx = \tan^{-1}\frac{u}{v}$$

$$\int \frac{u'v - uv'}{(u - v)^2}\, dx = -\frac{v}{u - v}$$

$$\int \frac{u'v - uv'}{u^2 - v^2}\, dx = \tfrac{1}{2}\ln\frac{u + v}{u - v}$$

$$\int u^{(n+1)}v\, dx = u^{(n)}v - u^{(n-1)}v' + u^{(n-2)}v'' - \cdots (-1)^{n+1}\int uv^{(n+1)}\, dx$$

In the following integral formulas, $u = f(x)$, a, m = constants. The constant of integration, C, is omitted.

$$\int u^m \, du = \frac{u^{m+1}}{m+1} \qquad m \neq -1 \qquad\qquad \int \frac{du}{u} = \ln |u| \qquad\qquad u \neq 0$$

$$\int \frac{1}{u^m} \, du = \frac{u^{1-m}}{1-m} \qquad m \neq 1 \qquad\qquad \int \sqrt[n]{u^m} \, du = \frac{n u \sqrt[n]{u^m}}{m+n} \qquad m \neq -n$$

$$\int \frac{du}{a^2 + u^2} = \frac{1}{a} \tan^{-1} \frac{u}{a} = -\frac{1}{a} \cot^{-1} \frac{u}{a} \qquad a \neq 0$$

$$\int \frac{du}{a^2 - u^2} = \frac{1}{a} \tanh^{-1} \frac{u}{a} = \frac{1}{2a} \ln \frac{a+u}{a-u} \qquad u^2 < a^2$$

$$\int \frac{du}{u^2 - a^2} = \frac{1}{a} \coth^{-1} \frac{u}{a} = \frac{1}{2a} \ln \frac{u+a}{u-a} \qquad u^2 > a^2$$

$$\int \frac{du}{\sqrt{a^2 + u^2}} = \sinh^{-1} \frac{u}{a} = \ln (u + \sqrt{u^2 + a^2})$$

$$\int \frac{du}{\sqrt{a^2 - u^2}} = \sin^{-1} \frac{u}{a} = -\cos^{-1} \frac{u}{a}$$

$$\int \frac{du}{\sqrt{u^2 - a^2}} = \cosh^{-1} \frac{u}{a} = \ln (u + \sqrt{u^2 - a^2})$$

$$\int \frac{du}{u\sqrt{a^2 + u^2}} = -\frac{1}{a} \operatorname{csch}^{-1} \frac{u}{a} = -\frac{1}{a} \sinh^{-1} \frac{a}{u} = -\frac{1}{a} \ln \frac{a + \sqrt{a^2 + u^2}}{u}$$

$$\int \frac{du}{u\sqrt{a^2 - u^2}} = -\frac{1}{a} \operatorname{sech}^{-1} \frac{u}{a} = -\frac{1}{a} \cosh^{-1} \frac{a}{u} = -\frac{1}{a} \ln \frac{a + \sqrt{a^2 - u^2}}{u}$$

$$\int \frac{du}{u\sqrt{u^2 - a^2}} = \frac{1}{a} \sec^{-1} \frac{u}{a} = \cos^{-1} \frac{a}{u}$$

$$\int \sqrt{a^2 + u^2} \, du = \frac{u}{2} \sqrt{a^2 + u^2} + \frac{a^2}{2} \sinh^{-1} \frac{u}{a} = \frac{u}{2} \sqrt{a^2 + u^2} + \frac{a^2}{2} \ln (u + \sqrt{a^2 + u^2})$$

$$\int \sqrt{a^2 - u^2} \, du = \frac{u}{2} \sqrt{a^2 - u^2} + \frac{a^2}{2} \sin^{-1} \frac{u}{a} = \frac{u}{2} \sqrt{a^2 - u^2} - \frac{a^2}{2} \cos^{-1} \frac{u}{a}$$

$$\int \sqrt{u^2 - a^2} \, du = \frac{u}{2} \sqrt{u^2 - a^2} - \frac{a^2}{2} \cosh^{-1} \frac{u}{a} = \frac{u}{2} \sqrt{u^2 - a^2} - \frac{a^2}{2} \ln (u + \sqrt{u^2 - a^2})$$

$$\int \frac{u \, du}{u^4 + a^4} = \frac{1}{2a^2} \tan^{-1} \frac{u^2}{a^2}$$

$$\int \frac{u \, du}{u^4 - a^4} = \frac{1}{4a^2} \ln \frac{u^2 - a^2}{u^2 + a^2}$$

[1]For additional cases refer to Secs. 19.02 through 19.40.

In the following integral formulas, $u = f(x)$ and a, b, m = constants. The constant of integration, C, is omitted.

$$\int a^u \, du = \frac{a^u}{\ln a} \qquad\qquad \int \frac{du}{a^u} = -\frac{1}{a^u \ln a}$$

$$\int e^u \, du = e^u \qquad\qquad \int \ln u \, du = u \ln u - u$$

$$\int \sin au \, du = \frac{-1}{a} \cos au \qquad\qquad \int \sinh au = \frac{1}{a} \cosh au$$

$$\int \cos au \, du = \frac{1}{a} \sin au \qquad\qquad \int \cosh au = \frac{1}{a} \sinh au$$

$$\int \tan au \, du = \frac{-1}{a} \ln \cos au \qquad\qquad \int \tanh au = \frac{1}{a} \ln \cosh au$$

$$\int \cot au \, du = \frac{1}{a} \ln \sin au \qquad\qquad \int \coth au = \frac{1}{a} \ln \sinh au$$

$$\int \sec au \, du = \frac{1}{2a} \ln \frac{1 + \sin au}{1 - \sin au} \qquad\qquad \int \operatorname{sech} au = \frac{1}{a} \tan^{-1} \sinh au$$

$$\int \csc au \, du = \frac{-1}{2a} \ln \frac{1 + \cos au}{1 - \cos au} \qquad\qquad \int \operatorname{csch} au = \frac{1}{a} \ln \tanh \frac{au}{2}$$

$$\int \frac{du}{\sin au} = \frac{-1}{2a} \ln \frac{1 + \cos au}{1 - \cos au} \qquad\qquad \int \frac{du}{\sinh au} = \frac{1}{a} \ln \tanh \frac{au}{2}$$

$$\int \frac{du}{\cos au} = \frac{1}{2a} \ln \frac{1 + \sin au}{1 - \sin au} \qquad\qquad \int \frac{du}{\cosh au} = \frac{1}{a} \tan^{-1} \sinh au$$

$$\int \frac{du}{\tan au} = \frac{1}{a} \ln \sin au \qquad\qquad \int \frac{du}{\tanh au} = \frac{1}{a} \ln \sinh au$$

$$\int \frac{du}{\cot au} = \frac{-1}{a} \ln \cos au \qquad\qquad \int \frac{du}{\coth au} = \frac{1}{a} \ln \cosh au$$

$$\int \frac{du}{\sec au} = \frac{1}{a} \sin au \qquad\qquad \int \frac{du}{\operatorname{sech} au} = \frac{1}{a} \sinh au$$

$$\int \frac{du}{\csc au} = \frac{-1}{a} \cos au \qquad\qquad \int \frac{du}{\operatorname{csch} au} = \frac{1}{a} \cosh au$$

$$\int e^{au} \sin bu \, du = -\frac{e^{au}}{r} \cos (bu + \omega) \qquad r = \sqrt{a^2 + b^2}$$

$$\int e^{au} \cos bu \, du = +\frac{e^{au}}{r} \sin (bu + \omega) \qquad \omega = \tan^{-1} \frac{b}{a}$$

[1]For additional cases refer to Secs. 19.41 through 19.86.

Integral	Substitution
$\displaystyle\int \frac{dx}{(x-a)^m}$	$\dfrac{1}{x-a} = t \qquad dx = -\dfrac{dt}{t^2}$
$\displaystyle\int \frac{f'(x)}{f(x)}\, dx$	$f(x) = t \qquad f'(x)\, dx = dt$
$\displaystyle\int x^m (a + bx^n)^p\, dx$	If $p = \alpha/\beta$ is a fraction and $(m+1)/n$ is an integer, use $$t = \sqrt[\beta]{a + bx^n}$$ If $p = \alpha/\beta$ is a fraction and $(m+1)/n + p$ is an integer, use $$t = \sqrt[\beta]{\frac{a + bx^n}{x^n}}$$ If p is an integer, use the binomial expansion.
$\displaystyle\int f(x^2)\, dx$	$x^2 = t \qquad 2x\, dx = dt$
$\displaystyle\int f\left[x;\ \left(\frac{ax+b}{cx+d}\right)^m\right] dx$	$\left(\dfrac{ax+b}{cx+d}\right)^m = t^m$
$\displaystyle\int f\left[x;\ \left(\frac{ax+b}{cx+d}\right)^m;\ \left(\frac{ax+b}{cx+d}\right)^n;\ \ldots\right] dx$	$\left(\dfrac{ax+b}{cx+d}\right)^r = t^r$ r is the least common multiple of m and n.
$\displaystyle\int f(x;\ \sqrt{ax^2 + bx + c})\, dx$	If $a > 0$, use $$x = \frac{t^2 - c}{b - 2t\sqrt{a}}$$ $$dx = 2\frac{-t^2\sqrt{a} + bt - c\sqrt{a}}{(b - 2t\sqrt{a})^2}\, dt$$ If $c > 0$, use $$x = \frac{2t\sqrt{c} - b}{a - t^2}$$ $$dx = \frac{2a\sqrt{c} - 2bt + 2t^2\sqrt{c}}{(a - t^2)^2}\, dt$$ If $ax^2 + bx + c = a(x - \alpha)(x - \beta)$, use $$x = \frac{t^2\alpha - a\beta}{t^2 - a} \qquad dx = \frac{2at(\beta - \alpha)}{(t^2 - a)^2}\, dt$$
$\displaystyle\int f(x;\ \sqrt[m]{ax + b})\, dx$	$\sqrt[m]{ax + b} = t \qquad dx = \dfrac{mt^{m-1}\, dt}{a}$

Integral and substitution	Transformations
$\int f(x;\ \sqrt{a^2-x^2})\,dx$	$\sin t = \dfrac{x}{a}\qquad\qquad \cos t = \dfrac{\sqrt{a^2-x^2}}{a}$
$x = a\sin t \qquad dx = a\cos t\,dt$	$\tan t = \dfrac{x}{\sqrt{a^2-x^2}}\qquad \cot t = \dfrac{\sqrt{a^2-x^2}}{x}$
$\int f(x;\ \sqrt{a^2-x^2})\,dx$	$\sinh t = \dfrac{x}{\sqrt{a^2-x^2}}\qquad \cosh t = \dfrac{a}{\sqrt{a^2-x^2}}$
$x = a\tanh t \qquad dx = \dfrac{a\,dt}{\cosh^2 t}$	$\tanh t = \dfrac{x}{a}\qquad\qquad \coth t = \dfrac{a}{x}$
$\int f(x;\ \sqrt{a^2+x^2})\,dx$	$\sin t = \dfrac{x}{\sqrt{a^2+x^2}}\qquad \cos t = \dfrac{a}{\sqrt{a^2+x^2}}$
$x = a\tan t \qquad dx = \dfrac{a\,dt}{\cos^2 t}$	$\tan t = \dfrac{x}{a}\qquad\qquad \cot t = \dfrac{a}{x}$
$\int f(x;\ \sqrt{a^2+x^2})\,dx$	$\sinh t = \dfrac{x}{a}\qquad\qquad \cosh t = \dfrac{\sqrt{a^2+x^2}}{a}$
$x = a\sinh t \qquad dx = a\cosh t\,dt$	$\tanh t = \dfrac{x}{\sqrt{a^2+x^2}}\qquad \coth t = \dfrac{\sqrt{a^2+x^2}}{x}$
$\int f(x;\ \sqrt{x^2-a^2})\,dx$	$\sin t = \dfrac{\sqrt{x^2-a^2}}{x}\qquad \cos t = \dfrac{a}{x}$
$x = \dfrac{a}{\cos t} \qquad dx = \dfrac{a\sin t\,dt}{\cos^2 t}$	$\tan t = \dfrac{\sqrt{x^2-a^2}}{a}\qquad \cot t = \dfrac{a}{\sqrt{x^2-a^2}}$
$\int f(x;\ \sqrt{x^2-a^2})\,dx$	$\sinh t = \dfrac{\sqrt{x^2-a^2}}{a}\qquad \cosh t = \dfrac{x}{a}$
$x = a\cosh t \qquad dx = a\sinh t\,dt$	$\tanh t = \dfrac{\sqrt{x^2-a^2}}{x}\qquad \coth t = \dfrac{x}{\sqrt{x^2-a^2}}$
$\int f(\sin x;\ \cos x;\ \tan x;\ \cot x)\,dx$	$\sin x = \dfrac{2t}{1+t^2}\qquad \cos x = \dfrac{1-t^2}{1+t^2}$
$\tan\dfrac{x}{2} = t \qquad dx = \dfrac{2dt}{1+t^2}$	$\tan x = \dfrac{2t}{1-t^2}\qquad \cot x = \dfrac{1-t^2}{2t}$
$\int f(\sinh x;\ \cosh x;\ \tanh x;\ \coth x)\,dx$	$\sinh x = \dfrac{2t}{1-t^2}\qquad \cosh x = \dfrac{1+t^2}{1-t^2}$
$\tanh\dfrac{x}{2} = t \qquad dx = \dfrac{2dt}{1-t^2}$	$\tanh x = \dfrac{2t}{1+t^2}\qquad \coth x = \dfrac{1+t^2}{2t}$
$\int f(e^x)\,dx$	$\int f(a^x)\,dx$
$e^x = t \qquad dx = \dfrac{dt}{t}$	$a^x = t \qquad dx = \dfrac{dt}{t\ln t}$

The calculation of a double integral is performed by successive evaluation of two definite integrals.

(1) Cartesian Coordinates

If P_1, P_2 and Q_1, Q_2 are points on a contour enclosing area A, selected so that they identify extreme coordinates x and y, respectively, then

$$A = \iint_A f(x, y) \, dx \, dy = \int_a^b \left[\int_{f_1(x)}^{f_2(x)} f(x, y) \, dy \right] dx$$

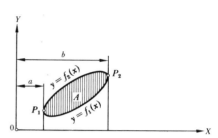

where the boundary of A consists of two continuous curves $y = f_1(x)$ and $y = f_2(x)$. The boundary of A is met by a line parallel to the Y axis in at most two points, and $x = a$, $x = b$ are the extreme values of x on A; or

$$A = \iint_A f(x, y) \, dx \, dy = \int_c^d \left[\int_{g_1(y)}^{g_2(y)} f(x, y) \, dx \right] dy$$

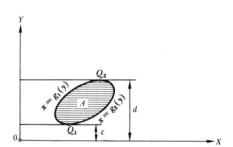

where the boundary of A consists of two continuous curves $x = g_1(y)$ and $x = g_2(y)$. The boundary of A is met by a line parallel to the X axis in at most two points, and $y = c$, $y = d$ are the extreme values of y on A.

If neither of these conditions is satisfied, the area A must be divided in two or more portions. Then

$$A = \iint_A f(x, y) \, dx \, dy$$

$$= \iint_{A_1} f(x, y) \, dx \, dy + \iint_{A_2} f(x, y) \, dx \, dy$$

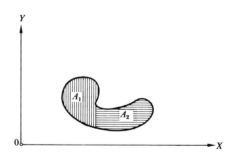

(2) Polar Coordinates

If S_1, S_2 are points on a contour enclosing area A, selected so that they identify extreme polar coordinates, then

$$A = \iint_A f(r, \theta) \, r \, dr \, d\theta = \int_\alpha^\beta \left[\int_{f_1(\theta)}^{f_2(\theta)} f(r, \theta) \, r \, dr \right] d\theta$$

where the boundary of A consists of two continuous curves $r = f_1(\theta)$ and $r = f_2(\theta)$. The boundary of A is met by two tangents from 0 in at most two points, and $\theta = \alpha$, $\theta = \beta$, are the extreme values of θ on A.

(3) Interpretation

If $f(x, y)$ has the same sign over A, the double integral may be interpreted as the *volume of a vertical cylinder* bounded below by the region A projected in the XY plane and above by the surface $z = f(x, y)$.

The calculation of a triple integral is performed by successive evaluation of three definite integrals.

(1) Cartesian Coordinates

$$V = \int\int\int_V f(x, y, z)\, dV = \int_a^b \int_{y_1(x)}^{y_2(x)} \int_{z_1(x, y)}^{z_2(x, y)} f(x, y, z)\, dz\, dy\, dx$$

(2) Cylindrical Coordinates

$$V = \int\int\int_V f(r, \theta, z)\, dV = \int_\alpha^\beta \int_{r_1(\theta)}^{r_2(\theta)} \int_{z_1(r, \theta)}^{z_2(r, \theta)} f(r, \theta, z)\, r\, dz\, d\theta\, dr$$

(3) Spherical Coordinates

$$V = \int\int\int_V f(r, \phi, \theta)\, dV = \int_\alpha^\beta \int_{\phi_1(\theta)}^{\phi_2(\theta)} \int_{r_1(\phi, \theta)}^{r_2(\phi, \theta)} f(r, \phi, \theta)\, r^2 \sin\phi\, dr\, d\phi\, d\theta$$

(4) Interpretation

If $f(x, y, z) = 1$, the triple integral may be interpreted as the *volume enclosed by the region* V.

(5) Curvilinear Coordinates

If $x = x(u, v, w)$, $y = y(u, v, w)$, $z = z(u, v, w)$, and $f(x, y, z) = g(u, v, w)$, then

$$V = \int\int\int_V f(x, y, z)\, dV = \int_{u_1}^{u_2} \int_{v_1(u)}^{v_2(u)} \int_{w_1(u, v)}^{w_2(u, v)} g(u, v, w)\underbrace{\frac{\partial(x, y, z)}{\partial(u, v, w)}}_{J}\, du\, dv\, dw$$

in which

$$J = \frac{\partial(x, y, z)}{\partial(u, v, w)} = \begin{vmatrix} \dfrac{\partial x}{\partial u} & \dfrac{\partial y}{\partial u} & \dfrac{\partial z}{\partial u} \\[2mm] \dfrac{\partial x}{\partial v} & \dfrac{\partial y}{\partial v} & \dfrac{\partial z}{\partial v} \\[2mm] \dfrac{\partial x}{\partial w} & \dfrac{\partial y}{\partial w} & \dfrac{\partial z}{\partial w} \end{vmatrix}$$

In the case of a double integral,

$$J = \frac{\partial(x, y)}{\partial(u, v)} = \begin{vmatrix} \dfrac{\partial x}{\partial u} & \dfrac{\partial y}{\partial u} \\[2mm] \dfrac{\partial x}{\partial v} & \dfrac{\partial y}{\partial v} \end{vmatrix}$$

Note: The order of integration is arbitrary; thus a *double integral* can be evaluated in *two ways*, and a *triple integral* can be evaluated in *six ways*.

(a) Area by single integration

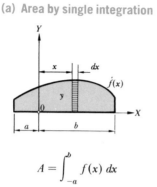

$$A = \int_{-a}^{b} f(x)\, dx$$

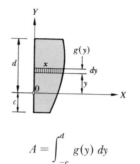

$$A = \int_{-c}^{d} g(y)\, dy$$

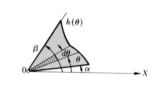

$$A = \int_{\alpha}^{\beta} \frac{h^2(\theta)}{2}\, d\theta$$

(b) Area by double integration

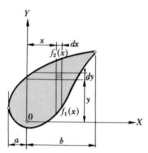

$$A = \int_{-a}^{b} \int_{f_1(x)}^{f_2(x)} dx\, dy$$

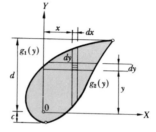

$$A = \int_{-c}^{d} \int_{g_1(y)}^{g_2(y)} dx\, dy$$

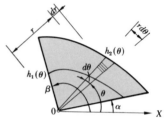

$$A = \int_{\alpha}^{\beta} \int_{h_1(\theta)}^{h_2(\theta)} r\, dr\, d\theta$$

(c) Length of plane curve

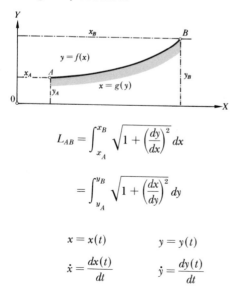

$$L_{AB} = \int_{x_A}^{x_B} \sqrt{1 + \left(\frac{dy}{dx}\right)^2}\, dx$$

$$= \int_{y_A}^{y_B} \sqrt{1 + \left(\frac{dx}{dy}\right)^2}\, dy$$

$$x = x(t) \qquad y = y(t)$$

$$\dot{x} = \frac{dx(t)}{dt} \qquad \dot{y} = \frac{dy(t)}{dt}$$

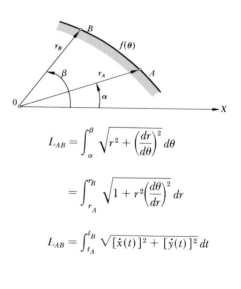

$$L_{AB} = \int_{\alpha}^{\beta} \sqrt{r^2 + \left(\frac{dr}{d\theta}\right)^2}\, d\theta$$

$$= \int_{r_A}^{r_B} \sqrt{1 + r^2\left(\frac{d\theta}{dr}\right)^2}\, dr$$

$$L_{AB} = \int_{t_A}^{t_B} \sqrt{[\dot{x}(t)]^2 + [\dot{y}(t)]^2}\, dt$$

(a) Length of space curve

$$x = x(t)$$
$$y = y(t) \qquad L_{AB} = \int_{t_A}^{t_B} \sqrt{\left(\frac{dx}{dt}\right)^2 + \left(\frac{dy}{dt}\right)^2 + \left(\frac{dz}{dt}\right)^2}\, dt$$
$$z = z(t)$$

(b) Area of surface of revolution generated by rotation of plane curve

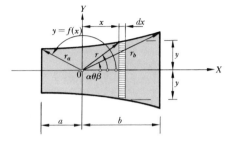

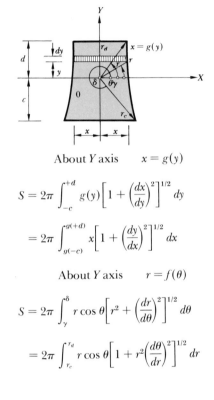

About X axis $y = f(x)$

$$S = 2\pi \int_{-a}^{+b} f(x)\left[1 + \left(\frac{dy}{dx}\right)^2\right]^{1/2} dx$$

$$= 2\pi \int_{f(-a)}^{f(+b)} y\left[1 + \left(\frac{dx}{dy}\right)^2\right]^{1/2} dy$$

About Y axis $x = g(y)$

$$S = 2\pi \int_{-c}^{+d} g(y)\left[1 + \left(\frac{dx}{dy}\right)^2\right]^{1/2} dy$$

$$= 2\pi \int_{g(-c)}^{g(+d)} x\left[1 + \left(\frac{dy}{dx}\right)^2\right]^{1/2} dx$$

About X axis $r = f(\theta)$

$$S = 2\pi \int_{\alpha}^{\beta} r\sin\theta\left[r^2 + \left(\frac{dr}{d\theta}\right)^2\right]^{1/2} d\theta$$

$$= 2\pi \int_{r_a}^{r_b} r\sin\theta\left[1 + r^2\left(\frac{d\theta}{dr}\right)^2\right]^{1/2} dr$$

About Y axis $r = f(\theta)$

$$S = 2\pi \int_{\gamma}^{\delta} r\cos\theta\left[r^2 + \left(\frac{dr}{d\theta}\right)^2\right]^{1/2} d\theta$$

$$= 2\pi \int_{r_c}^{r_d} r\cos\theta\left[1 + r^2\left(\frac{d\theta}{dr}\right)^2\right]^{1/2} dr$$

(c) Area of surface $z = f(x, y)$

$$S = \int\int_A \left[1 + \left(\frac{\partial z}{\partial x}\right)^2 + \left(\frac{\partial z}{\partial y}\right)^2\right]^{1/2} dx\, dy \qquad \text{integration over the projection } A \text{ on } XY \text{ plane}$$

(d) Volume of body of revolution generated by rotation of plane curve

About X axis $y = f(x)$

$$V = \pi \int_{-a}^{+b} [f(x)]^2\, dx$$

About Y axis $x = g(y)$

$$V = \pi \int_{-c}^{+d} [g(y)]^2\, dy$$

(e) Volume of body with known parallel cross section

$$V = \int_{-a}^{+b} A_x\, dx$$

$$V = \int_{-c}^{+d} A_y\, dy$$

(f) Volume of body of general shape

$$V = \int\int\int_V dx\, dy\, dz = \int\int\int_V r\, dr\, d\theta\, dz = \int\int\int_V r^2 \sin\phi\, d\theta\, d\phi\, dr$$

(1) System XY ($0 =$ **origin**, $C =$ **centroid**)

Area	$dA = dx\,dy$	$A = \displaystyle\iint_A dA$
Static moments	$M_x = \displaystyle\iint_A y\,dA$	$M_y = \displaystyle\iint_A x\,dA$
Coordinates of centroid	$x_C = \dfrac{M_y}{A}$	$y_C = \dfrac{M_x}{A}$

Moments of inertia

$$I_{xx} = \iint_A y^2\,dA$$

$$I_{yy} = \iint_A x^2\,dA$$

Products of inertia

$$I_{xy} = \iint_A xy\,dA$$

$$I_{yx} = \iint_A yx\,dA$$

Polar moment of inertia

$$J_0 = \iint_A r^2\,dA$$

$$= I_{xx} + I_{yy}$$

Radii of gyration

$$k_x = \sqrt{\frac{I_{xx}}{A}}$$

$$k_y = \sqrt{\frac{I_{yy}}{A}}$$

$$k_0 = \sqrt{\frac{J_0}{A}}$$

(2) System $X'Y'$ ($0' =$ **origin**, $C =$ **centroid**)

Static moments	$M'_x = M_x - bA$	$M'_y = M_y - aA$
Coordinates of centroid	$x'_C = x_C - a$	$y'_C = y_C - b$

Moments of inertia

$$I'_{xx} = I_{xx} - 2bM_x + b^2A$$

$$I'_{yy} = I_{yy} - 2aM_y + a^2A$$

Products of inertia

$$I'_{xy} = I'_{yx}$$

$$= I_{xy} - bM_x - aM_y + abA$$

Polar moment of inertia

$$J'_0 = J_0 + r^2A - 2(aM_x + bM_y)$$

Radii of gyration

$$k'_x = \sqrt{k_x^2 - 2by_C + b^2}$$

$$k'_y = \sqrt{k_y^2 - 2ax_C + a^2}$$

$$k'_0 = \sqrt{k_0^2 - 2(ax_C + by_C) + r^2}$$

(3) System U, V ($0 =$ **origin**, $C =$ **centroid**)

Static moments

$$M_u = M_x \cos\omega + M_y \sin\omega$$

$$M_v = -M_x \sin\omega + M_y \cos\omega$$

Moments and products of inertia

$$I_{uu} = I_{xx}\cos^2\omega + I_{yy}\sin^2\omega - I_{xy}\sin 2\omega$$

$$I_{vv} = I_{xx}\sin^2\omega + I_{yy}\cos^2\omega + I_{xy}\sin 2\omega$$

$$I_{uv} = I_{vu} = \frac{I_{xx} - I_{yy}}{2}\sin 2\omega + I_{xy}\cos 2\omega$$

Principal moments of inertia

Position of principal axes

$$\tan 2\omega = -\frac{2I_{xy}}{I_{xx} - I_{yy}}$$

Principal values

$$I_{1,2} = \frac{I_{xx} + I_{yy}}{2} \pm \frac{1}{2}\sqrt{(I_{xx} - I_{yy})^2 + 4I_{xy}^2}$$

| A = area | | $I_{AA}, I_{BB}, I_{CC}, I_{xx}, I_{yy}$ = moments of inertia |
| x_C, y_C = coordinates of centroid | | I_{AB}, I_{xy} = products of inertia |

Square			
$A = a^2$	$I_{AA} = \dfrac{a^4}{12}$	$I_{xx} = \dfrac{a^4}{3}$	
$x_C = \dfrac{a}{2}$	$I_{BB} = \dfrac{a^4}{12}$	$I_{yy} = \dfrac{a^4}{3}$	
$y_C = \dfrac{a}{2}$	$I_{AB} = 0$	$I_{xy} = \dfrac{a^4}{4}$	
Rectangle			
$A = ab$	$I_{AA} = \dfrac{ab^3}{12}$	$I_{xx} = \dfrac{ab^3}{3}$	
$x_C = \dfrac{a}{2}$	$I_{BB} = \dfrac{a^3b}{12}$	$I_{yy} = \dfrac{a^3b}{3}$	
$y_C = \dfrac{b}{2}$	$I_{AB} = 0$	$I_{xy} = \dfrac{a^2b^2}{4}$	
Triangle			
$A = \dfrac{bh}{2}$	$I_{TT} = \dfrac{bh^3}{4}$	$I_{xx} = \dfrac{bh^3}{12}$	
$x_C = \dfrac{v-u}{3}$	$I_{AA} = \dfrac{bh^3}{36}$	$I_{yy} = \dfrac{(u^3+v^3)h}{12}$	
$y_C = \dfrac{h}{3}$	$I_{AB} = \dfrac{(u^2-v^2)h^2}{72}$	$I_{xy} = \dfrac{(v^2-u^2)h^2}{24}$	

Triangle	x, y = coordinates in X, Y axes
$A = \dfrac{1}{2}\begin{vmatrix} x_1 & y_1 & 1 \\ x_2 & y_2 & 1 \\ x_3 & y_3 & 1 \end{vmatrix}$	$\bar{x}, \bar{y}$ = coordinates in A, B axes
	$I_{AA} = \dfrac{A}{12}(\bar{y}_1^2 + \bar{y}_2^2 + \bar{y}_3^2)$
$x_C = \dfrac{x_1 + x_2 + x_3}{3}$	$I_{BB} = \dfrac{A}{12}(\bar{x}_1^2 + \bar{x}_2^2 + \bar{x}_3^2)$
$y_C = \dfrac{y_1 + y_2 + y_3}{3}$	$I_{AB} = \dfrac{A}{12}(\bar{x}_1\bar{y}_1 + \bar{x}_2\bar{y}_2 + \bar{x}_3\bar{y}_3)$

Trapezoid		
$A = (a+b)h$	$I_{AA} = \dfrac{h^3}{18}\dfrac{a^2+4ab+b^2}{a+b}$	$I_{xx} = \dfrac{h^3}{6}(a+3b)$
$e = \dfrac{h}{3}\dfrac{a+2b}{a+b}$	$I_{BB} = \dfrac{h}{6}\dfrac{a^4-b^4}{a-b}$	$I_{TT} = \dfrac{h^3}{6}(3a+b)$
$f = \dfrac{h}{3}\dfrac{2a+b}{a+b}$	$I_{AB} = 0$	

Regular polygon	
$A = na^2 \cot \alpha$	$I_{AA} = \dfrac{A}{12}[3r^2 + a^2]$
$r = a \cot \alpha$	$= \dfrac{A}{12}[3R^2 - 2a^2]$
$R = \dfrac{a}{\sin \alpha}$	$= I_{BB} = I_{CC}$

A = area $I_{AA}, I_{BB}, I_{CC}, I_{xx}, I_{yy}$ = moments of inertia

x_C, y_C = coordinates of centroid I_{AB}, I_{xy} = products of inertia

Circle

$A = \pi a^2$

$x_C = a$

$y_C = a$

$I_{AA} = \dfrac{\pi a^4}{4}$

$I_{BB} = \dfrac{\pi a^4}{4}$

$I_{AB} = 0$

$I_{xx} = \dfrac{5\pi a^4}{4}$

$I_{yy} = \dfrac{5\pi a^4}{4}$

$I_{xy} = \pi a^4$

Hollow circle

$A = \pi(b^2 - a^2)$

$x_C = b$

$y_C = b$

$I_{AA} = \dfrac{\pi(b^4 - a^4)}{4}$

$I_{BB} = \dfrac{\pi(b^4 - a^4)}{4}$

$I_{AB} = 0$

$I_{xx} = I_{AA} + b^2 A$

$I_{yy} = I_{xx}$

$I_{xy} = \pi b^2(b^2 - a^2)$

Half hollow circle

$A = \dfrac{\pi(b^2 - a^2)}{2}$

$x_C = b$

$y_C = \dfrac{4(b^3 - a^3)}{3\pi(b^2 - a^2)}$

$I_{AA} = I_{xx} - y_C^2 A$

$I_{BB} = \dfrac{\pi(b^4 - a^4)}{8}$

$I_{AB} = 0$

$I_{xx} = \dfrac{\pi(b^4 - a^4)}{8}$

$I_{yy} = I_{xx} + b^2 A$

$I_{xy} = x_C y_C A$

Circular sector

$A = \alpha a^2$

$x_C = \dfrac{2a \sin \alpha}{3\alpha}$

$y_C = 0$

$I_{xx} = \dfrac{a^4}{4}(\alpha - \sin \alpha \cos \alpha)$

$I_{yy} = \dfrac{a^4}{4}(\alpha + \sin \alpha \cos \alpha)$

$I_{xy} = 0$

Circular segment

$A = a^2(\alpha - \sin \alpha \cos \alpha)$

$x_C = \dfrac{2a}{3}\dfrac{\sin^3 \alpha}{\alpha - \sin \alpha \cos \alpha}$

$y_C = 0$

$I_{xx} = \dfrac{Aa^2}{4}\left(1 - \dfrac{2\Delta}{3}\right)$

$I_{yy} = \dfrac{Aa^2}{4}(1 + 2\Delta)$

$\Delta = \dfrac{\sin^3 \alpha \cos \alpha}{\alpha - \sin \alpha \cos \alpha}$

Half circle

$A = \dfrac{\pi b^2}{2}$

$x_C = b$

$y_C = \dfrac{4b}{3\pi}$

$I_{AA} = 0.1098 b^4$

$I_{BB} = \dfrac{\pi b^4}{8}$

$I_{AB} = 0$

$I_{xx} = \dfrac{\pi b^4}{8}$

$I_{yy} = \dfrac{5\pi b^4}{8}$

$I_{xy} = \dfrac{2b^4}{3}$

A = area $I_{AA}, I_{BB}, I_{CC}, I_{xx}, I_{yy}$ = moments of inertia

x_C, y_C = coordinates of centroid I_{AB}, I_{xy} = products of inertia

Ellipse			
$A = \pi a b$	$I_{AA} = \dfrac{\pi a b^3}{4}$	$I_{xx} = \dfrac{5\pi a b^3}{4}$	
$x_C = a$	$I_{BB} = \dfrac{\pi a^3 b}{4}$	$I_{yy} = \dfrac{5\pi a^3 b}{4}$	
$y_C = b$	$I_{AB} = 0$	$I_{xy} = \pi a^2 b^2$	
Half ellipse			
$A = \dfrac{\pi a b}{2}$	$I_{AA} = \dfrac{\pi a b^3}{8}\left(1 - \dfrac{64}{9\pi^2}\right)$	$I_{xx} = \dfrac{\pi a b^3}{8}$	
$x_C = a$	$I_{BB} = \dfrac{\pi a^3 b}{8}$	$I_{yy} = \dfrac{5\pi a^3 b}{8}$	
$y_C = \dfrac{4b}{3\pi}$	$I_{AB} = 0$	$I_{xy} = \dfrac{2a^2 b^2}{3}$	
Quarter ellipse			
$A = \dfrac{\pi a b}{4}$	$I_{AA} = \dfrac{\pi a^3 b}{16}\left(1 - \dfrac{64}{9\pi^2}\right)$	$I_{xx} = \dfrac{\pi a b^3}{16}$	
$e = \dfrac{4a}{3\pi}$	$I_{BB} = \dfrac{\pi a b^3}{16}\left(1 - \dfrac{64}{9\pi^2}\right)$	$I_{yy} = \dfrac{\pi a^3 b}{16}$	
$f = \dfrac{4b}{3\pi}$	$I_{AB} = \dfrac{a^2 b^2}{8}\left(1 - \dfrac{32}{9\pi}\right)$	$I_{xy} = \dfrac{a^2 b^2}{8}$	
Parabola			
$A = \dfrac{4ab}{3} = \dfrac{2bl}{3}$	$I_{AA} = \dfrac{16ab^3}{175}$	$I_{xx} = \dfrac{32ab^3}{105}$	
	$I_{BB} = \dfrac{4a^3 b}{15}$	$I_{yy} = \dfrac{8a^3 b}{5}$	
$x_C = a$	$I_{AB} = 0$	$I_{xy} = \dfrac{8a^2 b^2}{15}$	
$y_C = \dfrac{2b}{5}$	$I_{TT} = \dfrac{4ab^3}{7}$		
Half parabola			
$A = \dfrac{2ab}{3}$	$I_{AA} = \dfrac{19ab^3}{480}$	$I_{xx} = \dfrac{2ab^3}{15}$	
$x_C = \dfrac{3a}{5}$	$I_{BB} = \dfrac{8a^3 b}{175}$	$I_{yy} = \dfrac{2a^3 b}{7}$	
$y_C = \dfrac{3b}{8}$	$I_{AB} = \dfrac{a^2 b^2}{60}$	$I_{xy} = \dfrac{a^2 b^2}{6}$	
Parabolic complement			
$A = \dfrac{ab}{3}$	$I_{AA} = \dfrac{37ab^3}{2,100}$	$I_{xx} = \dfrac{ab^3}{21}$	
$x_C = \dfrac{3a}{4}$	$I_{BB} = \dfrac{a^3 b}{80}$	$I_{yy} = \dfrac{a^3 b}{5}$	
$y_C = \dfrac{3b}{10}$	$I_{AB} = \dfrac{a^2 b^2}{120}$	$I_{xy} = \dfrac{a^2 b^2}{12}$	

(1) System X, Y $(0 = \text{origin}, C = \text{centroid})$

Volume
$\delta = \text{density}$
$$dV = dx\, dy\, dz\, \delta \qquad V = \iiint_V dV$$

Static moments
$$M_{xy} = \iiint_V z\, dV \qquad M_{yz} = \iiint_V x\, dV$$

$$M_{zx} = \iiint_V y\, dV$$

Coordinates of centroid
$$x_C = \frac{M_{yz}}{V} \qquad z_C = \frac{M_{xy}}{V}$$

$$y_C = \frac{M_{zx}}{V}$$

Moments of inertia
$$I_{xx} = \iiint_V (y^2 + z^2)\, dV$$

$$I_{yy} = \iiint_V (x^2 + z^2)\, dV$$

$$I_{zz} = \iiint_V (x^2 + y^2)\, dV$$

Products of inertia
$$I_{yz} = I_{zy} = \iiint_V yz\, dV$$

$$I_{xz} = I_{zx} = \iiint_V xz\, dV$$

$$I_{xy} = I_{yx} = \iiint_V xy\, dV$$

Polar moment of inertia
$$J_0 = \iiint_V (x^2 + y^2 + z^2)\, dV$$

$$= \frac{I_{xx} + I_{yy} + I_{zz}}{2}$$

Radii of gyration
$$k_x = \sqrt{\frac{I_{xx}}{V}} \qquad k_y = \sqrt{\frac{I_{yy}}{V}} \qquad k_z = \sqrt{\frac{I_{zz}}{V}} \qquad k_0 = \sqrt{\frac{J_0}{V}}$$

(2) System $X'Y'$ $(0' = \text{origin}, C = \text{centroid})$

Static moments
$$M'_{yz} = M_{yz} - aV \qquad M'_{zx} = M_{zx} - bV \qquad M'_{xy} = M_{xy} - cV$$

Coordinates of centroid
$$x'_C = x_C - a \qquad y'_C = y_C - b \qquad z'_C = z_C - c$$

Moments of inertia
$$I'_{xx} = I_{xx} - 2bM_{zx} - 2cM_{xy} + (b^2 + c^2)V$$

$$I'_{yy} = I_{yy} - 2aM_{yz} - 2cM_{xy} + (a^2 + c^2)V$$

$$I'_{zz} = I_{zz} - 2aM_{yz} - 2bM_{zx} + (a^2 + b^2)V$$

Products of inertia
$$I'_{yz} = I_{yz} - bM_{zx} - cM_{xy} + bcV$$

$$I'_{xz} = I_{xz} - aM_{yz} - cM_{xy} + acV$$

$$I'_{xy} = I_{xy} - aM_{yz} - bM_{zx} + abV$$

Polar moment of inertia $(r^2 = a^2 + b^2 + c^2)$
$$J'_0 = \frac{I_{xx} + I_{yy} + I_{zz}}{2} - 2aM_{yz} - 2bM_{zx} - 2cM_{xy} + r^2 V$$

Radii of gyration
$$k'_x = \sqrt{k_x^2 - 2by_C - 2cz_C + (b^2 + c^2)} \qquad k'_z = \sqrt{k_z^2 - 2ax_C - 2by_C + (a^2 + b^2)}$$

$$k'_y = \sqrt{k_y^2 - 2ax_C - 2cz_C + (a^2 + c^2)} \qquad k'_0 = \sqrt{k_0^2 - 2(ax_C + by_C + cz_C) + r^2}$$

$V =$ volume	$I_{xx}, I_{yy}, I_{zz}, I_{AA}, I_{BB}, I_{CC} =$ moments of inertia
$\delta =$ density per unit length	$I_{xy}, I_{yz}, I_{zx}, I_{AB}, I_{BC}, I_{CA} =$ products of inertia
$m = V\delta =$ mass	$x_C, y_C, z_C =$ coordinates of centroid

---

Straight bar

$m = l\delta$

$I_{AA} = \dfrac{ml^2}{12}$ $I_{xx} = \dfrac{ml^2}{3}$

$x_C = 0$ $I_{BB} = 0$ $I_{yy} = 0$

$y_C = l/2$ $I_{CC} = I_{AA}$ $I_{zz} = I_{xx}$

$z_C = 0$ $I_{AB} = 0, \ldots$ $I_{xy} = 0, \ldots$

Straight bar

$m = l\delta$

$I_{AA} = m\dfrac{b^2 + c^2}{12}$ $I_{xx} = m\dfrac{b^2 + c^2}{3}$

$x_C = \dfrac{a}{2}$ $I_{BB} = m\dfrac{c^2 + a^2}{12}$ $I_{yy} = m\dfrac{c^2 + a^2}{3}$

$y_C = \dfrac{b}{2}$ $I_{CC} = m\dfrac{a^2 + b^2}{12}$ $I_{zz} = m\dfrac{a^2 + b^2}{3}$

$z_C = \dfrac{c}{2}$ $I_{AB} = m\dfrac{ab}{12}, \ldots$ $I_{xy} = m\dfrac{ab}{3}, \ldots$

Circular ring

$m = 2\pi a\delta$

$I_{AA} = ma^2$ $I_{xx} = 3ma^2$

$x_C = 0$ $I_{BB} = \dfrac{ma^2}{2}$ $I_{yy} = \tfrac{3}{2}ma^2$

$y_C = a$ $I_{CC} = I_{BB}$ $I_{zz} = I_{yy}$

$z_C = a$ $I_{AB} = 0, \ldots$ $I_{xy} = I_{zx} = 0, I_{yz} = ma^2$

Circular bar

$m = 2\alpha a\delta$

$I_{AA} = ma^2\left(1 - \dfrac{\sin^2 \alpha}{\alpha^2}\right)$ $I_{xx} = ma^2$

$x_C = 0$ $I_{BB} = I_{yy} - mz_C^2$ $I_{yy} = \dfrac{ma^2}{2}\left(1 + \dfrac{\sin 2\alpha}{2\alpha}\right)$

$y_C = 0$ $I_{CC} = I_{zz}$ $I_{zz} = \dfrac{ma^2}{2}\left(1 - \dfrac{\sin 2\alpha}{2\alpha}\right)$

$z_C = \dfrac{a}{\alpha}\sin \alpha$ $I_{AB} = 0, \ldots$ $I_{xy} = 0, \ldots$

Parabolic bar

$\delta(\phi) = \dfrac{\delta}{\cos \phi}$

$I_{AA} = m\dfrac{15a^2 + 4h^2}{45}$ $I_{xx} = m\dfrac{5a^2 + 8h^2}{15}$

$m = 2a\delta$ $I_{BB} = \dfrac{4mh^2}{45}$ $I_{yy} = \dfrac{8mh^2}{15}$

$x_C = y_C = 0$ $I_{CC} = \dfrac{ma^2}{3}$ $I_{zz} = \dfrac{ma^2}{3}$

$z_C = \dfrac{2h}{3}$ $I_{AB} = 0, \ldots$ $I_{xy} = 0, \ldots$

V = volume $\qquad$ $I_{xx}, I_{yy}, I_{zz}, I_{AA}, I_{BB}, I_{CC}$ = moments of inertia

δ = density $\qquad$ $I_{xy}, I_{yz}, I_{zx}, I_{AB}, I_{BC}, I_{CA}$ = products of inertia

m = mass $\qquad\qquad\qquad\qquad$ x_C, y_C, z_C = coordinates of centroid

$m = V\delta$

Cube

$m = a^3\delta$

$x_C = \dfrac{a}{2}$

$y_C = \dfrac{a}{2}$

$z_C = \dfrac{a}{2}$

$I_{AA} = \dfrac{ma^2}{6}$

$I_{BB} = I_{CC} = I_{AA}$

$I_{TT} = \dfrac{5ma^2}{12}$

$I_{AB} = 0$

$I_{xx} = \dfrac{2ma^2}{3}$

$I_{yy} = I_{xx}$

$I_{zz} = I_{xx}$

$I_{xy} = I_{yz} = I_{zx} = \dfrac{ma^2}{4}$

Prism

$m = abc\delta$

$x_C = \dfrac{a}{2}$

$y_C = \dfrac{b}{2}$

$z_C = \dfrac{c}{2}$

$I_{AA} = m\dfrac{b^2 + c^2}{12}$

$I_{BB} = m\dfrac{a^2 + c^2}{12}$

$I_{TT} = m\dfrac{a^2 + 4c^2}{12}$

$I_{AB} = 0$

$I_{xx} = m\dfrac{b^2 + c^2}{3}$

$I_{yy} = m\dfrac{a^2 + c^2}{3}$

$I_{zz} = m\dfrac{a^2 + b^2}{3}$

$I_{xy} = \dfrac{mab}{4}$ $\qquad$ $\cdots$

Cylinder

$m = \pi a^2 h\delta$

$x_C = 0$

$y_C = 0$

$z_C = \dfrac{h}{2}$

$I_{AA} = m\dfrac{3a^2 + h^2}{12}$

$I_{BB} = m\dfrac{3a^2 + h^2}{12}$

$I_{CC} = \dfrac{ma^2}{2}$

$I_{AB} = 0$

$I_{xx} = m\dfrac{3a^2 + 4h^2}{12}$

$I_{yy} = m\dfrac{3a^2 + 4h^2}{12}$

$I_{zz} = \dfrac{ma^2}{2}$

$I_{xy} = I_{yz} = I_{zx} = 0$

Cone

$m = \dfrac{\pi a^2 h\delta}{3}$

$x_C = 0$

$y_C = 0$

$z_C = \dfrac{h}{4}$

$I_{AA} = \dfrac{3m}{80}(4a^2 + h^2)$

$I_{BB} = \dfrac{3m}{80}(4a^2 + h^2)$

$I_{TT} = \dfrac{3m}{20}(a^2 + 4h^2)$

$I_{AB} = 0$

$I_{xx} = \dfrac{m}{20}(3a^2 + 2h^2)$

$I_{yy} = \dfrac{m}{20}(3a^2 + 2h^2)$

$I_{zz} = \dfrac{3ma^2}{10}$

$I_{xy} = I_{yz} = I_{zx} = 0$

Rectangular pyramid

$m = \dfrac{abh\delta}{3}$

$x_C = 0$

$y_C = 0$

$z_C = \dfrac{h}{4}$

$I_{AA} = \dfrac{m}{80}(4b^2 + 3h^2)$

$I_{BB} = \dfrac{m}{80}(4a^2 + 3h^2)$

$I_{TT} = \dfrac{m}{20}(b^2 + 12h^2)$

$I_{AB} = 0$

$I_{xx} = \dfrac{m}{20}(b^2 + 2h^2)$

$I_{yy} = \dfrac{m}{20}(a^2 + 2h^2)$

$I_{zz} = \dfrac{m}{20}(a^2 + b^2)$

$I_{xy} = I_{yz} = I_{zx} = 0$

V = volume $I_{xx}, I_{yy}, I_{zz}, I_{AA}, I_{BB}, I_{CC}$ = moments of inertia

δ = density $I_{xy}, I_{yz}, I_{zx}, I_{AB}, I_{BC}, I_{CA}$ = products of inertia

m = mass x_C, y_C, z_C = coordinates of centroid

$m = V\delta$

Sphere			
$m = \dfrac{4\pi a^3 \delta}{3}$	$I_{TT} = \dfrac{7ma^2}{5}$	$I_{xx} = \dfrac{2ma^2}{5}$	
$x_C = 0$	$I_{NN} = I_{TT}$	$I_{xx} = I_{yy} = I_{zz}$	
$y_C = 0$			
$z_C = 0$	$I_{TN} = 0$	$I_{xy} = I_{yz} = I_{zx} = 0$	

Hemisphere			
$m = \dfrac{2\pi a^3 \delta}{3}$	$I_{AA} = \dfrac{83ma^2}{320}$	$I_{xx} = \dfrac{2ma^2}{5}$	
$x_C = 0$	$I_{TT} = \dfrac{208ma^2}{320}$	$I_{xx} = I_{yy} = I_{zz}$	
$y_C = 0$			
$z_C = \dfrac{3a}{8}$	$I_{TN} = 0$	$I_{xy} = I_{yz} = I_{zx} = 0$	

Ellipsoid			
$m = \dfrac{4\pi abc\delta}{3}$	$I_{TT} = m\dfrac{b^2 + 6c^2}{5}$	$I_{xx} = m\dfrac{b^2 + c^2}{5}$	
$x_C = 0$	$I_{NN} = m\dfrac{a^2 + 6c^2}{5}$	$I_{yy} = m\dfrac{a^2 + c^2}{5}$	
$y_C = 0$		$I_{zz} = m\dfrac{a^2 + b^2}{5}$	
$z_C = 0$	$I_{TN} = 0$	$I_{xy} = I_{yz} = I_{zx} = 0$	

Paraboloid of revolution			
$m = \dfrac{\pi a^2 h\delta}{2}$	$I_{AA} = \dfrac{m}{18}(3a^2 + h^2)$	$I_{xx} = \dfrac{m}{6}(a^2 + 3h^2)$	
$x_C = 0$	$I_{BB} = \dfrac{m}{18}(3a^2 + h^2)$	$I_{yy} = \dfrac{m}{6}(a^2 + 3h^2)$	
$y_C = 0$		$I_{zz} = \dfrac{ma^2}{3}$	
$z_C = \dfrac{2h}{3}$	$I_{AB} = 0$	$I_{xy} = I_{yz} = I_{zx} = 0$	

Elliptic paraboloid			
$m = \dfrac{\pi abh\delta}{2}$	$I_{AA} = \dfrac{m}{18}(3b^2 + h^2)$	$I_{xx} = \dfrac{m}{6}(b^2 + 3h^2)$	
$x_C = 0$	$I_{BB} = \dfrac{m}{18}(3a^2 + h^2)$	$I_{yy} = \dfrac{m}{6}(a^2 + 3h^2)$	
$y_C = 0$		$I_{zz} = \dfrac{m}{6}(a^2 + b^2)$	
$z_C = \dfrac{2h}{3}$	$I_{AB} = 0$	$I_{xy} = I_{yz} = I_{zx} = 0$	

$V = \text{volume}$ $I_{xx}, I_{yy}, I_{zz}, I_{AA}, I_{BB}, I_{CC} = \text{moments of inertia}$

$\delta = \text{density}$ $I_{xy}, I_{yz}, I_{zx}, I_{AB}, I_{BC}, I_{CA} = \text{products of inertia}$

$m = V\delta = \text{mass}$ $x_C, y_C, z_C = \text{coordinates of centroid}$

$t = 1 = \text{thickness}$

Spherical shell

$m = 4\pi a^2 \delta$	$I_{AA} = \dfrac{2ma^2}{3}$	$I_{xx} = \dfrac{5ma^2}{3}$
$x_C = 0$	$I_{BB} = I_{AA}$	$I_{yy} = I_{xx}$
$y_C = 0$	$I_{CC} = I_{AA}$	$I_{zz} = \dfrac{2ma^2}{3}$
$z_C = a$	$I_{AB} = 0, \ldots$	$I_{xy} = 0, \ldots$

Hemispherical shell

$m = 2\pi a^2 \delta$	$I_{AA} = \dfrac{5ma^2}{12}$	$I_{xx} = \dfrac{2ma^2}{3}$
$x_C = 0$	$I_{BB} = I_{AA}$	$I_{yy} = I_{xx}$
$y_C = 0$	$I_{CC} = \dfrac{2ma^2}{3}$	$I_{zz} = I_{yy}$
$z_C = -\dfrac{a}{2}$	$I_{AB} = 0, \ldots$	$I_{xy} = 0, \ldots$

Circular cylindrical shell

$m = 2\pi a h \delta$	$I_{AA} = \dfrac{m}{12}(6a^2 + h^2)$	$I_{xx} = \dfrac{m}{6}(3a^2 + 2h^2)$
$x_C = 0$	$I_{BB} = I_{AA}$	$I_{yy} = I_{xx}$
$y_C = 0$	$I_{CC} = ma^2$	$I_{zz} = ma^2$
$z_C = \dfrac{h}{2}$	$I_{AB} = 0, \ldots$	$I_{xy} = 0, \ldots$

Circular conical shell

$m = \pi a \sqrt{a^2 + h^2}\, \delta$	$I_{AA} = \dfrac{m}{18}(9a^2 + 10h^2)$	$I_{xx} = \dfrac{m}{2}(a^2 + 2h^2)$
$x_C = 0$	$I_{BB} = I_{AA}$	$I_{yy} = I_{xx}$
$y_C = 0$	$I_{CC} = \dfrac{ma^2}{2}$	$I_{zz} = \dfrac{ma^2}{2}$
$z_C = \dfrac{2h}{3}$	$I_{AB} = 0, \ldots$	$I_{xy} = 0, \ldots$

Semicircular cylindrical shell

$m = \pi a h \delta$	$I_{AA} = ma^2\left(1 - \dfrac{4}{\pi^2}\right)$	$I_{xx} = ma^2$
$x_C = \dfrac{h}{2}$	$I_{BB} = \dfrac{m}{12}\left(6a^2 - \dfrac{48a^2}{\pi^2} + h^2\right)$	$I_{yy} = \dfrac{m}{6}(3a^2 + 2h^2)$
$y_C = 0$	$I_{CC} = I_{NN} = \dfrac{m}{12}(6a^2 + h^2)$	$I_{zz} = I_{yy}$
$z_C = -\dfrac{2a}{\pi}$	$I_{AB} = 0, \ldots$	$I_{xy} = 0, \ldots$

10
VECTOR ANALYSIS

(1) Definitions

(a) Scalar is a quantity defined by magnitude (signed number) only and designated by ordinary letters such as $a, b, c, \ldots, r, \ldots, A, B, C, \ldots, R, \ldots, \alpha, \beta, \gamma, \ldots$. Examples of scalars are length, time, temperature, and mass.

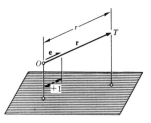

(b) Vector is a quantity defined by magnitude (scalar) and direction (line of action and sense) and designated by boldface letters such as $\mathbf{a}, \mathbf{b}, \mathbf{c}, \ldots, \mathbf{r}, \ldots, \mathbf{A}, \mathbf{B}, \mathbf{C}, \ldots, \mathbf{R}, \ldots$. Examples of vectors are force, moment, displacement, velocity, and acceleration.

(c) Graphical representation of a vector $\mathbf{r}$ is a directed segment given by its initial point O (origin of vector) and its end point T (terminus of vector). The magnitude r of the vector $\mathbf{r}$ is the length of this segment.

$$r = |\mathbf{r}| = \overline{OT}$$

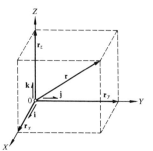

(d) Unit vector $\mathbf{e}$ is a vector of unit magnitude. Any vector $\mathbf{r}$ can be represented analytically as the product of its magnitude r and its unit vector $\mathbf{e}$.

$$\mathbf{r} = r\mathbf{e} \qquad \text{or} \qquad \mathbf{e} = \frac{\mathbf{r}}{r}$$

(2) Components and Magnitudes

(a) Resolution A vector $\mathbf{r}$ may be resolved into any number of components. In the right-handed cartesian coordinate system, $\mathbf{r}$ is resolved into three mutually perpendicular components, each parallel to the respective coordinate axis.

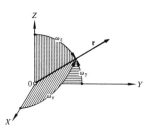

$$\mathbf{r} = \mathbf{r}_x + \mathbf{r}_y + \mathbf{r}_z = r_x\mathbf{i} + r_y\mathbf{j} + r_z\mathbf{k} = r\mathbf{e}$$

where $\mathbf{r}_x, \mathbf{r}_y, \mathbf{r}_z$ are the vector components, r_x, r_y, r_z are their magnitudes, $\mathbf{i}, \mathbf{j}, \mathbf{k}$ are the unit vectors in the X, Y, Z axes, respectively, r is the magnitude of $\mathbf{r}$ and $\mathbf{e}$ is its unit vector.

(b) Unit vectors $\mathbf{i}, \mathbf{j}, \mathbf{k}$, and $\mathbf{e}$ are inversely defined as

$$\mathbf{i} = \frac{\mathbf{r}_x}{r_x} \qquad \mathbf{j} = \frac{\mathbf{r}_y}{r_y} \qquad \mathbf{k} = \frac{\mathbf{r}_z}{r_z} \qquad \mathbf{e} = \frac{r_x}{r}\mathbf{i} + \frac{r_y}{r}\mathbf{j} + \frac{r_z}{k}\mathbf{k}$$

(c) Magnitude r is given by the magnitude of components r_x, r_y, r_z as

$$r = |\mathbf{r}| = \sqrt{r_x^2 + r_y^2 + r_z^2} = r_x\frac{\mathbf{i}}{\mathbf{e}} + r_y\frac{\mathbf{j}}{\mathbf{e}} + r_z\frac{\mathbf{k}}{\mathbf{e}} = \alpha r_x + \beta r_y + \gamma r_z$$

(d) Direction cosines α, β, γ of $\mathbf{r}$ defined as the cosines of the angles $\omega_x, \omega_y, \omega_z$ measured to $\mathbf{r}$ from the positive X, Y, Z axes, respectively, are

$$\cos \omega_x = \frac{r_x}{r} = \frac{\mathbf{i}}{\mathbf{e}} = \alpha \qquad \cos \omega_y = \frac{r_y}{r} = \frac{\mathbf{j}}{\mathbf{e}} = \beta \qquad \cos \omega_z = \frac{r_z}{r} = \frac{\mathbf{k}}{\mathbf{e}} = \gamma$$

and their relations are

$$\alpha^2 + \beta^2 + \gamma^2 = 1 \qquad \text{and} \qquad \mathbf{e} = \alpha\mathbf{i} + \beta\mathbf{j} + \gamma\mathbf{k}$$

(1) Vector Addition and Subtraction

(a) **Sum of vectors** **a** *and* **b** is a vector **c** formed by placing the initial point of **b** on the terminal point of **a** and joining the initial point of **a** to the terminal point of **b**.

$$\mathbf{a} + \mathbf{b} = \mathbf{c}$$

(b) **Difference of vectors** **a** *and* **b** is a vector **d** formed by placing the initial point of **b** on the initial point of **a** and joining the terminal point of **b** with the terminal point of **a**.

$$\mathbf{a} - \mathbf{b} = \mathbf{d}$$

(c) **Difference of vectors** **b** *and* **a** is a vector **e** formed by placing the initial point of **b** on the initial point of **a** and joining the terminal point of **a** with terminal point of **b**.

$$\mathbf{b} - \mathbf{a} = \mathbf{e}$$

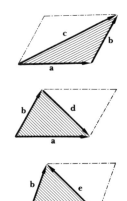

(2) Scalar – Vector Laws   (**a, b** = vectors; *m, n* = scalars)

$m\mathbf{a} = \mathbf{a}m$	Commutative law	$(m + n)\mathbf{a} = m\mathbf{a} + n\mathbf{a}$	Distributive law
$m(n\mathbf{a}) = (mn)\mathbf{a}$	Associative law	$m(\mathbf{a} + \mathbf{b}) = m\mathbf{a} + m\mathbf{b}$	Distributive law

(3) Vector Summation Laws

(a) **Commutative law**

$$\mathbf{a} + \mathbf{b} + \mathbf{c} = \mathbf{b} + \mathbf{c} + \mathbf{a} = \mathbf{c} + \mathbf{a} + \mathbf{b} = \mathbf{f}$$

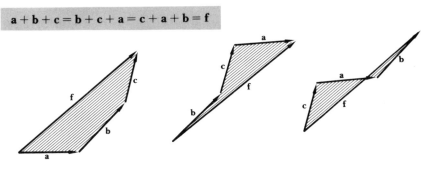

(b) **Associative law**

$$(\mathbf{a} + \mathbf{b}) + \mathbf{c} = \mathbf{a} + (\mathbf{b} + \mathbf{c}) = \mathbf{f}$$

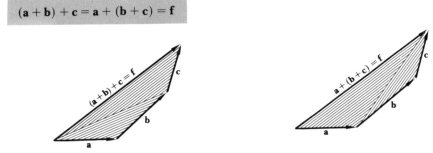

(1) Scalar Product

(a) **Dot product**(sclara product, inner product) of two vectors
a and **b** denoted as **a** · **b** or (**ab**) is defined as the product
of their magnitudes a, b and the cosine of the angle ω
between them. *The result is a scalar.*

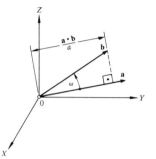

$$\mathbf{a} \cdot \mathbf{b} = ab \cos \omega = a_x b_x + a_y b_y + a_z b_z$$

$$\omega = \cos^{-1} \frac{\mathbf{a} \cdot \mathbf{b}}{ab} \qquad 0 < \omega < \pi$$

(b) **Product laws**   ($m, n =$ scalars)

Commutative: $\mathbf{a} \cdot \mathbf{b} = \mathbf{b} \cdot \mathbf{a}$
Associative: $m\mathbf{a} \cdot n\mathbf{b} = mn\,\mathbf{a} \cdot \mathbf{b}$
Distributive: $\mathbf{a} \cdot (\mathbf{b} + \mathbf{c}) = \mathbf{a} \cdot \mathbf{b} + \mathbf{a} \cdot \mathbf{c}$

(c) **Special cases**   ($a \neq 0, b \neq 0$)

Two vectors **a** and **b** are *normal* ($\omega = 90°$) if

$$\mathbf{a} \cdot \mathbf{b} = 0$$

From this,

$$\mathbf{i} \cdot \mathbf{j} = \mathbf{j} \cdot \mathbf{k} = \mathbf{k} \cdot \mathbf{i} = 0$$

Two vectors **a** and **b** are *parallel* ($\omega = 0°$) if

$$\mathbf{a} \cdot \mathbf{b} = ab$$

From this,

$$\mathbf{i} \cdot \mathbf{i} = \mathbf{j} \cdot \mathbf{j} = \mathbf{k} \cdot \mathbf{k} = 1$$

(2) Vector Product

(a) **Cross product** (vector product) of two vectors **a** and **b**
denoted as **a** × **b** or [**ab**] is defined as the product of their
magnitudes, the sine of the angle ω between them, and the
unit vector **n** normal to their plane. *The result is a vector.*

$$\mathbf{a} \times \mathbf{b} = ab \sin \omega\, \mathbf{n} = \begin{vmatrix} \mathbf{i} & \mathbf{j} & \mathbf{k} \\ a_x & a_y & a_z \\ b_x & b_y & b_z \end{vmatrix} = \begin{bmatrix} 0 & -a_z & a_y \\ a_z & 0 & -a_x \\ -a_y & a_x & 0 \end{bmatrix} \begin{bmatrix} b_x \\ b_y \\ b_z \end{bmatrix}$$

$$\omega = \sin^{-1} \frac{|\mathbf{a} \times \mathbf{b}|}{ab} \qquad 0 < \omega < \pi$$

(b) **Product laws**   ($m, n =$ scalars)

Noncommutative: $\mathbf{a} \times \mathbf{b} = -\mathbf{b} \times \mathbf{a}$
Associative: $(m\mathbf{a}) \times (n\mathbf{b}) = mn\,(\mathbf{a} \times \mathbf{b})$
Distributive: $\mathbf{a} \times (\mathbf{b} + \mathbf{c}) = \mathbf{a} \times \mathbf{b} + \mathbf{a} \times \mathbf{c}$

(c) **Special cases**   ($a \neq 0, b \neq 0$)

Two vectors $\vec{\mathbf{a}}$ and $\vec{\mathbf{b}}$ are *normal* ($\omega = 90°$) if

$$\mathbf{a} \times \mathbf{b} = ab\mathbf{n}$$

From this

$$\mathbf{i} \times \mathbf{j} = \mathbf{k} \qquad \mathbf{j} \times \mathbf{k} = \mathbf{i} \qquad \mathbf{k} \times \mathbf{i} = \mathbf{j}$$

Two vectors **a** and **b** are *parallel* ($\omega = 0°$) if

$$\mathbf{a} \times \mathbf{b} = 0$$

From this,

$$\mathbf{i} \times \mathbf{i} = \mathbf{j} \times \mathbf{j} = \mathbf{k} \times \mathbf{k} = 0$$

(1) Triple Products

(a) Scalar triple product (result a scalar)

$$\mathbf{a} \cdot (\mathbf{b} \times \mathbf{c}) = \mathbf{b} \cdot (\mathbf{c} \times \mathbf{a}) = \mathbf{c} \cdot (\mathbf{a} \times \mathbf{b}) = (\mathbf{a} \times \mathbf{b}) \cdot \mathbf{c} = (\mathbf{b} \times \mathbf{c}) \cdot \mathbf{a} = (\mathbf{c} \times \mathbf{a}) \cdot \mathbf{b} = [\mathbf{abc}]$$

$$[\mathbf{abc}] = \begin{vmatrix} a_x & a_y & a_z \\ b_x & b_y & b_z \\ c_x & c_y & c_z \end{vmatrix} = \begin{vmatrix} a_x & b_x & c_x \\ a_y & b_y & c_y \\ a_z & b_z & c_z \end{vmatrix} = [a_x \ a_y \ a_z] \begin{bmatrix} 0 & -b_z & b_y \\ b_z & 0 & -b_x \\ -b_y & b_x & 0 \end{bmatrix} \begin{bmatrix} c_x \\ c_y \\ c_z \end{bmatrix}$$

(b) Vector triple product (result a vector)

$$\mathbf{a} \times (\mathbf{b} \times \mathbf{c}) = (\mathbf{a} \cdot \mathbf{c})\mathbf{b} - (\mathbf{a} \cdot \mathbf{b})\mathbf{c} = \begin{vmatrix} \mathbf{b} & \mathbf{c} \\ \mathbf{a} \cdot \mathbf{b} & \mathbf{a} \cdot \mathbf{c} \end{vmatrix} = \begin{bmatrix} 0 & -a_z & a_y \\ a_z & 0 & -a_x \\ -a_y & a_x & 0 \end{bmatrix} \begin{bmatrix} 0 & -b_z & b_y \\ b_z & 0 & -b_x \\ -b_y & b_x & 0 \end{bmatrix} \begin{bmatrix} c_x \\ c_y \\ c_z \end{bmatrix}$$

$$(\mathbf{a} \times \mathbf{b}) \times \mathbf{c} = (\mathbf{a} \cdot \mathbf{c})\mathbf{b} - (\mathbf{b} \cdot \mathbf{c})\mathbf{a} = \begin{vmatrix} \mathbf{b} & \mathbf{a} \\ \mathbf{c} \cdot \mathbf{b} & \mathbf{c} \cdot \mathbf{a} \end{vmatrix} = \begin{bmatrix} 0 & -c_z & c_y \\ c_z & 0 & -c_x \\ -c_y & c_x & 0 \end{bmatrix} \begin{bmatrix} 0 & -b_z & b_y \\ b_z & 0 & -b_x \\ -b_y & b_x & 0 \end{bmatrix} \begin{bmatrix} a_x \\ a_y \\ a_z \end{bmatrix}$$

(2) Special Products

(a) Double scalar product (result a scalar)

$$(\mathbf{a} \times \mathbf{b}) \cdot (\mathbf{c} \times \mathbf{d}) = \begin{vmatrix} \mathbf{a} \cdot \mathbf{c} & \mathbf{b} \cdot \mathbf{c} \\ \mathbf{a} \cdot \mathbf{d} & \mathbf{b} \cdot \mathbf{d} \end{vmatrix} = [b_x \ b_y \ b_z] \begin{bmatrix} 0 & a_z & -a_y \\ -a_z & 0 & a_x \\ a_y & -a_y & 0 \end{bmatrix} \begin{bmatrix} 0 & -c_z & c_y \\ c_z & 0 & -c_x \\ -c_y & c_x & 0 \end{bmatrix} \begin{bmatrix} d_x \\ d_y \\ d_z \end{bmatrix}$$

$$(\mathbf{a} \times \mathbf{b})^2 = \begin{vmatrix} \mathbf{a}^2 & \mathbf{a} \cdot \mathbf{b} \\ \mathbf{a} \cdot \mathbf{b} & \mathbf{b}^2 \end{vmatrix} = [b_x \ b_y \ b_z] \begin{bmatrix} 0 & a_z & -a_y \\ -a_z & 0 & a_x \\ a_y & -a_x & 0 \end{bmatrix} \begin{bmatrix} 0 & -a_z & a_y \\ a_z & 0 & -a_x \\ -a_y & a_x & 0 \end{bmatrix} \begin{bmatrix} b_x \\ b_y \\ b_z \end{bmatrix}$$

(b) Double vector product (result a vector)

$$(\mathbf{a} \times \mathbf{b}) \times (\mathbf{c} \times \mathbf{d}) = \begin{vmatrix} \mathbf{i} & \mathbf{j} & \mathbf{k} \\ \begin{vmatrix} a_y & a_z \\ b_y & b_z \end{vmatrix} & \begin{vmatrix} a_z & a_x \\ b_z & b_x \end{vmatrix} & \begin{vmatrix} a_x & a_y \\ b_x & b_y \end{vmatrix} \\ \begin{vmatrix} c_y & c_z \\ d_y & d_z \end{vmatrix} & \begin{vmatrix} c_z & c_x \\ d_z & d_x \end{vmatrix} & \begin{vmatrix} c_x & c_y \\ d_x & d_y \end{vmatrix} \end{vmatrix} = -\mathbf{a}[\mathbf{bcd}] + \mathbf{b}[\mathbf{cda}] = \mathbf{c}[\mathbf{dab}] - \mathbf{d}[\mathbf{abc}]$$

(c) Double scalar triple product (result a scalar)

$$[\mathbf{a} \ \mathbf{b} \ \mathbf{c}][\mathbf{d} \ \mathbf{e} \ \mathbf{f}] = \begin{vmatrix} \mathbf{a} \cdot \mathbf{d} & \mathbf{a} \cdot \mathbf{e} & \mathbf{a} \cdot \mathbf{f} \\ \mathbf{b} \cdot \mathbf{d} & \mathbf{b} \cdot \mathbf{e} & \mathbf{b} \cdot \mathbf{f} \\ \mathbf{c} \cdot \mathbf{d} & \mathbf{c} \cdot \mathbf{e} & \mathbf{c} \cdot \mathbf{f} \end{vmatrix} = \begin{vmatrix} \mathbf{a} \\ \mathbf{b} \\ \mathbf{c} \end{vmatrix} [\mathbf{d} \ \mathbf{e} \ \mathbf{f}]$$

(d) Products of orthogonal vectors

If $\mathbf{a} \times (\mathbf{b} \times \mathbf{c}) = \mathbf{b} \times (\mathbf{c} \times \mathbf{a}) = \mathbf{c} \times (\mathbf{a} \times \mathbf{b})$ then $\mathbf{a}, \mathbf{b}, \mathbf{c}$ are *orthogonal*.
If $\mathbf{a} \times \mathbf{b} = \mathbf{c}$ $\mathbf{b} \times \mathbf{c} = \mathbf{a}$ $\mathbf{c} \times \mathbf{a} = \mathbf{b}$ then $\mathbf{a}, \mathbf{b}, \mathbf{c}$ are *orthogonal unit vectors*, and $\mathbf{a} \times (\mathbf{b} \times \mathbf{c}) =$
$\mathbf{b} \times (\mathbf{c} \times \mathbf{a}) = \mathbf{c} \times (\mathbf{a} \times \mathbf{b}) = 0$

(1) Ordinary Derivatives

(a) **Definitions** of limit, continuity, and vector function are formally identical with the definitions of scalar functions.

If $\mathbf{r} = \mathbf{r}(t) = x(t)\mathbf{i} + y(t)\mathbf{j} + z(t)\mathbf{k}$

   = vector function of scalar variable t

then

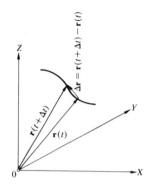

$$\frac{d\mathbf{r}}{dt} = \dot{\mathbf{r}} = \frac{dx(t)}{dt}\mathbf{i} + \frac{dy(t)}{dt}\mathbf{j} + \frac{dz(t)}{dt}\mathbf{k} = \dot{x}\mathbf{i} + \dot{y}\mathbf{j} + \dot{z}\mathbf{k}$$

$$\frac{d\dot{\mathbf{r}}}{dt} = \ddot{\mathbf{r}} = \frac{d\dot{x}(t)}{dt}\mathbf{i} + \frac{d\dot{y}(t)}{dt}\mathbf{j} + \frac{d\dot{z}(t)}{dt}\mathbf{k} = \ddot{x}\mathbf{i} + \ddot{y}\mathbf{j} + \ddot{z}\mathbf{k}$$

. . . .

(b) **Basic formulas** $[\alpha = \alpha(t) = \text{scalar function of } t]$

$$\frac{d}{dt}(\mathbf{r}_1 + \mathbf{r}_2) = \frac{d\mathbf{r}_1}{dt} + \frac{d\mathbf{r}_2}{dt} \qquad\qquad \frac{d}{dt}(\mathbf{r}_1 \cdot \mathbf{r}_2) = \mathbf{r}_1 \cdot \frac{d\mathbf{r}_2}{dt} + \frac{d\mathbf{r}_1}{dt} \cdot \mathbf{r}_2$$

$$\frac{d}{dt}(\alpha\mathbf{r}) = \alpha\frac{d\mathbf{r}}{dt} + \frac{d\alpha}{dt}\mathbf{r} \qquad\qquad \frac{d}{dt}(\mathbf{r}_1 \times \mathbf{r}_2) = \mathbf{r}_1 \times \frac{d\mathbf{r}_2}{dt} + \frac{d\mathbf{r}_1}{dt} \times \mathbf{r}_2$$

$$\frac{d}{dt}(\mathbf{r}_1 \cdot \mathbf{r}_2 \times \mathbf{r}_3) = \mathbf{r}_1 \cdot \mathbf{r}_2 \times \frac{d\mathbf{r}_3}{dt} + \mathbf{r}_1 \cdot \frac{d\mathbf{r}_2}{dt} \times \mathbf{r}_3 + \frac{d\mathbf{r}_1}{dt} \cdot \mathbf{r}_2 \times \mathbf{r}_3$$

$$\frac{d}{dt}[\mathbf{r}_1 \times (\mathbf{r}_2 \times \mathbf{r}_3)] = \mathbf{r}_1 \times \left(\mathbf{r}_2 \times \frac{d\mathbf{r}_3}{dt}\right) + \mathbf{r}_1 \times \left(\frac{d\mathbf{r}_2}{dt} \times \mathbf{r}_3\right) + \frac{d\mathbf{r}_1}{dt} \times (\mathbf{r}_2 \times \mathbf{r}_3)$$

(2) Partial Derivatives

If $\mathbf{r}(x, y, z)$ is a *vector function of several scalar variables* x, y, z, then partial derivatives are obtained by differentiating the magnitude of each component as a scalar function.

$$\frac{\partial\mathbf{r}}{\partial x} = \frac{\partial\mathbf{r}(x, y, z)}{\partial x} \qquad \frac{\partial\mathbf{r}}{\partial y} = \frac{\partial\mathbf{r}(x, y, z)}{\partial y} \qquad \frac{\partial\mathbf{r}}{\partial z} = \frac{\partial\mathbf{r}(x, y, z)}{\partial z}$$

$$\frac{\partial^2\mathbf{r}}{\partial x^2} = \frac{\partial}{\partial x}\left(\frac{\partial\mathbf{r}}{\partial x}\right) \qquad \frac{\partial^2\mathbf{r}}{\partial y^2} = \frac{\partial}{\partial y}\left(\frac{\partial\mathbf{r}}{\partial y}\right) \qquad \frac{\partial^2\mathbf{r}}{\partial z^2} = \frac{\partial}{\partial z}\left(\frac{\partial\mathbf{r}}{\partial z}\right)$$

$$\frac{\partial^2\mathbf{r}}{\partial x\partial y} = \frac{\partial}{\partial x}\left(\frac{\partial\mathbf{r}}{\partial y}\right) \qquad \frac{\partial^2\mathbf{r}}{\partial y\partial z} = \frac{\partial}{\partial y}\left(\frac{\partial\mathbf{r}}{\partial z}\right) \qquad \frac{\partial^2\mathbf{r}}{\partial z\partial x} = \frac{\partial}{\partial z}\left(\frac{\partial\mathbf{r}}{\partial x}\right)$$

$$\frac{\partial(\mathbf{r}_1 \cdot \mathbf{r}_2)}{\partial x} = \mathbf{r}_1 \cdot \frac{\partial\mathbf{r}_2}{\partial x} + \frac{\partial\mathbf{r}_1}{\partial x} \cdot \mathbf{r}_2$$

$$\frac{\partial^2(\mathbf{r}_1 \cdot \mathbf{r}_2)}{\partial x\partial y} = \mathbf{r}_1 \cdot \frac{\partial^2\mathbf{r}_2}{\partial x\partial y} + \frac{\partial\mathbf{r}_1}{\partial x} \cdot \frac{\partial\mathbf{r}_2}{\partial y} + \frac{\partial\mathbf{r}_1}{\partial y} \cdot \frac{\partial\mathbf{r}_2}{\partial x} + \frac{\partial^2\mathbf{r}_1}{\partial x\partial y} \cdot \mathbf{r}_2$$

$$\frac{\partial(\mathbf{r}_1 \times \mathbf{r}_2)}{\partial x} = \mathbf{r}_1 \times \frac{\partial\mathbf{r}_2}{\partial x} + \frac{\partial\mathbf{r}_1}{\partial x} \times \mathbf{r}_2$$

$$\frac{\partial^2(\mathbf{r}_1 \times \mathbf{r}_2)}{\partial x\partial y} = \mathbf{r}_1 \times \frac{\partial^2\mathbf{r}_2}{\partial x\partial y} + \frac{\partial\mathbf{r}_1}{\partial x} \times \frac{\partial\mathbf{r}_2}{\partial y} + \frac{\partial\mathbf{r}_1}{\partial y} \times \frac{\partial\mathbf{r}_2}{\partial x} + \frac{\partial^2\mathbf{r}_1}{\partial x\partial y} \times \mathbf{r}_2$$

(3) Partial Differentials

$$d(\mathbf{r}_1 \cdot \mathbf{r}_2) = \mathbf{r}_1 \cdot d\mathbf{r}_2 + d\mathbf{r}_1 \cdot \mathbf{r}_2$$

$$d(\mathbf{r}_1 \times \mathbf{r}_2) = \mathbf{r}_1 \times d\mathbf{r}_2 + d\mathbf{r}_1 \times \mathbf{r}_2$$

(4) Total Differentials

$$d\mathbf{r} = \frac{\partial\mathbf{r}}{\partial x}dx + \frac{\partial\mathbf{r}}{\partial y}dy + \frac{\partial\mathbf{r}}{\partial z}dz$$

$$d\mathbf{r} = \left(\frac{\partial\mathbf{r}}{\partial x}\frac{\partial x}{\partial t} + \frac{\partial\mathbf{r}}{\partial y}\frac{\partial y}{\partial t} + \frac{\partial\mathbf{r}}{\partial z}\frac{\partial z}{\partial t}\right)dt$$

(1) Position Vector

(a) Space curve is defined by the position vector

$$\mathbf{r} = x\mathbf{i} + y\mathbf{j} + z\mathbf{k} = r\mathbf{e}_r$$

where $x = x(s)$, $y = y(s)$, $z = z(s)$ are functions of the *curvilinear coordinate* s measured along the curve, or $x = x(t)$, $y = y(t)$, $z = z(t)$ are functions of parameter t (time), r is the magnitude of $\mathbf{r}$, and $\mathbf{e}_r$ is its unit vector.

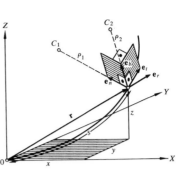

(b) Time derivatives of $\mathbf{r}$ are

$$\dot{\mathbf{r}} = \frac{d\mathbf{r}}{dt} = \dot{x}\mathbf{i} + \dot{y}\mathbf{j} + \dot{z}\mathbf{k} = \dot{r}\mathbf{e}_t$$

$$\ddot{\mathbf{r}} = \frac{d^2\mathbf{r}}{dt^2} = \ddot{x}\mathbf{i} + \ddot{y}\mathbf{j} + \ddot{z}\mathbf{k} = \ddot{r}\mathbf{e}_t + \dot{r}^2\kappa_1\mathbf{e}_n$$

where $\dot{r} = \sqrt{\dot{x}^2 + \dot{y}^2 + \dot{z}^2}$, $\ddot{r} = \sqrt{\ddot{x}^2 + \ddot{y}^2 + \ddot{z}^2}$,
$\mathbf{e}_t = $ *unit tangent vector*, $\mathbf{e}_n = $ *unit normal vector*, and $\kappa_1 = $ *curvature*, all defined below.

(c) Path derivatives of $\mathbf{r}$ are

$$\mathbf{r}' = \frac{d\mathbf{r}}{ds} = x'\mathbf{i} + y'\mathbf{j} + z'\mathbf{k} = r'\mathbf{e}_t = \frac{\dot{\mathbf{r}}}{\dot{r}} = \mathbf{e}_t$$

$$\mathbf{r}'' = \frac{d^2\mathbf{r}}{ds^2} = x''\mathbf{i} + y''\mathbf{j} + z''\mathbf{k} = r''\mathbf{e}_n = \frac{\dot{r}\ddot{\mathbf{r}} - \ddot{r}\dot{\mathbf{r}}}{\dot{r}^3} = \kappa_1\mathbf{e}_n$$

where $r' = \sqrt{x'^2 + y'^2 + z'^2}$ and $r'' = \sqrt{x''^2 + y''^2 + z''^2}$.

(2) Unit Vectors, Curvature, and Torsion

(a) Moving trihedral is formed at each point of the curve by three orthogonal unit vectors, $\mathbf{e}_t$, $\mathbf{e}_n$, and $\mathbf{e}_b = $ *unit binormal vector*, so that

$$\mathbf{e}_t = \mathbf{e}_n \times \mathbf{e}_b \qquad \mathbf{e}_n = \mathbf{e}_b \times \mathbf{e}_t \qquad \mathbf{e}_b = \mathbf{e}_t \times \mathbf{e}_n$$

(b) Transformation relations of $\mathbf{e}_t$, $\mathbf{e}_n$, $\mathbf{e}_b$ and their path derivatives $\mathbf{e}_t'$, $\mathbf{e}_n'$, $\mathbf{e}_b'$ to $\mathbf{i}$, $\mathbf{j}$, $\mathbf{k}$ are

$$\begin{bmatrix} \mathbf{e}_t \\ \mathbf{e}_n \\ \mathbf{e}_b \end{bmatrix} = \begin{bmatrix} x' & y' & z' \\ \dfrac{x''}{r''} & \dfrac{y''}{r''} & \dfrac{z''}{r''} \\ \dfrac{y'z'' - y''z'}{r''} & \dfrac{z'x'' - z''x'}{r''} & \dfrac{x'y'' - x''y'}{r''} \end{bmatrix} \begin{bmatrix} \mathbf{i} \\ \mathbf{j} \\ \mathbf{k} \end{bmatrix} \qquad \begin{bmatrix} \mathbf{e}_t' \\ \mathbf{e}_n' \\ \mathbf{e}_b' \end{bmatrix} = \begin{bmatrix} 0 & \kappa_1 & 0 \\ -\kappa_1 & 0 & \kappa_2 \\ 0 & -\kappa_2 & 0 \end{bmatrix} \begin{bmatrix} \mathbf{i} \\ \mathbf{j} \\ \mathbf{k} \end{bmatrix}$$

where $\mathbf{e}_t' = d\mathbf{e}_t/ds$, $\mathbf{e}_n' = d\mathbf{e}_n/ds$, $\mathbf{e}_b' = d\mathbf{e}_b/ds$.

(c) Curvature and torsion of the curve at the point given by $\mathbf{r}$ are, respectively,

$$\kappa_1 = r'' = \sqrt{(x'')^2 + (y'')^2 + (z'')^2} = 1/\rho_1$$

$$\kappa_2 = \frac{[\mathbf{r}' \cdot \mathbf{r}'' \times \mathbf{r}]}{(r')^2} = \frac{1}{\rho_2}$$

where $\rho_1 = $ *radius of curvature* and $\rho_2 = $ *radius of torsion*.

(1) Basic Equations

(a) Coordinates

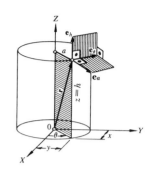

$$a = \sqrt{x^2 + y^2} \qquad\qquad x = a\cos\theta \qquad\qquad \dot{x} = \frac{dx}{dt}$$

$$\theta = \tan^{-1}\frac{y}{x} \qquad\qquad y = a\sin\theta \qquad\qquad \dot{y} = \frac{dy}{dt}$$

$$h = z \qquad\qquad z = h \qquad\qquad \dot{z} = \frac{dz}{dt}$$

where $a = a(t)$, $\theta = \theta(t)$, $h = h(t)$ are the time-dependent *cylindrical coordinates* and $x = x(t)$, $y = y(t)$, $z = z(t)$ are their time-dependent cartesian counterparts.

(b) Time derivatives

$$\dot{a} = \frac{da}{dt} = \frac{x\dot{x} + y\dot{y}}{a} \qquad \dot{\theta} = \frac{d\theta}{dt} = \frac{x\dot{y} - y\dot{x}}{a^2} \qquad \dot{h} = \frac{dh}{dt} = \dot{z}$$

(c) Position vector and its time derivatives

$$\mathbf{r} = a\mathbf{e}_a + h\mathbf{e}_h$$

$$\dot{\mathbf{r}} = \frac{d\mathbf{r}}{dt} = \dot{a}\mathbf{e}_a + a\dot{\theta}\mathbf{e}_\theta + \dot{h}\mathbf{e}_h$$

$$\ddot{\mathbf{r}} = \frac{d^2\mathbf{r}}{dt^2} = (\ddot{a} - a\dot{\theta}^2)\mathbf{e}_a + (a\ddot{\theta} + 2\dot{a}\dot{\theta})\mathbf{e}_\theta + \ddot{h}\mathbf{e}_h$$

where $\dot{a}, \dot{\theta}, \dot{h}$ are derived in (b), $\ddot{a} = d^2a/dt^2$, $\ddot{\theta} = d^2\theta/dt^2$, $\ddot{h} = d^2h/dt^2$, and $\mathbf{e}_a, \mathbf{e}_\theta, \mathbf{e}_h$ are the respective unit vectors.

(2) Matrix Transformations

$$C_\theta = \cos\theta \qquad\qquad S_\theta = \sin\theta \qquad\qquad \dot{C}_\theta = -\dot{\theta}\sin\theta \qquad\qquad \dot{S}_\theta = \dot{\theta}\cos\theta$$

(a) Time derivatives of coordinates

$$\begin{bmatrix} \dot{a} \\ a\dot{\theta} \\ \dot{h} \end{bmatrix} = \begin{bmatrix} C_\theta & S_\theta & 0 \\ -S_\theta & C_\theta & 0 \\ 0 & 0 & 1 \end{bmatrix} \begin{bmatrix} \dot{x} \\ \dot{y} \\ \dot{z} \end{bmatrix} \qquad\qquad \begin{bmatrix} \dot{x} \\ \dot{y} \\ \dot{z} \end{bmatrix} = \begin{bmatrix} C_\theta & -S_\theta & 0 \\ S_\theta & C_\theta & 0 \\ 0 & 0 & 1 \end{bmatrix} \begin{bmatrix} \dot{a} \\ a\dot{\theta} \\ \dot{h} \end{bmatrix}$$

(b) Transformations of unit vectors

$$\begin{bmatrix} \mathbf{e}_a \\ \mathbf{e}_\theta \\ \mathbf{e}_h \end{bmatrix} = \begin{bmatrix} C_\theta & S_\theta & 0 \\ -S_\theta & C_\theta & 0 \\ 0 & 0 & 1 \end{bmatrix} \begin{bmatrix} \mathbf{i} \\ \mathbf{j} \\ \mathbf{k} \end{bmatrix} \qquad\qquad \begin{bmatrix} \mathbf{i} \\ \mathbf{j} \\ \mathbf{k} \end{bmatrix} = \begin{bmatrix} C_\theta & -S_\theta & 0 \\ S_\theta & C_\theta & 0 \\ 0 & 0 & 1 \end{bmatrix} \begin{bmatrix} \mathbf{e}_a \\ \mathbf{e}_\theta \\ \mathbf{e}_h \end{bmatrix}$$

(c) Time derivatives of unit vectors

$$\begin{bmatrix} \dot{\mathbf{e}}_a \\ \dot{\mathbf{e}}_\theta \\ \dot{\mathbf{e}}_h \end{bmatrix} = \begin{bmatrix} 0 & \dot{\theta} & 0 \\ -\dot{\theta} & 0 & 0 \\ 0 & 0 & 0 \end{bmatrix} \begin{bmatrix} \mathbf{e}_a \\ \mathbf{e}_\theta \\ \mathbf{e}_h \end{bmatrix} = \begin{bmatrix} \dot{C}_\theta & \dot{S}_\theta & 0 \\ -\dot{S}_\theta & \dot{C}_\theta & 0 \\ 0 & 0 & 0 \end{bmatrix} \begin{bmatrix} \mathbf{i} \\ \mathbf{j} \\ \mathbf{k} \end{bmatrix}$$

(1) Basic Equations

(a) Coordinates

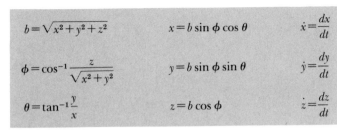

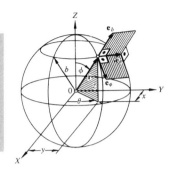

$$b=\sqrt{x^2+y^2+z^2} \qquad x=b\sin\phi\cos\theta \qquad \dot{x}=\frac{dx}{dt}$$

$$\phi=\cos^{-1}\frac{z}{\sqrt{x^2+y^2}} \qquad y=b\sin\phi\sin\theta \qquad \dot{y}=\frac{dy}{dt}$$

$$\theta=\tan^{-1}\frac{y}{x} \qquad z=b\cos\phi \qquad \dot{z}=\frac{dz}{dt}$$

where $b=b(t)$, $\phi=\phi(t)$, $\theta=\theta(t)$ are the time-dependent
spherical coordinates and $x=x(t)$, $y=y(t)$, $z=z(t)$ are their
time-dependent cartesian counterparts.

(b) Time derivatives $(a=\sqrt{x^2+y^2})$

$$\dot{b}=\frac{db}{dt}=\frac{x\dot{x}+y\dot{y}+z\dot{z}}{b} \qquad \dot{\phi}=\frac{d\phi}{dt}=\frac{z(x\dot{x}-y\dot{y})-\dot{z}a^2}{ab^2} \qquad \dot{\theta}=\frac{d\theta}{dt}=\frac{\dot{x}y-y\dot{x}}{a^2}$$

(c) Position vector and its time derivatives

$$\mathbf{r}=b\mathbf{e}_b \qquad\qquad\qquad \dot{\mathbf{r}}=\frac{d\mathbf{r}}{dt}=\dot{b}\mathbf{e}_b+b\dot{\phi}\mathbf{e}_\phi+b\dot{\theta}\sin\phi\,\mathbf{e}_\theta$$

$$\ddot{\mathbf{r}}=\frac{d^2\mathbf{r}}{dt^2}=(\ddot{b}-b\dot{\phi}^2-b\dot{\theta}^2\sin^2\phi)\mathbf{e}_b+(2\dot{b}\dot{\phi}+b\ddot{\phi}-b\dot{\theta}^2\sin\phi\cos\phi)\mathbf{e}_\phi$$
$$+(2\dot{b}\dot{\theta}\sin\phi+b\ddot{\theta}\sin\phi+2b\dot{\theta}\dot{\phi}\cos\phi)\mathbf{e}_\theta$$

where the single and double overdot and $\mathbf{e}_b$, $\mathbf{e}_\phi$, $\mathbf{e}_\theta$ have meanings similar to those in Sec. 10.07.

(2) Matrix Transformations

$$C_\phi=\cos\phi \qquad \dot{C}_\phi=-\dot{\phi}\sin\phi \qquad C_\theta=\cos\theta \qquad \dot{C}_\theta=-\dot{\theta}\sin\theta$$
$$S_\phi=\sin\phi \qquad \dot{S}_\phi=\dot{\phi}\cos\phi \qquad S_\theta=\sin\theta \qquad \dot{S}_\theta=\dot{\theta}\cos\theta$$

(a) Time derivatives of coordinates

$$\begin{bmatrix} \dot{b} \\ b\dot{\phi} \\ b\dot{\theta} \end{bmatrix}=\begin{bmatrix} S_\phi C_\theta & S_\phi S_\theta & C_\phi \\ C_\phi C_\theta & C_\phi S_\theta & -S_\phi \\ -S_\phi S_\theta & S_\phi C_\theta & 0 \end{bmatrix}\begin{bmatrix} \dot{x} \\ \dot{y} \\ \dot{z} \end{bmatrix} \qquad \begin{bmatrix} \dot{x} \\ \dot{y} \\ \dot{z} \end{bmatrix}=\begin{bmatrix} S_\phi C_\theta & C_\phi C_\theta & -S_\phi S_\theta \\ S_\phi S_\theta & C_\phi S_\theta & S_\phi C_\theta \\ C_\phi & -S_\phi & 0 \end{bmatrix}\begin{bmatrix} \dot{b} \\ b\dot{\phi} \\ b\dot{\theta} \end{bmatrix}$$

(b) Transformations of unit vectors

$$\begin{bmatrix} \mathbf{e}_b \\ \mathbf{e}_\phi \\ \mathbf{e}_\theta \end{bmatrix}=\begin{bmatrix} S_\phi C_\theta & S_\phi S_\theta & C_\phi \\ C_\phi C_\theta & C_\phi S_\theta & -S_\phi \\ -S_\theta & C_\theta & 0 \end{bmatrix}\begin{bmatrix} \mathbf{i} \\ \mathbf{j} \\ \mathbf{k} \end{bmatrix} \qquad \begin{bmatrix} \mathbf{i} \\ \mathbf{j} \\ \mathbf{k} \end{bmatrix}=\begin{bmatrix} S_\phi C_\theta & C_\phi C_\theta & -S_\theta \\ S_\phi S_\theta & C_\phi S_\theta & C_\theta \\ C_\phi & -S_\phi & 0 \end{bmatrix}\begin{bmatrix} \mathbf{e}_b \\ \mathbf{e}_\phi \\ \mathbf{e}_\theta \end{bmatrix}$$

(c) Time derivatives of unit vectors $(\lambda_1=\dot{\theta}\cos\phi,\ \lambda_2=\dot{\theta}\sin\phi)$

$$\begin{bmatrix} \dot{\mathbf{e}}_b \\ \dot{\mathbf{e}}_\phi \\ \dot{\mathbf{e}}_\theta \end{bmatrix}=\begin{bmatrix} 0 & \dot{\phi} & \lambda_2 \\ -\dot{\phi} & 0 & \lambda_1 \\ -\lambda_2 & -\lambda_1 & 0 \end{bmatrix}\begin{bmatrix} \mathbf{e}_b \\ \mathbf{e}_\phi \\ \mathbf{e}_\theta \end{bmatrix}=\begin{bmatrix} S_\phi\dot{C}_\theta+\dot{S}_\phi C_\theta & S_\phi\dot{S}_\theta+\dot{S}_\phi S_\theta & \dot{C}_\phi \\ C_\phi\dot{C}_\theta+\dot{C}_\phi C_\theta & C_\phi\dot{S}_\theta+\dot{C}_\phi S_\theta & -\dot{S}_\phi \\ -\dot{S}_\theta & \dot{C}_\theta & 0 \end{bmatrix}\begin{bmatrix} \mathbf{i} \\ \mathbf{j} \\ \mathbf{k} \end{bmatrix}$$

(1) Differential Elements

The *curvilinear coordinates* u_1, u_2, u_3 are said to be *orthogonal* if the coordinate curves are mutually perpendicular at every point. The differential element of length $ds_m = h_m \, du_m$ $(m = 1, 2, 3)$ is defined by the differential element of the respective coordinate and the corresponding *scaling factor*.

(2) Vector Function

The vector function $\mathbf{r} = r_1 \mathbf{e}_1 + r_2 \mathbf{e}_2 + r_3 \mathbf{e}_3$ is determined by *scalar functions* r_m and the respective *unit vector* $\mathbf{e}_m$, tangent to u_m.

$$\mathbf{e}_1 = \frac{1}{h_1} \frac{\partial \mathbf{r}}{\partial u_1} \qquad \mathbf{e}_2 = \frac{1}{h_2} \frac{\partial \mathbf{r}}{\partial u_2} \qquad \mathbf{e}_3 = \frac{1}{h_3} \frac{\partial \mathbf{r}}{\partial u_3}$$

(3) Table of h_m and ds_m

	x	y	z	a	θ	z	b	θ	ϕ
ds_m	dx	dy	dz	da	$a\,d\theta$	dz	db	$b \sin \phi \, d\theta$	$b\,d\phi$
h_m	1	1	1	1	a	1	1	$b \sin \phi$	b

(4) Differential Operators — General Formulas

(a) Gradient of scalar function $f = $ Grad f

$$\nabla f = \frac{1}{h_1} \frac{\partial f}{\partial u_1} \mathbf{e}_1 + \frac{1}{h_2} \frac{\partial f}{\partial u_2} \mathbf{e}_2 + \frac{1}{h_3} \frac{\partial f}{\partial u_3} \mathbf{e}_3$$

(b) Divergence of vector function $\mathbf{V} = $ Div $\mathbf{V}$

$$\nabla \cdot \mathbf{V} = \frac{1}{h_1 h_2 h_3} \left[\frac{\partial}{\partial u_1} (V_1 h_2 h_3) + \frac{\partial}{\partial u_2} (h_1 V_2 h_3) + \frac{\partial}{\partial u_3} (h_1 h_2 V_3) \right]$$

(c) Curl of vector function $\mathbf{V} = $ Curl $\mathbf{V} = $ Rot $\mathbf{V}$

$$\nabla \times \mathbf{V} = \frac{1}{h_1 h_2 h_3} \begin{vmatrix} h_1 \mathbf{e}_1 & \dfrac{\partial}{\partial u_1} & h_1 V_1 \\[2mm] h_2 \mathbf{e}_2 & \dfrac{\partial}{\partial u_2} & h_2 V_2 \\[2mm] h_3 \mathbf{e}_3 & \dfrac{\partial}{\partial u_3} & h_3 V_3 \end{vmatrix}$$

(d) Laplacian of scalar function f

$$\nabla^2 f = \frac{1}{h_1 h_2 h_3} \left[\frac{\partial}{\partial u_1} \left(\frac{h_2 h_3}{h_1} \frac{\partial f}{\partial u_1} \right) + \frac{\partial}{\partial u_2} \left(\frac{h_1 h_3}{h_2} \frac{\partial f}{u_2} \right) + \frac{\partial}{\partial u_3} \left(\frac{h_1 h_2}{h_3} \frac{\partial f}{\partial u_3} \right) \right]$$

(e) General operator's formulas

$$\nabla (f_1 + f_2) = \nabla f_1 + \nabla f_2 \qquad\qquad \nabla \cdot (f \mathbf{V}) = f (\nabla \cdot \mathbf{V}) + (\nabla f) \cdot \mathbf{V}$$

$$\nabla \cdot (\mathbf{V}_1 + \mathbf{V}_2) = \nabla \cdot \mathbf{V}_1 + \nabla \cdot \mathbf{V}_2 \qquad\qquad \nabla \times (f \mathbf{V}) = f (\nabla \times \mathbf{V}) + (\nabla f) \times \mathbf{V}$$

$$\nabla \times (\mathbf{V}_1 + \mathbf{V}_2) = \nabla \times \mathbf{V}_1 + \nabla \times \mathbf{V}_2 \qquad\qquad \nabla \cdot (\nabla f) = \nabla^2 f$$

(1) Cartesian Operators   (Sec. 10.05)   **(2) Cylindrical Operators**   (Sec. 10.07)

(a) Gradient

$$\nabla f = \frac{\partial f}{\partial x}\mathbf{i} + \frac{\partial f}{\partial y}\mathbf{j} + \frac{\partial f}{\partial z}\mathbf{k}$$

(a) Gradient

$$\nabla f = \frac{\partial f}{\partial a}\mathbf{e}_a + \frac{1}{a}\frac{\partial f}{\partial \theta}\mathbf{e}_\theta + \frac{\partial f}{\partial z}\mathbf{e}_z$$

(b) Divergence

$$\nabla \cdot \mathbf{V} = \frac{\partial V_x}{\partial x} + \frac{\partial V_y}{\partial y} + \frac{\partial V_z}{\partial z}$$

(b) Divergence

$$\nabla \cdot \mathbf{V} = \frac{1}{a}\left[\frac{\partial(aV_a)}{\partial a} + \frac{\partial V_\theta}{\partial \theta} + a\frac{\partial V_z}{\partial z}\right]$$

(c) Curl

$$\nabla \times \mathbf{V} = \begin{vmatrix} \mathbf{i} & \dfrac{\partial}{\partial x} & V_x \\[2mm] \mathbf{j} & \dfrac{\partial}{\partial y} & V_y \\[2mm] \mathbf{k} & \dfrac{\partial}{\partial z} & V_z \end{vmatrix}$$

(c) Curl

$$\nabla \times \mathbf{V} = \frac{1}{a}\begin{vmatrix} \mathbf{e}_a & \dfrac{\partial}{\partial a} & V_a \\[2mm] a\mathbf{e}_\theta & \dfrac{\partial}{\partial \theta} & aV_\theta \\[2mm] \mathbf{e}_z & \dfrac{\partial}{\partial z} & V_z \end{vmatrix}$$

(d) Laplacian

$$\nabla^2 f = \frac{\partial^2 f}{\partial x^2} + \frac{\partial^2 f}{\partial y^2} + \frac{\partial^2 f}{\partial z^2}$$

(d) Laplacian

$$\nabla^2 f = \frac{1}{a}\left[\frac{\partial}{\partial a}\left(a\frac{\partial f}{\partial a}\right) + \frac{1}{a}\frac{\partial}{\partial \theta}\left(\frac{\partial f}{\partial \theta}\right) + \frac{\partial}{\partial z}\left(a\frac{\partial f}{\partial z}\right)\right]$$

(3) Spherical Operators   (Sec. 10.08)

(a) Gradient

$$\nabla f = \frac{\partial f}{\partial b}\mathbf{e}_b + \frac{1}{b \sin \phi}\frac{\partial f}{\partial \theta}\mathbf{e}_\theta + \frac{1}{b}\frac{\partial f}{\partial \phi}\mathbf{e}_\phi$$

(b) Divergence

$$\nabla \cdot \mathbf{V} = \frac{1}{b^2 \sin \phi}\left[\frac{\partial}{\partial b}(b^2 \sin \phi V_b) + \frac{\partial}{\partial \theta}(bV_\theta) + \frac{\partial}{\partial \phi}(b \sin \phi V_\phi)\right]$$

(c) Curl

$$\nabla \times \mathbf{V} = \frac{1}{b^2 \sin \phi}\begin{vmatrix} \mathbf{e}_b & \dfrac{\partial}{\partial b} & V_b \\[2mm] b \sin \phi \mathbf{e}_\theta & \dfrac{\partial}{\partial \theta} & b \sin \phi V_\theta \\[2mm] b\mathbf{e}_\phi & \dfrac{\partial}{\partial \phi} & bV_\phi \end{vmatrix}$$

(d) Laplacian

$$\nabla^2 f = \frac{1}{b^2 \sin \phi}\left[\frac{\partial}{\partial b}\left(b^2 \sin \phi \frac{\partial f}{\partial b}\right) + \frac{\partial}{\partial \theta}\left(\frac{1}{\sin \phi}\frac{\partial f}{\partial \theta}\right) + \frac{\partial}{\partial \phi}\left(\sin \phi \frac{\partial f}{\partial \phi}\right)\right]$$

(1) Indefinite Integral

(a) Definition

The *indefinite integral of a vector function* $\mathbf{f}(t)$ of a single scalar variable t is the anti-derivative of that vector function.

$$\mathbf{f}(t) = \dot{\mathbf{F}}(t) = \frac{d\mathbf{F}(t)}{dt}$$

$$\int \mathbf{f}(t)\, dt = \int \dot{\mathbf{F}}(t)\, dt = \mathbf{F}(t) + \mathbf{c}$$

$\mathbf{c}$ = vector constant of integration

(b) General formulas

$\mathbf{E}$ = constant vector

m = constant scalar

t = variable scalar

$\mathbf{F}(t)$ = vector function of t

$\dot{\mathbf{F}}(t)$ = first derivative of $\mathbf{F}(t)$ with respect to t

$\ddot{\mathbf{F}}(t)$ = second derivative of $\mathbf{F}(t)$ with respect to t

$$\int m\dot{\mathbf{F}}(t)\, dt = m\mathbf{F}(t) + \mathbf{c}$$

$$\int \mathbf{E} \cdot \dot{\mathbf{F}}(t)\, dt = \mathbf{E} \cdot \mathbf{F}(t) + c$$

$$\int \mathbf{E} \times \dot{\mathbf{F}}(t)\, dt = \mathbf{E} \times \mathbf{F}(t) + \mathbf{c}$$

$$\int \mathbf{F}(t) \cdot \dot{\mathbf{F}}(t)\, dt = \frac{\mathbf{F}^2(t)}{2} + c$$

$$\int \mathbf{F}(t) \times \ddot{\mathbf{F}}(t)\, dt = \mathbf{F}(t) \times \dot{\mathbf{F}}(t) + \mathbf{c}$$

$$\int \int \ddot{\mathbf{F}}(t)\, dt\, dt = \mathbf{F}(t) + \mathbf{c}_1 t + \mathbf{c}_2$$

(2) Definite Integral

(a) Definition

The *definite integral of a vector function* $\mathbf{f}(t)$ of a single scalar variable t is the limit of a sum between the lower and the upper limit.

$$\int_{t=a}^{t=b} \mathbf{f}(t)\, dt = \int_{t=a}^{t=b} \dot{\mathbf{F}}(t)\, dt = \mathbf{F}(b) - \mathbf{F}(a)$$

a, b = scalar constants

(b) Interchange of limits

$$\int_{t=a}^{t=b} \mathbf{f}(t)\, dt = -\int_{t=b}^{t=a} \mathbf{f}(t)\, dt = \mathbf{F}(b) - \mathbf{F}(a)$$

(c) Decomposition of limits

$$\int_{t=a}^{t=b} \mathbf{f}(t)\, dt = \int_{t=a}^{t=c} \mathbf{f}(t)\, dt + \int_{t=c}^{t=b} \mathbf{f}(t)\, dt$$

$$= \mathbf{F}(b) - \mathbf{F}(a)$$

(d) Sum of several functions

$$\int_{t=a}^{t=b} [\mathbf{f}_1(t) + \mathbf{f}_2(t) + \mathbf{f}_3(t)]\, dt = \int_{t=a}^{t=b} \mathbf{f}_1(t)\, dt + \int_{t=a}^{t=b} \mathbf{f}_2(t)\, dt + \int_{t=a}^{t=b} \mathbf{f}_3(t)\, dt$$

(e) Constant factors $\mathbf{E}$, m

$$\int_{t=a}^{t=b} \mathbf{E} \times \mathbf{f}(t)\, dt = \mathbf{E} \times \int_{t=a}^{t=b} \mathbf{f}(t)\, dt$$

$$\int_{t=a}^{t=b} m\mathbf{f}(t)\, dt = m \int_{t=a}^{t=b} \mathbf{f}(t)\, dt$$

$$\int_{t=a}^{t=b} \mathbf{E} \cdot \mathbf{f}(t)\, dt = \mathbf{E} \cdot \int_{t=a}^{t=b} \mathbf{f}(t)\, dt$$

(1) Line Integral

(a) Definition

The *line integral of a vector* **F** over a path C is defined as the integral of the scalar product of **F** and the elemental path $d\mathbf{r}$ along C. **r** defines a curve which has continuous derivatives, and **F** is continuous along C.

$$\int_a^b \mathbf{F} \cdot d\mathbf{r} = \int_a^b F_x \, dx + \int_a^b F_y \, dy + \int_a^b F_z \, dz$$

(b) Theorems

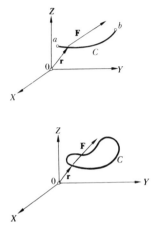

If $\mathbf{F} = \nabla\phi = \dfrac{\partial\phi}{\partial x}\mathbf{i} + \dfrac{\partial\phi}{\partial y}\mathbf{j} + \dfrac{\partial\phi}{\partial z}\mathbf{k}$ and C is *an open curve*, then

$$\int_a^b \nabla\phi \cdot d\mathbf{r} \text{ is } \textit{independent of the path} \text{ between } a \text{ and } b.$$

If $\mathbf{F} = \nabla\phi = \dfrac{\partial\phi}{\partial x}\mathbf{i} + \dfrac{\partial\phi}{\partial y}\mathbf{j} + \dfrac{\partial\phi}{\partial z}\mathbf{k}$ and C is *a closed curve*, then

$$\oint \nabla\phi \cdot d\mathbf{r} \text{ is } \textit{equal to zero.}$$

(2) Surface Integral

(a) A normal surface vector is a vector whose length is equal to the area bounded by a closed curve C and whose direction is perpendicular to the plane of the area.

$$dS = ds\,\mathbf{n}$$

(b) Flow of a scalar field

$$P = \int\int_S \phi \, dS = \mathbf{i} \int\int_{\Sigma yz} \phi \, dy \, dz + \mathbf{j} \int\int_{\Sigma xz} \phi \, dx \, dz + \mathbf{k} \int\int_{\Sigma xy} \phi \, dx \, dy$$

(c) Scalar flow of a vector field

$$Q = \int\int_S \mathbf{F} \cdot dS = \int\int_{\Sigma yz} F_x \, dy \, dz + \int\int_{\Sigma xz} F_y \, dx \, dz + \int\int_{\Sigma xy} F_z \, dx \, dy$$

(d) Vector flow of a vector field

$$\mathbf{R} = \int\int_S \mathbf{F} \times dS = \int\int_{\Sigma yz} (F_z\mathbf{j} - F_y\mathbf{k}) \, dy \, dz + \int\int_{\Sigma xz} (F_x\mathbf{k} - F_z\mathbf{i}) \, dx \, dz + \int\int_{\Sigma xy} (F_y\mathbf{i} - F_x\mathbf{j}) \, dx \, dy$$

(e) Closed surface integral notation

$$P = \oint_S \phi \, dS \qquad Q = \oint_S \mathbf{F} \cdot dS \qquad R = \oint_S \mathbf{F} \times dS$$

Note: In integrals **P**, **Q**, **R** each integral is taken over projection of the surface on the respective coordinate plane.

(1) Gauss' Theorem

A *scalar flow of a vector function* **F** through a closed surface S is equal to the integral of $\nabla \cdot \mathbf{F}$ over the volume V bounded by S.

$$\underbrace{\oint_S \mathbf{F} \cdot d\mathbf{S}}_{\substack{\text{Closed surface} \\ \text{integral}}} = \underbrace{\int_V \nabla \cdot \mathbf{F} \, dV}_{\substack{\text{Closed volume} \\ \text{integral}}}$$

In cartesian coordinates

$$\oint_S \mathbf{F} \cdot d\mathbf{S} = \underset{\Sigma yz}{\int\int} F_x \, dy \, dz + \underset{\Sigma zx}{\int\int} F_y \, dz \, dx + \underset{\Sigma xy}{\int\int} F_z \, dx \, dy = \underset{\Sigma xyz}{\int\int\int} \left(\frac{\partial F_x}{\partial x} + \frac{\partial F_y}{\partial y} + \frac{\partial F_z}{\partial z} \right) dx \, dy \, dz$$

Thus a *closed volume integral* can be reduced to a *closed surface integral*.

(2) Stokes' Theorem

A *circulation of a vector function* **F** about a closed path C is equal to a vector flow of the same vector function over an arbitrary surface bounded by C.

$$\underbrace{\oint_C \mathbf{F} \cdot d\mathbf{r}}_{\substack{\text{Closed path} \\ \text{integral}}} = \underbrace{\int_S (\nabla \times \mathbf{F}) \cdot d\mathbf{S}}_{\substack{\text{Surface} \\ \text{integral}}}$$

In *cartesian coordinates*

$$\oint_C \mathbf{F} \cdot d\mathbf{r} = \int_C (F_x \, dx + F_y \, dy + F_z \, dz)$$

$$= \underset{\Sigma yz}{\int\int} \left(\frac{\partial F_z}{\partial y} - \frac{\partial F_y}{\partial z} \right) dy \, dz + \underset{\Sigma zx}{\int\int} \left(\frac{\partial F_x}{\partial z} - \frac{\partial F_z}{\partial x} \right) dz \, dx + \underset{\Sigma xy}{\int\int} \left(\frac{\partial F_y}{\partial x} - \frac{\partial F_x}{\partial y} \right) dx \, dy$$

Thus a *surface integral* can be reduced to a *line integral*.

(3) Green's Theorem

If in Gauss' theorem $\mathbf{F} = \alpha \nabla \beta$, where α, β are scalar functions of x, y, z, then

$$\underbrace{\oint_S (\alpha \nabla \beta) \cdot d\mathbf{S}}_{\substack{\text{Closed surface} \\ \text{integral}}} = \underbrace{\int_V \nabla \cdot (\alpha \nabla \beta) \, dV}_{\substack{\text{Closed volume} \\ \text{integral}}} = \underbrace{\int_V (\alpha \nabla^2 \beta) + \nabla \alpha \cdot \nabla \beta) \, dV}_{\substack{\text{From operator's} \\ \text{formula}}}$$

If $\alpha = +1$,

$$\oint_S \nabla \beta \cdot d\mathbf{S} = \int_V \nabla^2 \beta \, dV$$

and in *cartesian coordinates*

$$\underset{\Sigma yz}{\int\int} \frac{\partial \beta}{\partial x} \, dy \, dz + \underset{\Sigma zx}{\int\int} \frac{\partial \beta}{\partial y} \, dz \, dx + \underset{\Sigma xy}{\int\int} \frac{\partial \beta}{\partial z} \, dx \, dy = \underset{\Sigma xyz}{\int\int\int} \left(\frac{\partial^2 \beta}{\partial x^2} + \frac{\partial^2 \beta}{\partial y^2} + \frac{\partial^2 \beta}{\partial z^2} \right) dx \, dy \, dz$$

11
FUNCTIONS OF A COMPLEX VARIABLE

$$p = a + bi \qquad r = \sqrt{a^2 + b^2} \qquad \phi = \tan^{-1}\frac{b}{a} \qquad a, b, c, d = \text{real numbers}$$

(1) Algebraic Forms

(a) **Imaginary number.** The second root of a negative number,

$$\sqrt{-b^2} = b\sqrt{-1} = bi$$

is called the *imaginary number.* The basis of imaginary numbers is the *imaginary unit i.*

(b) **Complex number** consists of a real part and an imaginary part,

$$p = a + bi \qquad q = a - bi$$

where q is the *conjugate* of p.

$$\begin{aligned}
\sqrt{-1} &= i & i^{4k+1} &= i \\
i^2 &= -1 & i^{4k+2} &= -1 \\
i^3 &= -i & i^{4k+3} &= -i \\
i^4 &= 1 & i^{4k+4} &= 1 \\
k &= 0, 1, 2, \ldots
\end{aligned}$$

(c) **Basic operations**

$$p = a + bi \qquad q = a - bi \qquad\qquad p = a + bi \qquad q = c + di$$

$$p + q = 2a \qquad p - q = 2bi \qquad\qquad p \pm q = (a \pm c) + (b \pm d)i$$

$$pq = a^2 + b^2 \qquad\qquad\qquad pq = (ac - bd) + (ad + bc)i$$

$$\frac{p}{q} = \frac{a^2 + 2abi - b^2}{a^2 + b^2} \qquad\qquad \frac{p}{q} = \frac{(ac + bd) + (bc - ad)i}{c^2 + d^2}$$

(d) **Complex surds**

$$\sqrt{\pm bi} = \sqrt{\frac{b}{2}} \pm i\sqrt{\frac{b}{2}} = \frac{\sqrt{2b}}{2}(1 \pm i) \qquad \sqrt{a + bi} \pm \sqrt{a - bi} = \sqrt{2(a \pm \sqrt{a^2 + b^2})}$$

$$\sqrt{\pm i} = \sqrt{\frac{1}{2}} \pm i\sqrt{\frac{1}{2}} = \frac{\sqrt{2}}{2}(1 \pm i) \qquad \sqrt{a \pm bi} = \sqrt{\frac{\sqrt{a^2 + b^2} + a}{2}} \pm i\sqrt{\frac{\sqrt{a^2 + b^2} - a}{2}}$$

(2) Transcendent Forms

(a) **Complex number** $p = a + bi$ can be represented as a point in the *complex plane* (Argand or Gauss plane).

$$p = a + bi = re^{i\phi} = r(\cos\phi + i\sin\phi)$$

where $e = 2.718\,281\,828 \ldots$ (Sec. 1.04).

(b) **Basic operations** $(m, n = 1, 2, 3, \ldots)(k = 0, 1, 2, \ldots, n-1)$

$$p_1 = a_1 + b_1 i = r_1(\cos\phi_1 + i\sin\phi_1) \qquad\qquad p_2 = a_2 + b_2 i = r_2(\cos\phi_2 + i\sin\phi_2)$$

$$p_1 p_2 = r_1 r_2[\cos(\phi_1 + \phi_2) + i\sin(\phi_1 + \phi_2)] \qquad p_1 : p_2 = (r_1 : r_2)[\cos(\phi_1 - \phi_2) + i\sin(\phi_1 - \phi_2)]$$

$$p^{m/n} = r^{m/n}(\cos\phi + i\sin\phi)^{m/n} = r^{m/n}\left[\cos\frac{m}{n}(\phi + 2k\pi) + i\sin\frac{m}{n}(\phi + 2k\pi)\right]$$

$$\sqrt[n]{1} = \cos\frac{2k\pi}{n} + i\sin\frac{2k\pi}{n} \qquad\qquad \sqrt[n]{-1} = \cos\frac{(2k+1)\pi}{n} + i\sin\frac{(2k+1)\pi}{n}$$

$$z = x + iy \qquad r = \sqrt{x^2 + y^2} \qquad \phi = \tan^{-1}\frac{y}{x} \qquad x, y = \text{real independent variables}$$

(1) Function of Complex Variable

(a) **Complex variable.** If x and y are two independent real variables, then

$$z = x + iy \qquad w = f(z)$$

are called the *complex variable z* and the *function of complex variable w*, respectively.

(b) **Function of complex variable** is said to be *analytic* (regular, holomorphic) at a point z_0, if its derivative with respect to z exists at that point and at every point in some neighborhood of that point.

(c) **Rules of operations** with analytic functions are formally identical with those of functions of real variables, including their representation by power series.

(2) Exponential Functions

(a) **Basic relations** $(m, n = 1, 2, 3, \ldots)\ (k = 0, \pm1, \pm2, \ldots)$

$$z^{m/n} = r^{m/n}\left(\cos\frac{m\phi}{n} + i\sin\frac{m\phi}{n}\right) = r^{m/n}e^{im\phi/n}$$

$$e^{ix} = 1 + \frac{ix}{1!} + \frac{(ix)^2}{2!} + \frac{(ix)^3}{3!} + \cdots + \frac{(ix)^n}{n!} + \cdots = \cos x + i \sin x$$

$$e^{-ix} = 1 - \frac{ix}{1!} + \frac{(ix)^2}{2!} - \frac{(ix)^3}{3!} + \cdots \pm \frac{(ix)^n}{n!} + \cdots = \cos x - i \sin x$$

$$e^{2k\pi i} = 1$$
$$e^{(2k+1)\pi i} = -1$$
$$e^{\phi + 2k\pi i} = e^{\phi}$$
$$e^{\phi + (2k+1)\pi i} = -e^{\phi}$$
$$i^i = (0.20788\ldots)e^{2k\pi}$$

(b) **Derivative and integral**

$$\frac{d(e^{(a+bi)x})}{dx} = (a + bi)e^{(a+bi)x} = (a + bi)e^{ax}(\cos bx + i \sin bx)$$

$$\int e^{(a+bi)x}\,dx = \frac{e^{(a+bi)x}}{a + bi} = \frac{e^{ax}}{a^2 + b^2}[(a \cos bx + b \sin bx) + i(a \sin bx - b \cos bx)]$$

(3) Logarithmic Functions

(a) **Basic relations** $(k = 0, \pm1, \pm2, \ldots)$

$$\ln z = \ln re^{i(\phi + 2k\pi)} = \ln r + i(\phi + 2k\pi)$$

$$\ln a = \ln a + 2k\pi i \qquad\qquad \ln bi = \ln b + i\left(\frac{\pi}{2} + 2k\pi\right)$$

$$\ln (-a) = \ln a + i(2k+1)\pi \qquad \ln (-bi) = \ln b + i\left(\frac{3\pi}{2} + 2k\pi\right)$$

$$\ln 1 = 2k\pi i$$
$$\ln (-1) = (2k+1)\pi i$$
$$\ln i = (4k+1)\frac{\pi i}{2}$$
$$\ln (-i) = (4k+3)\frac{\pi i}{2}$$

(b) **Derivative and Integral**

$$\frac{d[\ln (a + bi)x]}{dx} = \frac{d[\ln (a + bi)]}{dx} + \frac{d(\ln x)}{dx} = \frac{1}{x}$$

$$\int [\ln (a + bi)x]\,dx = x[\ln (a + bi)x - 1] = x\left(\ln \sqrt{a^2 + b^2} - 1 + i \tan^{-1}\frac{b}{a} + \ln x\right)$$

$$z = x + iy \qquad\qquad a, b = \text{real numbers} \qquad\qquad k = 0, 1, 2, \ldots \qquad n = 1, 2, 3, \ldots$$

$$p = a + bi \qquad\qquad A = \tfrac{1}{2}\sqrt{(1+a)^2 + b^2} \qquad\qquad B = \tfrac{1}{2}\sqrt{(1-a)^2 + b^2}$$

(1) Trigonometric Functions

(a) Basic relations

$$\sin ix = \sin (ix \pm 2k\pi) = i \sinh x \qquad\qquad \sin x = \sin (x \pm 2k\pi) = -i \sinh ix$$

$$\cos ix = \cos (ix \pm 2k\pi) = \cosh x \qquad\qquad \cos x = \cos (x \pm 2k\pi) = \cosh ix$$

$$\tan ix = \tan (ix \pm k\pi) = i \tanh x \qquad\qquad \tan x = \tan (x \pm k\pi) = -i \tanh ix$$

$$\cot ix = \cot (ix \pm k\pi) = -i \coth x \qquad\qquad \cot x = \cot (x \pm k\pi) = i \coth ix$$

$$\sec ix = \sec (ix \pm 2k\pi) = \operatorname{sech} x \qquad\qquad \sec x = \sec (x \pm 2k\pi) = \operatorname{sech} ix$$

$$\csc ix = \csc (ix \pm 2k\pi) = -i \operatorname{csch} x \qquad\qquad \csc x = \csc (x \pm 2k\pi) = i \operatorname{csch} ix$$

(b) Complex variable argument

$$\sin z = \sin (z \pm 2k\pi) = -i \sinh iz = \sin x \cosh y + i \cos x \sinh y$$

$$\cos z = \cos (z \pm 2k\pi) = \cosh iz = \cos x \cosh y - i \sin x \sinh y$$

$$\tan z = \tan (z \pm k\pi) = -i \tanh iz = \frac{\sin 2x + i \sinh 2y}{\cos 2x + \cosh 2y}$$

(c) Identities $\qquad (k = 0, 1, 2, \ldots, n-1)$

$$\sin x = \frac{e^{ix} - e^{-ix}}{2i} \qquad\qquad (\cos x \pm i \sin x)^n = \cos nx \pm i \sin nx$$

$$\cos x = \frac{e^{ix} + e^{-ix}}{2} \qquad\qquad \sqrt[n]{\cos x \pm i \sin x} = \cos \frac{x + 2k\pi}{n} \pm i \sin \frac{x + 2k\pi}{n}$$

(2) Inverse Trigonometric Functions

(a) Basic relations

$$\sin^{-1} ix = i \sinh^{-1} x \qquad \cos^{-1} x = \pm i \cosh^{-1} x \qquad \tan^{-1} ix = i \tanh^{-1} x$$

$$\csc^{-1} ix = -i \cosh^{-1} x \qquad \sec^{-1} x = \pm i \operatorname{sech}^{-1} x \qquad \cot^{-1} ix = -i \coth^{-1} x$$

(b) Complex variable argument

$$\sin^{-1} z = -\cos^{-1} z + \frac{\pi}{2}(1 \pm 4k) = -i \sinh^{-1} iz = -i \ln (iz \pm \sqrt{1 - z^2}) \pm 2k\pi$$

$$\cos^{-1} z = -\sin^{-1} z + \frac{\pi}{2}(1 \pm 4k) = -i \cosh^{-1} z = -i \ln (iz \pm \sqrt{z^2 - 1}) \pm 2k\pi$$

$$\tan^{-1} z = -\cot^{-1} z + \frac{\pi}{2}(1 \pm 2k) = -i \tanh^{-1} iz = -i \ln \sqrt{\frac{1 + iz}{1 - iz}} \pm k\pi$$

(c) Complex number argument[1]

$$\sin^{-1} p = (-1)^k [\sin^{-1} (A - B) + i\theta \cosh^{-1} (A + B)] \pm k\pi$$

$$\cos^{-1} p = \pm [\cos^{-1} (A - B) - i\theta \cosh^{-1} (A + B) \pm 2k\pi]$$

$$\tan^{-1} p = \frac{i}{2}\left[(1 \pm 2k)\pi - \tan^{-1} \frac{1 + b}{a} - \tan^{-1} \frac{1 - b}{a} \right] + i\theta \ln \sqrt{\frac{C}{D}}$$

$$\theta = \begin{cases} +1 & \text{if } b \geq 0 \\[2mm] -1 & \text{if } b < 0 \end{cases}$$

[1]For C, D see opposite page.

$z = x + iy$	$a, b =$ real numbers	$k = 0, 1, 2, \ldots$	$n = 1, 2, 3, \ldots$
$p = a + bi$	$C = \frac{1}{2}\sqrt{(1+b)^2 + a^2}$	$D = \frac{1}{2}\sqrt{(1-b)^2 + a^2}$	

(1) Hyperbolic Functions

(a) Basic relations

$$\sinh ix = \sinh (ix \pm 2k\pi i) = i \sin x \qquad \sinh x = \sinh (x \pm 2k\pi i) = -i \sin ix$$

$$\cosh ix = \cosh (ix \pm 2k\pi i) = \cos x \qquad \cosh x = \cosh (x \pm 2k\pi i) = \cos ix$$

$$\tanh ix = \tanh (ix \pm k\pi i) = i \tan x \qquad \tanh x = \tanh (x \pm k\pi i) = -i \tan ix$$

$$\coth ix = \coth (ix \pm k\pi i) = -i \cot x \qquad \coth x = \coth (x \pm k\pi i) = i \cot ix$$

$$\operatorname{sech} ix = \operatorname{sech} (ix \pm 2k\pi i) = \sec x \qquad \operatorname{sech} x = \operatorname{sech} (x \pm 2k\pi i) = \sec ix$$

$$\operatorname{csch} ix = \operatorname{csch} (ix \pm 2k\pi i) = -i \csc x \qquad \operatorname{csch} x = \operatorname{csch} (x \pm 2k\pi i) = i \csc ix$$

(b) Complex variable argument

$$\sinh z = \sinh (z \pm 2k\pi i) = -i \sin iz = \sinh x \cos y + i \cosh x \sin y$$

$$\cosh z = \cosh (z \pm 2k\pi i) = \cos iz = \cosh x \cos y + i \sinh x \sin y$$

$$\tanh z = \tanh (z \pm k\pi i) = -i \tan iz = \frac{\sinh 2x + i \sin 2y}{\cosh 2x + \cos 2y}$$

(c) Identities $(k = 0, 1, 2, \ldots, n-1)$

$$\sinh ix = \frac{e^{ix} - e^{-ix}}{2} \qquad\qquad (\cosh x \pm \sinh x)^n = \cosh nx \pm \sinh nx$$

$$\cosh ix = \frac{e^{ix} + e^{-ix}}{2} \qquad\qquad \sqrt[n]{\cosh x \pm \sinh x} = \cosh \frac{x + 2k\pi i}{n} \pm \sinh \frac{x + 2k\pi i}{n}$$

(2) Inverse Hyperbolic Functions

(a) Basic relations

$\sinh^{-1} ix = i \sin^{-1} x$	$\cosh^{-1} x = \pm i \cos^{-1} x$	$\tanh^{-1} ix = i \tan^{-1} x$
$\operatorname{csch}^{-1} ix = -i \csc^{-1} x$	$\operatorname{sech}^{-1} x = \pm i \sec^{-1} x$	$\coth^{-1} ix = -i \cot^{-1} x$

(b) Complex variable argument

$$\sinh^{-1} z = -i \sin^{-1} iz = \ln (\sqrt{z^2 + 1} + z) \pm 2k\pi i = -\ln (\sqrt{x^2 + 1} - z) \pm 2k\pi i$$

$$\cosh^{-1} z = i \cos^{-1} z = \ln (z + \sqrt{z^2 - 1}) \pm 2k\pi i = -\ln (z - \sqrt{z^2 - 1}) \pm 2k\pi i$$

$$\tanh^{-1} z = -i \tan^{-1} iz = \ln \sqrt{\frac{1+z}{1-z}} \pm k\pi i$$

(c) Complex number argument[1]

$$\sinh^{-1} p = (-1)^k [\cosh^{-1} (C+D) + i\theta \sin^{-1} (C-D)] + k\pi i$$

$$\cosh^{-1} p = \pm [\cosh^{-1} (A+B) + i\theta \cos^{-1} (A-B) + k\pi i]$$

$$\tanh^{-1} p = \frac{i}{2}\left[(1 \pm 2k) \pi - \tan^{-1}\frac{1+a}{b} - \tan^{-1}\frac{1-a}{b} \right] + \ln \sqrt{\frac{A}{B}}$$

$$\theta = \begin{cases} +1 & \text{if } b \geq 0 \\ -1 & \text{if } b < 0 \end{cases}$$

[1]For A, B see opposite page.

$[A], [B]$ = real matrices of order $m \times m$ $[I]$ = unit matrix of order $m \times m$

$[\ ]^T$ = transpose (Sec. 1.10) $[\ ]^{-1}$ = inverse (Sec. 1.11)

(a) Complex matrix $[M]$, its conjugate $[\overline{M}]$, and its associate $[M*]$ are, respectively,

$$[M] = [A] + i[B] = [A + iB] \qquad [\overline{M}] = [A] - i[B] = [A - iB]$$
$$[M*] = [\overline{M}]^T = [A]^T - i[B]^T = [A - iB]^T$$

(b) Hermitian matrix is the complex generalization of the *normal matrix* (Sec. 1.11).

If $[M] = [M*]$ then $[M][M*] = [M*][M] = [M]^2$

(c) Unitary matrix is the complex generalization of the *orthogonal matrix* (Sec. 1.11).

If $[M*] = [M]^{-1}$ then $[M][M*] = [M*][M] = [I]$

(d) Involutory matrix is the complex generalization of the *orthonormal matrix* (Sec. 1.11).

If $[M] = [M*] = [M]^{-1}$ then $[M][M] = [I]$

(2) Operations

(a) Summation and resolution

$$[M] + [\overline{M}] = 2[A] \qquad [M] - [\overline{M}] = 2i[B] \qquad [M] = \tfrac{1}{2}[M + M*] + \tfrac{1}{2}[M - M*]$$

(b) Product of two hermitian matrices is a hermitian matrix, product of two unitary matrices is a unitary matrix, and product of two involutory matrices is an involutory matrix.

(3) Classification

Real matrices $B = 0$			Complex matrices $B \neq 0$		
Normal	$A = A^T$		Hermitian	$M = M*$	
Antinormal	$A = -A^T$		Antihermitian	$M = -M*$	
Orthogonal		$A^T = A^{-1}$	Unitary		$M* = M^{-1}$
Antiorthogonal		$A^T = -A^{-1}$	Antiunitary		$M* = -M^{-1}$
Orthonormal	$A = A^T = A^{-1}$		Involutory	$M = M* = M^{-1}$	
Antiorthonormal	$A = -A^T = -A^{-1}$		Anti-involutory	$M = -M* = -M^{-1}$	

12

FOURIER SERIES

(1) Basic Case

(a) Definition

Any *single-valued function* $f(\theta)$ that is *continuous* except for a *finite number of discontinuities* in an interval $-\pi < \theta < +\pi$, and has a finite number of maxima and minima in this interval may be represented by a *convergent Fourier series*.

$$f(\theta) = \frac{a_0}{2} + a_1 \cos\theta + a_2 \cos 2\theta + a_3 \cos 3\theta + \cdots + b_1 \sin\theta + b_2 \sin 2\theta + b_3 \sin 3\theta + \cdots$$

$$= \frac{a_0}{2} + \sum_{n=1}^{\infty} (a_n \cos n\theta + b_n \sin n\theta)$$

If $f(\theta)$ is a *periodic function* of θ with *period* 2π,

$$a_n = \frac{1}{\pi} \int_{-\pi}^{+\pi} f(\theta) \cos n\theta \, d\theta \qquad n = 0, 1, 2, \ldots \qquad b_n = \frac{1}{\pi} \int_{-\pi}^{+\pi} f(\theta) \sin n\theta \, d\theta \qquad n = 1, 2, \ldots$$

(b) Phase angles α and β

The cosine and sine terms in the Fourier series may be combined in a single cosine or sine series with phase angles α or β, respectively.

$$f(\theta) = \frac{A_0}{2} + \sum_{n=1}^{\infty} A_n \cos(n\theta + \alpha_n)$$

$$A_n = \sqrt{a_n^2 + b_n^2}$$

$$\alpha_n = \tan^{-1}\left(-\frac{b_n}{a_n}\right)$$

$$f(\theta) = \frac{B_0}{2} + \sum_{n=1}^{\infty} B_n \sin(n\theta + \beta_n)$$

$$B_n = \sqrt{a_n^2 + b_n^2}$$

$$\beta_n = \tan^{-1}\frac{a_n}{b_n}$$

(2) Special Cases

(a) Change in variable $\quad \left(\theta = \frac{\pi x}{l} \text{ and } -l < x < +l\right)$

$$f(x) = \frac{\bar{a}_0}{2} + \sum_{n=1}^{\infty}\left(\bar{a}_n \cos\frac{n\pi x}{l} + \bar{b}_n \sin\frac{n\pi x}{l}\right)$$

$$\bar{a}_n = \frac{1}{l} \int_{-l}^{+l} f(x) \cos\frac{n\pi x}{l} \, dx$$

$$\bar{b}_n = \frac{1}{l} \int_{-l}^{+l} f(x) \sin\frac{n\pi x}{l} \, dx$$

(b) Change in variable $\quad \left(\theta = \frac{2\pi t}{T} \text{ and } -\frac{T}{2} < t < +\frac{T}{2}\right)$

$$f(t) = \frac{a_0^*}{2} + \sum_{n=1}^{\infty}\left(a_n^* \cos\frac{2n\pi t}{T} + b_n^* \sin\frac{2n\pi t}{T}\right)$$

$$a_n^* = \frac{2}{T} \int_{-T/2}^{+T/2} f(t) \cos\frac{2n\pi t}{T} \, dt$$

$$b_n^* = \frac{2}{T} \int_{-T/2}^{+T/2} f(t) \sin\frac{2n\pi t}{T} \, dt$$

(1) Change in Limits

In the development of Fourier series the *limits of integral may be changed* (shifting of interval) as shown.

	$-2\pi < \theta < 0$	$\phi < \theta < \phi + 2\pi$	$0 < \theta < 2\pi$
a_n	$\dfrac{1}{\pi}\displaystyle\int_{-2\pi}^{0} f(\theta)\cos n\theta\, d\theta$	$\dfrac{1}{\pi}\displaystyle\int_{\phi}^{\phi+2\pi} f(\theta)\cos n\theta\, d\theta$	$\dfrac{1}{\pi}\displaystyle\int_{0}^{2\pi} f(\theta)\cos n\theta\, d\theta$
b_n	$\dfrac{1}{\pi}\displaystyle\int_{-2\pi}^{0} f(\theta)\sin n\theta\, d\theta$	$\dfrac{1}{\pi}\displaystyle\int_{\phi}^{\phi+2\pi} f(\theta)\sin n\theta\, d\theta$	$\dfrac{1}{\pi}\displaystyle\int_{0}^{2\pi} f(\theta)\sin n\theta\, d\theta$
	$-2l < x < 0$	$a < x < a+2l$	$0 < x < 2l$
$\bar{a}_n$	$\dfrac{1}{l}\displaystyle\int_{-2l}^{0} f(x)\cos\dfrac{n\pi x}{l}\, dx$	$\dfrac{1}{l}\displaystyle\int_{a}^{a+2l} f(x)\cos\dfrac{n\pi x}{l}\, dx$	$\dfrac{1}{l}\displaystyle\int_{0}^{2l} f(x)\cos\dfrac{n\pi x}{l}\, dx$
$\bar{b}_n$	$\dfrac{1}{l}\displaystyle\int_{-2l}^{0} f(x)\sin\dfrac{n\pi x}{l}\, dx$	$\dfrac{1}{l}\displaystyle\int_{a}^{a+2l} f(x)\sin\dfrac{n\pi x}{l}\, dx$	$\dfrac{1}{l}\displaystyle\int_{0}^{2l} f(x)\cos\dfrac{n\pi x}{l}\, dx$
	$-T < t < 0$	$C < t < C+T$	$0 < t < T$
a_n^*	$\dfrac{2}{T}\displaystyle\int_{-T}^{0} f(t)\cos\dfrac{2n\pi t}{T}\, dt$	$\dfrac{2}{T}\displaystyle\int_{C}^{C+T} f(t)\cos\dfrac{2n\pi t}{T}\, dt$	$\dfrac{2}{T}\displaystyle\int_{0}^{T} f(t)\cos\dfrac{2n\pi t}{T}\, dt$
b_n^*	$\dfrac{2}{T}\displaystyle\int_{-T}^{0} f(t)\sin\dfrac{2n\pi t}{T}\, dt$	$\dfrac{2}{T}\displaystyle\int_{C}^{C+T} f(t)\sin\dfrac{2n\pi t}{T}\, dt$	$\dfrac{2}{T}\displaystyle\int_{0}^{T} f(t)\sin\dfrac{2n\pi t}{T}\, dt$

(2) Identities

In the Fourier series expansion the following identities are useful:

	n	n even	n odd	$\dfrac{n}{2}$ odd	$\dfrac{n}{2}$ even
$\sin n\pi$	0	0	0	0	0
$\cos n\pi$	$(-1)^n$	$+1$	-1	$+1$	$+1$
$\sin\dfrac{n\pi}{2}$		0	$(-1)^{n-1/2}$	0	0
$\cos\dfrac{n\pi}{2}$		$(-1)^{n/2}$	0	-1	$+1$

If all derivatives are finite and the series is convergent, then

$$\frac{f'(2\pi) - f'(0)}{n^2} - \frac{f'''(2\pi) - f'''(0)}{n^4} + \cdots = \int_{0}^{2\pi} f(\theta)\cos n\theta\, d\theta$$

$$1 - \frac{1}{3} + \frac{1}{5} - \frac{1}{7} + \cdots = \frac{\pi}{4}$$

$$1 + \frac{1}{2^2} + \frac{1}{3^2} + \frac{1}{4^2} + \cdots = \frac{\pi^2}{6}$$

$$1 - \frac{1}{3^3} + \frac{1}{5^3} - \frac{1}{7^3} + \cdots = \frac{\pi^3}{32}$$

$$1 - \frac{1}{2^2} + \frac{1}{3^2} - \frac{1}{4^2} + \cdots = \frac{\pi^2}{12}$$

$$1 + \frac{1}{3^4} + \frac{1}{5^4} + \frac{1}{7^4} + \cdots = \frac{\pi^4}{96}$$

$$1 + \frac{1}{2^4} + \frac{1}{3^4} + \frac{1}{4^4} + \cdots = \frac{\pi^4}{90}$$

(1) Closed Form   ($s=$ constant; $n=1,2,\ldots$)

$$\sum_{n=1}^{\infty} s^n \sin nx = \frac{s \sin x}{1 - 2s \cos x + s^2} \qquad s^2 < 1$$

$$\sum_{n=0}^{\infty} s^n \cos nx = \frac{1 - s \cos x}{1 - 2s \cos x + s^2} \qquad s^2 < 1$$

$$\sum_{n=1}^{\infty} \frac{s^n}{n} \sin nx = \tan^{-1} \frac{s \sin x}{1 - s \cos x} \qquad s^2 \leqslant 1$$

$$\sum_{n=1}^{\infty} \frac{s^n}{n} \cos nx = \ln \frac{1}{\sqrt{1 - 2s \cos x + s^2}} \qquad s^2 \leqslant 1$$

$$\sum_{n=1}^{\infty} \frac{\sin nx}{n} = \frac{\pi - x}{2} \qquad\qquad \sum_{n=1}^{\infty} \frac{\cos nx}{n} = \frac{1}{2} \ln \frac{1}{2(1 - \cos x)}$$

$$\sum_{n=1}^{\infty} \frac{\sin nx}{n^3} = \frac{\pi^2 x}{6} - \frac{\pi x^2}{4} + \frac{x^3}{12} \quad 0 < x < 2\pi \qquad \sum_{n=1}^{\infty} \frac{\cos nx}{n^2} = \frac{\pi^2}{6} - \frac{\pi x}{2} + \frac{x^2}{4} \quad 0 < x < 2\pi$$

$$\sum_{n=1}^{\infty} \frac{\sin nx}{n^5} = \frac{\pi^4 x}{90} - \frac{\pi^2 x^3}{36} + \frac{\pi x^4}{48} - \frac{x^5}{240} \qquad \sum_{n=1}^{\infty} \frac{\cos nx}{n^4} = \frac{\pi^4}{90} - \frac{\pi^2 x^2}{12} + \frac{\pi x^3}{12} - \frac{x^4}{48}$$

(2) Complex Form

Since the *exponential* and *trigonometric functions* are *connected* by

$$\cos \theta = \frac{e^{i\theta} + e^{-i\theta}}{2} \qquad \sin \theta = \frac{e^{i\theta} - e^{-i\theta}}{2i}$$

with $\omega_n = \dfrac{n\pi}{l}$ and $n = 0, \pm 1, \pm 2, \ldots$, then

$$f(x) = \frac{1}{2}\left(C_0 + \sum_{n=1}^{\infty} C_n e^{i\omega_n x} + \sum_{n=1}^{\infty} D_n e^{-i\omega_n x}\right)$$

in which

$$C_n = \frac{1}{l} \int_{-l}^{l} f(x) e^{-i\omega_n x}\, dx \qquad\qquad D_n = \frac{1}{l} \int_{-l}^{l} f(x) e^{i\omega_n x}\, dx$$

and

$$C_n = \bar{a}_n - i\bar{b}_n \qquad\qquad D_n = \bar{a}_n + i\bar{b}_n$$

or in simpler form

$$f(x) = \frac{1}{2} \sum_{n=-\infty}^{+\infty} C_n e^{-i\omega_n x}$$

The set of coefficients $\{C_n\}$ is called the *spectrum* of $f(x)$.

(1) Even Functions

$$\bar{a}_n = \frac{2}{l} \int_0^l f(x) \cos \frac{n\pi x}{l} \, dx$$

$$\bar{b}_n = 0 \qquad n = 0, 1, 2, 3$$

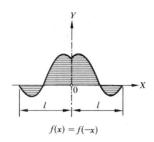

$$f(x) = f(-x)$$

$$\bar{a}_{2n} = \frac{2}{l} \int_0^l f(x) \cos \frac{2n\pi x}{l} \, dx$$

$$\bar{b}_{2n} = \frac{2}{l} \int_0^l f(x) \sin \frac{2n\pi x}{l} \, dx$$

$$\bar{a}_{2n+1} = 0 \qquad \bar{b}_{2n+1} = 0 \qquad n = 0, 1, 2, \ldots$$

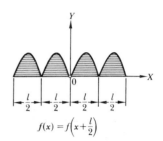

$$f(x) = f\left(x + \frac{l}{2}\right)$$

(2) Odd Functions

$$\bar{b}_n = \frac{2}{l} \int_0^l f(x) \sin \frac{n\pi x}{l} \, dx$$

$$\bar{a}_n = 0 \qquad n = 1, 2, 3, \ldots$$

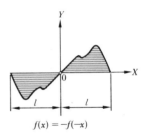

$$f(x) = -f(-x)$$

$$\bar{a}_{2n+1} = \frac{2}{l} \int_0^l f(x) \cos \frac{(2n+1)\pi x}{l} \, dx$$

$$\bar{b}_{2n+1} = \frac{2}{l} \int_0^l f(x) \sin \frac{(2n+1)\pi x}{l} \, dx$$

$$\bar{a}_{2n} = 0 \qquad \bar{b}_{2n} = 0 \qquad n = 0, 1, 2, \ldots$$

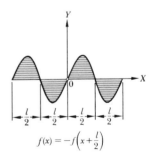

$$f(x) = -f\left(x + \frac{l}{2}\right)$$

$$\bar{b}_{2n+1} = \frac{4}{l} \int_0^{l/2} f(x) \sin \frac{(2n+1)\pi x}{l} \, dx$$

$$\bar{a}_{2n} = 0 \qquad \bar{b}_n = 0 \qquad n = 0, 1, 2, \ldots$$

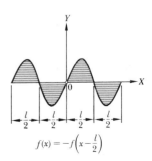

$$f(x) = -f\left(x - \frac{l}{2}\right)$$

$$f(x)=\frac{A_0}{2}+\sum_{n=1}^{\infty}\left(A_n\cos\frac{n\pi x}{L}+B_n\sin\frac{n\pi x}{L}\right)$$

$\alpha=n\pi a$

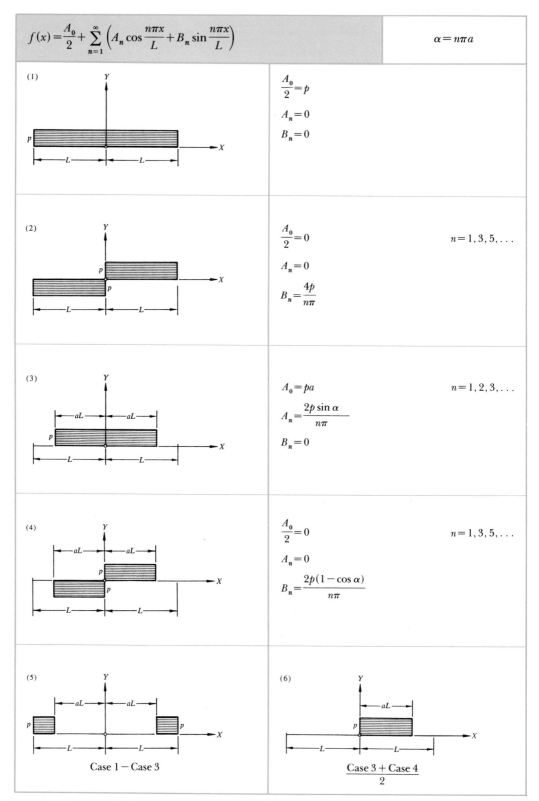

(1)

$$\frac{A_0}{2}=p$$

$$A_n=0$$

$$B_n=0$$

(2)

$$\frac{A_0}{2}=0$$

$$A_n=0$$

$$B_n=\frac{4p}{n\pi}$$

$n=1,3,5,\ldots$

(3)

$$A_0=pa$$

$$A_n=\frac{2p\sin\alpha}{n\pi}$$

$$B_n=0$$

$n=1,2,3,\ldots$

(4)

$$\frac{A_0}{2}=0$$

$$A_n=0$$

$$B_n=\frac{2p(1-\cos\alpha)}{n\pi}$$

$n=1,3,5,\ldots$

(5)

Case 1 − Case 3

(6)

$$\frac{\text{Case 3}+\text{Case 4}}{2}$$

[1]In Secs. 12.05 through 12.13 Fourier coefficients $\bar{a}_0$, $\bar{a}_n$, $\bar{b}_n$ introduced in Sec. 12.01 $(2a)$ are denoted as A_0, A_n, B_n, respectively, to eliminate the potential conflict with the position factors a and b.

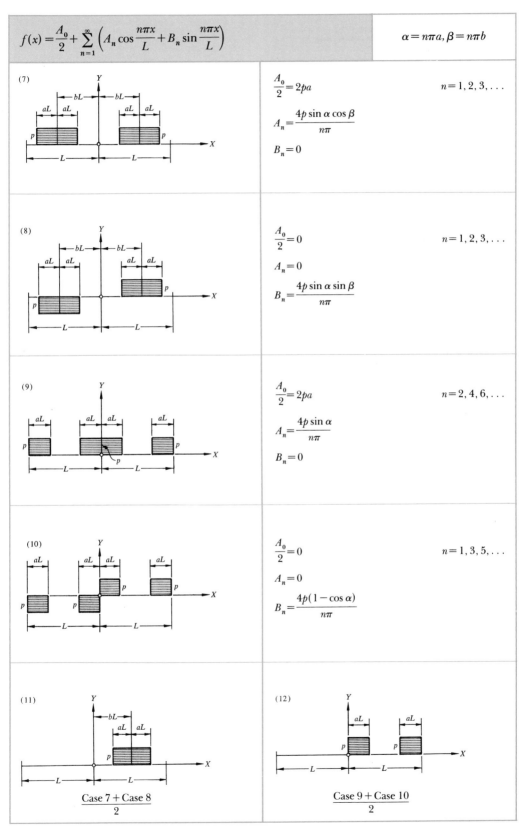

$$f(x) = \frac{A_0}{2} + \sum_{n=1}^{\infty} \left(A_n \cos \frac{n\pi x}{L} + B_n \sin \frac{n\pi x}{L} \right)$$

$$\alpha = n\pi a, \ \beta = n\pi b$$

(7)

$$\frac{A_0}{2} = 2pa \qquad n = 1, 2, 3, \ldots$$

$$A_n = \frac{4p \sin \alpha \cos \beta}{n\pi}$$

$$B_n = 0$$

(8)

$$\frac{A_0}{2} = 0 \qquad n = 1, 2, 3, \ldots$$

$$A_n = 0$$

$$B_n = \frac{4p \sin \alpha \sin \beta}{n\pi}$$

(9)

$$\frac{A_0}{2} = 2pa \qquad n = 2, 4, 6, \ldots$$

$$A_n = \frac{4p \sin \alpha}{n\pi}$$

$$B_n = 0$$

(10)

$$\frac{A_0}{2} = 0 \qquad n = 1, 3, 5, \ldots$$

$$A_n = 0$$

$$B_n = \frac{4p(1 - \cos \alpha)}{n\pi}$$

(11)

$$\frac{\text{Case 7} + \text{Case 8}}{2}$$

(12)

$$\frac{\text{Case 9} + \text{Case 10}}{2}$$

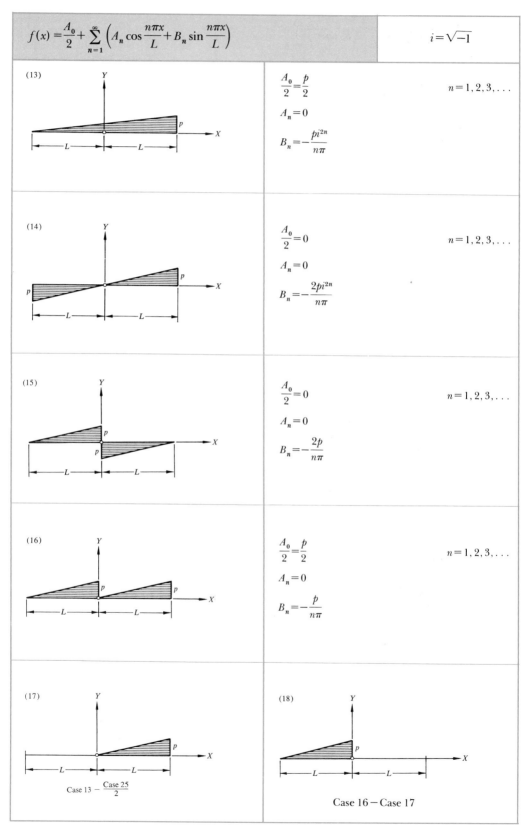

$$f(x) = \frac{A_0}{2} + \sum_{n=1}^{\infty} \left(A_n \cos \frac{n\pi x}{L} + B_n \sin \frac{n\pi x}{L} \right)$$

$i = \sqrt{-1}$

(13)

$\dfrac{A_0}{2} = \dfrac{p}{2}$ $n = 1, 2, 3, \ldots$

$A_n = 0$

$B_n = -\dfrac{pi^{2n}}{n\pi}$

(14)

$\dfrac{A_0}{2} = 0$ $n = 1, 2, 3, \ldots$

$A_n = 0$

$B_n = -\dfrac{2pi^{2n}}{n\pi}$

(15)

$\dfrac{A_0}{2} = 0$ $n = 1, 2, 3, \ldots$

$A_n = 0$

$B_n = -\dfrac{2p}{n\pi}$

(16)

$\dfrac{A_0}{2} = \dfrac{p}{2}$ $n = 1, 2, 3, \ldots$

$A_n = 0$

$B_n = -\dfrac{p}{n\pi}$

(17)

Case 13 $- \dfrac{\text{Case 25}}{2}$

(18)

Case 16 $-$ Case 17

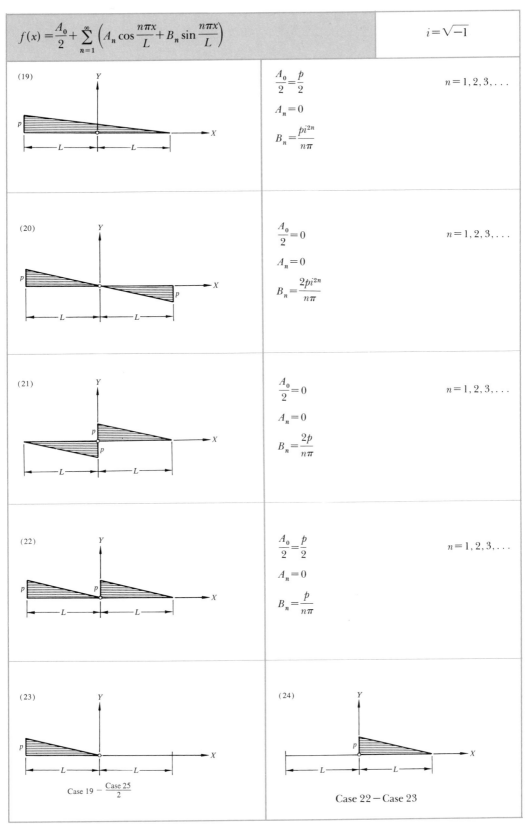

$$f(x) = \frac{A_0}{2} + \sum_{n=1}^{\infty} \left(A_n \cos \frac{n\pi x}{L} + B_n \sin \frac{n\pi x}{L} \right)$$

$i = \sqrt{-1}$

(19)

$\dfrac{A_0}{2} = \dfrac{p}{2}$

$A_n = 0$

$B_n = \dfrac{pi^{2n}}{n\pi}$

$n = 1, 2, 3, \ldots$

(20)

$\dfrac{A_0}{2} = 0$

$A_n = 0$

$B_n = \dfrac{2pi^{2n}}{n\pi}$

$n = 1, 2, 3, \ldots$

(21)

$\dfrac{A_0}{2} = 0$

$A_n = 0$

$B_n = \dfrac{2p}{n\pi}$

$n = 1, 2, 3, \ldots$

(22)

$\dfrac{A_0}{2} = \dfrac{p}{2}$

$A_n = 0$

$B_n = \dfrac{p}{n\pi}$

$n = 1, 2, 3, \ldots$

(23)

Case 19 $- \dfrac{\text{Case 25}}{2}$

(24)

Case 22 $-$ Case 23

$$f(x) = \frac{A_0}{2} + \sum_{n=1}^{\infty} \left(A_n \cos \frac{n\pi x}{L} + B_n \sin \frac{n\pi x}{L} \right)$$

$\alpha = n\pi a$

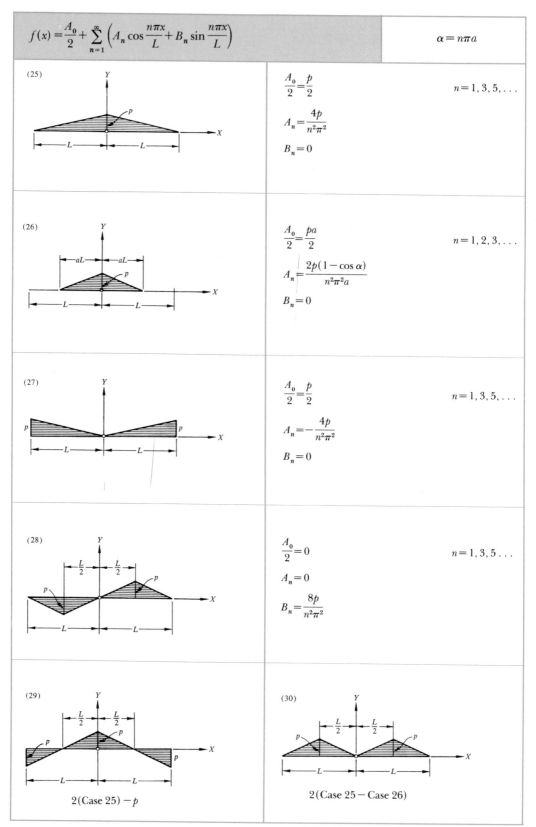

(25)

$$\frac{A_0}{2} = \frac{p}{2}$$

$$A_n = \frac{4p}{n^2\pi^2}$$

$$B_n = 0$$

$n = 1, 3, 5, \ldots$

(26)

$$\frac{A_0}{2} = \frac{pa}{2}$$

$$A_n = \frac{2p(1 - \cos \alpha)}{n^2\pi^2 a}$$

$$B_n = 0$$

$n = 1, 2, 3, \ldots$

(27)

$$\frac{A_0}{2} = \frac{p}{2}$$

$$A_n = -\frac{4p}{n^2\pi^2}$$

$$B_n = 0$$

$n = 1, 3, 5, \ldots$

(28)

$$\frac{A_0}{2} = 0$$

$$A_n = 0$$

$$B_n = \frac{8p}{n^2\pi^2}$$

$n = 1, 3, 5 \ldots$

(29)

$2(\text{Case } 25) - p$

(30)

$2(\text{Case } 25 - \text{Case } 26)$

$$f(x) = \frac{A_0}{2} + \sum_{n=1}^{\infty} \left(A_n \cos \frac{n\pi x}{L} + B_n \sin \frac{n\pi x}{L} \right)$$

$\alpha = n\pi a, \beta = n\pi b$

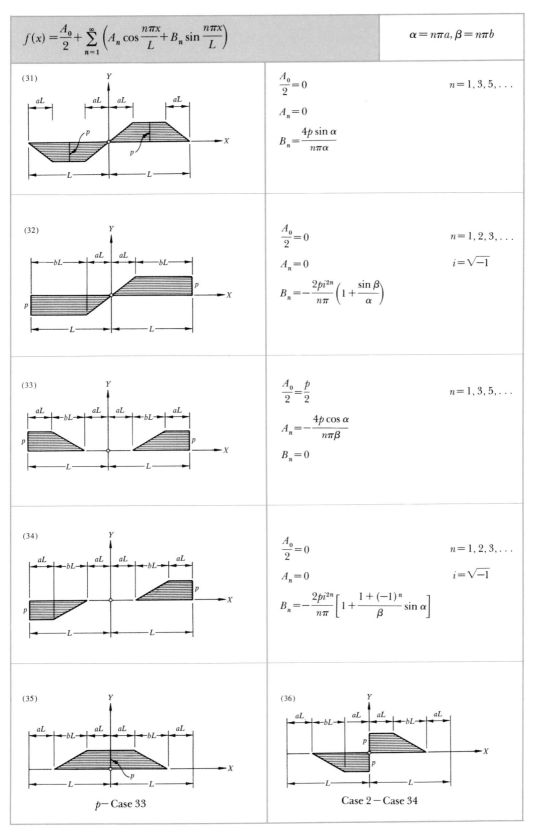

(31)

$\dfrac{A_0}{2} = 0$ $n = 1, 3, 5, \ldots$

$A_n = 0$

$B_n = \dfrac{4p \sin \alpha}{n\pi\alpha}$

(32)

$\dfrac{A_0}{2} = 0$ $n = 1, 2, 3, \ldots$

$A_n = 0$ $i = \sqrt{-1}$

$B_n = -\dfrac{2pi^{2n}}{n\pi} \left(1 + \dfrac{\sin \beta}{\alpha} \right)$

(33)

$\dfrac{A_0}{2} = \dfrac{p}{2}$ $n = 1, 3, 5, \ldots$

$A_n = -\dfrac{4p \cos \alpha}{n\pi\beta}$

$B_n = 0$

(34)

$\dfrac{A_0}{2} = 0$ $n = 1, 2, 3, \ldots$

$A_n = 0$ $i = \sqrt{-1}$

$B_n = -\dfrac{2pi^{2n}}{n\pi} \left[1 + \dfrac{1 + (-1)^n}{\beta} \sin \alpha \right]$

(35)

p – Case 33

(36)

Case 2 – Case 34

$$f(x) = \frac{A_0}{2} + \sum_{n=1}^{\infty} \left(A_n \cos \frac{n\pi x}{L} + B_n \sin \frac{n\pi x}{L} \right)$$

$i = \sqrt{-1}$

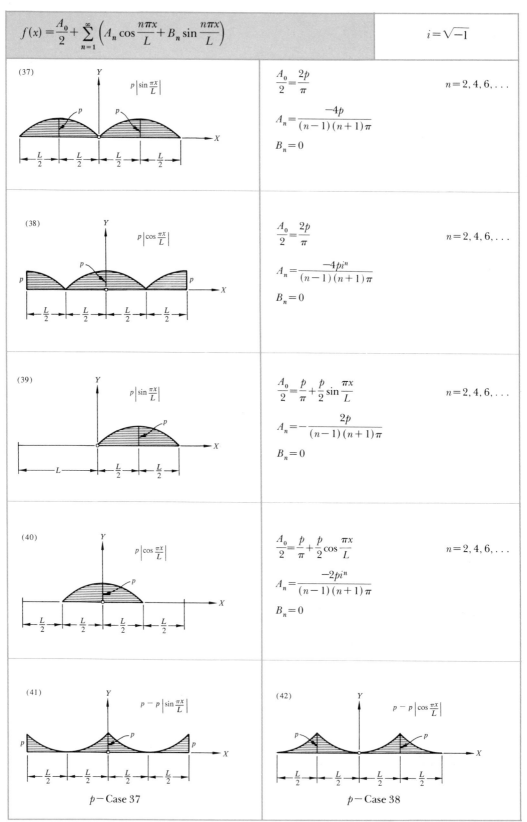

(37)
$$\frac{A_0}{2} = \frac{2p}{\pi}$$
$$n = 2, 4, 6, \ldots$$
$$A_n = \frac{-4p}{(n-1)(n+1)\pi}$$
$$B_n = 0$$

(38)
$$\frac{A_0}{2} = \frac{2p}{\pi}$$
$$n = 2, 4, 6, \ldots$$
$$A_n = \frac{-4pi^n}{(n-1)(n+1)\pi}$$
$$B_n = 0$$

(39)
$$\frac{A_0}{2} = \frac{p}{\pi} + \frac{p}{2} \sin \frac{\pi x}{L}$$
$$n = 2, 4, 6, \ldots$$
$$A_n = -\frac{2p}{(n-1)(n+1)\pi}$$
$$B_n = 0$$

(40)
$$\frac{A_0}{2} = \frac{p}{\pi} + \frac{p}{2} \cos \frac{\pi x}{L}$$
$$n = 2, 4, 6, \ldots$$
$$A_n = \frac{-2pi^n}{(n-1)(n+1)\pi}$$
$$B_n = 0$$

(41)

p − Case 37

(42)

p − Case 38

$$f(x) = \frac{A_0}{2} + \sum_{n=1}^{\infty} \left(A_n \cos \frac{n\pi x}{L} + B_n \sin \frac{n\pi x}{L} \right)$$

$i = \sqrt{-1}$

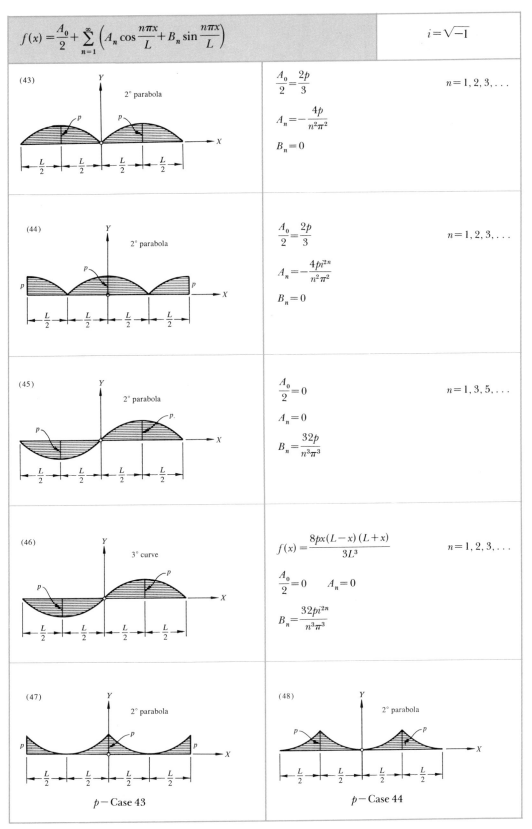

(43)   2° parabola

$$\frac{A_0}{2} = \frac{2p}{3} \qquad n = 1, 2, 3, \ldots$$

$$A_n = -\frac{4p}{n^2\pi^2}$$

$$B_n = 0$$

(44)   2° parabola

$$\frac{A_0}{2} = \frac{2p}{3} \qquad n = 1, 2, 3, \ldots$$

$$A_n = -\frac{4pi^{2n}}{n^2\pi^2}$$

$$B_n = 0$$

(45)   2° parabola

$$\frac{A_0}{2} = 0 \qquad n = 1, 3, 5, \ldots$$

$$A_n = 0$$

$$B_n = \frac{32p}{n^3\pi^3}$$

(46)   3° curve

$$f(x) = \frac{8px(L-x)(L+x)}{3L^3} \qquad n = 1, 2, 3, \ldots$$

$$\frac{A_0}{2} = 0 \qquad A_n = 0$$

$$B_n = \frac{32pi^{2n}}{n^3\pi^3}$$

(47)   2° parabola

$p -$ Case 43

(48)   2° parabola

$p -$ Case 44

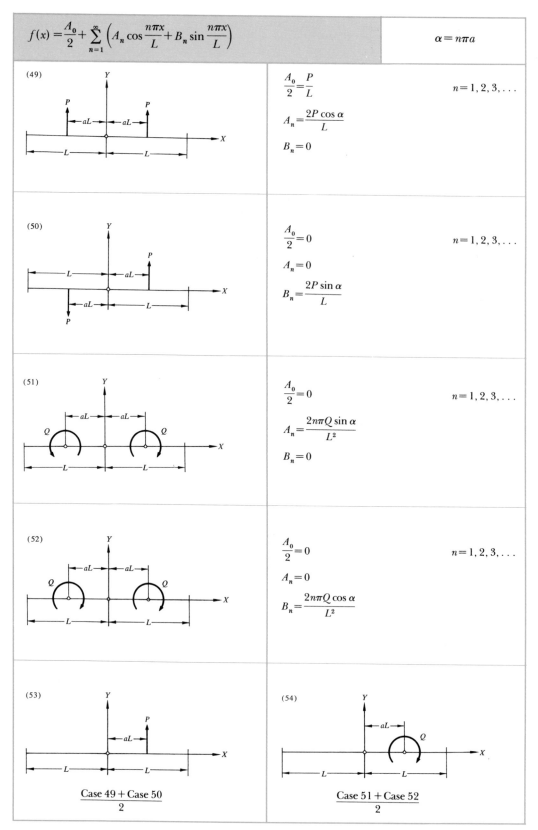

$$f(x) = \frac{A_0}{2} + \sum_{n=1}^{\infty} \left(A_n \cos \frac{n\pi x}{L} + B_n \sin \frac{n\pi x}{L} \right)$$

$\alpha = n\pi a$

(49)

$\dfrac{A_0}{2} = \dfrac{P}{L}$ $n = 1, 2, 3, \ldots$

$A_n = \dfrac{2P \cos \alpha}{L}$

$B_n = 0$

(50)

$\dfrac{A_0}{2} = 0$ $n = 1, 2, 3, \ldots$

$A_n = 0$

$B_n = \dfrac{2P \sin \alpha}{L}$

(51)

$\dfrac{A_0}{2} = 0$ $n = 1, 2, 3, \ldots$

$A_n = \dfrac{2n\pi Q \sin \alpha}{L^2}$

$B_n = 0$

(52)

$\dfrac{A_0}{2} = 0$ $n = 1, 2, 3, \ldots$

$A_n = 0$

$B_n = \dfrac{2n\pi Q \cos \alpha}{L^2}$

(53)

$$\frac{\text{Case 49} + \text{Case 50}}{2}$$

(54)

$$\frac{\text{Case 51} + \text{Case 52}}{2}$$

13
HIGHER TRANSCENDENT FUNCTIONS

(1) Definition

Integrals which cannot be evaluated as finite combinations of elementary functions are called *integral functions*. The most typical functions in this group evaluated by *series expansion* (Sec. 9.01) are given below.

(2) Integral–Sine, –Cosine, and –Exponential Functions

$$\text{Si}(x) = \int_0^x \frac{\sin x}{x}\,dx = x - \frac{1}{3}\frac{x^3}{3!} + \frac{1}{5}\frac{x^5}{5!} - \frac{1}{7}\frac{x^7}{7!} + \cdots \qquad \text{Si}(\infty) = \frac{\pi}{2}$$

$$\text{Ci}(x) = \int_{+\infty}^x \frac{\cos x}{x}\,dx = C + \ln x - \frac{1}{2}\frac{x^2}{2!} + \frac{1}{4}\frac{x^4}{4!} + \cdots \qquad \text{Ci}(\infty) = 0$$

$$\text{Ei}(x) = \int_{+\infty}^x \frac{e^{-x}}{x}\,dx = C + \ln x - x + \frac{1}{2}\frac{x^2}{2!} - \frac{1}{3}\frac{x^3}{3!} + \cdots \qquad \text{Ei}(\infty) = 0$$

$$\bar{\text{Ei}}(x) = \int_{-\infty}^x \frac{e^x}{x}\,dx = C + \ln x + x + \frac{1}{2}\frac{x^2}{2!} + \frac{1}{3}\frac{x^3}{3!} + \cdots \qquad \bar{\text{Ei}}(\infty) = \infty$$

$$C = \int_{+\infty}^0 e^{-x}\ln x\,dx = 0.577\,215\,665 = \text{Euler's constant} \qquad \text{(Sec. 7.02)}$$

(3) Fresnel Integrals

$$\int_0^x \frac{\sin x}{\sqrt{x}}\,dx = 2\sqrt{x}\left(\frac{1}{3}\frac{x}{1!} - \frac{1}{7}\frac{x^3}{3!} + \frac{1}{11}\frac{x^5}{5!} - \cdots\right) \qquad \int_0^{+\infty} \frac{\sin x}{\sqrt{x}}\,dx = \sqrt{\frac{\pi}{2}}$$

$$\int_0^x \frac{\cos x}{\sqrt{x}}\,dx = 2\sqrt{x}\left(1 - \frac{1}{5}\frac{x^2}{2!} + \frac{1}{9}\frac{x^4}{4!} - \frac{1}{13}\frac{x^6}{6!} + \cdots\right) \qquad \int_0^{+\infty} \frac{\cos x}{\sqrt{x}}\,dx = \sqrt{\frac{\pi}{2}}$$

$$S(x) = \sqrt{\frac{2}{\pi}}\int_0^x \sin x^2\,dx \qquad S(-x) = -S(x)$$

$$S(0) = 0$$

$$= \sqrt{\frac{2}{\pi}}\left(\frac{1}{1!}\frac{x^3}{3} - \frac{1}{3!}\frac{x^7}{7} + \frac{1}{5!}\frac{x^{11}}{11} - \cdots\right) \qquad S(\infty) = \tfrac{1}{2}$$

$$C(x) = \sqrt{\frac{2}{\pi}}\int_0^x \cos x^2\,dx \qquad C(-x) = -C(x)$$

$$C(0) = 0$$

$$= \sqrt{\frac{2}{\pi}}\left(\frac{1}{0!}\frac{x}{1} - \frac{1}{2!}\frac{x^5}{5} + \frac{1}{4!}\frac{x^9}{9} - \cdots\right) \qquad C(\infty) = \tfrac{1}{2}$$

(4) Error Function

$$\text{erf}(x) = \frac{2}{\sqrt{\pi}}\int_0^x e^{-x^2}\,dx \qquad \text{erf}(-x) = -\text{erf}(x)$$

$$\text{erf}(0) = 0$$

$$= \frac{2}{\sqrt{\pi}}\left(\frac{1}{0!}\frac{x}{1} - \frac{1}{1!}\frac{x^3}{3} + \frac{1}{2!}\frac{x^5}{5} - \frac{1}{3!}\frac{x^7}{7} + \cdots\right) \qquad \text{erf}(\infty) = 1$$

(a) Sine integral

$$\mathrm{Si}(x) = \int_0^x \frac{\sin x}{x}\,dx$$

x	0	1	2	3	4	5	6	7	8	9	x
0.	0.0000	0.0999	0.1996	0.2985	0.3965	0.4931	0.5881	0.6812	0.7721	0.8605	0.
1.	0.9461	1.0287	1.1080	1.1840	1.2562	1.3247	1.3892	1.4496	1.5058	1.5578	1.
2.	1.6054	1.6487	1.6876	1.7222	1.7525	1.7785	1.8004	1.8182	1.8321	1.8422	2.
3.	1.8487	1.8517	1.8514	1.8481	1.8419	1.8331	1.8219	1.8086	1.7934	1.7765	3.
4.	1.7582	1.7387	1.7184	1.6973	1.6758	1.6541	1.6325	1.6110	1.5900	1.5696	4.

(b) Cosine integral

$$\mathrm{Ci}(x) = \int_{+\infty}^x \frac{\cos x}{x}\,dx$$

x	0	1	2	3	4	5	6	7	8	9	x
0.	$-\infty$	-1.7279	-1.0422	-0.6492	-0.3788	-0.1778	-0.0223	$+0.1051$	$+0.1983$	$+0.2761$	0.
1.	$+0.3374$	$+0.3849$	$+0.4205$	$+0.4457$	$+0.4620$	$+0.4704$	$+0.4717$	$+0.4670$	$+0.4568$	$+0.4419$	1.
2.	$+0.4230$	$+0.4005$	$+0.3751$	$+0.3472$	$+0.3173$	$+0.2859$	$+0.2533$	$+0.2201$	$+0.1865$	$+0.1529$	2.
3.	$+0.1196$	$+0.0870$	$+0.0553$	$+0.0247$	-0.0045	-0.0321	-0.0580	-0.0819	-0.1038	-0.1235	3.
4.	-0.1410	-0.1562	-0.1690	-0.1795	-0.1877	-0.1935	-0.1970	-0.1984	-0.1976	-0.1948	4.

(c) Exponential integral

$$\mathrm{Ei}(x) = \int_{+\infty}^x \frac{e^{-x}}{x}\,dx$$

x	0	1	2	3	4	5	6	7	8	9	x
0.	$-\infty$	-1.8229	-1.2227	-0.9057	-0.7024	-0.5598	-0.4544	-0.3738	-0.3106	-0.2602	0.
1.	-0.2194	-0.1860	-0.1584	-0.1355	-0.1162	-0.1000	-0.0863	-0.0747	-0.0647	-0.0562	1.
2.	-0.0489	-0.0426	-0.0372	-0.0325	-0.0284	-0.0249	-0.0219	-0.0192	-0.0169	-0.0148	2.

(d) Error integral

$$\mathrm{erf}(x) = \frac{2}{\sqrt{\pi}} \int_0^x e^{-x^2}\,dx$$

x	0	1	2	3	4	5	6	7	8	9	x
0.	0.0000	0.1125	0.2227	0.3286	0.4284	0.5205	0.6039	0.6778	0.7421	0.7969	0.
1.	0.8427	0.8802	0.9103	0.9340	0.9523	0.9661	0.9764	0.9838	0.9891	0.9928	1.
2.	0.9953	0.9970	0.9981	0.9989	0.9994	0.9996	0.9998	0.9999	0.9999	1.0000	2.

$x=$ signed real number $(|x| \leq \infty)$ $m, n, r = 0, 1, 2, \ldots$

$u=$ signed real number $(|u| \leq 1)$ $n! = n$ factorial (Sec. 1.03)

(1) Gamma Function Γ

(a) Definition

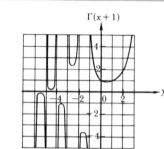

$\Gamma(x+1)$

$$\Gamma(x) = \int_0^\infty t^{x-1} e^{-t}\, dt \qquad x > 0$$

$$\Gamma(x) = \lim_{n \to \infty} \frac{n^x n!}{x(x+1)(x+2) \cdots (x+n)}$$

(b) Functional equations   (Sec. 1.03)

$$\Gamma(x+r) = (x+r-1)(x+r-2) \cdots (x+1)\Gamma(x+1)$$

$$\Gamma(x-r) = \frac{\Gamma(x+1)}{(x-r)(x-r+1)(x-r+2) \cdots (x-1)x}$$

$$\Gamma(n+1) = n(n-1)(n-2) \cdots 2 \cdot 1 = n!$$

$$\Gamma(n) = (n-1)(n-2)(n-3) \cdots 2 \cdot 1 = (n-1)!$$

(c) Reflection equations   (Sec. A.02)

$$\Gamma(u)\Gamma(-u) = \frac{-\pi}{u \sin u\pi} \qquad \Gamma(u)\Gamma(1-u) = \frac{\pi}{\sin u\pi}$$

$$\Gamma(1+u)\Gamma(-u) = \frac{-\pi}{\sin u\pi} \qquad \Gamma(1+u)\Gamma(1-u) = \frac{u\pi}{\sin u\pi}$$

(d) Special values   (Sec. A.01)

$$\Gamma(-n) = \infty \qquad \Gamma(0) = \infty \qquad \Gamma(1) = 1 \qquad \Gamma(2) = 1$$

$$\Gamma(n+\tfrac{1}{2}) = \frac{(2n)!\sqrt{\pi}}{n!2^{2n}} \qquad \Gamma(\tfrac{1}{2}) = \sqrt{\pi} \qquad \Gamma(\tfrac{3}{2}) = \frac{\sqrt{\pi}}{2} \qquad \Gamma(-n+\tfrac{1}{2}) = \frac{(-1)^n n! 2^{2n}\sqrt{\pi}}{(2n)!}$$

(2) Pi Function $-\Pi$

(a) Definition

$$\Pi(x) = \Gamma(x+1) = \int_0^\infty t^x e^{-t}\, dt \qquad x > 0$$

(b) Recursion formulas

$$\Pi(n) = n(n-1)(n-2) \cdots 1 = n! \qquad \Pi(n) = \frac{\Pi(n+1)}{n+1} \qquad \Pi(n) = n\Pi(n-1)$$

(3) Beta Function $-\mathrm{B}$

(a) Definition

$$\mathrm{B}(x, y) = \int_0^1 t^{x-1}(1-t)^{y-1}\, dt = \frac{\Gamma(x)\Gamma(y)}{\Gamma(x+y)} \qquad x > 0, y > 0$$

(b) Relations
$$\mathrm{B}(m, n) = \mathrm{B}(n, m)$$

$$\mathrm{B}(m, n) = \frac{(m-1)!(n-1)!}{(m+n-1)!}$$

$C = 0.577\ 215\ 665 = $ Euler's constant (Sec. 7.02) $k, m, n, r = 0, 1, 2, \ldots$

$\Gamma(\ ) = $ gamma function (Sec. 13.03) $Z(\ ) = $ zeta function (Sec. A.06)

(1) Digamma Function ψ

(a) Definition

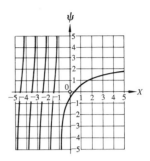

$$\psi(x) = \frac{d}{dx}\big\{\ln[\Gamma(x+1)]\big\} = \frac{\dfrac{d\Gamma(x+1)}{dx}}{\Gamma(x+1)}$$

$$\psi(x) = \lim_{n \to \infty}\left(\ln n - \sum_{k=1}^{n}\frac{1}{x+k}\right) = \sum_{k=1}^{\infty}\left(\frac{1}{k} - \frac{1}{x+k}\right) - C$$

(b) Functional equations   (Sec. 8.03)

$$\psi(x+r) = \psi(x) + \sum_{k=1}^{r}\frac{1}{x+k} \qquad\qquad \psi(n) = \sum_{k=1}^{n}\frac{1}{k} - C$$

$$\psi(x-r) = \psi(x) - \sum_{k=1}^{r}\frac{1}{x+1-k} \qquad\qquad \psi(n-\tfrac{1}{2}) = 2\sum_{k=1}^{n}\frac{1}{2k-1} - \ln 4 - C$$

(c) Reflection equations

$$\psi(u) = 1 + \frac{1}{2u} - \frac{1}{(1+u)(1-u)} - \frac{\pi}{2\tan u\pi} - C - \sum_{k=1}^{8} b_{2k}u^{2k} - (|\varepsilon| \le 5 \times 10^{-11}) \qquad |u| \le 0.5$$

$b_2 = 0.202\ 056\ 903$	$b_4 = 0.036\ 927\ 755$	$b_6 = 0.008\ 349\ 277$	$b_8 = 0.002\ 008\ 393$
$b_{10} = 0.000\ 494\ 189$	$b_{12} = 0.000\ 122\ 713$	$b_{14} = 0.000\ 030\ 588$	$b_{16} = 0.000\ 007\ 637$

(d) Special values

$$\psi(-n) = \pm\infty \qquad \psi(0) = -C \qquad \psi(1) = 1 - C \qquad \psi(n) = 1 + \frac{1}{2} + \frac{1}{3} + \cdots + \frac{1}{n} - C$$

$$\psi(N) = \ln N + \frac{1}{2N} - \frac{N^{-2}}{12}\left(1 - \frac{N^{-2}}{10}\left(1 - \frac{10N^{-2}}{21}\left(1 - \frac{21N^{-2}}{20}\right)\right)\right) - (|\varepsilon| \le 5 \times 10^{-11}) \qquad N \ge 5$$

(2) Polygamma Function $\psi^{(m)}$

(a) Definition

$$\psi^{(m)}(x) = \frac{d^m}{dx^m}[\psi(x)] = \frac{d^{m+1}}{dx^{m+1}}\big\{\ln[\Gamma(x+1)]\big\} = (-1)^{m+1}m!\sum_{k=1}^{\infty}\frac{1}{(x+k)^{m+1}}$$

(b) Integral argument

$$\psi^{(1)}(n) = Z(2) - \sum_{k=1}^{n}\frac{1}{k^2} \qquad\qquad \psi^{(m)}(n) = (-1)^{m+1}m!\left[Z(m+1) - \sum_{k=1}^{n}\frac{1}{k^{m+1}}\right]$$

(c) Reflection equation

$$\psi^{(m)}(r \pm u) = \psi^{(m)}(\pm u) \mp (-1)^{m+1}m!\sum_{k=1}^{r}\frac{1}{(k \pm u)^{m+1}} \qquad |u| \le 0.5$$

(1) Elliptic Integrals, Normal Form, Formulas

$$F(k, x) = \int_0^x \frac{dx}{\sqrt{(1 - x^2)(1 - k^2 x^2)}} = F(k, \phi) = \int_0^\phi \frac{d\phi}{\sqrt{1 - k^2 \sin^2 \phi}}$$

$$E(k, x) = \int_0^x \sqrt{\frac{1 - k^2 x^2}{1 - x^2}} \, dx = E(k, \phi) = \int_0^\phi \sqrt{1 - k^2 \sin^2 \phi} \, d\phi$$

where $k = \sin \omega =$ modulus (given constant) in the interval $0 \leq k \leq +1$, $x =$ independent variable in interval $-1 \leq x \leq +1$.

(2) Elliptic Integrals, Complete Form, Formulas

$$F\left(k, \frac{\pi}{2}\right) = K = \frac{\pi}{2}\left\{1 + \left(\frac{1}{2}\right)^2 k^2 + \left[\frac{(1)(3)}{(2)(4)}\right]^2 k^4 + \left[\frac{(1)(3)(5)}{(2)(4)(6)}\right]^2 k^6 + \cdots\right\}$$

$$E\left(k, \frac{\pi}{2}\right) = E = \frac{\pi}{2}\left\{1 - \left(\frac{1}{2}\right)^2 \frac{k^2}{1} - \left[\frac{(1)(3)}{(2)(4)}\right]^2 \frac{k^4}{3} - \left[\frac{(1)(3)(5)}{(2)(4)(6)}\right]^2 \frac{k^6}{5} - \cdots\right\}$$

(3) Elliptic Integrals, Degenerated Form

$$F(0, x) = \sin^{-1} x \qquad K = \frac{\pi}{2} \qquad\qquad F(1, x) = \tanh^{-1} x \qquad K = \infty$$

$$E(0, x) = \sin^{-1} x \qquad E = \frac{\pi}{2} \qquad\qquad E(1, x) = x \qquad E = 1$$

13.06 ELLIPTIC FUNCTIONS

(1) Definition

If $u = F(k, \phi)$, the inverse function is designated as $\phi = \text{am } u$ and is called the *elliptic function of Jacobi.*

$$x = \text{sn } u = \sin (\text{am } u) \qquad\qquad \text{sn}^2 u + \text{cn}^2 u = 1$$
$$\sqrt{1 - x^2} = \text{cn } u = \cos (\text{am } u) \qquad\qquad \text{dn}^2 u + k^2 \text{sn}^2 u = 1$$
$$\sqrt{1 - k^2 x^2} = \text{dn } u = \sqrt{1 - k^2 \text{sn}^2 u} \qquad\qquad \text{dn}^2 u - k^2 \text{cn}^2 u = 1 - k^2$$

$$\text{sn } u = u - \frac{(1 + k^2) u^3}{3!} + \frac{(1 + 14k^2 + k^4) u^5}{5!} - \cdots$$

$$\text{cn } u = 1 - \frac{u^2}{2!} + \frac{(1 + 4k^2) u^4}{4!} - \frac{(1 + 44k^2 + 16k^4) u^6}{6!} + \cdots$$

$$\text{dn } u = 1 - \frac{k^2 u^2}{2!} + \frac{k^2 (4 + k^2) u^4}{4!} - \frac{k^2 (16 + 44k^2 + k^4) u^6}{6!} + \cdots$$

$$\text{sn } (k = 0) = \sin u \qquad\qquad \text{cn } (k = 0) = \cos u \qquad\qquad \text{dn } (k = 0) = 1$$

$$\text{sn } (k = 1) = \tanh u \qquad\qquad \text{cn } (k = 1) = \frac{1}{\cosh u} \qquad\qquad \text{dn } (k = 1) = \frac{1}{\cosh u}$$

(2) Derivatives

$$\frac{d}{dx}(\text{sn } u) = \text{cn } u \text{ dn } u$$

$$\frac{d}{dx}(\text{cn } u) = - \text{sn } u \text{ dn } u$$

$$\frac{d}{dx}(\text{dn } u) = - k^2 \text{ sn } u \text{ cn } u$$

(3) Integrals

$$\int \text{sn } u \, du = \frac{1}{k} \ln (\text{dn } u - k \text{ cn } u)$$

$$\int \text{cn } u \, du = \frac{1}{k} \cos^{-1} (\text{dn } u)$$

$$\int \text{dn } u \, du = \sin^{-1} (\text{sn } u)$$

(a) First kind

$$F(k, \phi) = \int_0^{\phi} \frac{d\phi}{\sqrt{1 - k^2 \sin^2 \phi}} \qquad k = \sin \omega$$

ω	0°	10°	20°	30°	40°	50°	60°	70°	80°	90°	ω
k / φ	0	0.1737	0.3420	0.5000	0.6428	0.7660	0.8660	0.9397	0.9848	1.0000	k / φ
0°	0.0000	0.0000	0.0000	0.0000	0.0000	0.0000	0.0000	0.0000	0.0000	0.0000	0°
10°	0.1745	0.1746	0.1746	0.1748	0.1749	0.1751	0.1752	0.1753	0.1754	0.1754	10°
20°	0.3491	0.3493	0.3499	0.3508	0.3520	0.3533	0.3545	0.3555	0.3561	0.3564	20°
30°	0.5236	0.5243	0.5263	0.5294	0.5334	0.5379	0.5422	0.5459	0.5484	0.5493	30°
40°	0.6981	0.6997	0.7043	0.7117	0.7213	0.7323	0.7436	0.7535	0.7604	0.7629	40°
50°	0.8727	0.8756	0.8842	0.8982	0.9173	0.9401	0.9647	0.9876	1.0044	1.0107	50°
60°	1.0472	1.0519	1.0660	1.0896	1.1226	1.1643	1.2126	1.2619	1.3014	1.3170	60°
70°	1.2217	1.2286	1.2495	1.2853	1.3372	1.4068	1.4944	1.5959	1.6918	1.7354	70°
80°	1.3963	1.4057	1.4344	1.4846	1.5597	1.6660	1.8125	2.0119	2.2653	2.4363	80°
90°	1.5708	1.5828	1.6200	1.6858	1.7868	1.9356	2.1565	2.5046	3.1534	∞	90°

(b) Second kind

$$E(k, \phi) = \int_0^{\phi} \sqrt{1 - k^2 \sin^2 \phi}\, d\phi \qquad k = \sin \omega$$

ω	0°	10°	20°	30°	40°	50°	60°	70°	80°	90°	ω
k / φ	0	0.1737	0.3420	0.5000	0.6428	0.7660	0.8660	0.9397	0.9848	1.0000	k / φ
0°	0.0000	0.0000	0.0000	0.0000	0.0000	0.0000	0.0000	0.0000	0.0000	0.0000	0°
10°	0.1745	0.1745	0.1744	0.1743	0.1742	0.1740	0.1739	0.1738	0.1737	0.1736	10°
20°	0.3491	0.3489	0.3483	0.3473	0.3462	0.3450	0.3438	0.3429	0.3422	0.3420	20°
30°	0.5236	0.5229	0.5209	0.5179	0.5141	0.5100	0.5061	0.5029	0.5007	0.5000	30°
40°	0.6981	0.6966	0.6921	0.6851	0.6763	0.6667	0.6575	0.6497	0.6446	0.6428	40°
50°	0.8727	0.8698	0.8614	0.8483	0.8317	0.8134	0.7954	0.7801	0.7697	0.7660	50°
60°	1.0472	1.0426	1.0290	1.0076	0.9801	0.9493	0.9184	0.8914	0.8728	0.8660	60°
70°	1.2217	1.2149	1.1949	1.1632	1.1221	1.0750	1.0266	0.9830	0.9514	0.9397	70°
80°	1.3963	1.3870	1.3597	1.3161	1.2590	1.1926	1.1225	1.0565	1.0054	0.9848	80°
90°	1.5708	1.5589	1.5238	1.4675	1.3931	1.3055	1.2111	1.1184	1.0401	1.0000	90°

13.08 OTHER ELLIPTIC INTEGRALS, NORMAL FORM

$$\lambda = \sqrt{1 - k^2 \sin^2 \phi} \qquad\qquad k = \sin \omega$$

$$\kappa = \frac{F - E}{k^2} \qquad\qquad l = \cos \omega$$

$$\int_0^{\phi} \frac{\sin^2 \phi \, d\phi}{\lambda} = \kappa \qquad\qquad \int_0^{\phi} \frac{\cos^2 \phi \, d\phi}{\lambda} = F - \kappa$$

$$\int_0^{\phi} \frac{\sin^2 \phi \, d\phi}{\lambda^2} = \frac{(F - \kappa)\lambda - \sin \phi \cos \phi}{l^2 \lambda} \qquad\qquad \int_0^{\phi} \frac{\cos^2 \phi \, d\phi}{\lambda^2} = \frac{\kappa \lambda + \sin \phi \cos \phi}{\lambda}$$

ω	$F\left(k,\frac{\pi}{2}\right)$	$E\left(k,\frac{\pi}{2}\right)$	ω	$F\left(k,\frac{\pi}{2}\right)$	$E\left(k,\frac{\pi}{2}\right)$	ω	$F\left(k,\frac{\pi}{2}\right)$	$E\left(k,\frac{\pi}{2}\right)$
0°	1.5708	1.5708	50°	1.9356	1.3055	82°0′	3.3699	1.0278
1°	1.5709	1.5707	51°	1.9539	1.2963	82°12′	3.3946	1.0267
2°	1.5713	1.5703	52°	1.9729	1.2870	82°24′	3.4199	1.0256
3°	1.5719	1.5697	53°	1.9927	1.2776	82°36′	3.4460	1.0245
4°	1.5727	1.5689	54°	2.0133	1.2682	82°48′	3.4728	1.0234
5°	1.5738	1.5678	55°	2.0347	1.2587	83°0′	3.5004	1.0223
6°	1.5751	1.5665	56°	2.0571	1.2492	83°12′	3.5288	1.0213
7°	1.5767	1.5650	57°	2.0804	1.2397	83°24′	3.5581	1.0202
8°	1.5785	1.5632	58°	2.1047	1.2301	83°36′	3.5884	1.0192
9°	1.5805	1.5611	59°	2.1300	1.2206	83°48′	3.6196	1.0182
10°	1.5828	1.5589	60°	2.1565	1.2111	84°0′	3.6519	1.0172
11°	1.5854	1.5564	61°	2.1842	1.2015	84°12′	3.6853	1.0163
12°	1.5882	1.5537	62°	2.2132	1.1921	84°24′	3.7198	1.0153
13°	1.5913	1.5507	63°	2.2435	1.1826	84°36′	3.7557	1.0144
14°	1.5946	1.5476	64°	2.2754	1.1732	84°48′	3.7930	1.0135
15°	1.5981	1.5442	65°	2.3088	1.1638	85°0′	3.8317	1.0127
16°	1.6020	1.5405	66°	2.3439	1.1546	85°12′	3.8721	1.0118
17°	1.6061	1.5367	67°	2.3809	1.1454	85°24′	3.9142	1.0110
18°	1.6105	1.5326	68°	2.4198	1.1362	85°36′	3.9583	1.0102
19°	1.8151	1.5283	69°	2.4610	1.1273	85°48′	4.0044	1.0094
20°	1.6200	1.5238	70°0′	2.5046	1.1184	86°0′	4.0528	1.0087
21°	1.6252	1.5191	70°30′	2.5273	1.1140	86°12′	4.1037	1.0079
22°	1.6307	1.5142	71°0′	2.5507	1.1096	86°24′	4.1574	1.0072
23°	1.6365	1.5090	71°30′	2.5749	1.1053	86°36′	4.2142	1.0065
24°	1.6426	1.5037	72°0′	2.5998	1.1011	86°48′	4.2746	1.0059
25°	1.6490	1.4981	72°30′	2.6256	1.0968	87°0′	4.3387	1.0053
26°	1.6557	1.4924	73°0′	2.6521	1.0927	87°12′	4.4073	1.0047
27°	1.6627	1.4864	73°30′	2.6796	1.0885	87°24′	4.4812	1.0041
28°	1.6701	1.4803	74°0′	2.7081	1.0844	87°36′	4.5609	1.0036
29°	1.6777	1.4740	74°30′	2.7375	1.0804	87°48′	4.6477	1.0031
30°	1.6858	1.4675	75°0′	2.7681	1.0764	88°0′	4.7427	1.0026
31°	1.6941	1.4608	75°30′	2.7998	1.0725	88°12′	4.8479	1.0022
32°	1.7028	1.4539	76°0′	2.8327	1.0686	88°24′	4.9654	1.0017
33°	1.7119	1.4469	76°30′	2.8669	1.0648	88°36′	5.0988	1.0014
34°	1.7214	1.4397	77°0′	2.9026	1.0611	88°48′	5.2527	1.0010
35°	1.7313	1.4323	77°30′	2.9397	1.0574	89°0′	5.4349	1.0008
36°	1.7415	1.4248	78°0′	2.9786	1.0538	89°6′	5.5402	1.0006
37°	1.7522	1.4171	78°30′	3.0192	1.0502	89°12′	5.6579	1.0005
38°	1.7633	1.4092	79°0′	3.0617	1.0468	89°18′	5.7914	1.0005
39°	1.7748	1.4013	79°30′	3.1064	1.0434	89°24′	5.9455	1.0003
40°	1.7868	1.3931	80°0′	3.1534	1.0401	89°30′	6.1278	1.0002
41°	1.7992	1.3849	80°12′	3.1729	1.0388	89°36′	6.3509	1.0001
42°	1.8122	1.3765	80°24′	3.1928	1.0375	89°42′	6.6385	1.0001
43°	1.8256	1.3680	80°36′	3.2132	1.0363	89°48′	7.0440	1.0000
44°	1.8396	1.3594	80°48′	3.2340	1.0350	89°54′	7.7371	1.0000
45°	1.8541	1.3506	81°0′	3.2553	1.0338	90°0′	∞	1.0000
46°	1.8692	1.3418	81°12′	3.2771	1.0326			
47°	1.8848	1.3329	81°24′	3.2995	1.0313			
48°	1.9011	1.3238	81°36′	3.3223	1.0302			
49°	1.9180	1.3147	81°48′	3.3458	1.0290			

(Ref. 13.03, p. 85)

14
ORDINARY DIFFERENTIAL EQUATIONS

(1) Definitions

A differential equation is an *algebraic* or *transcendent equality* involving differentials or derivatives. The *order* of a differential equation is the order of the highest derivative. The *degree* of a differential equation is the algebraic degree of the highest derivative. The *number of independent variables* defines the differential equation as an *ordinary differential equation* (single independent variable) or a *partial differential equation* (two or more independent variables). A *homogeneous* differential equation has all terms of the same degree in the independent variable and its derivatives. If one or more terms do not involve the independent variable, the differential equation is *nonhomogeneous*. A homogeneous equation obtained from a nonhomogeneous equation by setting the nonhomogeneous terms equal to zero is the *reduced equation*. A *linear differential equation* consists of linear terms. A linear term is one which is of the first degree in the independent variables and their derivatives.

(2) Solution

An ordinary differential equation of order n given in general form as

$$F[x, y(x), y'(x), y''(x), \ldots, y^{(n)}(x)] = 0$$

has a *general solution* (general integral)

$$y = y(x, C_1, C_2, \ldots, C_n)$$

where $C_1, C_2, \ldots, C_n$ are *arbitrary* and *independent constants*. A general solution of a reduced differential equation is the *complementary solution* (complementary function). A *particular solution* is a special case of the general solution with definite values assigned to the constants. A *singular solution* is a solution which cannot be obtained from the general solution by specifying the values of the arbitrary constants.

14.02 SPECIAL FIRST-ORDER DIFFERENTIAL EQUATIONS

(1) Direct Separation

Equation	Solution
$y' = f(x)$	$y = \int f(x) \, dx + C$
$y' = g(y)$	$x = \int \dfrac{dy}{g(y)} + C$
$y' = f(x)g(y)$	$\int \dfrac{dy}{g(y)} = \int f(x) \, dx + C$
$y' = \dfrac{f(x)}{g(y)}$	$\int g(y) \, dy = \int f(x) \, dx + C$
$y' = \dfrac{g(y)}{f(x)}$	$\int \dfrac{dy}{g(y)} = \int \dfrac{dx}{f(x)} + C$

(2) Separation by Substitution

Equation, substitution	Solution
$yf(x, y) \, dx + xg(x, y) \, dy = 0$	
$y = rx$	$\int \dfrac{dx}{x} = \int \dfrac{g(r) \, dr}{f(r) + rg(r)} + C$
$yf(x, y) \, dx + xg(x, y) \, dy = 0$	
$y = \dfrac{r}{x}$	$\int \dfrac{dx}{x} = \int \dfrac{f(r) \, dr}{r[g(r) - f(r)]} + C$

(1) Direct Solution $(a = \text{constant})$

Equation	Solution
$y'' = a$	$y = \frac{1}{2}ax^2 + C_1 x + C_2$
$y'' = f(x)$	$y = \int\int f(x)\, dx\, dx + C_1 x + C_2$

(2) Substitution $(y' = \psi)$

Equation

Solution

$y'' = f(y)$

$$\int \psi\, d\psi = \int f(y)\, dy + C$$

$$x = \pm \int \frac{dy}{\sqrt{2\int f(y)\, dy + C_1}} + C_2$$

$y'' = f(y')$

$$\int \frac{d\psi}{f(\psi)} = \int dx + C$$

$$x = \int \frac{d\psi}{f(\psi)} + C_1 \qquad\qquad y = \int \frac{\psi\, d\psi}{f(\psi)} + C_2$$

$y'' = f(x, y')$

$$\psi' = f(x, \psi)$$

$$\psi = f(x, C_1) \qquad\qquad y = \int f(x, C_1)\, dx + C_2$$

$y'' = f(y, y')$

$$\psi \frac{d\psi}{dy} = f(y, \psi)$$

$$\psi = f(y, C_1) \qquad\qquad x = \int \frac{dy}{f(y, C_1)} + C_2$$

14.04 nth-ORDER DIFFERENTIAL EQUATION, SPECIAL CASE

Equation

Solution

$\dfrac{d^{(n)}y}{dx^n} = f(x)$

$$y = \frac{1}{(n-1)!} \int_0^x f(\tau)(x - \tau)^{n-1}\, d\tau + g(x)$$

$$g(x) = C_0 + C_1 x + C_2 x^2 + \cdots + C_{n-1} x^{n-1}$$

14.05 EXACT DIFFERENTIAL EQUATION

If M and N are functions of (x, y) and $\partial M/\partial y = \partial N/\partial x$, then $M\, dx + N\, dy = 0$ is an exact differential equation, the solution of which is

$$\int M\, dx + \int \left(N - \int \frac{\partial M}{\partial y}\, dx\right) dy + C = 0 \qquad \text{or} \qquad \int N\, dy + \int \left(M - \int \frac{\partial N}{\partial x}\, dy\right) dx + C = 0$$

If the condition $\partial M/\partial y = \partial N/\partial x$ is not satisfied, there exists a function $\psi(x, y) = \psi$ (*integrating factor*) such that

$$\frac{\partial(\psi M)}{\partial y} = \frac{\partial(\psi N)}{\partial x}$$

(1) Variable Coefficients $y' + P(x)y = Q(x)$

Condition

$Q(x) = 0$

$Q(x) \neq 0$

Solution

$y = C \exp\left[-\int P(x)\,dx\right]$

$y = \exp\left[-\int P(x)\,dx\right]$

$\times \left\{\int Q(x) \exp\left[\int P(x)\,dx\right] dx + C\right\}$

(2) Constant Coefficients $(A, B, \alpha, \beta = \text{constants})$ $y' + By = Q(x)$

Condition

$Q(x) = 0$

$Q(x) = A$

$Q(x) = Ax$

$Q(x) = Ax^2$

$Q(x) = Af(x)$

$Q(x) = Ae^{\alpha x}$

$Q(x) = A \sin \beta x$

$Q(x) = A \cos \beta x$

$Q(x) = Ae^{\alpha x} \sin \beta x$

$Q(x) = Ae^{\alpha x} \cos \beta x$

Solution

$y = Ce^{-Bx}$

$y = Ce^{-Bx} + \dfrac{A}{B}$

$y = Ce^{-Bx} + \dfrac{A}{B}\left(x - \dfrac{1}{B}\right)$

$y = Ce^{-Bx} + \dfrac{A}{B}\left(x^2 - \dfrac{2x}{B} + \dfrac{2}{B^2}\right)$

$y = Ce^{-Bx} + \dfrac{A}{B}\left[f(x) - \dfrac{f'(x)}{B} + \dfrac{f''(x)}{B^2} - \cdots\right]$

$y = Ce^{-Bx} + \dfrac{A}{\alpha + B}e^{\alpha x}$

$y = Ce^{-Bx} + \dfrac{A(B \cos \beta x - \beta \sin \beta x)}{\beta^2 + B^2}$

$y = Ce^{-Bx} + \dfrac{A(B \cos \beta x + \beta \sin \beta x)}{\beta^2 + B^2}$

$y = Ce^{-Bx} - \dfrac{A[(\alpha + B) \sin \beta x + \beta \cos \beta x]}{\beta^2 + (\alpha + B)^2}$

$y = Ce^{-Bx} + \dfrac{A[\beta \sin \beta x + (\alpha + B) \cos \beta x]}{\beta^2 + (\alpha + B)^2}$

(3) Bernoulli's Equation $y' + P(x)y = Q(x)y^n$

Substitution:

$y = z^{1/(1-n)}$ $y' = \dfrac{1}{1-n} z^{n/(1-n)} z'$

Reduced equation:

$z' + (1 - n)P(x)z = (1 - n)Q(x)$

Solution:

$y^{1-n} = \exp\left[(n - 1)\int P(x)\,dx\right]\left\{(1 - n)\int Q(x) \exp\left[(1 - n)\int P(x)\,dx\right] dx + C\right\}$

(1) Standard Form

A linear differential equation of order n with constant coefficients is given as

$$y^{(n)} + a_1 y^{(n-1)} + a_2 y^{(n-2)} + \cdots + a_n y = f(x)$$

where $f(x)$ is an *arbitrary function* of x.

(2) General Solution

For such an equation, the general solution is

$$y = y_C + y_P$$

where y_C = *complementary function* and y_P = *particular solution*.

(3) Complementary Function

By the substitution $y = e^{\lambda x}$ the reduced differential equation transforms into

$$f(\lambda)e^{\lambda x} = (\lambda^n + a_1\lambda^{n-1} + a_2\lambda^{n-2} + \cdots + a_n)e^{\lambda x} = 0$$

where $f(\lambda) = 0$ is the *characteristic equation*, the roots of which take one of the forms given below (or their combinations) and yield the *coefficients of the complementary functions* given below.

Roots	Complementary function
Real, distinct $\lambda_1 \neq \lambda_2 \neq \cdots \neq \lambda_{n-1} \neq \lambda_n$	$y_C = C_1 e^{\lambda_1 x} + C_2 e^{\lambda_2 x} + \cdots + C_n e^{\lambda_n x}$
Real, repeated $\lambda_1 = \lambda_2 = \cdots = \lambda_{n-1} = \lambda_n$	$y_C = (C_1 + C_2 x + C_3 x + \cdots + C_n x^{n-1})e^{\lambda x}$
Complex, distinct $\lambda_1 = \alpha + \beta i \qquad \lambda_2 = \alpha - \beta i$ $\cdots\cdots\cdots\cdots\cdots\cdots\cdots\cdots\cdots$ $\lambda_{n-1} = \gamma + \delta i \qquad \lambda_n = \gamma - \delta i$	$y_C = e^{\alpha x}(C_1 \cos \beta x + C_2 \sin \beta x)$ $\qquad + \cdots + e^{\gamma x}(C_{n-1} \cos \delta x + C_n \sin \delta x)$
Complex, repeated $\lambda_1 = \lambda_3 = \cdots \lambda_{n-1} = \alpha + \beta i$ $\lambda_2 = \lambda_4 = \cdots \lambda_n = \alpha - \beta i$	$y_C = e^{\alpha x}(C_1 + C_3 x + \cdots + C_{n-1}x^{n/2-1}) \cos \beta x$ $\qquad + e^{\alpha x}(C_2 + C_4 x + \cdots + C_n x^{n/2}) \sin \beta x$

(4) Particular Solution

$$y_P = e^{\lambda_n x} \int e^{(\lambda_{n-1}-\lambda_n)x} \int e^{(\lambda_{n-2}-\lambda_{n-1})x} \cdots \int e^{(\lambda_1-\lambda_2)x} \int e^{-\lambda_1 x} f(x) \, (dx)^n$$

$$y'' + ay = f(x) \qquad \lambda^2 + a = 0 \qquad \boxed{a \neq 0}$$

(1) Complementary Solution (a = real number)

$$a < 0$$
$$\lambda_{1,2} = \pm\sqrt{-a} = \pm\alpha$$
$$y_C = A\cosh\alpha x + B\sinh\alpha x$$

$$a > 0$$
$$\lambda_{1,2} = \pm\sqrt{-a} = \pm i\alpha$$
$$y_C = A\cos\alpha x + B\sin\alpha x$$

(2) Particular Solution (c, b, ω = real numbers)

$$y_P = \frac{1}{a}\left[f(x) - \frac{f''(x)}{a} + \frac{f^{1v}(x)}{a^2} - \frac{f^{v1}(x)}{a^3} + \cdots\right]$$

$$u = a + b^2 - \omega^2$$
$$v = 2b\omega$$

$f(x)$	y_P	$f(x)$	y_P
c	$\dfrac{c}{a}$	cx^2	$\dfrac{c}{a}\left(x^2 - \dfrac{2}{a}\right)$
cx	$\dfrac{cx}{a}$	ce^{bx}	$\dfrac{ce^{bx}}{a + b^2}$
$c\cos\omega x$	$\dfrac{c\cos\omega x}{a - \omega^2}$	$ce^{bx}\cos\omega x$	$c\,\dfrac{u\cos\omega x + v\sin\omega x}{u^2 + v^2}e^{bx}$
$c\sin\omega x$	$\dfrac{c\sin\omega x}{a - \omega^2}$	$ce^{bx}\sin\omega x$	$c\,\dfrac{u\sin\omega x - v\cos\omega x}{u^2 + v^2}e^{bx}$

14.09 THIRD-ORDER DIFFERENTIAL EQUATION

$$y''' + py'' + qy' + ry = f(x) \qquad \lambda^3 + p\lambda^2 + q\lambda + r = 0 \qquad \boxed{p, q, r \neq 0}$$

(1) Complementary Solution (p, q, r = real numbers, D given in Sec. 1.05)

$D < 0$	$\lambda_1 = \alpha,\ \lambda_2 = \beta,\ \lambda_3 = \gamma$	$y_C = Ae^{\alpha x} + Be^{\beta x} + Ce^{\gamma x}$
$D = 0$	$\lambda_1 = \lambda_2 = \alpha,\ \lambda_3 = \gamma$	$y_C = (A + Bx)e^{\alpha x} + Ce^{\gamma x}$
$D = 0$	$\lambda_1 = \lambda_2 = \lambda_3 = \alpha$	$y_C = (A + Bx + Cx^2)e^{\alpha x}$
$D > 0$	$\lambda_1, \lambda_2 = \alpha \pm i\beta,\ \lambda_3 = \gamma$	$y_C = (A\cos\beta x + B\sin\beta x)e^{\alpha x} + Ce^{\gamma x}$

(2) Particular Solution (b, c, ω = real numbers)

$$y_P = \sum_{k=0}^{\infty} s_k f^{(k)}(x) \qquad s_0 = \frac{1}{r} \qquad s_1 = -\frac{q}{r^2} \qquad s_2 = -\frac{pr - q^2}{r^3}$$

$$s_k = -\frac{1}{r}\left(qs_{k-1} + ps_{k-2} + s_{k-3}\right)$$

$$u = r + qb + pb^2 + b^3 - 2(p + 3b)\omega^2$$
$$v = (q + 2pb + 3b^2)\omega - \omega^3$$

$f(x)$	y_P	$f(x)$	y_P
c	$\dfrac{c}{r}$	cx^2	$\dfrac{c}{r}\left[x^2 - \dfrac{2q}{r}x + \dfrac{2(q^2 - pr)}{r^2}\right]$
cx	$\dfrac{c}{r}\left(x - \dfrac{q}{r}\right)$	ce^{bx}	$\dfrac{ce^{bx}}{b^3 + pb^2 + qb + r}$
$c\cos\omega x$	$c\,\dfrac{u\cos\omega x + v\sin\omega x*}{u^2 + v^2}$	$ce^{bx}\cos\omega x$	$c\,\dfrac{u\cos\omega x + v\sin\omega x}{u^2 + v^2}e^{bx}$
$c\sin\omega x$	$c\,\dfrac{u\sin\omega x - v\cos\omega x*}{u^2 + v^2}$	$ce^{bx}\sin\omega x$	$c\,\dfrac{u\sin\omega x - v\cos\omega x}{u^2 + v^2}e^{bx}$

*$b = 0$ in u and v.

$$y'' + py' + qy = f(x) \qquad \lambda^2 + p\lambda + q = 0 \qquad \boxed{p, q \neq 0}$$

(1) Complementary Solution $\quad (p, q = \text{real numbers}, D = p^2 - 4q)$

$D > 0$	$D = 0$	$D < 0$
$\lambda_{1,2} = \dfrac{-p \pm \sqrt{D}}{2} = \alpha, \beta$	$\lambda_{1,2} = -\dfrac{p}{2} = \lambda$	$\lambda_{1,2} = \dfrac{-p \pm \sqrt{D}}{2} = \alpha \pm i\beta$
$y_C = Ae^{\alpha x} + Be^{\beta x}$	$y_C = (A + Bx)e^{\lambda x}$	$y_C = (A \cos \beta x + B \sin \beta x)e^{\alpha x}$

(2) Particular solution $\quad (b, c, \omega = \text{real numbers})$

$$y_P = \frac{1}{q}\left\{ f(x) + \sum_{k=1}^{\infty} \left(\frac{-1}{q}\right)^k \left[(p - q)^k + q^k\right] f^{(k)}(x) \right\}$$

$$\boxed{\begin{aligned} u &= q + bp + b^2 - \omega^2 \\ v &= (2b + p)\omega \end{aligned}}$$

$f(x)$	y_P	$f(x)$	y_P
c	$\dfrac{c}{q}$	cx^2	$\dfrac{c}{q}\left[x^2 - \dfrac{2px}{q} + \dfrac{2(p^2 - q)}{q^2}\right]$
cx	$\dfrac{c}{q}\left(x - \dfrac{p}{q}\right)$	ce^{bx}	$\dfrac{ce^{bx}}{q + pb + b^2}$
$c \cos \omega x$	$c\,\dfrac{u \cos \omega x + v \sin \omega x}{u^2 + v^2}*$	$ce^{bx} \cos \omega x$	$c\,\dfrac{u \cos \omega x + v \sin \omega x}{u^2 + v^2}e^{bx}$
$c \sin \omega x$	$c\,\dfrac{u \sin \omega x - v \cos \omega x}{u^2 + v^2}*$	$ce^{bx} \sin \omega x$	$c\,\dfrac{u \sin \omega x - v \cos \omega x}{u^2 + v^2}e^{bx}$

$\quad *b = 0$ in u and v.

14.11 FOURTH-ORDER DIFFERENTIAL EQUATION

$$y^{iv} + ay = f(x) \qquad \lambda^4 + a = 0 \qquad \boxed{a \neq 0}$$

(1) Complementary Solution $\quad (a = \text{real number})$

$a > 0 \qquad\qquad \lambda_1 = (1 + i)\alpha \qquad \lambda_2 = (1 - i)\alpha \qquad \lambda_3 = -(1 + i)\alpha \qquad \lambda_4 = -(1 - i)\alpha$

$\alpha = \sqrt[4]{\dfrac{a}{4}} \qquad\qquad y_C = e^{\alpha x}(A \cos \alpha x + B \sin \alpha x) + e^{-\alpha x}(C \cos \alpha x + D \sin \alpha x)$

$a < 0 \qquad\qquad \lambda_1 = \beta \qquad\qquad\quad \lambda_2 = -\beta \qquad\qquad \lambda_3 = i\beta \qquad\qquad\quad \lambda_4 = -i\beta$

$\beta = \sqrt[4]{-a} \qquad\qquad y_C = A \cosh \beta x + B \sinh \beta x + C \cos \beta x + D \sin \beta x$

(2) Particular Solution $\quad (b, c, \omega = \text{real numbers})$

$$y_P = \frac{1}{a}\left[f(x) - \frac{f^{iv}(x)}{a} + \frac{f^{viii}(x)}{a^2} - \frac{f^{xxii}(x)}{a^3} + \cdots \right]$$

$$\boxed{\begin{aligned} u &= a + (b^2 - \omega^2)^2 \\ v &= 4b(b^2 - \omega^2) \end{aligned}}$$

$f(x)$	y_P	$f(x)$	y_P
c	$\dfrac{c}{a}$	cx^2	$\dfrac{cx^2}{a}$
cx	$\dfrac{cx}{a}$	ce^{bx}	$\dfrac{ce^{bx}}{a + b^4}$
$c \cos \omega x$	$\dfrac{c \cos \omega}{a + \omega^4}$	$ce^{bx} \cos \omega x$	$c\,\dfrac{u \cos \omega x + v \sin \omega x}{u^2 + v^2}e^{bx}$
$c \sin \omega x$	$\dfrac{c \sin \omega x}{a + \omega^4}$	$ce^{bx} \sin \omega x$	$c\,\dfrac{u \sin \omega x - v \cos \omega x}{u^2 + v^2}e^{bx}$

$$y^{iv} + py'' + qy = f(x) \qquad \lambda^4 + p\lambda^2 + q = 0 \qquad \boxed{p, q \neq 0}$$

(1) Complementary Solution $(p, q = \text{real numbers}, D = p^2 - 4q)$

Conditions		Solution
$p > 0$	$D > 0$	$\lambda_{1,\,2} = \pm i \sqrt{\tfrac{1}{2}(p + \sqrt{D})} = \pm i\alpha \qquad \lambda_{3,\,4} = \pm i \sqrt{\tfrac{1}{2}(p - \sqrt{D})} = \pm i\beta$ $y_C = A \cos \alpha x + B \sin \alpha x + C \cos \beta x + D \sin \beta x$
	$D = 0$	$\lambda_1 = \lambda_2 = i\sqrt{\dfrac{p}{2}} = i\lambda \qquad\qquad \lambda_3 = \lambda_4 = -i\sqrt{\dfrac{p}{2}} = -i\lambda$ $y_C = (A + Bx)\cos \lambda x + (C + Dx)\sin \lambda x$
	$D < 0$	$\lambda_{1,2} = \pm i \sqrt{\dfrac{p}{2} + \dfrac{i}{2}\sqrt{-D}} = \pm \dfrac{i}{\sqrt{2}}\left(\sqrt{\sqrt{q}+\dfrac{p}{2}} + i\sqrt{\sqrt{q}-\dfrac{p}{2}}\right) = \pm(\bar{\alpha}i - \bar{\beta})$ $\lambda_{3,\,4} = \pm i \sqrt{\dfrac{p}{2} + \dfrac{i}{2}\sqrt{-D}} = \pm \dfrac{i}{\sqrt{2}}\left(\sqrt{\sqrt{q}+\dfrac{p}{2}} + i\sqrt{\sqrt{q}-\dfrac{p}{2}}\right) = \pm(\bar{\alpha}i + \bar{\beta})$ $y_C = \cos \bar{\alpha}x(A \cosh \bar{\beta}x + B \sinh \bar{\beta}x) + \sin \bar{\alpha}x(C \cosh \bar{\beta}x + D \sinh \bar{\beta}x)$
$p < 0$	$D > 0$	$\lambda_{1,2} = \pm\sqrt{\tfrac{1}{2}(-p + \sqrt{D})} = \pm\alpha \qquad \lambda_{3,4} = \pm\sqrt{\tfrac{1}{2}(-p - \sqrt{D})} = \pm\beta$ $y_C = A \cosh \alpha x + B \sinh \alpha x + C \cosh \beta x + D \sinh \beta x$
	$D = 0$	$\lambda_1 = \lambda_2 = \sqrt{-\dfrac{p}{2}} = \lambda \qquad\qquad \lambda_3 = \lambda_4 = -\sqrt{-\dfrac{p}{2}} = -\lambda$ $y_C = (A + Bx)\cosh \lambda x + (C + Dx)\sinh \lambda x$
	$D < 0$	$\lambda_{1,\,2} = \pm\sqrt{-\dfrac{p}{2} - \dfrac{i}{2}\sqrt{-D}} = \pm\dfrac{1}{\sqrt{2}}\left(\sqrt{\sqrt{q}-\dfrac{p}{2}} + i\sqrt{\sqrt{q}+\dfrac{p}{2}}\right) = \pm(\bar{\alpha} - i\bar{\beta})$ $\lambda_{3,4} = \pm\sqrt{-\dfrac{p}{2} - \dfrac{i}{2}\sqrt{-D}} = \pm\dfrac{1}{\sqrt{2}}\left(\sqrt{\sqrt{q}-\dfrac{p}{2}} - i\sqrt{\sqrt{q}+\dfrac{p}{2}}\right) = \pm(\bar{\alpha} - i\bar{\beta})$ $y_C = \cosh \bar{\alpha}x(A \cos \bar{\beta}x + B \sin \bar{\beta}x) + \sinh \bar{\alpha}x(C \cos \bar{\beta}x + D \sin \bar{\beta}x)$

(2) Particular Solution $(b, c, \omega = \text{real numbers})$

$$y_P = \sum_{k=0}^{\infty} s_{2k} f^{(2k)}(x) \qquad f^{(0)} = f(x) \qquad f^{(2k)}(x) = d^{2k}f(x)/dx^{2k}$$

$$s_0 = \frac{1}{q} \qquad s_2 = -\frac{p}{q^2} \qquad s_4 = \frac{p^2 - q}{q^3} \qquad s_6 = -\frac{p^3 - 2pq}{q^4} \qquad \boxed{s_{2k} = -\frac{1}{q}(p\,s_{2k-2} + s_{2k-4})}$$

$f(x)$	y_P	$f(x)$	y_P
c	$\dfrac{c}{q}$	cx^2	$\dfrac{c}{q}\left(x^2 - \dfrac{2p}{q}\right)$
cx	$\dfrac{cx}{q}$	ce^{bx}	$\dfrac{ce^{bx}}{q + pb^2 + b^4}$
$c \cos \omega x$	$c\dfrac{\cos \omega x}{q - p\omega^2 + \omega^4}$	$c \sin \omega x$	$c\dfrac{\sin \omega x}{q - p\omega^2 + \omega^4}$

(1) Standard Form

Euler's differential equation of order n is given as

$$a_0 x^n y^{(n)} + a_1 x^{n-1} y^{(n-1)} + a_2 x^{n-2} y^{(n-2)} + \cdots + a_n y = f(x)$$

where $f(x)$ is an arbitrary function of x.

(2) Complementary Solution

By the substitution $y = x^\lambda$ the reduced differential equation transforms into

$$f(\lambda) x^\lambda = \left[a_0 \frac{\lambda!}{(\lambda - n)!} + a_1 \frac{\lambda!}{(\lambda - n + 1)!} + a_2 \frac{\lambda!}{(\lambda - n + 2)!} + \cdots + a_n \right] x^\lambda = 0$$

where $f(\lambda) = 0$ is the *characteristic equation*, the roots of which take one of the forms (or their combinations) given in Sec. 14.07 (3).

(3) Particular Solution

In general, there is no *general method of finding a particular solution* of this differential equation. A method which sometimes yields a solution is to assume a series

$$y_P = \frac{1}{a_n} [f(x) + A_1 f'(x) + A_2 f''(x) + \cdots].$$

where $A_1, A_2, \ldots$ are functions given by the following conditions:

$$A_1 = -x \bar{a}_{n-1}$$

$$A_2 = -x^2 \bar{a}_{n-2} - x \bar{a}_{n-1} A_1$$

$$A_3 = -x^3 \bar{a}_{n-3} - x^2 \bar{a}_{n-2} A_1 - x a_{n-1} A_2$$

$$\cdots \cdots \cdots \cdots \cdots \cdots \cdots \cdots \cdots \cdots \cdots \cdots \cdots$$

$$A_n = -x^n \bar{a}_0 - x^{n-1} \bar{a}_1 A_1 - x^{n-2} \bar{a}_2 A_2 - x^{n-3} \bar{a}_3 A_3 - \cdots$$

$$A_{n+1} = 0 - x^n \bar{a}_0 A_1 - x^{n-1} \bar{a}_1 A_2 - x^{n-2} \bar{a}_2 A_3 - \cdots$$

$$\cdots \cdots \cdots \cdots \cdots \cdots \cdots \cdots \cdots \cdots \cdots \cdots \qquad \bar{a}_j = \frac{a_j}{a_n}$$

14.14   SECOND-ORDER EULER'S DIFFERENTIAL EQUATION

$$x^2 y'' + bxy' + cy = f(x) \qquad\qquad \lambda^2 + p\lambda + q = 0$$

(1) Complementary Solution $(D = p^2 - 4q)$

$D > 0$	$D = 0$	$D < 0$
$\lambda_{1,2} = \dfrac{-p \pm \sqrt{D}}{2} = \alpha, \beta$	$\lambda_{1,2} = -\dfrac{p}{2} = \lambda$	$\lambda_{1,2} = \dfrac{-p \pm \sqrt{D}}{2} = \alpha \pm i\beta$
$y_C = A x^\alpha + B x^\beta$	$y_C = x^\lambda (A \ln |x| + B)$	$y_C = x^\alpha [A \cos (\beta \ln |x|) + B \sin (\beta \ln |x|)]$

(2) Particular Solution

The method of function series [Sec. 14.12 (3)] frequently yields a particular solution.

(1) Concept

Some differential equations, particularly homogeneous linear equations with variable coefficients, can be solved by assuming a solution in the form of an *infinite power series* such as

$$y = b_0 + b_1 x + b_2 x^2 + \cdots + b_r x^r + \cdots = \sum_{r=0}^{\infty} b_r x^r$$

the first and higher derivatives of which are

$$y' = \sum_{r=0}^{\infty} r b_r x^{r-1} \qquad\qquad y'' = \sum_{r=0}^{\infty} r(r-1) b_r x^{r-2}, \ldots$$

After these expressions have been substituted in the given differential equation, the terms in like power of x are combined, and the coefficients of each power of x are set equal to zero. The system of equations thus obtained yields the values of the constants of the series of which n (corresponding to the order) must remain unknown (arbitrary) as the constants of integration.

(2) Transformation

Frequently it serves to an advantage to transform the given differential equation

$$y'' + a(x)y' + b(x)y = 0$$

by the substitution of

$$y = y(x) = u(x) \exp\left[-\tfrac{1}{2}\int a(x)\, dx\right] \qquad \text{to} \qquad u'' + \left[b(x) - \frac{a(x)'}{2} - \left(\frac{a(x)}{2}\right)^2\right] u = 0$$

(3) Orthogonal Polynomials

A *set of polynomials* $p_n(x)$ $(n = 0, 1, 2, \ldots)$ of degree n in x is orthogonal in the interval (a, b) with respect to the weight function $w(x)$ if

$$\int_a^b w(x) p_m(x) p_n(x)\, dx = \begin{cases} 0 & \text{for } m \neq n \\ a_n & \text{for } m = n \end{cases} \qquad m, n = 0, 1, 2, \ldots$$

Under certain conditions, this relationship admits the representation of a function in the form

$$f(x) = \sum_{n=0}^{\infty} C_n p_n(x) \qquad \text{with} \qquad C_n = \frac{1}{a_n} \int_a^b f(x) w(x) p_n(x)\, dx$$

(4) Classical Orthogonal Polynomials

Classical orthogonal polynomials of particular interest are designated by the names of their discoverers. The *Legendre* (Sec. 14.17), *Chebyshev* (Sec. 14.18), *Laguerre* (Sec. 14.19), and *Hermite polynomials* (Sec. 14.20) have *two typical properties*:

(a) $p_n(x)$ *satisfies the differential equation*

$$a(x)y'' + b(x)y' + cy = 0$$

where $a(x), b(x)$ are independent of n, c, and n, c are independent of x.

(b) The polynomial *can be represented by the generalized Rodrigues formula as*

$$p_n(x) = \frac{1}{k_n w(x)} \frac{d^n}{dx^n} \{w(x)[\phi(x)]^n\}$$

where $\phi(x)$ is a polynomial of the first or second degree.

(1) Gauss' Differential Equation　　　$(\alpha, \beta, \gamma = \text{constants})$

$$x(1-x)y'' - [(\alpha + \beta + 1)x - \gamma]y' - \alpha\beta y = 0$$

(2) Solution　　　$(y = A_1 y_1 + A_2 y_2)$

$y_1 = F(\alpha, \beta, \gamma, x)$

$$= 1 + \frac{\alpha\beta}{\gamma}\frac{x}{1!} + \frac{\alpha(\alpha+1)\beta(\beta+1)}{\gamma(\gamma+1)}\frac{x^2}{2!} + \frac{\alpha(\alpha+1)(\alpha+2)\beta(\beta+1)(\beta+2)}{\gamma(\gamma+1)(\gamma+2)}\frac{x^3}{3!} + \cdots$$

$y_2 = x^{1-\gamma}F(\alpha-\gamma+1, \beta-\gamma+1, 2-\gamma, x)$

$$= x^{1-\gamma}\left[1 + \frac{(\alpha-\gamma+1)(\beta-\gamma+1)}{(2-\gamma)}\frac{x}{1!}\right.$$

$$\left. + \frac{(\alpha-\gamma+1)(\alpha-\gamma+2)(\beta-\gamma+1)(\beta-\gamma+2)}{(2-\gamma)(3-\gamma)}\frac{x^2}{2!} + \cdots\right] \qquad |x| < 1, \gamma \neq 0, 1, 2, \ldots$$

$y_1 = F(\alpha, \beta, \alpha+\beta-\gamma+1, 1-x)$

$y_2 = (1-x)^{\gamma-\alpha-\beta}F(\gamma-\beta, \gamma-\alpha, \gamma-\alpha-\beta+1, 1-x)$ 　　　$|x-1| < 1, \alpha+\beta-\gamma \neq 0, 1, 2, \ldots$

$y_1 = x^{-\alpha}F(\alpha, \alpha-\gamma+1, \alpha-\beta+1, x^{-1})$

$y_2 = x^{-\beta}F(\beta, \beta-\gamma+1, \beta-\alpha+1, x^{-1})$ 　　　$|x| > 1, \alpha-\beta \neq 0, 1, 2, \ldots$

14.17　CONFLUENT HYPERGEOMETRIC DIFFERENTIAL EQUATION

(1) Kummer's Differential Equation　　　$(\beta, \gamma = \text{constants})$

$$xy'' + (\gamma - x)y' - \beta y = 0$$

(2) Solution　　　$(y = A_1 y_1 + A_2 y_2)$

$y_1 = F(\beta, \gamma, x)$

$$= 1 + \frac{\beta}{\gamma}\frac{x}{1!} + \frac{\beta(\beta+1)}{\gamma(\gamma+1)}\frac{x^2}{2!} + \cdots$$

$y_2 = x^{1-\gamma}F(\beta-\gamma+1, 2-\gamma, x)$

$$= x^{1-\gamma}\left[1 + \frac{\beta-\gamma+1}{2-\gamma}\frac{x}{1!} + \frac{(\beta-\gamma+1)(\beta-\gamma+2)}{(2-\gamma)(3-\gamma)}\frac{x^2}{2!} + \cdots\right] \qquad \gamma \neq 0, 1, 2, \ldots$$

$\bar{P}_n(x), \bar{Q}_n(x) =$ Legendre functions

$P_n(x), Q_n(x) =$ Legendre polynomials

$\Lambda[\ ] =$ nested sum (Sec. 8.11)

$(\ )!! =$ double factorial (Sec. 1.03)

(1) Differential Equation

$$(1 - x^2)y'' - 2xy' + n(n+1)y = 0$$

$-\infty < n < +\infty$

Interval: $[-1, +1]$
Weight: $w(x) = 1$

(2) Solution for Àll n: $y = A_1\bar{P}_n(x) + A_2\bar{Q}_n(x)$

$$\bar{P}_n(x) = \bigwedge_{k=0}^{\infty}\left[1 - \frac{(n-2k)(n+2k+1)}{(2k+1)(2k+2)}S\right]$$

$S = x^2$

$$= \left(1 - \frac{n(n+1)}{1 \cdot 2}S\left(1 - \frac{(n-2)(n+3)}{3 \cdot 4}S\left(1 - \frac{(n-4)(n+5)}{5 \cdot 6}S\left(1 - \cdots\right)\right)\right)\right)$$

$$\bar{Q}_n(x) = x \bigwedge_{k=0}^{\infty}\left[1 - \frac{(n-2k-1)(n+2k+2)}{(2k+2)(2k+3)}S\right]$$

$$= x\left(1 - \frac{(n-1)(n+2)}{2 \cdot 3}S\left(1 - \frac{(n-3)(n+4)}{4 \cdot 5}S\left(1 - \frac{(n-5)(n+6)}{6 \cdot 7}S\left(1 - \cdots\right)\right)\right)\right)$$

(3) Solution for $n = 0, 1, 2, \ldots$: $y = A_1 P_n(x) + A_2 Q_n(x)$

$$P_n(x) = \frac{(2n-1)!!}{n!}x^n \bigwedge_{k=0}^{t}\left[1 - \frac{(n-2k)(n-2k-1)}{(2k+2)(2n-2k-1)}R\right]$$

$R = x^{-2}$

$$= \frac{(2n-1)!!}{n!}x^n\left(1 - \frac{n(n-1)}{2(2n-1)}R\left(1 - \frac{(n-2)(n-3)}{4(2n-3)}R\left(1 - \frac{(n-4)(n-5)}{6(2n-5)}R\left(1 - \cdots\right)\right)\right)\right)$$

$$Q_n(x) = \left(\ln\sqrt{\frac{x+1}{x-1}}\right)P_n(x) - \sum_{k=1}^{n}\frac{P_{k-1}(x)P_{n-k}(x)}{k}$$

where $P(x)$ series terminates with $t = n/2$ for even n and with $t = (n-1)/2$ for odd n.

(4) Special Values

$$P_n(0) = \begin{cases} (-1)^{n/2}\dfrac{(n-1)!!}{n!!} & n \text{ even} \\ 0 & n \text{ odd} \end{cases}$$

$$Q_n(0) = \begin{cases} 0 & n \text{ even} \\ (-1)^{(n+1)/2}\dfrac{(n-1)!!}{n!!} & n \text{ odd} \end{cases}$$

$$P_n(1) = 1 \qquad P_n(-1) = (-1)^n$$

$$Q_n(1) = \infty \qquad Q_n(-1) = \infty(-1)^{n+1}$$

(5) Relations   (Sec. A.14)

$$P_{n+1}(x) = \frac{2n+1}{n+1}xP_n(x) - \frac{n}{n+1}P_{n-1}(x)$$

$$Q_{n+1}(x) = \frac{2n+1}{n+1}xQ_n(x) - \frac{n}{n+1}Q_{n-1}(x)$$

$$\int_{-1}^{+1}P_m(x)P_n(x)\,dx = \begin{cases} 0 & m \neq n \\ \dfrac{2}{2n+1} & m = n \end{cases}$$

$$\int_{-1}^{+1}Q_m(x)Q_n(x)\,dx = \begin{cases} 0 & m \neq n \\ \dfrac{2}{2n+1} & m = n \end{cases}$$

(1) Graphs of $P_n(x)$

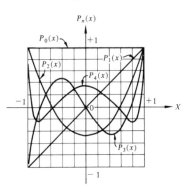

(2) Graphs of $Q_n(x)$

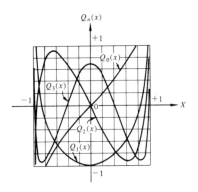

$$P_n(-x) = (-1)^n P_n(x) \qquad |x| \le 1$$

(3) Table of $P_n(x)$

n \ x	0.0	0.2	0.4	0.6	0.8
0	+1.000 000 0	+1.000 000 0	+1.000 000 0	+1.000 000 0	+1.000 000 0
1	0.000 000 0	+0.200 000 0	+0.400 000 0	+0.600 000 0	+0.800 000 0
2	−0.500 000 0	−0.440 000 0	−0.260 000 0	+0.040 000 0	+0.460 000 0
3	0.000 000 0	−0.280 000 0	−0.440 000 0	−0.360 000 0	+0.080 000 0
4	+0.375 000 0	+0.232 000 0	+0.113 000 0	−0.408 000 0	−0.233 000 0
5	0.000 000 0	+0.307 520 0	+0.270 640 0	−0.152 640 0	−0.399 520 0
6	−0.312 500 0	−0.080 576 0	+0.292 636 0	+0.172 096 0	−0.391 796 0
7	0.000 000 0	−0.293 516 8	−0.014 590 4	+0.322 598 4	−0.239 651 2
8	+0.273 437 5	−0.039 564 8	−0.266 999 3	+0.212 339 2	−0.016 655 3
9	0.000 000 0	+0.245 957 1	−0.188 763 6	−0.046 103 0	+0.187 855 3
10	−0.246 093 8	+0.129 072 0	+0.096 835 1	−0.243 662 7	+0.300 529 8
11	0.000 000 0	−0.174 346 4	+0.245 553 1	−0.237 192 7	+0.288 213 4
12	+0.225 585 9	−0.185 136 9	+0.099 488 2	−0.049 414 1	+0.166 441 6

$$Q_n(-x) = (-1)^{n+1} Q_n(x) \qquad |x| \le 1$$

(4) Table of $Q_n(x)$

n \ x	0.0	0.2	0.4	0.6	0.8
0	0.000 000 0	+0.202 732 6	+0.423 648 9	+0.693 147 2	+1.098 612 3
1	−1.000 000 0	−0.959 953 5	−0.830 540 4	−0.584 111 7	−0.121 110 2
2	0.000 000 0	−0.389 202 3	−0.710 148 7	−0.872 274 1	−0.694 638 3
3	+0.666 666 7	+0.509 901 6	+0.080 261 1	−0.482 866 3	−0.845 444 4
4	0.000 000 0	+0.470 367 3	+0.588 794 3	+0.147 195 0	−0.662 643 4
5	−0.533 333 3	−0.238 589 1	+0.359 723 0	+0.545 264 7	−0.277 851 0
6	0.000 000 0	−0.479 455 4	−0.226 865 1	+0.477 127 8	+0.144 688 8
7	+0.457 142 9	+0.026 421 5	−0.476 862 4	+0.064 286 9	+0.453 123 1
8	0.000 000 0	+0.429 431 5	−0.159 139 8	−0.345 164 0	+0.553 081 9
9	−0.406 349 2	+0.138 743 9	+0.303 638 7	−0.448 329 9	+0.432 993 1
10	0.000 000 0	−0.333 765 6	+0.373 991 2	−0.200 448 5	+0.160 375 2
11	+0.369 408 4	−0.253 568 6	+0.009 580 9	+0.177 967 6	−0.148 694 4
12	0.000 000 0	+0.208 750 5	−0.335 479 9	+0.388 407 2	−0.375 007 2

$\bar{T}_n(x), \bar{U}_n(x) =$ Chebyshev functions

$T_n(x), U_n(x) =$ Chebyshev polynomials

$\Lambda[\ ] =$ nested sum (Sec. 8.11)

(1) Differential Equation

$$(1-x^2)y'' - xy' + n^2 y = 0 \qquad -\infty < n < +\infty$$

Interval: $[-1, +1]$

Weight: $w(x) = \dfrac{1}{\sqrt{1-x^2}}$

(2) Solution for All n: $y = A_1 \bar{T}_n(x) + A_2 \bar{U}_n(x)$

$$\bar{T}_n(x) = \bigwedge_{k=0}^{\infty}\left[1 - \frac{n^2-(2k)^2}{(2k+1)(2k+2)}S\right]$$

$S = x^2$

$$= \left(1 - \frac{n^2}{1\cdot 2}S\left(1 - \frac{n^2-2^2}{3\cdot 4}S\left(1 - \frac{n^2-4^2}{5\cdot 6}S\left(1-\cdots\right)\right)\right)\right)$$

$$\bar{U}_n(x) = x\bigwedge_{k=0}^{\infty}\left[1 - \frac{n^2-(2k+1)^2}{(2k+2)(2k+3)}S\right]$$

$$= x\left(1 - \frac{n^2-1}{2\cdot 3}S\left(1 - \frac{n^2-3^2}{4\cdot 5}S\left(1 - \frac{n^2-5^2}{6\cdot 7}S\left(1-\cdots\right)\right)\right)\right)$$

(3) Solution for $n = 0, 1, 2, \ldots$: $y = A_1 T_n(x) + A_2 U_n(x)$

$$T_n(x) = x^n\bigwedge_{k=0}^{t}\left[1 - \frac{(n-2k)(n-2k-1)}{(2k+1)(2k+2)}R\right]$$

$R = \dfrac{1-x^2}{x^2}$

$$= x^n\left(1 - \frac{n(n-1)}{1\cdot 2}R\left(1 - \frac{(n-2)(n-3)}{3\cdot 4}R\left(1 - \frac{(n-4)(n-5)}{5\cdot 6}R\left(1-\cdots\right)\right)\right)\right)$$

$$U_n(x) = \frac{1}{n}\sqrt{1-x^2}\frac{dT_n(x)}{dx} \qquad U_0(x) = \sin^{-1}x$$

where $T_n(x)$ series terminates with $t = n/2$ for even n and with $t = (n-1)/2$ for odd n.

(4) Special Values

$$T_n(0) = \begin{cases}(-1)^{n/2} & n\text{ even}\\ 0 & n\text{ odd}\end{cases} \qquad U_n(0) = \begin{cases}0 & n\text{ even}\\ (-1)^{n+1} & n\text{ odd}\end{cases}$$

$$T_n(1) = 1 \qquad T_n(-1) = (-1)^n \qquad U_n(1) = 0 \qquad U_n(-1) = 0$$

(5) Relations   (Sec. A.14)

$$T_{n+1}(x) = 2xT_n(x) - T_{n-1}(x) \qquad\qquad U_{n+1}(x) = 2xU_n(x) - U_{n-1}(x)$$

$$\int_{-1}^{+1}\frac{T_m(x)T_n(x)}{\sqrt{1-x^2}}dx = \begin{cases}0 & m\neq n\\ \dfrac{\pi}{2} & m=n\neq\\ \pi & m=n=0\end{cases} \qquad \int_{-1}^{+1}\frac{U_m(x)U_n(x)}{\sqrt{1-x^2}}dx = \begin{cases}0 & m\neq n\\ \dfrac{\pi}{2} & m=n\neq 0\\ 0 & m=n=0\end{cases}$$

(1) Graphs of $T_n(x)$

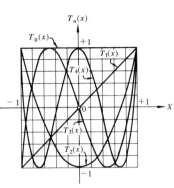

(2) Graphs of $U_n(x)$

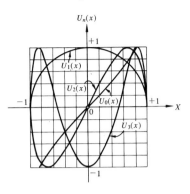

(3) Table of $T_n(x)$

$$T_n(-x) = (-1)^n T_n(x) \qquad |x| \le 1$$

n \ x	0.0	0.2	0.4	0.6	0.8
0	+1.000 000 0	+1.000 000 0	+1.000 000 0	+1.000 000 0	+1.000 000 0
1	0.000 000 0	+0.200 000 0	+0.400 000 0	+0.600 000 0	+0.800 000 0
2	−1.000 000 0	−0.920 000 0	−0.680 000 0	−0.280 000 0	+0.280 000 0
3	0.000 000 0	−0.568 000 0	−0.944 000 0	−0.936 000 0	−0.352 000 0
4	+1.000 000 0	+0.692 800 0	−0.075 200 0	−0.843 200 0	−0.843 200 0
5	0.000 000 0	+0.845 120 0	+0.883 840 0	−0.075 840 0	−0.997 120 0
6	−1.000 000 0	−0.354 752 0	+0.782 272 0	+0.752 192 0	−0.752 192 0
7	0.000 000 0	−0.987 020 8	−0.258 022 4	+0.978 470 4	−0.206 387 2
8	+1.000 000 0	−0.040 056 3	−0.988 689 9	+0.421 972 5	+0.421 972 5
9	0.000 000 0	+0.970 998 3	−0.532 929 5	−0.472 103 4	+0.881 543 2
10	−1.000 000 0	+0.428 455 6	+0.562 346 3	−0.988 496 6	+0.988 496 6
11	0.000 000 0	−0.799 616 0	+0.982 806 6	−0.714 092 5	+0.700 051 4
12	+1.000 000 0	−0.748 302 4	+0.223 899 0	+0.131 585 6	+0.131 585 6

(4) Table of $U_n(x)$

$$U_n(-x) = (-1)^{n+1} U_n(x) \qquad |x| \le 1$$

n \ x	0.0	0.2	0.4	0.6	0.8
0	0.000 000 0	+0.201 357 9	+0.411 516 8	+0.643 501 1	+0.927 295 2
1	+1.000 000 0	+0.979 795 9	+0.916 515 1	+0.800 000 0	+0.600 000 0
2	0.000 000 0	+0.391 918 4	+0.733 212 1	+0.960 000 0	+0.960 000 0
3	−1.000 000 0	−0.823 028 6	−0.329 945 5	+0.352 000 0	+0.936 000 0
4	0.000 000 0	−0.721 129 8	−0.997 168 5	−0.537 600 0	+0.537 600 0
5	+1.000 000 0	+0.534 576 7	−0.467 789 3	−0.997 120 0	−0.075 840 0
6	0.000 000 0	+0.934 960 5	+0.622 937 1	−0.658 944 0	−0.658 944 0
7	−1.000 000 0	−0.160 592 5	+0.966 138 9	+0.206 387 2	−0.978 470 4
8	0.000 000 0	−0.999 197 5	+0.149 740 6	+0.906 608 6	−0.906 608 6
9	+1.000 000 0	−0.239 086 5	−0.846 159 4	+0.881 543 2	−0.472 103 4
10	0.000 000 0	+0.903 562 9	−0.826 668 1	+0.151 243 7	+0.151 243 1
11	−1.000 000 0	+0.600 511 7	+0.184 824 9	−0.700 051 4	+0.714 092 4
12	0.000 000 0	−0.663 358 2	+0.974 528 0	−0.991 304 8	+0.991 304 7

$\bar{L}_n(x), \bar{L}_n^*(x) =$ Laguerre functions

$L_n(x) =$ Laguerre polynomial

$L_n^*(x) =$ Laguerre auxiliary function

$\Lambda\,[\,] =$ nested sum (Sec. 8.11)

(1) Differential Equation

$$xy'' + (1-x)y' + ny = 0 \qquad -\infty < n < +\infty$$

Interval: $[0, \infty]$
Weight: $W(x) = e^{-x}$

(2) Solution for All n: $y = A_1 \bar{L}_n(x) + A_2 \bar{L}_n^*(x)$

$$\bar{L}_n(x) = (-x)^n \sum_{k=0}^{\infty}\Lambda\left[1 - \frac{(n-k)^2}{(k+1)x}\right]$$

$$= (-x)^n\left(1 - \frac{n^2}{x}\left(1 - \frac{(n-1)^2}{2x}\left(1 - \frac{(n-2)^2}{3x}\left(1 - \cdots\right)\right)\right)\right)$$

$$\bar{L}_n^*(x) = \frac{e^x}{x^{n+1}} \sum_{k=0}^{\infty}\Lambda\left[1 + \frac{(n+k+1)^2}{(k+1)x}\right]$$

$$= \frac{e^x}{x^{n+1}}\left(1 + \frac{(n+1)^2}{x}\left(1 + \frac{(n+2)^2}{2x}\left(1 + \frac{(n+3)^3}{3x}\left(1 + \cdots\right)\right)\right)\right)$$

(3) Solution for $n = 0, 1, 2, \ldots$: $y = A_1 L_n(x) + A_2 L_n^*(x)$

$$L_n(x) = n! \sum_{k=0}^{n}\Lambda\left[1 - \frac{(n-k)x}{(k+1)^2}\right]$$

$$= n!\left(1 - \frac{nx}{1 \cdot 1}\left(1 - \frac{(n-1)x}{2 \cdot 2}\left(1 - \frac{(n-2)x}{3 \cdot 3}\left(1 - \cdots\right)\right)\right)\right)$$

$$L_n^*(x) = L_n(x)\ln x + \sum_{k=1}^{\infty} C_{n,k}x^k$$

where $C_{n,1} = n+1$ and $C_{n,k} = \frac{1}{k^2}\left[(k-1)C_{n,k-1} + \frac{n+1}{k!}\binom{-n}{k-1}\right]$.

(4) Special Values and Relations   Sec. A.14

$$L_{n+1}(x) = (2n+1-x)L_n(x) - n^2 L_{n-1}(x) \qquad L_n(0) = n! \qquad L_n^*(0) = \infty$$

$$\int_0^{\infty} e^{-x}L_m(x)L_n(x)\,dx = \begin{cases} 0 & m \neq n \\ (n!)^2 & m = n \end{cases}$$

$$\int_0^{\infty} x^p e^{-x}L_n(x)\,dx = \begin{cases} 0 & p < n \\ (-1)^n(n!)^2 & p = n \end{cases}$$

(5) Table[1] of $L_n(x)$

$0 \leq x \leq \infty$

n	x 0.5	1.0	3.0	5.0
0	+1.00000 00000	+1.00000 00000	+1.00000 00000	+1.00000 00000
1	+0.50000 00000	0.00000 00000	−2.00000 00000	−4.00000 00000
2	+0.12500 00000	−0.50000 00000	−0.50000 00000	+3.50000 00000
3	−0.14583 33333	−0.66666 66667	+1.00000 00000	+2.66666 66667
4	−0.33072 91667	−0.62500 00000	+1.37500 00000	−1.29166 66667
5	−0.44557 29167	−0.46666 66667	+0.85000 00000	−3.16666 66667
6	−0.50414 49653	−0.25694 44444	−0.01250 00000	−2.09027 77778
7	−0.51833 92237	−0.04047 61905	−0.74642 85714	+0.32539 68254
8	−0.49836 29984	+0.15399 30556	−1.10870 53571	+2.23573 90873
9	−0.45291 95204	+0.30974 42681	−1.06116 07143	+2.69174 38272
10	−0.38937 44141	+0.41894 59325	−0.70002 23214	+1.75627 61795
11	−0.31390 72988	+0.48013 41791	−0.18079 95130	+0.10754 36909
12	−0.23164 96389	+0.49621 22235	+0.34035 46063	−1.44860 42948

[1](Ref. 14.01, p. 800.)

$\bar{H}_n(x), \bar{H}_n^*(x) =$ Hermite functions	$\Lambda[\ ] =$ nested sum (Sec. 8.11)
$H_n(x), H_n^*(x) =$ Hermite polynomial	$(\ )!! =$ double factorial (Sec. 1.03)

(1) Differential Equation

$$y'' - 2xy' + 2ny = 0 \qquad -\infty < n < +\infty \qquad \text{Interval: } [-\infty, \infty]$$
$$\text{Weight: } w(x) = e^{-x^2}$$

(2) Solution for All n: $y = A_1 \bar{H}_n(x) + A_2 \bar{H}_n^*(x)$

$$\bar{H}_n(x) = \mathop{\Lambda}_{k=0}^{\infty}\left[1 - \frac{2(n-2k)}{(1+2k)(2+2k)} S\right] = \left(1 - \frac{2n}{1\cdot 2} S\left(1 - \frac{2(n-2)}{3\cdot 4} S\left(1 - \frac{2(n-4)}{5\cdot 6} S\left(1 - \cdots\right)\right)\right)\right)$$

$$\bar{H}_n^*(x) = x \mathop{\Lambda}_{k=0}^{\infty}\left[1 - \frac{2(n-1-2k)}{(2+2k)(3+3k)} S\right] \qquad\qquad S = x^2$$

$$= x\left(1 - \frac{2(n-1)}{2\cdot 3} S\left(1 - \frac{2(n-3)}{4\cdot 5} S\left(1 - \frac{2(n-5)}{6\cdot 7} S\left(1 - \cdots\right)\right)\right)\right)$$

(3) Solution for $n = 0, 1, 2, \ldots$: $y = A_1 H_n(x) + A_2 H_n^*(x)$

$$H_n(x) = (2x)^n \mathop{\Lambda}_{k=0}^{n}\left[1 - \frac{(n-k)(n-k-1)}{k+1} R\right] = H_n^*(x) \qquad R = \frac{1}{4x^2}$$

$$= (2x)^n\left(1 - \frac{n(n-1)}{1} R\left(1 - \frac{(n-2)(n-3)}{2} R\left(1 - \frac{(n-4)(n-5)}{3} R\left(1 - \cdots\right)\right)\right)\right)$$

where $A_1 = \begin{cases} i^n \dfrac{n!C}{(n/2)!} & n \text{ even} \\ 0 & n \text{ odd} \end{cases} \qquad A_2 = \begin{cases} 0 & n \text{ even} \\ 2i^{n-1}\dfrac{n!C}{[(n-1)/2]!} & n \text{ odd} \end{cases}$

and $C =$ arbitrary constant.

(4) Special Values and Relations   Sec. A.14

$$H_{n+1}(x) = 2xH_n(x) - 2n H_{n-1}(x) \qquad\qquad \frac{d}{dx}[H_{n+1}(x)] = 2(n+1)H_n(x)$$

$$H_n(0) = \begin{cases} (\sqrt{-2})^n (n-1)!! & n \text{ even} \\ 0 & n \text{ odd} \end{cases} \qquad \int_{-\infty}^{+\infty} e^{-x^2} H_m(x)H_n(x)\, dx = \begin{cases} 0 & m \neq n \\ 2^n n! \sqrt{\pi} & m = n \end{cases}$$

(5) Table[1] of $H_n(x)$

$$H_n(-x) = (-1)^n H_n(x) \qquad |x| \leq \infty$$

n \ x	0.5		1.0		3.0		5.0	
0	+1.00000		+1.00000		+1.00000 00		1.00000 00000	
1	+1.00000		+2.00000		+6.00000 00		1.00000 00000	(+01)
2	−1.00000		+2.00000		+3.40000 00	(+01)	9.80000 00000	(+01)
3	−5.00000		−4.00000		+1.80000 00	(+02)	9.40000 00000	(+02)
4	+1.00000		−2.00000	(+01)	+8.76000 00	(+02)	8.81200 00000	(+03)
5	+4.10000	(+01)	−8.00000	(+00)	+3.81600 00	(+03)	8.06000 00000	(+04)
6	+3.10000	(+01)	+1.84000	(+02)	+1.41360 00	(+04)	7.17880 00000	(+05)
7	−4.61000	(+02)	+4.64000	(+02)	+3.90240 00	(+04)	6.21160 00000	(+06)
8	−8.95000	(+02)	−1.64800	(+03)	+3.62400 00	(+04)	5.20656 80000	(+07)
9	+6.48100	(+03)	−1.07200	(+04)	−4.06944 00	(+05)	4.21271 20000	(+08)
10	+2.25910	(+04)	+8.22400	(+03)	−3.09398 40	(+06)	3.27552 97600	(+09)
11	−1.07029	(+05)	+2.30848	(+05)	−1.04250 24	(+07)	2.43298 73600	(+10)
12	−6.04031	(+05)	+2.80768	(+05)	+5.51750 40	(+06)	1.71237 08128	(+11)

[1](Ref. 14.01, p. 802.)

(1) Differential Equation

$$x^2 y'' + xy' + (x^2 - n^2)y = 0$$

(2) Solution

$$y = A_1 J_n(x) + A_2 J_{-n}(x) \qquad n \neq 0, 1, 2, \ldots \qquad\qquad y = A_1 J_n(x) + A_2 Y_n(x) \qquad \text{all } n$$

(3) Bessel Functions of the First Kind of Order n

$$J_n(x) = \sum_{k=0}^{\infty} \frac{(-1)^k (x/2)^{2k+n}}{k!\,\Gamma(k+1+n)}$$

$$= \frac{x^n}{2^n \Gamma(1+n)}\left[1 - \frac{x^2}{2(2+2n)} + \frac{x^4}{(2)(4)(2+2n)(4+2n)} - \cdots\right]$$

$$J_{-n}(x) = \sum_{k=0}^{\infty} \frac{(-1)^k (x/2)^{2k-n}}{k!\,\Gamma(k+1-n)}$$

$$= \frac{x^{-n}}{2^{-n} \Gamma(1-n)}\left[1 - \frac{x^2}{2(2-2n)} + \frac{x^4}{(2)(4)(2-2n)(4-2n)} - \cdots\right]$$

$$= (-1)^n J_n(x) \qquad n = 0, 1, 2, \ldots$$

$$J_0(x) = 1 - \frac{(x/2)^2}{(1!)^2} + \frac{(x/2)^4}{(2!)^2} - \frac{(x/2)^6}{(3!)^2} + \cdots$$

$$J_1(x) = \frac{x}{2}\left[1 - \frac{(x/2)^2}{2(1!)^2} + \frac{(x/2)^4}{3(2!)^2} - \frac{(x/2)^6}{4(3!)^2} + \cdots\right] = -\frac{d}{dx}[J_0(x)]$$

(4) Bessel Functions of the Second Kind of Order n

$$Y_n(x) = \begin{cases} \dfrac{J_n(x) \cos n\pi - J_{-n}(x)}{\sin n\pi} & n \neq 0, 1, 2, \ldots \\[4mm] \lim\limits_{p \to n} \dfrac{J_p(x) \cos p\pi - J_{-p}(x)}{\sin p\pi} & n = 0, 1, 2, \ldots \end{cases}$$

$$Y_{-n}(x) = (-1)^n Y_n(x) \qquad n = 0, 1, 2, \ldots$$

$$Y_0(x) = \frac{2}{\pi}\left[\left(\ln\frac{x}{2} + C\right)J_0(x) + \frac{2}{1}J_2(x) - \frac{2}{2}J_4(x) + \frac{2}{3}J_6(x) - \cdots\right]$$

$$Y_1(x) = \frac{2}{\pi}\left[\left(\ln\frac{x}{2} + C\right)J_1(x) - \frac{1}{x} - \frac{1}{2}J_1(x) + \frac{9}{4}J_3(x) - \cdots\right] = -\frac{d}{dx}[Y_0(x)]$$

where $C = 0.577\,215\,665$ (Sec. 7.02)

(1) Recurrence Relations

$$J_{n+1}(x) = \frac{2n}{x} J_n(x) - J_{n-1}(x)$$

$$= \frac{n}{x} J_n(x) - \frac{d}{dx}[J_n(x)]$$

$$= -x^n \frac{d}{dx}[x^{-n} J_n(x)]$$

$$J_{n-1}(x) = \frac{2n}{x} J_n(x) - J_{n+1}(x)$$

$$= \frac{n}{x} J_n(x) + \frac{d}{dx}[J_n(x)]$$

$$= x^{-n} \frac{d}{dx}[x^n J_n(x)]$$

$$\frac{d}{dx}[J_n(x)] = \tfrac{1}{2}[J_{n-1}(x) - J_{n+1}(x)]$$

$$\int x^{n+1} J_n(x)\, dx = x^{n+1} J_{n+1}(x)$$

$$Y_{n+1}(x) = \frac{2n}{x} Y_n(x) - Y_{n-1}(x)$$

$$= \frac{n}{x} Y_n(x) - \frac{d}{dx}[Y_n(x)]$$

$$= -x^n \frac{d}{dx}[x^{-n} Y_n(x)]$$

$$Y_{n-1}(x) = \frac{2n}{x} Y_n(x) - Y_{n+1}(x)$$

$$= \frac{n}{x} Y_n(x) + \frac{d}{dx}[Y_n(x)]$$

$$= x^{-n} \frac{d}{dx}[x^n Y_n(x)]$$

$$\frac{d}{dx}[Y_n(x)] = \tfrac{1}{2}[Y_{n-1}(x) - Y_{n+1}(x)]$$

$$\int x^{n+1} Y_n(x)\, dx = x^{n+1} Y_{n+1}(x)$$

(2) Half-Odd Integers   (See also Sec. 14.28–3)

$$J_{1/2}(x) = \sqrt{\frac{2}{\pi x}} \sin x$$

$$J_{3/2}(x) = \sqrt{\frac{2}{\pi x}} \left(\frac{\sin x}{x} - \cos x \right)$$

$$J_{5/2}(x) = \sqrt{\frac{2}{\pi x}} \left[\left(\frac{3}{x^2} - 1 \right) \sin x - \frac{3}{x} \cos x \right]$$

$$J_{-1/2}(x) = \sqrt{\frac{2}{\pi x}} \cos x$$

$$J_{-3/2}(x) = -\sqrt{\frac{2}{\pi x}} \left(\frac{\cos x}{x} + \sin x \right)$$

$$J_{-5/2}(x) = \sqrt{\frac{2}{\pi x}} \left[\left(\frac{3}{x^2} - 1 \right) \cos x + \frac{3}{x} \sin x \right]$$

$$Y_{1/2}(x) = -\sqrt{\frac{2}{\pi x}} \cos x$$

$$Y_{3/2}(x) = -\sqrt{\frac{2}{\pi x}} \left(\frac{\cos x}{x} + \sin x \right)$$

$$Y_{5/2}(x) = -\sqrt{\frac{2}{\pi x}} \left[\left(\frac{3}{x^2} - 1 \right) \cos x + \frac{3}{x} \sin x \right]$$

$$Y_{-1/2}(x) = \sqrt{\frac{2}{\pi x}} \sin x$$

$$Y_{-3/2}(x) = -\sqrt{\frac{2}{\pi x}} \left(\frac{\sin x}{x} + \cos x \right)$$

$$Y_{-5/2}(x) = \sqrt{\frac{2}{\pi x}} \left[\left(\frac{3}{x^2} - 1 \right) \sin x - \frac{3}{x} \cos x \right]$$

(3) Hankel Functions of Order n

$$H_n^{(1)}(x) = J_n(x) + i Y_n(x)$$

$$J_n(x) = \frac{H_n^{(1)}(x) + H_n^{(2)}(x)}{2}$$

$$H_n^{(2)}(x) = J_n(x) - i Y_n(x)$$

$$Y_n(x) = \frac{H_n^{(1)}(x) - H_n^{(2)}(x)}{2i}$$

(1) Asymptotic Approximation

$$J_n(x) \approx \sqrt{\frac{2}{\pi x}} \cos\left(x - \frac{n\pi}{2} - \frac{\pi}{4}\right) \qquad x > 25$$

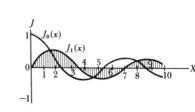

(2) Numerical Values $J_0(x)$ $x = 0 - 10$

x	0	1	2	3	4	5	6	7	8	9	x
0.	1.0000	0.9975	0.9900	0.9776	0.9604	0.9385	0.9120	0.8812	0.8463	0.8075	0.
1.	0.7652	0.7196	0.6711	0.6201	0.5669	0.5118	0.4554	0.3980	0.3400	0.2818	1.
2.	0.2239	0.1667	0.1104	0.0555	0.0025	−0.0484	−0.0968	−0.1424	−0.1850	−0.2243	2.
3.	−0.2601	−0.2921	−0.3202	−0.3443	−0.3643	−0.3801	−0.3918	−0.3992	−0.4026	−0.4018	3.
4.	−0.3971	−0.3887	−0.3766	−0.3610	−0.3423	−0.3205	−0.2961	−0.2693	−0.2404	−0.2097	4.
5.	−0.1776	−0.1443	−0.1103	−0.0758	−0.0412	−0.0068	0.0270	0.0599	0.0917	0.1220	5.
6.	0.1506	0.1773	0.2017	0.2238	0.2433	0.2601	0.2740	0.2851	0.2931	0.2981	6.
7.	0.3001	0.2991	0.2951	0.2882	0.2786	0.2663	0.2516	0.2346	0.2154	0.1944	7.
8.	0.1717	0.1475	0.1222	0.0960	0.0692	0.0419	0.0146	−0.0125	−0.0392	−0.0653	8.
9.	−0.0903	−0.1142	−0.1367	−0.1577	−0.1768	−0.1939	−0.2090	−0.2218	−0.2323	−0.2403	9.
10.	−0.2459										

(3) Numerical Values $J_1(x)$ $x = 0 - 10$

x	0	1	2	3	4	5	6	7	8	9	x
0.	0.0000	0.0499	0.0995	0.1483	0.1960	0.2423	0.2867	0.3290	0.3688	0.4059	0.
1.	0.4401	0.4709	0.4983	0.5220	0.5419	0.5579	0.5699	0.5778	0.5815	0.5812	1.
2.	0.5767	0.5683	0.5560	0.5399	0.5202	0.4971	0.4708	0.4416	0.4097	0.3754	2.
3.	0.3391	0.3009	0.2613	0.2207	0.1792	0.1374	0.0955	0.0538	0.0128	−0.0272	3.
4.	−0.0660	−0.1033	−0.1386	−0.1719	−0.2028	−0.2311	−0.2566	−0.2791	−0.2985	−0.3147	4.
5.	−0.3276	−0.3371	−0.3432	−0.3460	−0.3453	−0.3414	−0.3343	−0.3241	−0.3110	−0.2951	5.
6.	−0.2767	−0.2559	−0.2329	−0.2081	−0.1816	−0.1538	−0.1250	−0.0953	−0.0652	−0.0349	6.
7.	−0.0047	0.0252	0.0543	0.0826	0.1096	0.1352	0.1592	0.1813	0.2014	0.2192	7.
8.	0.2346	0.2476	0.2580	0.2657	0.2708	0.2731	0.2728	0.2697	0.2641	0.2559	8.
9.	0.2453	0.2324	0.2174	0.2004	0.1816	0.1613	0.1395	0.1166	0.0928	0.0684	9.
10.	0.0435										

(4) Asymptotic Series for Large x

$$J_n(x) \approx \sqrt{\frac{2}{\pi x}}\left[\cos\psi\left(1 - \frac{\alpha_1\alpha_3}{2!} + \frac{\alpha_1\alpha_3\alpha_5\alpha_7}{4!} - \cdots\right) - \sin\psi\left(\frac{\alpha_1}{1!} - \frac{\alpha_1\alpha_3\alpha_5}{3!} + \cdots\right)\right]$$

$$\psi = x - \frac{n\pi}{2} - \frac{\pi}{4} \qquad \alpha_k = \frac{4n^2 - k^2}{8x} \qquad k = 1, 2, \ldots \qquad x > 15$$

(1) Asymptotic Approximation

$$Y_n(x) \approx \sqrt{\frac{2}{\pi x}} \sin\left(x - \frac{n\pi}{2} - \frac{\pi}{4}\right) \qquad x > 25$$

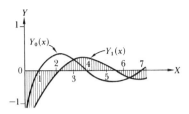

(2) Numerical Values $Y_0(x)$ $x = 0 - 10$

x	0	1	2	3	4	5	6	7	8	9	x
0.	$-\infty$	-1.5342	-1.0811	-0.8073	-0.6060	-0.4445	-0.3085	-0.1907	-0.0868	0.0056	0.
1.	0.0883	0.1622	0.2281	0.2865	0.3379	0.3824	0.4204	0.4520	0.4774	0.4968	1.
2.	0.5104	0.5183	0.5208	0.5181	0.5104	0.4981	0.4813	0.4605	0.4359	0.4079	2.
3.	0.3769	0.3431	0.3071	0.2691	0.2296	0.1890	0.1477	0.1061	0.0645	0.0234	3.
4.	-0.0169	-0.0561	-0.0938	-0.1296	-0.1633	-0.1947	-0.2235	-0.2494	-0.2723	-0.2921	4.
5.	-0.3085	-0.3216	-0.3313	-0.3374	-0.3402	-0.3395	-0.3354	-0.3282	-0.3177	-0.3044	5.
6.	-0.2882	-0.2694	-0.2483	-0.2251	-0.1999	-0.1732	-0.1452	-0.1162	-0.0864	-0.0563	6.
7.	-0.0259	0.0042	0.0339	0.0628	0.0907	0.1173	0.1424	0.1658	0.1872	0.2065	7.
8.	0.2235	0.2381	0.2501	0.2595	0.2662	0.2702	0.2715	0.2700	0.2659	0.2592	8.
9.	0.2499	0.2383	0.2245	0.2086	0.1907	0.1712	0.1502	0.1279	0.1045	0.0804	9.
10.	0.0557										10.

(3) Numerical Values $Y_1(x)$ $x = 0 - 10$

x	0	1	2	3	4	5	6	7	8	9	x
0.	$-\infty$	-6.4590	-3.3238	-2.2931	-1.7809	-1.4715	-1.2604	-1.1032	-0.9781	-0.8731	0.
1.	-0.7812	-0.6981	-0.6211	-0.5485	-0.4791	-0.4123	-0.3476	-0.2847	-0.2237	-0.1644	1.
2.	-0.1070	-0.0517	0.0015	0.0523	0.1005	0.1459	0.1884	0.2276	0.2635	0.2959	2.
3.	0.3247	0.3496	0.3707	0.3879	0.4010	0.4102	0.4154	0.4167	0.4141	0.4078	3.
4.	0.3979	0.3846	0.3680	0.3484	0.3260	0.3010	0.2737	0.2445	0.2136	0.1812	4.
5.	0.1479	0.1137	0.0792	0.0445	0.0101	-0.0238	-0.0568	-0.0887	-0.1192	-0.1481	5.
6.	-0.1750	-0.1998	-0.2223	-0.2422	-0.2596	-0.2741	-0.2857	-0.2945	-0.3002	-0.3029	6.
7.	-0.3027	-0.2995	-0.2934	-0.2846	-0.2731	-0.2591	-0.2428	-0.2243	-0.2039	-0.1817	7.
8.	-0.1581	-0.1331	-0.1072	-0.0806	-0.0535	-0.0262	0.0011	0.0280	0.0544	0.0799	8.
9.	0.1043	0.1275	0.1491	0.1691	0.1871	0.2032	0.2171	0.2287	0.2379	0.2447	9.
10.	0.2490										10.

(4) Asymptotic Series for Large x

$$Y_n(x) \approx \sqrt{\frac{2}{\pi x}}\left[\sin\psi\left(1 - \frac{\alpha_1\alpha_3}{2!} + \frac{\alpha_1\alpha_3\alpha_5\alpha_7}{4!} - \cdots\right) + \cos\psi\left(\frac{\alpha_1}{1!} - \frac{\alpha_1\alpha_3\alpha_5}{3!} + \cdots\right)\right]$$

$$\psi = x - \frac{n\pi}{2} - \frac{\pi}{4} \qquad \alpha_k = \frac{4n^2 - k^2}{8x} \qquad k = 1, 2, \ldots \qquad x > 15$$

(1) Differential Equation

$$x^2 y'' + xy' - (x^2 + n^2)y = 0$$

(2) Solution

$$y = A_1 I_n(x) + A_2 I_{-n}(x) \qquad n \neq 0, 1, 2, \dots \qquad y = A_1 I_n(x) + A_2 K_n(x) \qquad \text{all } n$$

(3) Modified Bessel Functions of the First Kind of Order n

$$I_n(x) = i^{-n} J_n(ix) = \sum_{k=0}^{\infty} \frac{(x/2)^{2k+n}}{k! \Gamma(k+1+n)}$$

$$= \frac{x^n}{2^n \Gamma(1+n)} \left[1 + \frac{x^2}{2(2+2n)} + \frac{x^4}{(2)(4)(2+2n)(4+2n)} + \cdots \right]$$

$$I_{-n}(x) = i^n J_{-n}(ix) = \sum_{k=0}^{\infty} \frac{(x/2)^{2k-n}}{k! \Gamma(k+1-n)}$$

$$= \frac{x^{-n}}{2^{-n} \Gamma(1-n)} \left[1 + \frac{x^2}{2(2-2n)} + \frac{x^4}{(2)(4)(2-2n)(4-2n)} + \cdots \right]$$

$$I_{-n}(x) = I_n(x) \qquad n = 0, 1, 2, \dots$$

$$I_0(x) = 1 + \frac{(x/2)^2}{(1!)^2} + \frac{(x/2)^4}{(2!)^2} + \frac{(x/2)^6}{(3!)^2} + \cdots$$

$$I_1(x) = \frac{x}{2} \left[1 + \frac{(x/2)^2}{2(1!)^2} + \frac{(x/2)^4}{3(2!)^2} + \frac{(x/2)^6}{4(3!)^2} + \cdots \right] = \frac{d}{dx} [I_0(x)]$$

(4) Modified Bessel Functions of the Second Kind of Order n

$$K_n(x) = \begin{cases} \dfrac{\pi}{2} \dfrac{I_{-n}(x) - I_n(x)}{\sin n\pi} & n \neq 0, 1, 2, \dots \\[3mm] \lim_{p \to n} \dfrac{\pi}{2} \dfrac{I_{-p}(x) - I_p(x)}{\sin p\pi} & n = 0, 1, 2, \dots \end{cases}$$

$$K_{-n}(x) = K_n(x) \qquad n = 0, 1, 2, \dots$$

$$K_0(x) = -\left[\left(\ln \frac{x}{2} + C \right) I_0(x) - \frac{2}{1} I_2(x) - \frac{2}{2} I_4(x) - \frac{2}{3} I_6(x) - \cdots \right]$$

$$-K_1(x) = -\left[\left(\ln \frac{x}{2} + C \right) I_1(x) - \frac{1}{x} - \frac{1}{2} I_1(x) - \frac{9}{4} I_3(x) - \cdots \right] = \frac{d}{dx} [K_0(x)]$$

where $C = 0.577\,215\,665$ (Sec. 7.02)

(1) Recurrence Relations

$$I_{n+1}(x) = -\frac{2n}{x}I_n(x) + I_{n-1}(x)$$

$$= -\frac{n}{x}I_n(x) + \frac{d}{dx}[I_n(x)]$$

$$= x^n\frac{d}{dx}[x^{-n}I_n(x)]$$

$$I_{n-1}(x) = \frac{2n}{x}I_n(x) + I_{n+1}(x)$$

$$= \frac{n}{x}I_n(x) + \frac{d}{dx}[I_n(x)]$$

$$= x^{-n}\frac{d}{dx}[x^nI_n(x)]$$

$$\frac{d}{dx}[I_n(x)] = \tfrac{1}{2}[I_{n+1}(x) + I_{n-1}(x)]$$

$$\int x^{n+1}I_n(x)\,dx = x^{n+1}I_{n+1}(x)$$

$$K_{n+1}(x) = \frac{2n}{x}K_n(x) + K_{n-1}(x)$$

$$= \frac{n}{x}K_n(x) - \frac{d}{dx}[K_n(x)]$$

$$= -x^n\frac{d}{dx}[x^{-n}K_n(x)]$$

$$K_{n-1}(x) = -\frac{2n}{x}K_n(x) + K_{n+1}(x)$$

$$= -\frac{n}{x}K_n(x) - \frac{d}{dx}[K_n(x)]$$

$$= -x^{-n}\frac{d}{dx}[x^nK_n(x)]$$

$$\frac{d}{dx}[K_n(x)] = -\tfrac{1}{2}[K_{n+1}(x) + K_{n-1}(x)]$$

$$\int x^{n+1}K_n(x)\,dx = -x^{n+1}K_{n+1}(x)$$

(2) Half-Odd Integers   (See also Sec. 14.29-3)

$$I_{1/2}(x) = \sqrt{\frac{2}{\pi x}}\sinh x$$

$$I_{3/2}(x) = -\sqrt{\frac{2}{\pi x}}\left(\frac{\sinh x}{x} - \cosh x\right)$$

$$I_{5/2}(x) = \sqrt{\frac{2}{\pi x}}\left[\left(\frac{3}{x^2} + 1\right)\sinh x - \frac{3}{x}\cosh x\right]$$

$$I_{-1/2}(x) = \sqrt{\frac{2}{\pi x}}\cosh x$$

$$I_{-3/2}(x) = -\sqrt{\frac{2}{\pi x}}\left(\frac{\cosh x}{x} - \sinh x\right)$$

$$I_{-5/2}(x) = \sqrt{\frac{2}{\pi x}}\left[\left(\frac{3}{x^2} + 1\right)\cosh x - \frac{3}{x}\sinh x\right]$$

$$K_{1/2}(x) = e^{-x}\sqrt{\frac{\pi}{2x}}$$

$$K_{3/2}(x) = e^{-x}\sqrt{\frac{\pi}{2x}}\left(\frac{1}{x} + 1\right)$$

$$K_{5/2}(x) = e^{-x}\sqrt{\frac{\pi}{2x}}\left(\frac{2}{x^2} + \frac{2}{x} + 1\right)$$

$$K_{-1/2}(x) = e^{-x}\sqrt{\frac{\pi}{2x}}$$

$$K_{-3/2}(x) = e^{-x}\sqrt{\frac{\pi}{2x}}\left(\frac{1}{x} + 1\right)$$

$$K_{-5/2}(x) = e^{-x}\sqrt{\frac{\pi}{2x}}\left(\frac{2}{x^2} + \frac{2}{x} + 1\right)$$

(3) Transformations $(n = p \pm 1/2,\ p = 0, 1, 2, \ldots)$

$$J_{p+1/2}(x) = (-1)^pY_{-p-1/2}(x) = (-1)^p\sqrt{\frac{2}{\pi x}}x^{p+1}\left(\frac{1}{x}\frac{d}{dx}\right)^p\frac{\sin x}{x}$$

$$Y_{p+1/2}(x) = -(-1)^pJ_{-p-1/2}(x) = -(-1)^p\sqrt{\frac{2}{\pi x}}x^{p+1}\left(\frac{1}{x}\frac{d}{dx}\right)^p\frac{\cos x}{x}$$

$$I_{p+1/2}(x) = \sqrt{\frac{2}{\pi x}}x^{p+1}\left(\frac{1}{x}\frac{d}{dx}\right)^p\frac{\sinh x}{x}$$

$$K_{p+1/2}(x) = K_{-p-1/2}(x) = (-1)^p\sqrt{\frac{\pi}{2x}}x^{p+1}\left(\frac{1}{x}\frac{d}{dx}\right)^p\frac{e^{-x}}{x}$$

(1) Asymptotic Approximation

$$I_n(x) \approx \sqrt{\frac{1}{2\pi x}}\, e^x \qquad x > 25$$

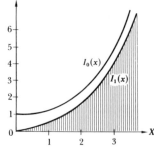

(2) Numerical Values

$I_0(x)$ $x = 0 - 10$

x	0	1	2	3	4	5	6	7	8	9	x
0.	1.000	1.003	1.010	1.023	1.040	1.063	1.092	1.126	1.167	1.213	0.
1.	1.266	1.326	1.394	1.469	1.553	1.647	1.750	1.864	1.990	2.128	1.
2.	2.280	2.446	2.629	2.830	3.049	3.290	3.553	3.842	4.157	4.503	2.
3.	4.881	5.294	5.747	6.243	6.785	7.378	8.028	8.739	9.517	10.37	3.
4.	11.30	12.32	13.44	14.67	16.01	17.48	19.09	20.86	22.79	24.91	4.
5.	27.24	29.79	32.58	35.65	39.01	42.69	46.74	51.17	56.04	61.38	5.
6.	67.23	73.66	80.72	88.46	96.96	106.3	116.5	127.8	140.1	153.7	6.
7.	168.6	185.0	202.9	222.7	244.3	268.2	294.3	323.1	354.7	389.4	7.
8.	427.6	469.5	515.6	566.3	621.9	683.2	750.5	824.4	905.8	995.2	8.
9.	1,094	1,202	1,321	1,451	1,595	1,753	1,927	2,119	2,329	2,561	9.
10.	2,816										

(3) Numerical Values

$I_1(x)$ $x = 0 - 10$

x	0	1	2	3	4	5	6	7	8	9	x
0.	0.0000	0.0501	0.1005	0.1517	0.2040	0.2579	0.3137	0.3719	0.4329	0.4971	0.
1.	0.5652	0.6375	0.7147	0.7973	0.8861	0.9817	1.085	1.196	1.317	1.448	1.
2.	1.591	1.745	1.914	2.098	2.298	2.517	2.755	3.016	3.301	3.613	2.
3.	3.953	4.326	4.734	5.181	5.670	6.206	6.793	7.436	8.140	8.913	3.
4.	9.759	10.69	11.71	12.82	14.05	15.39	16.86	18.48	20.25	22.20	4.
5.	24.34	26.68	29.25	32.08	35.18	38.59	42.33	46.44	50.95	55.90	5.
6.	61.34	67.32	73.89	81.10	89.03	97.74	107.3	117.8	129.4	142.1	6.
7.	156.0	171.4	188.3	206.8	227.2	249.6	274.2	301.3	381.1	363.9	7.
8.	399.9	439.5	483.0	531.0	583.7	641.6	705.4	775.5	852.7	937.5	8.
9.	1,031	1,134	1,247	1,371	1,508	1,658	1,824	2,006	2,207	2,428	9.
10.	2,671										

(4) Asymptotic Series for Large x

$$I_n(x) \approx \frac{e^x}{\sqrt{2\pi x}}\left(1 - \frac{\alpha_1}{1!} + \frac{\alpha_1 \alpha_3}{2!} - \cdots\right)$$

$$\alpha_k = \frac{4n^2 - k^2}{8x} \qquad k = 1, 2, \ldots \qquad x > 15$$

(1) Asymptotic Approximation

$$K_n(x) \approx \sqrt{\frac{\pi}{2x}} e^{-x} \qquad x > 25$$

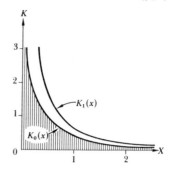

(2) Numerical Values $K_0(x)$ $x = 0 - 10$

x	0	1	2	3	4	5	6	7	8	9	x
0.	∞	2.4271	1.7527	1.3725	1.1145	0.9244	0.7775	0.6605	0.5653	0.4867	0.
1.	0.4210	0.3656	0.3185	0.2782	0.2437	0.2138	0.1880	0.1655	0.1459	0.1288	1.
2.	0.1139	0.1008	0.08927	0.07914	0.07022	0.06235	0.05540	0.04926	0.04382	0.03901	2.
3.	0.03474	0.03095	0.02759	0.02461	0.02196	0.01960	0.01750	0.01563	0.01397	0.01248	3.
4.	0.01116	$0.0^2 9980$	$0.0^2 8927$	$0.0^2 7988$	$0.0^2 7149$	$0.0^2 6400$	$0.0^2 5730$	$0.0^2 5132$	$0.0^2 4597$	$0.0^2 4119$	4.
5.	$0.0^2 3691$	$0.0^2 3308$	$0.0^2 2966$	$0.0^2 2659$	$0.0^2 2385$	$0.0^2 2139$	$0.0^2 1918$	$0.0^2 1721$	$0.0^2 1544$	$0.0^2 1386$	5.
6.	$0.0^2 1244$	$0.0^2 1117$	$0.0^2 1003$	$0.0^3 9001$	$0.0^3 8083$	$0.0^3 7259$	$0.0^3 6520$	$0.0^3 5857$	$0.0^3 5262$	$0.0^3 4728$	6.
7.	$0.0^3 4248$	$0.0^3 3817$	$0.0^3 3431$	$0.0^3 3084$	$0.0^3 2772$	$0.0^3 2492$	$0.0^3 2240$	$0.0^3 2014$	$0.0^3 1811$	$0.0^3 1629$	7.
8.	$0.0^3 1465$	$0.0^3 1317$	$0.0^3 1185$	$0.0^3 1066$	$0.0^4 9588$	$0.0^4 8626$	$0.0^4 7761$	$0.0^4 6983$	$0.0^4 6283$	$0.0^4 5654$	8.
9.	$0.0^4 5088$	$0.0^4 4579$	$0.0^4 4121$	$0.0^4 3710$	$0.0^4 3339$	$0.0^4 3006$	$0.0^4 2706$	$0.0^4 2436$	$0.0^4 2193$	$0.0^4 1975$	9.
10.	$0.0^4 1778$										

(3) Numerical Values $K_1(x)$ $x = 0 - 10$

x	0	1	2	3	4	5	6	7	8	9	x
0.	∞	9.8538	4.7760	3.0560	2.1844	1.6564	1.3028	1.0503	0.8618	0.7165	0.
1.	0.6019	0.5098	0.4346	0.3725	0.3208	0.2774	0.2406	0.2094	0.1826	0.1597	1.
2.	0.1399	0.1227	0.1079	0.09498	0.08372	0.07389	0.06528	0.05774	0.05111	0.04529	2.
3.	0.04016	0.03563	0.03164	0.02812	0.02500	0.02224	0.01979	0.01763	0.01571	0.01400	3.
4.	0.01248	0.01114	$0.0^2 9938$	$0.0^2 8872$	$0.0^2 7923$	$0.0^2 7078$	$0.0^2 6325$	$0.0^2 5654$	$0.0^2 5055$	$0.0^2 4521$	4.
5.	$0.0^2 4045$	$0.0^2 3619$	$0.0^2 3239$	$0.0^2 2900$	$0.0^2 2597$	$0.0^2 2326$	$0.0^2 2083$	$0.0^2 1866$	$0.0^2 1673$	$0.0^2 1499$	5.
6.	$0.0^2 1344$	$0.0^2 1205$	$0.0^2 1081$	$0.0^3 9691$	$0.0^3 8693$	$0.0^3 7799$	$0.0^3 6998$	$0.0^3 6280$	$0.0^3 5636$	$0.0^3 5059$	6.
7.	$0.0^3 4542$	$0.0^3 4078$	$0.0^3 3662$	$0.0^3 3288$	$0.0^3 2953$	$0.0^3 2653$	$0.0^3 2383$	$0.0^3 2141$	$0.0^3 1924$	$0.0^3 1729$	7.
8.	$0.0^3 1554$	$0.0^3 1396$	$0.0^3 1255$	$0.0^3 1128$	$0.0^3 1014$	$0.0^4 9120$	$0.0^4 8200$	$0.0^4 7374$	$0.0^4 6631$	$0.0^4 5964$	8.
9.	$0.0^4 5364$	$0.0^4 4825$	$0.0^4 4340$	$0.0^4 3904$	$0.0^4 3512$	$0.0^4 3160$	$0.0^4 2843$	$0.0^4 2559$	$0.0^4 2302$	$0.0^4 2072$	9.
10.	$0.0^4 1865$										

(4) Asymptotic Series for Large x

$$K_n(x) \approx \sqrt{\frac{\pi}{2x}} e^{-x} \left(1 + \frac{\alpha_1}{1!} + \frac{\alpha_1 \alpha_3}{2!} + \cdots \right)$$

$$\alpha_k = \frac{4n^2 - k^2}{8x} \qquad k = 1, 2, \ldots \qquad x > 15$$

(1) Differential Equation

$$x^2 y'' + xy' \pm ia^2 x^2 y = 0$$

(2) Solution

$$y = A_1[\text{Ber}\,(ax) \mp i\,\text{Bei}\,(ax)] + A_2[\text{Ker}\,(ax) \mp i\,\text{Kei}\,(ax)]$$

(3) Ber, Bei **Functions**

$$\text{Ber}\,(ax) = 1 - \frac{(ax/2)^4}{(2!)^2} + \frac{(ax/2)^8}{(4!)^2} - \cdots$$

$$\text{Bei}\,(ax) = \frac{(ax/2)^2}{(1!)^2} - \frac{(ax/2)^6}{(3!)^2} + \frac{(ax/2)^{10}}{(5!)^2} - \cdots$$

(4) Ker, Kei **Functions**

$$\text{Ker}\,(ax) = -\left(\ln\frac{ax}{2} + C\right)\text{Ber}\,(ax) + \frac{\pi}{4}\text{Bei}\,(ax) - \lambda_2 + \lambda_4 - \cdots$$

$$\text{Kei}\,(ax) = -\left(\ln\frac{ax}{2} + C\right)\text{Bei}\,(ax) - \frac{\pi}{4}\text{Ber}\,(ax) + \lambda_1 - \lambda_3 - \cdots$$

where $C = 0.577\,215\,665$ $\lambda_k = \dfrac{(ax/2)^{2k}}{(k!)^2}\left(1 + \dfrac{1}{2} + \dfrac{1}{3} + \cdots + \dfrac{1}{k}\right)$ (Sec. 13.04)

(5) Ber', Bei' **Functions**

$$\text{Ber}'\,(ax) = -\frac{2a(ax/2)^3}{(2!)^2} + \frac{4a(ax/2)^7}{(4!)^2} - \cdots = \frac{d}{dx}[\text{Ber}\,(ax)]$$

$$\text{Bei}'\,(ax) = \frac{a(ax/2)}{(1!)^2} - \frac{3a(ax/2)^5}{(3!)^2} + \cdots = \frac{d}{dx}[\text{Bei}\,(ax)]$$

(6) Ker', Kei' **Functions**

$$\text{Ker}'\,(ax) = -\left(\ln\frac{ax}{2} + C\right)\text{Ber}'\,(ax) - \frac{1}{x}\text{Ber}\,(ax) + \frac{\pi}{4}\text{Bei}'\,(ax) - \lambda_2' + \lambda_4' - \cdots = \frac{d}{dx}[\text{Ker}\,(ax)]$$

$$\text{Kei}'\,(ax) = -\left(\ln\frac{ax}{2} + C\right)\text{Bei}'\,(ax) - \frac{1}{x}\text{Bei}\,(ax) - \frac{\pi}{4}\text{Ber}'\,(ax) + \lambda_1' - \lambda_3' + \cdots = \frac{d}{dx}[\text{Kei}\,(ax)]$$

where $\lambda_k' = \dfrac{d\lambda_k}{dx}$

(7) Relations

$$\text{Ber}\,(x) + i\,\text{Bei}\,(x) = J_0(xi\sqrt{i}) = I_0(x\sqrt{i})$$

$$\text{Ker}\,(x) + i\,\text{Kei}\,(x) = K_0(x\sqrt{i})$$

$$\int x\,\text{Ber}\,(x)\,dx = x\,\text{Bei}'\,(x) \qquad\qquad \int x\,\text{Bei}\,(x)\,dx = -x\,\text{Ber}'\,(x)$$

$$\int x\,\text{Ker}\,(x)\,dx = x\,\text{Kei}'\,(x) \qquad\qquad \int x\,\text{Kei}\,(x)\,dx = -x\,\text{Ker}'\,(x)$$

(1) Differential Equation

$$x^2 y'' + xy' \pm (ix^2 + n^2)y = 0$$

(2) Solution

$$y = A_1[\text{Ber}_n(x) \mp i\,\text{Bei}_n(x)] + A_2[\text{Ker}_n(x) \mp i\,\text{Kei}_n(x)]$$

(3) Ber$_n$, Bei$_n$ Functions

$$\text{Ber}_n(x) = \sum_{k=0}^{\infty} \frac{(-1)^{n+k}(x/2)^{n+2k}}{k!(n+k)!} \cos\frac{(n+2k)\pi}{4}$$

$$\text{Bei}_n(x) = \sum_{k=0}^{\infty} \frac{(-1)^{n+k+1}(x/2)^{n+2k}}{k!(n+k)!} \sin\frac{(n+2k)\pi}{4} \qquad n = 0, 1, 2, \ldots \qquad k = 0, 1, 2, \ldots$$

(4) Ker$_n$, Kei$_n$ Functions

$$\text{Ker}_n(x) = -\left(\ln\frac{x}{2} + C\right)\text{Ber}_n(x) + \frac{\pi}{4}\text{Bei}_n(x) + \frac{1}{2}\sum_{k=0}^{n-1}\frac{(n-k-1)!(x/2)^{2k-n}}{k!}\cos\frac{(3n+2k)\pi}{4}$$

$$+ \frac{1}{2}\sum_{k=0}^{\infty}\frac{(x/2)^{2k+n}}{k!(n+k)!}(\tau_k + \tau_{k+n})\cos\frac{(3n+2k)\pi}{4}$$

$$\text{Kei}_n(x) = -\left(\ln\frac{x}{2} + C\right)\text{Bei}_n(x) - \frac{\pi}{4}\text{Ber}_n(x) - \frac{1}{2}\sum_{k=0}^{n-1}\frac{(n-k-1)!(x/2)^{2k-n}}{k!}\sin\frac{(3n+2k)\pi}{4}$$

$$+ \frac{1}{2}\sum_{k=0}^{\infty}\frac{(x/2)^{2k+n}(\tau_k + \tau_{k+n})}{k!(n+k)!}\sin\frac{(3n+2k)\pi}{4}$$

where $\qquad \tau_k = 1 + \dfrac{1}{2} + \dfrac{1}{3} + \cdots + \dfrac{1}{k} \qquad \tau_{k+n} = 1 + \dfrac{1}{2} + \dfrac{1}{3} + \cdots + \dfrac{1}{k+n}$

$$C = 0.577\,215\,665 \qquad\qquad k, n = 0, 1, 2, \ldots \qquad\qquad \text{(Sec. 13.04)}$$

(5) Relations

$$\text{Ber}_n(x) + i\,\text{Bei}_n(x) = J_n(xi\sqrt{i}) = i^n I_n(x\sqrt{i})$$

$$\text{Ber}_n(x) + \text{Bei}_n(x) = -\frac{x}{n\sqrt{2}}[\text{Bei}_{n-1}(x) + \text{Bei}_{n+1}(x)]$$

$$\text{Ber}_n(x) - \text{Bei}_n(x) = -\frac{x}{n\sqrt{2}}[\text{Ber}_{n-1}(x) + \text{Ber}_{n+1}(x)]$$

$$\text{Ker}_n(x) + i\,\text{Kei}_n(x) = i^{-n} K_n(x\sqrt{i})$$

$$\text{Ker}_n(x) + \text{Kei}_n(x) = -\frac{x}{n\sqrt{2}}[\text{Kei}_{n-1}(x) + \text{Kei}_{n+1}(x)]$$

$$\text{Ker}_n(x) - \text{Kei}_n(x) = -\frac{x}{n\sqrt{2}}[\text{Ker}_{n-1}(x) + \text{Kei}_{n+1}(x)]$$

(1) Asymptotic Approximation

$$\text{Ber}(x) \approx \frac{e^{x/\sqrt{2}}}{\sqrt{2\pi x}}\cos\left(\frac{x}{\sqrt{2}} - \frac{\pi}{8}\right)$$

$$\text{Bei}(x) \approx \frac{e^{x/\sqrt{2}}}{\sqrt{2\pi x}}\sin\left(\frac{x}{\sqrt{2}} - \frac{\pi}{8}\right)$$

$$x > 25$$

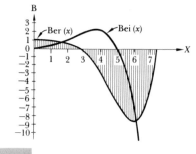

(2) Numerical Values

Ber(x) $x = 0 - 10$

x	0	1	2	3	4	5	6	7	8	9	x
0.	1.0000	1.0000	1.0000	0.9999	0.9996	0.9990	0.9980	0.9962	0.9936	0.9890	0.
1.	0.9844	0.9771	0.9676	0.9554	0.9401	0.9211	0.8979	0.8700	0.8367	0.7975	1.
2.	0.7517	0.6987	0.6377	0.5680	0.4890	0.4000	0.3001	0.1887	0.0651	−0.0714	2.
3.	−0.2214	−0.3855	−0.5644	−0.7584	−0.9680	−1.1936	−1.4353	−1.6933	−1.9674	−2.2576	3.
4.	−2.5634	−2.8843	−3.3295	−3.5679	−3.9283	−4.2991	−4.6784	−5.0639	−5.4531	−5.8429	4.
5.	−6.2301	−6.6107	−6.9803	−7.3344	−7.6674	−7.9736	−8.2466	−8.4794	−8.6644	−8.7937	5.
6.	−8.8583	−8.8491	−8.7561	−8.5688	−8.2762	−7.8669	−7.3287	−6.6492	−5.8155	−4.8146	6.
7.	−3.6329	−2.2571	−0.6737	1.1308	3.1695	5.4550	7.9994	10.814	13.909	17.293	7.
8.	20.974	24.957	29.245	33.840	38.738	43.936	49.423	55.187	61.210	67.469	8.
9.	73.936	80.576	87.350	94.208	101.10	107.95	114.70	121.26	127.54	133.43	9.
10.	138.84										

(3) Numerical Values

Bei(x) $x = 0 - 10$

x	0	1	2	3	4	5	6	7	8	9	x
0.	0.0000	0.0²2500	0.01000	0.02250	0.04000	0.06249	0.08998	0.1224	0.1599	0.2023	0.
1.	0.2496	0.3017	0.3587	0.4204	0.4867	0.5576	0.6327	0.7120	0.7953	0.8821	1.
2.	0.9723	1.0654	1.1610	1.2585	1.3575	1.4572	1.5569	1.6557	1.7529	1.8472	2.
3.	1.9376	2.0228	2.1016	2.1723	2.2334	2.2832	2.3199	2.3413	2.3454	2.3300	3.
4.	2.2927	2.2309	2.1422	2.0236	1.8726	1.6860	1.4610	1.1946	0.8837	0.5251	4.
5.	0.1160	−0.3467	−0.8658	−1.4443	−2.0845	−2.7890	−3.5597	−4.3986	−5.3068	−6.2854	5.
6.	−7.3347	−8.4545	−9.6437	−10.901	−12.223	−13.607	−15.047	−16.538	−18.074	−19.644	6.
7.	−21.239	−22.848	−24.456	−26.049	−27.609	−29.116	−30.548	−31.882	−33.092	−34.147	7.
8.	−35.017	−35.667	−36.061	−36.159	−35.920	−35.298	−34.246	−32.714	−30.651	−28.003	8.
9.	−24.713	−20.724	−15.976	−10.412	−3.9693	3.4106	11.787	21.218	31.758	43.459	9.
10.	56.371										

(4) Asymptotic Series for Large x

$$\text{Ber}(x) \approx \frac{e^{x/\sqrt{2}}}{\sqrt{2\pi x}}\{\cos\phi[1 + (1^2)\beta_1 + (1^2)(3^2)\beta_2 + \cdots] + \sin\phi[(1^2)\gamma_1 + (1^2)(3^2)\gamma_2 + \cdots]\}$$

$$\text{Bei}(x) \approx \frac{e^{x/\sqrt{2}}}{\sqrt{2\pi x}}\{\sin\phi[1 + (1^2)\beta_1 + (1^2)(3^2)\beta_2 + \cdots] - \cos\phi[(1^2)\gamma_1 + (1^2)(3^2)\gamma_2 + \cdots]\}$$

$$\phi = \frac{x}{\sqrt{2}} - \frac{\pi}{8} \qquad \beta_k = \frac{\cos(k\pi/4)}{k!(8x)^k} \qquad \gamma_k = \frac{\sin(k\pi/4)}{k!(8x)^k} \qquad k = 1, 2, \ldots \qquad x > 15$$

(1) Asymptotic Approximation

$$\text{Ker}\,(x) \approx \sqrt{\frac{\pi}{2x}}\, e^{-x/\sqrt{2}} \cos\left(\frac{x}{\sqrt{2}} + \frac{\pi}{8}\right)$$

$x > 25$

$$\text{Kei}\,(x) \approx -\sqrt{\frac{\pi}{2x}}\, e^{-x/\sqrt{2}} \sin\left(\frac{x}{\sqrt{2}} + \frac{\pi}{8}\right)$$

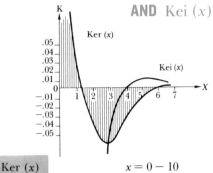

(2) Numerical Values

Ker (x) $x = 0 - 10$

x	0	1	2	3	4	5	6	7	8	9	x
0.	∞	2.4205	1.7331	1.3372	1.0626	0.8559	0.6931	0.5614	0.4529	0.3625	0.
1.	0.2867	0.2228	0.1689	0.1235	0.08513	0.05293	0.02603	$0.0^2 3691$	-0.01470	-0.02966	1.
2.	-0.04166	-0.05111	-0.05834	-0.06367	-0.06737	-0.06969	-0.07083	-0.07097	-0.07030	-0.06894	2.
3.	-0.06703	-0.06468	-0.06198	-0.05903	-0.05590	-0.05264	-0.04932	-0.04597	-0.04265	-0.03937	3.
4.	-0.03618	-0.03308	-0.03011	-0.02726	-0.02456	-0.02200	-0.01960	-0.01734	-0.01525	-0.01330	4.
5.	-0.01151	$-0.0^2 9865$	$-0.0^2 8359$	$-0.0^2 6989$	$-0.0^2 5749$	$-0.0^2 4632$	$-0.0^2 3632$	$-0.0^2 2740$	$-0.0^2 1952$	$-0.0^2 1258$	5.
6.	$-0.0^2 6530$	$-0.0^3 1295$	$-0.0^3 3191$	$-0.0^3 6991$	$-0.0^2 1017$	$-0.0^2 1278$	$-0.0^2 1488$	$-0.0^2 1653$	$-0.0^2 1777$	$-0.0^2 1866$	6.
7.	$0.0^2 1922$	$0.0^2 1951$	$0.0^2 1956$	$0.0^2 1940$	$0.0^2 1907$	$0.0^2 1860$	$0.0^2 1800$	$0.0^2 1731$	$0.0^2 1655$	$0.0^2 1572$	7.
8.	$0.0^2 1486$	$0.0^2 1397$	$0.0^2 1306$	$0.0^2 1216$	$0.0^2 1126$	$0.0^2 1037$	$0.0^3 9511$	$0.0^3 8675$	$0.0^3 7871$	$0.0^3 7102$	8.
9.	$0.0^3 6372$	$0.0^3 5681$	$0.0^3 5030$	$0.0^3 4422$	$0.0^3 3855$	$0.0^3 3330$	$0.0^3 2846$	$0.0^3 2402$	$0.0^3 1996$	$0.0^3 1628$	9.
10.	$0.0^3 1295$										

(3) Numerical Values

Kei (x) $x = 0 - 10$

x	0	1	2	3	4	5	6	7	8	9	x
0.	-0.7854	-0.7769	-0.7581	-0.7331	-0.7038	-0.6716	-0.6374	-0.6022	-0.5664	-0.5305	0.
1.	-0.4950	-0.4601	-0.4262	-0.3933	-0.3617	-0.3314	-0.3026	-0.2752	-0.2494	-0.2251	1.
2.	-0.2024	-0.1812	-0.1614	-0.1431	-0.1262	-0.1107	-0.09644	-0.08342	-0.07157	-0.06083	2.
3.	-0.05112	-0.04240	-0.03458	-0.02762	-0.02145	-0.01600	-0.01123	$0.0^2 7077$	$-0.0^2 3487$	$0.0^3 4108$	3.
4.	$0.0^2 2198$	$0.0^2 4386$	$0.0^2 6194$	$0.0^2 7661$	$0.0^2 8826$	$0.0^2 9721$	0.01038	0.01083	0.01110	0.01121	4.
5.	0.01119	0.01105	0.01082	0.01051	0.01014	$0.0^2 9716$	$0.0^2 9255$	$0.0^2 8766$	$0.0^2 8258$	$0.0^2 7739$	5.
6.	$0.0^2 7216$	$0.0^2 6696$	$0.0^2 6183$	$0.0^2 5681$	$0.0^2 5194$	$0.0^2 4724$	$0.0^2 4274$	$0.0^2 3846$	$0.0^2 3440$	$0.0^2 3058$	6.
7.	$0.0^2 2700$	$0.0^2 2366$	$0.0^2 2057$	$0.0^2 1770$	$0.0^2 1507$	$0.0^2 1267$	$0.0^2 1048$	$0.0^3 8498$	$0.0^3 6714$	$0.0^3 5117$	7.
8.	$0.0^3 3696$	$0.0^3 2440$	$0.0^3 1339$	$0.0^4 3809$	$-0.0^4 4449$	$-0.0^3 1149$	$-0.0^3 1742$	$-0.0^3 2233$	$-0.0^3 2632$	$-0.0^3 2949$	8.
9.	$-0.0^3 3192$	$-0.0^3 3368$	$-0.0^3 3486$	$-0.0^3 3552$	$-0.0^3 3574$	$-0.0^3 3557$	$-0.0^3 3508$	$-0.0^3 3430$	$-0.0^3 3329$	$-0.0^3 3210$	9.
10.	$-0.0^3 3075$										

(4) Asymptotic Series for Large x

$$\text{Ker}\,(x) \approx \sqrt{\frac{\pi}{2x}}\, e^{-x/\sqrt{2}}\{\cos\eta\,[1 - (1^2)\beta_1 + (1^2)(3^2)\beta_2 - \cdots] + \sin\eta\,[(1^2)\gamma_1 - (1^2)(3^2)\gamma_2 + \cdots]\}$$

$$\text{Kei}\,(x) \approx \sqrt{\frac{\pi}{2x}}\, e^{-x/\sqrt{2}}\{-\cos\eta\,[(1^2)\gamma_1 - (1^2)(3^3)\gamma_2 + \cdots] - \sin\eta\,[1 - (1^2)\beta_1 + (1^2)(3^2)\beta_2 - \cdots]\}$$

$$\eta = \frac{x}{\sqrt{2}} + \frac{\pi}{8} \qquad \beta_k = \frac{\cos\,(k\pi/4)}{k!\,(8x)^k} \qquad \gamma_k = \frac{\sin\,(k\pi/4)}{k!\,(8x)^k} \qquad k = 1, 2, \ldots \qquad x > 15$$

$$J_n(x+y) = \sum_{k=-\infty}^{+\infty} J_k(x) J_{n-k}(y) \qquad n = 0, \pm 1, \pm 2, \ldots$$

$$1 = J_0(x) + 2J_2(x) + \cdots + 2J_{2n}(x) + \cdots$$

$$x = 2[J_1(x) + 3J_3(x) + \cdots + (2n+1)J_{2n+1}(x) + \cdots]$$

$$x^2 = 8[J_2(x) + 4J_4(x) + \cdots + n^2 J_{2n}(x) + \cdots]$$

$$\sin x = 2[J_1(x) - J_3(x) + J_5(x) - \cdots]$$

$$\cos x = J_0(x) - 2[J_2(x) - J_4(x) + \cdots]$$

$$\sinh x = 2[I_1(x) + I_3(x) + I_5(x) + \cdots]$$

$$\cosh x = I_0(x) + 2[I_2(x) + I_4(x) + \cdots]$$

$$\sin(x \sin \omega) = 2[J_1(x) \sin \omega + J_3(x) \sin 3\omega + \cdots]$$

$$\sin(x \cos \omega) = 2[J_1(x) \cos \omega - J_3(x) \cos 3\omega - \cdots]$$

$$\cos(x \sin \omega) = J_0(x) + 2[J_2(x) \cos 2\omega + J_4(x) \cos 4\omega + \cdots]$$

$$\cos(x \cos \omega) = J_0(x) - 2[J_2(x) \cos 2\omega - J_4(x) \cos 4\omega + \cdots]$$

14.36   DEFINITE INTEGRALS INVOLVING BESSEL FUNCTIONS

$$\int_0^\infty J_n(\alpha x)\, dx = \frac{1}{\alpha} \qquad n > -1 \qquad\qquad \int_0^\infty J_n(\alpha x) \frac{dx}{x} = \frac{1}{n} \qquad n = 1, 2, \ldots$$

$$\int_0^\infty e^{-\alpha x} J_0(\beta x)\, dx = \frac{1}{\sqrt{\alpha^2 + \beta^2}} \qquad\qquad \int_0^\infty e^{-\alpha x} J_1(\beta x)\, dx = \frac{1}{\beta}\left(1 - \frac{\alpha}{\sqrt{\alpha^2 + \beta^2}}\right)$$

$$\int_0^\infty J_n(\alpha x) \sin \beta x\, dx = \begin{cases} \dfrac{\sin(n \sin^{-1}(\beta/\alpha))}{\sqrt{\alpha^2 - \beta^2}} & 0 < \beta < \alpha \\[4mm] \dfrac{\alpha^n \cos(n\pi/2)}{\sqrt{b^2 - \alpha^2}(\beta + \sqrt{\beta^2 - \alpha^2})^n} & 0 < \alpha < \beta \end{cases} \quad n > -2$$

$$\int_0^\infty J_n(\alpha x) \cos \beta x\, dx = \begin{cases} \dfrac{\cos[n \cos^{-1}(\beta/\alpha)]}{\sqrt{\alpha^2 - \beta^2}} & 0 < \beta < \alpha \\[4mm] \dfrac{-\alpha^n \sin(n\pi/2)}{\sqrt{\beta^2 - \alpha^2}(\beta + \sqrt{\beta^2 - \alpha^2})^n} & 0 < \alpha < \beta \end{cases} \quad n > -1$$

$$\int_0^\infty \frac{J_m(x) J_n(x)}{x}\, dx = \begin{cases} \dfrac{2}{(m^2 - n^2)\pi} \sin \dfrac{(m-n)\pi}{2} & m \ne n \\[4mm] \dfrac{1}{2m} & m = n \end{cases} \quad m + n > 0$$

15
PARTIAL
DIFFERENTIAL
EQUATIONS

(1) Definition

A partial differential equation of order n is *a functional equation* involving at least one nth partial derivative of the unknown function $\Phi(x_1, x_2, \ldots, x_m)$ of two or more variables $x_1, x_2, \ldots, x_m$.

(2) Solution

A partial differential equation of order n given in general form as

$$F\left(x_1, x_2, \ldots, x_m, \Phi; \frac{\partial \Phi}{\partial x_1}, \frac{\partial \Phi}{\partial x_2}, \ldots, \frac{\partial \Phi}{\partial x_m}, \frac{\partial^2 \Phi}{\partial x_1^2}, \frac{\partial^2 \Phi}{\partial x_2^2}, \ldots, \frac{\partial^2 \Phi}{\partial x_m^2}, \ldots\right) = 0$$

has a *general solution* (general integral), which involves *arbitrary functions*. A *particular solution* (particular integral) is a special case of the general solution with *specific functions* substituted for the arbitrary functions. These specific functions are generated by the given conditions (boundary conditions, initial conditions). Many partial differential equations admit *singular solutions* (singular integrals) unrelated to the general integral.

(3) Separation of Variables

Although the separation-of-variables method is not universally applicable, it offers a convenient and simple solution of many partial differential equations in engineering and the physical sciences. The underlying idea is to *separate* the given partial differential equation into *a set of ordinary differential equations*, the solution of which is known.
The *assumed solution* is

$$\Phi(x_1, x_2, \ldots, x_m) = \Phi_1(x_1)\Phi_2(x_2) \cdots \Phi_m(x_m)$$

the linear combination of which is the general solution. This procedure is applicable when the substitution of the assumed solution in the given partial differential equation transforms this equation into

$$\psi_1\left(x_1; \frac{d\Phi_1}{\partial x_1}; \ldots\right) + \psi_2\left(x_2, x_3, \ldots, x_m; \frac{d\Phi_2}{dx_2}; \ldots\right) = 0$$

A setting $\psi_1 = C$ and $\psi_2 = -C$ breaks the equation $\psi_1 + \psi_2 = 0$ into an ordinary differential equation and a partial differential equation, the second of which (if necessary) can be the subject of a repeated separation process.

(4) Classification

Partial differential equations are classified according to the type of conditions imposed by the problem. Linear differential equations of the second order in the two variables given as

$$Au_{xx} + Bu_{xy} + Cu_{yy} + Du_x + Eu_y + Fu + G = 0$$

where $u_{xx}, u_{xy}, \ldots$ are the partial derivatives with respect to x and/or y and $A, B, \ldots$, are functions of x and y, are classified as *elliptic, hyperbolic, or parabolic* according to whether

$$\Delta = \begin{vmatrix} A & B \\ B & C \end{vmatrix}$$

is *positive, negative, or zero*. The concept of these definitions can be extended for differential equations with more than two variables.

		$A, B, C =$ integration constants
$a, b, c =$ given constants	$u_x = \dfrac{du}{dx}$	$u_y = \dfrac{du}{dy}$
$u = u(x, y)$		
$P = P(x, y, u)$	$Q = Q(x, y, u)$	$R = R(x, y, u)$

(1) Quasilinear Equation

For

$$P(x, y, u)\frac{du}{dx} + Q(x, y, u)\frac{du}{dy} = R(x, y, z)$$

the *characteristic equation* is

$$dx : dy : du = P : Q : R \qquad \text{or} \qquad \frac{dx}{P} = \frac{dy}{Q} = \frac{du}{R}$$

from which

$$f(x, y, u) = A \qquad\qquad g(x, y, u) = B$$

and the *general solution* is

$$F[\, f(x, y, u)\, ; g(x, y, u)\,] = 0$$

where $F[\;\;]$ is an *arbitrary function*.

(2) Special Linear Equations

Differential equation	Solution
$a\,u_x \pm b\,u_y = 0$	$u = f(ay \mp bx)$
$y\,u_x - x\,u_y = 0$	$u = f(x^2 + y^2)$
$y\,u_x + x\,u_y = u$	$u = (x + y)f(x^2 - y^2)$
$xu\,u_y + yu\,u_x = cx$	$u^2 = 2cy + f(x^2 + y^2)$
$a\,u_x + b\,u_y = c$	$F\left[\left(\dfrac{x}{a} - \dfrac{u}{c}\right); \left(\dfrac{y}{b} - \dfrac{u}{c}\right)\right] = 0$
$x\,u_x + y\,u_y = u$	$F\left(\dfrac{x}{u}; \dfrac{y}{u}\right) = 0$
$(x - a)\,u_x + (y - b)\,u_y = u - c$	$F\left(\dfrac{x - a}{u - c}; \dfrac{y - b}{u - c}\right) = 0$
$yu\,u_x + xu\,u_y = xy$	$F[\,(x^2 - y^2); (x^2 - u^2)\,] = 0$

(3) Special Nonlinear Equations

Differential equation	Solution	
$u_x\,u_y = c$	$u = Ax + By + C$	$B = \dfrac{c}{A}$
$u_x\,u_y = xy$	$u = Ax^2 + By^2 + C$	$B = \dfrac{1}{4A}$
$u_x\,u_y = u_x + u_y$	$u = Ax + By + C$	$B = \dfrac{A}{A - 1}$
$u_x\,u_y = x\,u_x + y\,u_y$	$u = Ax^2 + xy + By^2 + C$	$B = \dfrac{1}{4A}$

A, B, C = integration constants α, β, γ = separation constants

$J(\ ), Y(\ )$ = Bessel functions (Sec. 14.24)

(1) Two-dimensional Cases

(a) Rectangular coordinates $\dfrac{\partial^2 \Phi}{\partial x^2} + \dfrac{\partial^2 \Phi}{\partial y^2} = 0$

Solution:

$$\Phi_\alpha = (A_1 e^{i\alpha x} + A_2 e^{-i\alpha x})(B_1 e^{\alpha y} + B_2 e^{-\alpha y}) \qquad\qquad \alpha \neq 0$$

$$\Phi_0 = (A_1 + A_2 x)(B_1 + B_2 y) \qquad\qquad \alpha = 0$$

(b) Polar coordinates $\dfrac{\partial^2 \Phi}{\partial r^2} + \dfrac{1}{r}\dfrac{\partial \Phi}{\partial r} + \dfrac{1}{r^2}\dfrac{\partial^2 \Phi}{\partial \phi^2} = 0$

Solution:

$$\Phi_\alpha = (A_1 r^\alpha + A_2 r^{-\alpha})(B_1 e^{\alpha i\phi} + B_2 e^{-\alpha i\phi}) \qquad\qquad \alpha \neq 0$$

$$\Phi_0 = (A_1 + A_2 \ln r)(B_1 + B_2 \phi) \qquad\qquad \alpha = 0$$

(2) Three-dimensional Cases

(a) Rectangular coordinates $\dfrac{\partial^2 \Phi}{\partial x^2} + \dfrac{\partial^2 \Phi}{\partial y^2} + \dfrac{\partial^2 \Phi}{\partial z^2} = 0$

Solution: $(\alpha^2 + \beta^2 + \gamma^2 = 0)$

$$\phi_{\alpha\beta\gamma} = (A_1 e^{\alpha x} + A_2 e^{-\alpha x})(B_1 e^{\beta y} + B_2 e^{-\beta y})(C_1 e^{\gamma z} + C_2 e^{-\gamma z}) \qquad \alpha \neq 0, \beta \neq 0, \gamma \neq 0$$

$$\phi_{000} = (A_1 + A_2 x)(B_1 + B_2 y)(C_1 + C_2 z) \qquad \alpha = 0, \beta = 0, \gamma = 0$$

(b) Cylindrical coordinates $\dfrac{\partial^2 \Phi}{\partial r^2} + \dfrac{1}{r}\dfrac{\partial \Phi}{\partial r} + \dfrac{1}{r^2}\dfrac{\partial^2 \Phi}{\partial \phi^2} + \dfrac{\partial^2 \Phi}{\partial z^2} = 0$

Solution:

$$\phi_{\alpha\beta} = [A_1 J_\alpha(i\beta r) + A_2 Y_\alpha(i\beta r)](B_1 e^{i\alpha\phi} + B_2 e^{-i\alpha\phi})(C_1 e^{i\beta z} + C_2 e^{-i\beta z}) \qquad \alpha \neq 0, \beta \neq 0$$

$$\phi_{00} = (A_1 + A_2 \ln r)(B_1 + B_2 \phi)(C_1 + C_2 z) \qquad \alpha = 0, \beta = 0$$

$$\phi_{\alpha 0} = (A_1 + A_2 \ln r)(B_1 e^{i\alpha\phi} + B_2 e^{-i\alpha\phi})(C_1 + C_2 z) \qquad \alpha \neq 0, \beta = 0$$

$$\phi_{0\beta} = J_0(i\beta r)(B_1 + B_2 \phi)(C_1 e^{i\beta z} + C_2 e^{-i\beta z}) \qquad \alpha = 0, \beta \neq 0$$

$A, B, C =$ integration constants $\alpha, \beta, \gamma =$ separation constants

$J(\ ), Y(\ ) =$ Bessel functions (Sec. 14.24) $k =$ given constant

(1) Two-dimensional Cases

(a) Rectangular coordinates
$$\frac{\partial^2 \Phi}{\partial x^2} + \frac{\partial^2 \Phi}{\partial y^2} + k^2 \Phi = 0$$

Solution: $(\alpha^2 + \beta^2 = k^2)$

$$\Phi_{\alpha\beta} = (A_1 e^{i\alpha x} + A_2 e^{-i\alpha x})(B_1 e^{i\beta y} + B_2 e^{-i\beta y}) \qquad \alpha \neq 0$$

$$\Phi_{0k} = (A_1 + A_2 x)(B_1 e^{iky} + B_2 e^{-iky}) \qquad \alpha = 0$$

(b) Polar coordinates
$$\frac{\partial^2 \Phi}{\partial r^2} + \frac{1}{r}\frac{\partial \Phi}{\partial r} + \frac{1}{r^2}\frac{\partial^2 \Phi}{\partial \phi^2} + k^2 \Phi = 0$$

Solution:

$$\Phi_{\alpha k} = [A_1 J_\alpha(ikr) + A_2 Y_\alpha(ikr)](B_1 e^{i\alpha\phi} + B_2 e^{-i\alpha\phi}) \qquad \alpha \neq 0$$

$$\Phi_{0k} = J_0(ikr)(B_1 + B_2 \phi) \qquad \alpha = 0$$

(2) Three-dimensional Cases

(a) Rectangular coordinates
$$\frac{\partial^2 \Phi}{\partial x^2} + \frac{\partial^2 \Phi}{\partial y^2} + \frac{\partial^2 \Phi}{\partial z^2} + k^2 \Phi = 0$$

Solution: $(\alpha^2 + \beta^2 + \gamma^2 = k^2)$

$$\Phi_{\alpha\beta\gamma} = (A_1 e^{i\alpha x} + A_2 e^{-i\alpha x})(B_1 e^{i\beta y} + B_2 e^{-i\beta y})(C_1 e^{i\gamma z} + C_2 e^{-i\gamma z}) \qquad \alpha \neq 0, \beta \neq 0, \gamma \neq 0$$

$$\Phi_{00k} = (A_1 + A_2 x)(B_1 + B_2 y)(C_1 e^{ikz} + C_2 e^{-ikz}) \qquad \alpha = 0, \beta = 0$$

(b) Cylindrical coordinates
$$\frac{\partial^2 \Phi}{\partial r^2} + \frac{1}{r}\frac{\partial \Phi}{\partial r} + \frac{1}{r^2}\frac{\partial^2 \Phi}{\partial \phi^2} + \frac{\partial^2 \Phi}{\partial z^2} + k^2 \Phi = 0$$

Solution:

$$\Phi_{\alpha\beta k} = [A_1 J_\alpha(i\gamma r) + A_2 Y_\alpha(i\gamma r)](B_1 e^{i\alpha\phi} + B_2 e^{-i\alpha\phi})(C_1 e^{i\beta z} + C_2 e^{-i\beta z}) \qquad \alpha \neq 0, \beta \neq 0$$

$$\Phi_{00k} = J_0(ikr)(B_1 + B_2 \phi)(C_1 + C_2 z) \qquad \alpha = 0, \beta = 0$$

$$\gamma = \sqrt{k^2 - \beta^2}$$

$A, B, C =$ integration constants $\qquad \alpha, \beta, \gamma, \lambda, \omega =$ separation constants

$c =$ given constant $\qquad J(\ ), Y(\ ) =$ Bessel functions (Sec. 14.24)

(1) Rectangular Coordinates

(a) One-dimensional case

$$\frac{\partial^2 \Phi}{\partial x^2} - \frac{1}{c^2} \frac{\partial \Phi}{\partial t} = 0$$

Solution: $(\alpha = \lambda, \omega = c^2 \lambda^2)$

$\Phi_\alpha = (A_1 e^{i\alpha x} + A_2 e^{-i\alpha x}) e^{-\omega t} \qquad\qquad \alpha \neq 0$

$\Phi_0 = A_1 + A_2 x \qquad\qquad\qquad\qquad \alpha = 0$

(b) Two-dimensional case

$$\frac{\partial^2 \Phi}{\partial x^2} + \frac{\partial^2 \Phi}{\partial y^2} - \frac{1}{c^2} \frac{\partial \Phi}{\partial t} = 0$$

Solution: $(\alpha^2 + \beta^2 = \lambda^2, \omega = c^2 \lambda^2)$

$\Phi_{\alpha\beta} = (A_1 e^{i\alpha x} + A_2 e^{-i\alpha x})(B_1 e^{i\beta y} + B_2 e^{-i\beta y}) e^{-\omega t} \qquad \alpha \neq 0, \beta \neq 0$

$\Phi_{0\lambda} = (A_1 + A_2 x)(B_1 e^{i\lambda y} + B_2 e^{-i\lambda y}) e^{-\omega t} \qquad \alpha = 0, \beta = \lambda$

(c) Three-dimensional case

$$\frac{\partial^2 \Phi}{\partial x^2} + \frac{\partial^2 \Phi}{\partial y^2} + \frac{\partial^2 \Phi}{\partial z^2} - \frac{1}{c^2} \frac{\partial \Phi}{\partial t} = 0$$

Solution: $(\alpha^2 + \beta^2 + \gamma^2 = \lambda^2, \omega = c^2 \lambda^2)$

$\Phi_{\alpha\beta\gamma} = (A_1 e^{i\alpha x} + A_2 e^{-i\alpha x})(B_1 e^{i\beta y} + B_2 e^{-i\beta y})(C_1 e^{i\gamma z} + C_2 e^{-i\gamma z}) e^{-\omega t} \qquad \alpha \neq 0, \beta \neq 0, \gamma \neq 0$

$\Phi_{00\lambda} = (A_1 + A_2 x)(B_1 + B_2 y)(C_1 e^{i\lambda z} + C_2 e^{-i\lambda z}) e^{-\omega t} \qquad \alpha = 0, \beta = 0, \gamma = \lambda$

(2) Cylindrical Coordinates[1]

(a) Two-dimensional case

$$\frac{\partial^2 \Phi}{\partial r^2} + \frac{1}{r} \frac{\partial \Phi}{\partial r} + \frac{\partial^2 \Phi}{\partial \phi^2} - \frac{1}{c^2} \frac{\partial \Phi}{\partial t} = 0$$

Solution: $(\alpha = 1, 2, 3, \ldots)(\omega = c^2 \lambda^2)$

$\Phi_{\alpha\beta} = [A_1 J_\alpha(i\lambda r) + A_2 Y_\alpha(i\lambda r)](B_1 e^{i\alpha \phi} + B_2 e^{-i\alpha \phi}) e^{-\omega t} \qquad \alpha \neq 0, \lambda \neq 0$

$\Phi_{0\lambda} = J_0(i\lambda r)(B_1 + B_2 \phi) e^{-\omega t} \qquad \alpha = 0, \lambda \neq 0$

(b) Three-dimensional case

$$\frac{\partial^2 \Phi}{\partial r^2} + \frac{1}{r} \frac{\partial \Phi}{\partial r} + \frac{1}{r^2} \frac{\partial^2 \Phi}{\partial \phi^2} + \frac{\partial^2 \Phi}{\partial z^2} - \frac{1}{c^2} \frac{\partial \Phi}{\partial t} = 0$$

Solution: $(\alpha = 1, 2, 3, \ldots)(\omega = c^2 \lambda^2)$

$\Phi_{\alpha\beta\gamma} = [A_1 J_\alpha(i\gamma r) + A_2 Y_\alpha(i\gamma r)](B_1 e^{i\alpha \phi} + B_2 e^{-i\alpha \phi})(C_1 e^{i\beta z} + C_2 e^{-i\beta z}) e^{-\omega t}$

$\qquad\qquad\qquad\qquad\qquad\qquad\qquad\qquad\qquad\qquad \alpha \neq 0, \beta \neq 0, \gamma = \sqrt{\lambda^2 - \beta^2}$

$\Phi_{00\lambda} = J_0(i\lambda r)(B_1 + B_2 \phi)(C_1 + C_2 z) e^{-\omega t} \qquad\qquad \alpha = 0, \beta = 0, \gamma = \lambda$

[1] In all axially symmetrical cases $\alpha = 0$.

A, B, C, D = integration constants	$\alpha, \beta, \gamma, \lambda, \omega$ = separation constants
c = given constant	$J(\ ), Y(\ )$ = Bessel functions (Sec. 14.24)

(1) Rectangular Coordinates

(a) One-dimensional case

$$\frac{\partial^2\Phi}{\partial x^2} - \frac{1}{c^2}\frac{\partial^2\Phi}{\partial t^2} = 0$$

Solution: $(\alpha = \lambda,\ \omega = c\lambda)$

$$\Phi_\alpha = (A_1 e^{i\alpha x} + A_2 e^{-i\alpha x})(B_1 e^{i\omega t} + B_2 e^{-i\omega t}) \qquad \alpha \neq 0$$

$$\Phi_0 = A_1 + A_2 x \qquad \alpha = 0$$

(b) Two-dimensional case

$$\frac{\partial^2\Phi}{\partial x^2} + \frac{\partial^2\Phi}{\partial y^2} - \frac{1}{c^2}\frac{\partial^2\Phi}{\partial t^2} = 0$$

Solution: $(\alpha^2 + \beta^2 = \lambda^2,\ \omega = c\lambda)$

$$\Phi_{\alpha\beta} = (A_1 e^{i\alpha x} + A_2 e^{-i\alpha x})(B_1 e^{i\beta y} + B_2 e^{-i\beta y})(C_1 e^{i\omega t} + C_2 e^{-i\omega t}) \qquad \alpha \neq 0, \beta \neq 0$$

$$\Phi_{0\lambda} = (A_1 + A_2 x)(B_1 e^{i\lambda y} + B_2 e^{-i\lambda y})(C_1 e^{i\omega t} + C_2 e^{-i\omega t}) \qquad \alpha = 0, \beta = \lambda$$

(c) Three-dimensional case

$$\frac{\partial^2\Phi}{\partial x^2} + \frac{\partial^2\Phi}{\partial y^2} + \frac{\partial^2\Phi}{\partial z^2} - \frac{1}{c^2}\frac{\partial^2\Phi}{\partial t^2} = 0$$

Solution: $(\alpha^2 + \beta^2 + \gamma^2 = \lambda^2,\ \omega = c\lambda)$

$$\Phi_{\alpha\beta\gamma} = (A_1 e^{i\alpha x} + A_2 e^{-i\alpha x})(B_1 e^{i\beta y} + B_2 e^{-i\beta y})(C_1 e^{i\gamma z} + C_2 e^{-i\gamma z})(D_1 e^{i\omega t} + D_2 e^{-i\omega t})$$

$$\alpha \neq 0, \beta \neq 0, \gamma \neq 0$$

$$\Phi_{00\lambda} = (A_1 + A_2 x)(B_1 + B_2 y)(C_1 e^{i\lambda z} + C_2 e^{-i\lambda z})(D_1 e^{i\omega t} + D_2 e^{-i\omega t}) \qquad \alpha = 0, \beta = 0, \gamma = \lambda$$

(2) Cylindrical Coordinates[1]

(a) Two-dimensional case

$$\frac{\partial^2\Phi}{\partial r^2} + \frac{1}{r}\frac{\partial\Phi}{\partial r} + \frac{1}{r^2}\frac{\partial^2\Phi}{\partial\phi^2} - \frac{1}{c^2}\frac{\partial^2\Phi}{\partial t^2} = 0$$

Solution: $(\alpha = 1, 2, 3, \ldots)(\omega = c\lambda)$

$$\Phi_{\alpha\beta} = [A_1 J_\alpha(i\lambda r) + A_2 Y_\alpha(i\lambda r)](B_1 e^{i\alpha\phi} + B_2 e^{-i\alpha\phi})(C_1 e^{i\omega t} + C_2 e^{-i\omega t}) \qquad \alpha \neq 0, \lambda \neq 0$$

$$\Phi_{0\lambda} = J_0(i\lambda r)(B_1 + B_2\phi)(C_1 e^{i\omega t} + C_2 e^{-i\omega t}) \qquad \alpha = 0, \lambda \neq 0$$

(b) Three-dimensional case

$$\frac{\partial^2\Phi}{\partial r^2} + \frac{1}{r}\frac{\partial\Phi}{\partial r} + \frac{1}{r^2}\frac{\partial^2\Phi}{\partial\phi^2} + \frac{\partial^2\Phi}{\partial z^2} - \frac{1}{c^2}\frac{\partial^2\Phi}{\partial t^2} = 0$$

Solution: $(\alpha = 1, 2, 3, \ldots)(\omega = c\lambda)$

$$\Phi_{\alpha\beta\gamma} = [A_1 J_\alpha(i\gamma r) + A_2 Y_\alpha(i\gamma r)](B_1 e^{i\alpha\phi} + B_2 e^{-i\alpha\phi})(C_1 e^{i\beta z} + C_2 e^{-i\beta z})(D_1 e^{i\omega t} + D_2 e^{-i\omega t})$$

$$\alpha \neq 0, \beta \neq 0, \gamma = \sqrt{\lambda^2 - \beta^2}$$

$$\Phi_{00\lambda} = J_0(i\lambda r)(B_1 + B_2\phi)(C_1 + C_2 z)(D_1 e^{i\omega t} + D_2 e^{-i\omega t}) \qquad \alpha = 0, \beta = 0, \gamma = \lambda$$

[1] In all axially symmetrical cases $\alpha = 0$.

$u = u(x, t) =$ longitudinal displacement	$v = v(x, t) =$ transverse displacement
$a, b, l =$ given constants	$A, B, C, D, \ldots =$ integration constants
$f(x, t) =$ forcing function	$\omega =$ angular frequency (eigenvalue)
$X(x), Y(x) =$ eigenvectors (Sec. 1.14)	$Q(t), T(t) =$ time functions

(1) Longitudinal Vibration

(a) General solution

$$a\frac{\partial^2 u}{\partial x^2} - b\frac{\partial^2 u}{\partial t^2} = -f(x, t) \qquad \lambda_k = \omega_k \sqrt{\frac{b}{a}} \qquad k = 1, 2, 3, \ldots$$

$$u(x, t) = \sum_{k=1}^{\infty} X_k(x) T_k(t)$$

$$X_k(x) = A_k \cos \lambda_k x + B_k \frac{\sin \lambda_k x}{\lambda_k} \qquad Q_k(t) = \int_0^l f(x, t) X_k(x)\, dx$$

$$T_k(t) = C_k \frac{\cos \omega_k t}{\Delta_k} + D_k \frac{\sin \omega_k t}{\Delta_k \omega_k} + \frac{b}{\Delta_k \omega_k} \int_0^t Q_k(\tau) \sin \omega_k(t - \tau)\, d\tau$$

(b) Integration constants

$$A_k = X_k(+0) \qquad B_k = X_k'(+0) \qquad \Delta_k = \int_0^l X_k^2(x)\, dx$$

$$C_k = \int_0^l u(x, 0) X_k(x)\, dx \qquad D_k = \int_0^l \dot{u}(x, 0) X_k(x)\, dx$$

$$\text{where } X_k'(+0) = \frac{dX(x)}{dx}\bigg|_{x=0}, \quad \dot{u}(x, 0) = \frac{du(x, t)}{dt}\bigg|_{t=0}.$$

(2) Transverse Vibration

(a) General solution

$$a\frac{\partial^4 v}{\partial x^4} + b\frac{\partial^2 v}{\partial t^2} = f(x, t) \qquad \lambda_k = \sqrt[4]{\frac{b\omega_k^2}{a}} \qquad k = 1, 2, 3, \ldots$$

$$v(x, t) = \sum_{k=1}^{\infty} Y_k(x) T_k(t)$$

$$Y_k(x) = A_k \Psi_1 + B_k \Psi_2 + C_k \Psi_3 + D_k \Psi_4 \qquad Q_k(t) = \int_0^l f(x, t) Y_k(x)\, dx$$

$$\Psi_1 = \frac{\cosh \lambda_k x + \cos \lambda_k x}{2} \qquad \Psi_2 = \frac{\sinh \lambda_k x + \sin \lambda_k x}{2\lambda_k}$$

$$\Psi_3 = \frac{\cosh \lambda_k x - \cos \lambda_k x}{2\lambda_k^2} \qquad \Psi_4 = \frac{\sinh \lambda_k x - \sin \lambda_k x}{2\lambda_k^3}$$

$$T_k(t) = E_k \frac{\cos \omega_k t}{\Delta_k} + F_k \frac{\sin \omega_k t}{\Delta_k \omega_k} + \frac{b}{\Delta_k \omega_k} \int_0^t Q_k(\tau) \sin \omega_k(t - \tau)\, d\tau$$

(b) Integration constants

$$A_k = Y_k(+0) \qquad B_k = Y_k'(+0) \qquad C_k = Y''(+0) \qquad D_k = Y'''(+0)$$

$$\Delta_k = \int_0^l Y_k^2(x)\, dx \qquad E_k = \int_0^l v(x, 0) Y_k(x)\, dx \qquad F_k = \int_0^l \dot{v}(x, 0) Y_k(x)\, dx$$

$$\text{where } Y'(+0) = \frac{dY(x)}{dx}\bigg|_{x=0}, \quad Y''(+0) = \frac{d^2 Y(x)}{dx^2}\bigg|_{x=0}, \ldots, \dot{v}(x, 0) = \frac{dv(x, t)}{dt}\bigg|_{t=0}.$$

16
LAPLACE TRANSFORMS

(1) Definitions

If $F(t)$ is a *piecewise-continuous function* of the real variable $t \geqslant 0$, then the *Laplace transform* of $F(t)$ is

$$\mathscr{L}\{F(t)\} = \int_0^\infty e^{-st}F(t)\, dt = f(s)$$

where s = real or complex and $e = 2.71828\cdots$.
The *inverse Laplace transform* of $f(s)$ is

$$\mathscr{L}^{-1}\{f(s)\} = F(t)$$

Not every function $f(s)$ has an inverse Laplace transform.

(2) Basic Relationships $(a, b = \text{real numbers}, n = 1, 2, 3, \ldots)$

$$\mathscr{L}(a) = \frac{a}{s} \qquad \mathscr{L}(t) = \frac{1}{s^2} \qquad \mathscr{L}(t^n) = \frac{n!}{s^{n+1}}$$

$$\mathscr{L}\{aF(t)\} = af(s) \qquad\qquad \mathscr{L}\{tF(t)\} = -f'(s)$$

$$\mathscr{L}\{F(at)\} = \frac{1}{a}f\left(\frac{s}{a}\right) \qquad\qquad \mathscr{L}\left\{F\left(\frac{t}{a}\right)\right\} = af(as)$$

$$\mathscr{L}\left\{\frac{1}{t}F(t)\right\} = \int_s^\infty f(s)\, ds \qquad\qquad \mathscr{L}\{t^nF(t)\} = (-1)^n f^{(n)}(s)$$

$$\mathscr{L}\{e^{at}F(t)\} = f(s-a) \qquad\qquad \mathscr{L}\{F(t-a)\} = e^{-as}f(s)$$

$$\mathscr{L}\{aF_1(t) + bf_2(t)\} = \mathscr{L}\{aF_1(t)\} + \mathscr{L}\{bF_2(t)\} = af_1(s) + bf_2(s)$$

(3) Derivatives

$$\mathscr{L}\{F'(t)\} = sf(s) - F(+0)$$

$$\mathscr{L}\{F''(t)\} = s^2f(s) - sF(+0) - F'(+0)$$

$$\mathscr{L}\{F^{(n)}(t)\} = s^nf(s) - s^{n-1}F(+0) - s^{n-2}F'(+0) - \cdots - F^{(n-1)}(+0)$$

(4) Integrals $[u = u(t), v = v(t)]$

$$\mathscr{L}\left\{\int_0^t F(u)\, du\right\} = \frac{f(s)}{s} \qquad\qquad \mathscr{L}\left\{\int_0^t \int_0^u F(v)\, dv\, du\right\} = \frac{f(s)}{s^2}$$

$$\mathscr{L}\left\{\int_0^t F(t-\tau)G(\tau)\, d\tau\right\} = \mathscr{L}\left\{\int_0^t F(t)G(t-\tau)\, d\tau\right\} = f(s) \bullet g(s)$$

(5) Periodic Functions $[G(T+t) = G(t), H(T+t) = -H(t)]$

$$\mathscr{L}\{G(T+t)\} = \frac{\int_0^T e^{-st}G(t)\, dt}{1 - e^{-sT}} \qquad\qquad \mathscr{L}\{H(T+t)\} = \frac{\int_0^T e^{-st}H(t)\, dt}{1 + e^{-sT}}$$

$f(s)$	$F(t)$	$f(s)$	$F(t)$
$\dfrac{1}{s}$	1	$\dfrac{1}{s-a}$	e^{at}
$\dfrac{1}{s^2}$	t	$\dfrac{1}{(s-a)^2}$	te^{at}
$\dfrac{1}{s^n}$	$\dfrac{t^{n-1}}{(n-1)!}$	$\dfrac{1}{(s-a)^n}$	$\dfrac{t^{n-1}e^{at}}{(n-1)!}$
$\dfrac{1}{\sqrt{s}}$	$\dfrac{1}{\sqrt{\pi t}}$	$\dfrac{1}{s(s-a)}$	$\dfrac{1-e^{at}}{a}$
$\dfrac{1}{\sqrt{s^3}}$	$2\sqrt{\dfrac{t}{\pi}}$	$\dfrac{1}{s^2(s-a)}$	$\dfrac{e^{at}-at-1}{a^2}$
$\dfrac{1}{s^2+a^2}$	$\dfrac{\sin at}{a}$	$\dfrac{1}{s^2-a^2}$	$\dfrac{\sinh at}{a}$
$\dfrac{s}{s^2+a^2}$	$\cos at$	$\dfrac{s}{s^2-a^2}$	$\cosh at$
$\dfrac{1}{(s-b)^2+a^2}$	$\dfrac{e^{bt}\sin at}{a}$	$\dfrac{1}{(s-b)^2-a^2}$	$\dfrac{e^{bt}\sinh at}{a}$
$\dfrac{s-b}{(s-b)^2+a^2}$	$e^{bt}\cos at$	$\dfrac{s-b}{(s-b)^2-a^2}$	$e^{bt}\cosh at$
$\dfrac{1}{(s-a)(s-b)}\quad a\neq b$	$\dfrac{e^{bt}-e^{at}}{b-a}$	$\dfrac{s}{(s-a)(s-b)}\quad a\neq b$	$\dfrac{be^{bt}-ae^{at}}{b-a}$
$\dfrac{1}{(s^2+a^2)^2}$	$\dfrac{\sin at - at\cos at}{2a^3}$	$\dfrac{1}{(s^2-a^2)^2}$	$-\dfrac{\sinh at - at\cosh at}{2a^3}$
$\dfrac{s}{(s^2+a^2)^2}$	$\dfrac{t\sin at}{2a}$	$\dfrac{s}{(s^2-a^2)^2}$	$\dfrac{t\sinh at}{2a}$
$\dfrac{s^2}{(s^2+a^2)^2}$	$\dfrac{\sin at + at\cos at}{2a}$	$\dfrac{s^2}{(s^2-a^2)^2}$	$\dfrac{\sinh at + at\cosh at}{2a}$
$\dfrac{s^3}{(s^2+a^2)^2}$	$\cos at - \dfrac{at\sin at}{2}$	$\dfrac{s^3}{(s^2-a^2)^2}$	$\cosh at + \dfrac{at\sinh at}{2}$
$\dfrac{1}{s(s^2+a^2)}$	$\dfrac{1-\cos at}{a^2}$	$\dfrac{1}{s(s^2-a^2)}$	$\dfrac{\cosh at - 1}{a^2}$
$\dfrac{1}{s^2(s^2+a^2)}$	$\dfrac{at-\sin at}{a^3}$	$\dfrac{1}{s^2(s^2-a^2)}$	$\dfrac{\sinh at - at}{a^3}$

a, b = constants		$n, k = 1, 2, 3, \ldots$	
erf = error function (Sec. 13.01)		Γ = gamma function (Sec. 13.03)	
J = Bessel function (Sec. 14.24)		I = Bessel function (Sec. 14.28)	

$f(s)$	$F(t)$	$f(s)$	$F(t)$
$\dfrac{1}{\sqrt{s+a} + \sqrt{s+b}}$	$\dfrac{e^{-bt} - e^{-at}}{2(b-a)t\sqrt{\pi t^3}}$	$\sqrt{s-a} + \sqrt{s-b}$	$\dfrac{e^{bt} - e^{at}}{2t\sqrt{\pi t^3}}$
$\dfrac{1}{s\sqrt{s+a}}$	$\dfrac{\operatorname{erf}\sqrt{at}}{\sqrt{a}}$	$\dfrac{1}{\sqrt{s}(s-a)}$	$e^{at}\dfrac{\operatorname{erf}\sqrt{at}}{\sqrt{a}}$
$\dfrac{1}{(s^2 + a^2)^{1/2}}$	$J_0(at)$	$\dfrac{1}{(s^2 - a^2)^{1/2}}$	$I_0(at)$
$\dfrac{1}{(s^2 + a^2)^{3/2}}$	$\dfrac{tJ_1(at)}{a}$	$\dfrac{1}{(s^2 - a^2)^{3/2}}$	$\dfrac{tI_1(at)}{a}$
$\dfrac{1}{(s^2 + a^2)^{k}}$ $k > 0$	$\dfrac{\sqrt{\pi}}{\Gamma(k)}\left(\dfrac{t}{2a}\right)^{k-1/2} J_{k-1/2}(at)$	$\dfrac{1}{(s^2 - a^2)^{k}}$ $k > 0$	$\dfrac{\sqrt{\pi}}{\Gamma(k)}\left(\dfrac{t}{2a}\right)^{k-1/2} I_{k-1/2}(at)$
$\dfrac{s}{(s^2 + a^2)^{3/2}}$	$tJ_0(at)$	$\dfrac{s}{(s^2 - a^2)^{3/2}}$	$tI_0(at)$
$\dfrac{s^2}{(s^2 + a^2)^{3/2}}$	$J_0(at) - at J_1(at)$	$\dfrac{s^2}{(s^2 - a^2)^{3/2}}$	$I_0(at) + at I_1(at)$
$(\sqrt{s^2 + a^2} - s)^k$ $k > 0$	$\dfrac{ka^k J_k(at)}{t}$	$(s - \sqrt{s^2 - a^2})^k$ $k > 0$	$\dfrac{ka^k I_k(at)}{t}$
$\dfrac{(\sqrt{s^2 + a^2} - s)^k}{\sqrt{s^2 + a^2}}$ $k > -1$	$a^k J_k(at)$	$\dfrac{(s - \sqrt{s^2 - a^2})^k}{\sqrt{s^2 - a^2}}$ $k > -1$	$a^k I_k(at)$
$\dfrac{1 + e^{-ks}}{s}$	$\begin{cases} 1 & \text{if } 0 < t < k \\ 2 & \text{if } t > k \end{cases}$	$\dfrac{1 - e^{-ks}}{s}$	$\begin{cases} 1 & \text{if } 0 < t < k \\ 0 & \text{if } t > k \end{cases}$
$\dfrac{e^{-ks}}{s^{n+1}}$ $n + 1 > 0$	$\begin{cases} 0 & \text{if } 0 < t < k \\ \dfrac{(t-k)^n}{\Gamma(n+1)} & \text{if } t > k \end{cases}$	$\dfrac{e^{-(k/s)}}{\sqrt{s}}$	$\dfrac{\cos 2\sqrt{kt}}{\sqrt{\pi t}}$
$\dfrac{e^{-(k/s)}}{s}$	$J_0(2\sqrt{kt})$	$\dfrac{e^{k/s}}{s}$	$I_0(2\sqrt{kt})$
$\dfrac{e^{-(k/s)}}{s^{n+1}}$ $k > -1$	$\left(\dfrac{t}{k}\right)^{n/2} J_n(2\sqrt{kt})$	$\dfrac{e^{k/s}}{s^{n+1}}$ $k > -1$	$\left(\dfrac{t}{k}\right)^{n/2} I_n(2\sqrt{kt})$

$f(s)$	$F(t)$
$\dfrac{1}{s^3 + a^3}$	$\dfrac{e^{at/2}}{3a^2}\left(\sqrt{3}\sin\dfrac{at\sqrt{3}}{2} - \cos\dfrac{at\sqrt{3}}{2} + e^{-(3at/2)}\right)$
$\dfrac{s}{s^3 + a^3}$	$\dfrac{e^{at/2}}{3a}\left(\sqrt{3}\sin\dfrac{at\sqrt{3}}{2} + \cos\dfrac{at\sqrt{3}}{2} - e^{-(3at/2)}\right)$
$\dfrac{s^2}{s^3 + a^3}$	$\dfrac{1}{3}\left(2e^{at/2}\cos\dfrac{at\sqrt{3}}{2} + e^{-at}\right)$
$\dfrac{1}{s^3 - a^3}$	$-\dfrac{e^{-(at/2)}}{3a^2}\left(\sqrt{3}\sin\dfrac{at\sqrt{3}}{2} + \cos\dfrac{at\sqrt{3}}{2} - e^{3at/2}\right)$
$\dfrac{s}{s^3 - a^3}$	$\dfrac{e^{-(at/2)}}{3a}\left(\sqrt{3}\sin\dfrac{at\sqrt{3}}{2} - \cos\dfrac{at\sqrt{3}}{2} + e^{3at/2}\right)$
$\dfrac{s^2}{s^3 - a^3}$	$\dfrac{1}{3}\left(2e^{-(at/2)}\cos\dfrac{at\sqrt{3}}{2} + e^{at}\right)$
$\dfrac{s}{(s^2 + a^2)(s^2 + b^2)} \qquad a \neq b$	$\dfrac{\cos at - \cos bt}{b^2 - a^2}$
$\dfrac{1}{s^4 + 4a^4}$	$\dfrac{1}{4a^3}(\sin at \cosh at - \cos at \sinh at)$
$\dfrac{s}{s^4 + 4a^4}$	$\dfrac{1}{2a^2}\sin at \sinh at$
$\dfrac{s^2}{s^4 + 4a^4}$	$\dfrac{1}{2a}(\sin at \cosh at + \cos at \sinh at)$
$\dfrac{s^3}{s^4 + 4a^4}$	$\cos at \cosh at$
$\dfrac{1}{s^4 - a^4}$	$\dfrac{1}{2a^3}(\sinh at - \sin at)$
$\dfrac{s}{s^4 - a^4}$	$\dfrac{1}{2a^2}(\cosh at - \cos at)$
$\dfrac{s^2}{s^4 - a^4}$	$\dfrac{1}{2a}(\sinh at + \sin at)$
$\dfrac{s^3}{s^4 - a^4}$	$\dfrac{1}{2}(\cosh at + \cos at)$

$f(s)$	$F(t)$
$\dfrac{e^{-as}}{s}$	Unit function
$\dfrac{e^{-as}(1 - e^{-bs})}{s}$	Pulse function
$\dfrac{1}{s(1 - e^{-as})}$	Step function
$\dfrac{\tan\alpha\, e^{-bs}}{s^2}$	$\tan\alpha$
$\dfrac{\tan\beta(1 - e^{-bs})}{s^2}$	$\tan\beta$

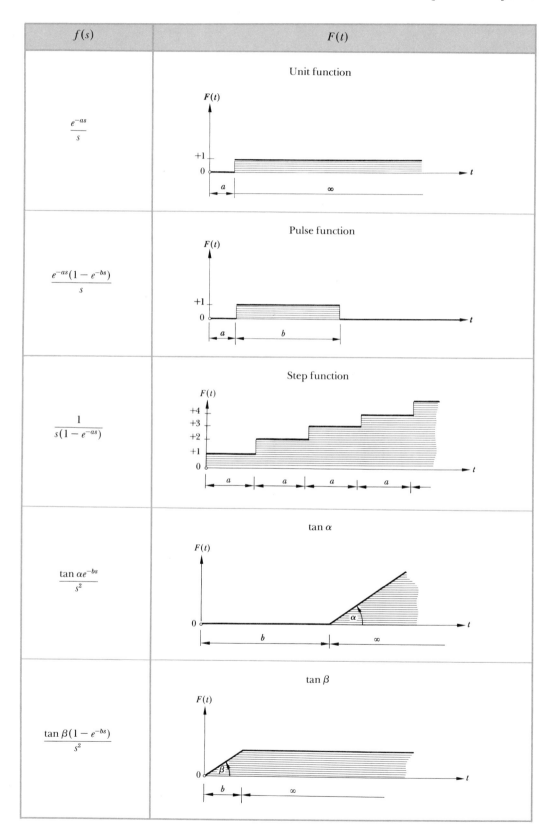

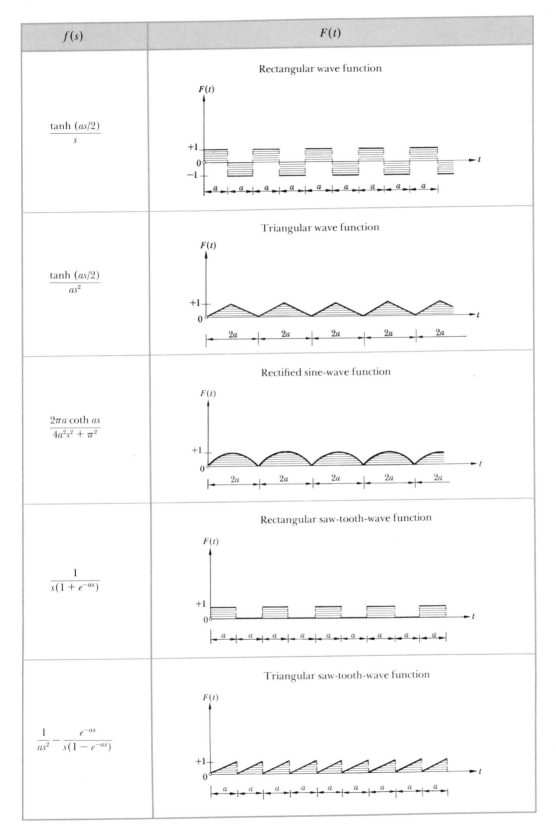

$f(s)$	$F(t)$
$\dfrac{\tanh\,(as/2)}{s}$	Rectangular wave function
$\dfrac{\tanh\,(as/2)}{as^2}$	Triangular wave function
$\dfrac{2\pi a \coth as}{4a^2 s^2 + \pi^2}$	Rectified sine-wave function
$\dfrac{1}{s(1 + e^{-as})}$	Rectangular saw-tooth-wave function
$\dfrac{1}{as^2} - \dfrac{e^{-as}}{s(1 - e^{-as})}$	Triangular saw-tooth-wave function

$$y'' + ay = f(x)$$

$y(+0), y'(+0) =$ initial values $\qquad a =$ real number $\neq 0$ $\qquad \lambda = \sqrt{|a|}$

(1) General Solution

$$y(x) = y(+0)A_r(x) + y'(+0)B_r(x) + F_r(x)$$

$\qquad r = 1, 2$

(2) Solution of the First Kind

$\qquad a < 0 \qquad r = 1$

(a) Shape functions and convolution integral

$$A_1(x) = \cosh \lambda x \qquad B_1(x) = \frac{\sinh \lambda x}{\lambda}$$

$$F_1(x) = \int_0^x f(x-\tau)B_1(\tau)\,d\tau \qquad \text{(Sec. 16.08)}$$

(b) Transport matrix equation

$$\begin{bmatrix} y(x) \\ y'(x) \end{bmatrix} = \begin{bmatrix} A_1(x) & B_1(x) \\ -aB_1(x) & A_1(x) \end{bmatrix} \begin{bmatrix} y(+0) \\ y'(+0) \end{bmatrix} + \begin{bmatrix} F_1(x) \\ F_1'(x) \end{bmatrix}$$

(3) Solution of the Second Kind

$\qquad a > 0 \qquad r = 2$

(a) Shape functions and convolution integral

$$A_2(x) = \cos \lambda x \qquad B_2(x) = \frac{\sin \lambda x}{\lambda}$$

$$F_2(x) = \int_0^x f(x-\tau)B_2(\tau)\,d\tau \qquad \text{(Sec. 16.08)}$$

(b) Transport matrix equation

$$\begin{bmatrix} y(x) \\ y'(x) \end{bmatrix} = \begin{bmatrix} A_2(x) & B_2(x) \\ -aB_2(x) & A_2(x) \end{bmatrix} \begin{bmatrix} y(+0) \\ y'(+0) \end{bmatrix} + \begin{bmatrix} F_2(x) \\ F_2'(x) \end{bmatrix}$$

(4) Applications

$y'' - 4y = 10 \sin 5x$	$a = -4,$	$r = 1,$	$\lambda = \sqrt{|-4|} = 2,$	$w = 10,$	$\beta = 5$

$$A_1(x) = \cosh 2x \qquad B_1(x) = \tfrac{1}{2}\sinh 2x \qquad F_1(x) = \frac{10}{-4-25}(\sin 5x - \tfrac{5}{2}\sinh 2x)$$

$$y(x) = y(+0)\cosh 2x + y'(+0)\frac{\sinh 2x}{2} - \tfrac{10}{29}(\sin 5x - \tfrac{5}{2}\sinh 2x)$$

$y'' + 4y = 10 \sin 5x$	$a = 4,$	$r = 2,$	$\lambda = \sqrt{4} = 2,$	$w = 10,$	$\beta = 5$

$$A_2(x) = \cos 2x \qquad B_2(x) = \tfrac{1}{2}\sin 2x \qquad F_2(x) = \frac{10}{4-25}(\sin 5x - \tfrac{5}{2}\sin 2x)$$

$$y(x) = y(+0)\cos 2x + y'(+0)\frac{\sin 2x}{2} - \tfrac{10}{21}(\sin 5x - \tfrac{5}{2}\sin 2x)$$

$$y'' + ay = f(x)$$

The analytical expressions for $F_r(x)$ below apply for $F_1(x)$ and also for $F_2(x)$ in Sec. 16.07 (if the numerical value of a is used with its proper sign and the respective shape functions are applied).

$a, b, c, w, \alpha, \beta, P = $ real numbers

$f(x)$	$F_r(x) = \int_0^x f(x-\tau)B_r(\tau)d\tau$	$r = 1, 2$
	$\dfrac{w}{a}[1 - A_r(x)]$	
	$\dfrac{w}{a}[x - B_r(x)]$	
	$\dfrac{w}{a^2}[ax^2 - 2 + 2A_r(x)]$	
	$\dfrac{w}{a + \alpha^2}[e^{-\alpha x} - A_r(x) + \alpha B_r(x)]$	
	$\dfrac{w}{a}[b - x - bA_r(x) + B_r(x)]$	
	$\dfrac{w}{a - \beta^2}[\sin \beta x - \beta B_r(x)]$	
	$\dfrac{w}{a - \beta^2}[\cos \beta x - A_r(x)]$	
	$0 \qquad\qquad$ for $x < c$ $PB_r(x - c) \qquad$ for $x \geq c$	

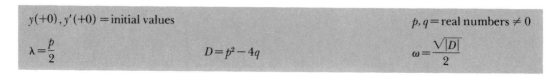

$$y'' + py' + qy = f(x)$$

$y(+0), y'(+0) =$ initial values

$p, q =$ real numbers $\neq 0$

$$\lambda = \frac{p}{2} \qquad\qquad D = p^2 - 4q \qquad\qquad \omega = \frac{\sqrt{|D|}}{2}$$

(1) General Solution

$$y(x) = y(+0)A_r(x) + y'(+0)B_r(x) + F_r(x) \qquad \boxed{r = 1, 2, 3}$$

(2) Solution of the First Kind

$\boxed{D < 0, r = 1}$

(a) Shape functions and convolution integral

$$A_1(x) = e^{-\lambda x}\left(\cos \omega x + \frac{\lambda}{\omega}\sin \omega x\right) \qquad B_1(x) = e^{-\lambda x}\frac{\sin \omega x}{\omega}$$

$$F_1(x) = \int_0^x f(x-\tau)B_1(\tau)\, dt \qquad (\text{Sec. 16.10})$$

(b) Transport matrix equation

$$\begin{bmatrix} y(x) \\ y'(x) \end{bmatrix} = \begin{bmatrix} A_1(x) & B_1(x) \\ -qB_1(x) & A_1(x) - pB_1(x) \end{bmatrix}\begin{bmatrix} y(+0) \\ y'(+0) \end{bmatrix} + \begin{bmatrix} F_1(x) \\ F_1'(x) \end{bmatrix}$$

(3) Solution of the Second Kind

$\boxed{D = 0, r = 2}$

(a) Shape functions and convolution integral

$$A_2(x) = (1 + \lambda x)e^{-\lambda x} \qquad B_2(x) = xe^{-\lambda x}$$

$$F_2(x) = \int_0^x f(x-\tau)B_2(\tau)\, d\tau \qquad (\text{Sec. 16.10})$$

(b) Transport matrix equation

$$\begin{bmatrix} y(x) \\ y'(x) \end{bmatrix} = \begin{bmatrix} A_2(x) & B_2(x) \\ -qB_2(x) & A_2(x) - pB_2(x) \end{bmatrix}\begin{bmatrix} y(+0) \\ y'(+0) \end{bmatrix} + \begin{bmatrix} F_2(x) \\ F_2'(x) \end{bmatrix}$$

(4) Solution of the Third Kind

$\boxed{D > 0, r = 3}$

(a) Shape functions and convolution integral

$$A_3(x) = e^{-\lambda x}\left(\cosh \omega x + \frac{\lambda}{\omega}\sinh \omega x\right) \qquad B_3(x) = e^{-\lambda x}\frac{\sinh \omega x}{\omega}$$

$$F_3(x) = \int_0^x f(x-\tau)B_3(\tau)\, d\tau \qquad (\text{Sec. 16.10})$$

(b) Transport matrix equation

$$\begin{bmatrix} y(x) \\ y'(x) \end{bmatrix} = \begin{bmatrix} A_3(x) & B_3(x) \\ -qB_3(x) & A_3(x) - pB_3(x) \end{bmatrix}\begin{bmatrix} y(+0) \\ y'(+0) \end{bmatrix} + \begin{bmatrix} F_3(x) \\ F_3'(x) \end{bmatrix}$$

$$y'' + py' + qy = f(x)$$

The analytical expressions for $F_r(x)$ below apply for $F_1(x)$, $F_2(x)$, and also for $F_3(x)$ in Sec. 16.09 (if the numerical values of p, q are used with their proper signs and the respective shape functions are applied).

$b, c, p, q, w, \alpha, \beta, P$ = real numbers

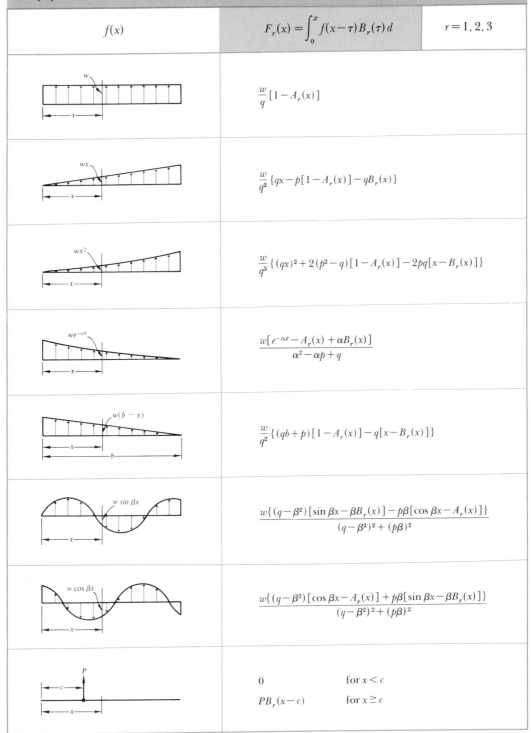

$f(x)$	$F_r(x) = \displaystyle\int_0^x f(x-\tau) B_r(\tau)\, d$	$r = 1, 2, 3$
w	$\dfrac{w}{q}[1 - A_r(x)]$	
wx	$\dfrac{w}{q^2}\{qx - p[1 - A_r(x)] - qB_r(x)\}$	
wx^2	$\dfrac{w}{q^3}\{(qx)^2 + 2(p^2 - q)[1 - A_r(x)] - 2pq[x - B_r(x)]\}$	
$we^{-\alpha x}$	$\dfrac{w[e^{-\alpha x} - A_r(x) + \alpha B_r(x)]}{\alpha^2 - \alpha p + q}$	
$w(b - x)$	$\dfrac{w}{q^2}\{(qb + p)[1 - A_r(x)] - q[x - B_r(x)]\}$	
$w \sin \beta x$	$\dfrac{w\{(q - \beta^2)[\sin \beta x - \beta B_r(x)] - p\beta[\cos \beta x - A_r(x)]\}}{(q - \beta^2)^2 + (p\beta)^2}$	
$w \cos \beta x$	$\dfrac{w\{(q - \beta^2)[\cos \beta x - A_r(x)] + p\beta[\sin \beta x - \beta B_r(x)]\}}{(q - \beta^2)^2 + (p\beta)^2}$	
P	$\begin{array}{ll} 0 & \text{for } x < c \\ PB_r(x - c) & \text{for } x \geq c \end{array}$	

$$y^{iv} + ay = f(x)$$

$y(+0), y'(+0), y''(+0), y'''(+0) =$ initial values

$a =$ real number $\neq 0$

(1) General Solution

$$y(x) = y(+0)A_r(x) + y'(+0)B_r(x) + y''(+0)C_r(x) + y'''(+0)D_r(x) + F_r(x)$$

$r = 1, 2$

(2) Solution of the First Kind

$a < 0, r = 1, \lambda = \sqrt[4]{|a|}$

(a) Shape functions

$$A_1(x) = \frac{\cosh \lambda x + \cos \lambda x}{2}$$

$$B_1(x) = \frac{\sinh \lambda x + \sin \lambda x}{2\lambda}$$

$$C_1(x) = \frac{\cosh \lambda x - \cos \lambda x}{2\lambda^2}$$

$$D_1(x) = \frac{\sinh \lambda x - \sin \lambda x}{2\lambda^3}$$

(b) Convolution integral

$$F_1(x) = \int_0^x f(x-\tau)D_1(\tau)\, d\tau \qquad \text{(Sec. 16.12)}$$

(c) Transport matrix equation

$$
\left[\begin{array}{c} y(x) \\ y'(x) \\ \hline y''(x) \\ y'''(x) \end{array}\right]
=
\left[\begin{array}{cc|cc}
A_1(x) & B_1(x) & C_1(x) & D_1(x) \\
-aD_1(x) & A_1(x) & B_1(x) & C_1(x) \\
\hline
-aC_1(x) & -aD_1(x) & A_1(x) & B_1(x) \\
-aB_1(x) & -aC_1(x) & -aD_1(x) & A_1(x)
\end{array}\right]
\left[\begin{array}{c} y(+0) \\ y'(+0) \\ \hline y''(+0) \\ y'''(+0) \end{array}\right]
+
\left[\begin{array}{c} F_1(x) \\ F_1'(x) \\ \hline F_1''(x) \\ F_1'''(x) \end{array}\right]
$$

(3) Solution of the Second Kind

$a > 0, r = 2, \lambda = \sqrt[4]{\dfrac{a}{4}}$

(a) Shape functions

$$A_2(x) = \cosh \lambda x \cos \lambda x$$

$$B_2(x) = \frac{1}{2\lambda}[\cosh \lambda x \sin \lambda x + \sinh \lambda x \cos \lambda x]$$

$$C_2(x) = \frac{1}{2\lambda^2}\sinh \lambda x \sin \lambda x$$

$$D_2(x) = \frac{1}{4\lambda^3}[\cosh \lambda x \sin \lambda x - \sinh \lambda x \cos \lambda x]$$

(b) Convolution integral

$$F_2(x) = \int_0^x f(x-\tau)D_2(\tau)\, d\tau \qquad \text{(Sec. 16.12)}$$

(c) Transport matrix equation

$$
\left[\begin{array}{c} y(x) \\ y'(x) \\ \hline y''(x) \\ y'''(x) \end{array}\right]
=
\left[\begin{array}{cc|cc}
A_2(x) & B_2(x) & C_2(x) & D_2(x) \\
-aD_2(x) & A_2(x) & B_2(x) & C_2(x) \\
\hline
-aC_2(x) & -aD_2(x) & A_2(x) & B_2(x) \\
-aB_2(x) & -aC_2(x) & -aD_2(x) & A_2(x)
\end{array}\right]
\left[\begin{array}{c} y(+0) \\ y'(+0) \\ \hline y''(+0) \\ y'''(+0) \end{array}\right]
+
\left[\begin{array}{c} F_2(x) \\ F_2'(x) \\ \hline F_2''(x) \\ F_2'''(x) \end{array}\right]
$$

$$y^{iv} + ay = f(x)$$

The analytical expressions for $F_r(x)$ below apply for $F_1(x)$ and also for $F_2(x)$ in Sec. 16.11 (if the numerical value of a is used with its proper sign and the respective shape functions are applied).

$a, c, w, \alpha, \beta, P, Q =$ real numbers

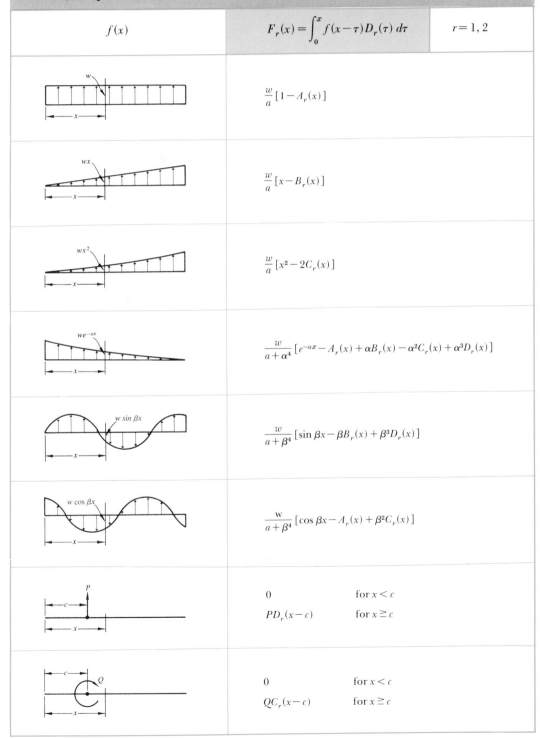

$f(x)$	$F_r(x) = \int_0^x f(x-\tau)D_r(\tau)\,d\tau$	$r = 1, 2$
	$\dfrac{w}{a}[1 - A_r(x)]$	
	$\dfrac{w}{a}[x - B_r(x)]$	
	$\dfrac{w}{a}[x^2 - 2C_r(x)]$	
	$\dfrac{w}{a+\alpha^4}[e^{-\alpha x} - A_r(x) + \alpha B_r(x) - \alpha^2 C_r(x) + \alpha^3 D_r(x)]$	
	$\dfrac{w}{a+\beta^4}[\sin\beta x - \beta B_r(x) + \beta^3 D_r(x)]$	
	$\dfrac{w}{a+\beta^4}[\cos\beta x - A_r(x) + \beta^2 C_r(x)]$	
	$0 \qquad\qquad$ for $x < c$ $PD_r(x-c) \qquad$ for $x \geq c$	
	$0 \qquad\qquad$ for $x < c$ $QC_r(x-c) \qquad$ for $x \geq c$	

$$y^{iv} + ay'' = f(x)$$

$y(+0), y'(+0), y''(+0), y'''(+0) = $ initial values

$a = $ real number $\neq 0$

(1) General Solution

$$y(x) = y(+0) + y'(+0)x + y''(+0)C_r(x) + y'''(+0)D_r(x) + F_r(x)$$

$r = 1, 2$

(2) Solution of the First Kind

$a < 0, r = 1, \lambda = \sqrt{|a|}$

(a) Shape functions

$$A_1(x) = \cosh \lambda x \qquad\qquad B_1(x) = \frac{\sinh \lambda x}{\lambda}$$

$$C_1(x) = \frac{\cosh \lambda x - 1}{\lambda^2} \qquad D_1(x) = \frac{\sinh \lambda x - \lambda x}{\lambda^3}$$

(b) Convolution integral

$$F_1(x) = \int_0^x f(x-\tau)D_1(\tau)\,d\tau \qquad\qquad \text{(Sec. 16.14)}$$

(c) Transport matrix equation

$$
\begin{bmatrix} y(x) \\ y'(x) \\ \hline y''(x) \\ y'''(x) \end{bmatrix}
=
\left[\begin{array}{cc|cc}
1 & x & C_1(x) & D_1(x) \\
0 & 1 & B_1(x) & C_1(x) \\
\hline
0 & 0 & A_1(x) & B_1(x) \\
0 & 0 & -aB_1(x) & A_1(x)
\end{array}\right]
\begin{bmatrix} y(+0) \\ y'(+0) \\ \hline y''(+0) \\ y'''(+0) \end{bmatrix}
+
\begin{bmatrix} F_1(x) \\ F_1'(x) \\ \hline F_1''(x) \\ F_1'''(x) \end{bmatrix}
$$

(3) Solution of the Second Kind

$a > 0, r = 2, \lambda = \sqrt{a}$

(a) Shape functions

$$A_2(x) = \cos \lambda x \qquad\qquad B_2(x) = \frac{\sin \lambda x}{\lambda}$$

$$C_2(x) = \frac{1 - \cos \lambda x}{\lambda^2} \qquad D_2(x) = \frac{\lambda x - \sin \lambda x}{\lambda^3}$$

(b) Convolution integral

$$F_2(x) = \int_0^x f(x-\tau)D_2(\tau)\,d\tau \qquad\qquad \text{(Sec. 16.14)}$$

(c) Transport matrix equation

$$
\begin{bmatrix} y(x) \\ y'(x) \\ \hline y''(x) \\ y'''(x) \end{bmatrix}
=
\left[\begin{array}{cc|cc}
1 & x & C_2(x) & D_2(x) \\
0 & 1 & B_2(x) & C_2(x) \\
\hline
0 & 0 & A_2(x) & B_2(x) \\
0 & 0 & -aB_2(x) & A_2(x)
\end{array}\right]
\begin{bmatrix} y(+0) \\ y'(+0) \\ \hline y''(+0) \\ y'''(+0) \end{bmatrix}
+
\begin{bmatrix} F_2(x) \\ F_2'(x) \\ \hline F_2''(x) \\ F_2'''(x) \end{bmatrix}
$$

$$y^{iv} + ay'' = f(x)$$

The analytical expressions for $F_r(x)$ below apply for $F_1(x)$ and also for $F_2(x)$ in Sec. 16.13 (if the numerical value of a is used with its proper sign and the respective shape functions are applied).

$a, c, w, \alpha, \beta, P, Q$ = real numbers

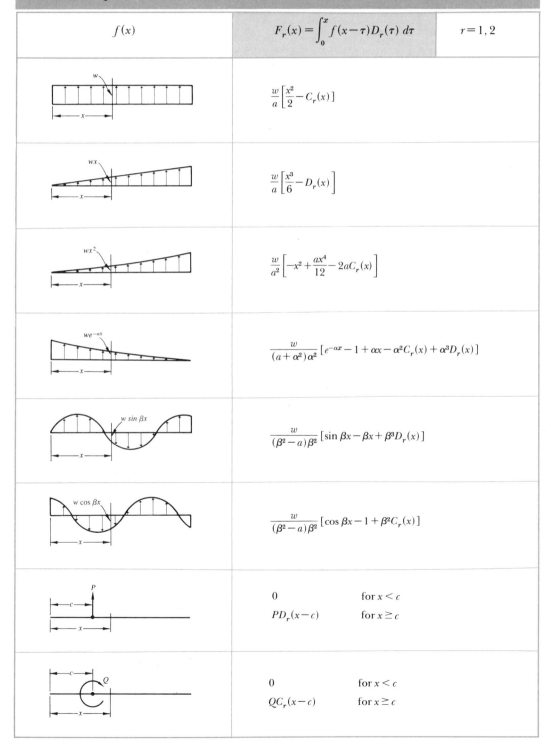

$f(x)$	$F_r(x) = \int_0^x f(x-\tau)D_r(\tau)\,d\tau$	$r = 1, 2$
w	$\dfrac{w}{a}\left[\dfrac{x^2}{2} - C_r(x)\right]$	
wx	$\dfrac{w}{a}\left[\dfrac{x^3}{6} - D_r(x)\right]$	
wx^2	$\dfrac{w}{a^2}\left[-x^2 + \dfrac{ax^4}{12} - 2aC_r(x)\right]$	
$we^{-\alpha x}$	$\dfrac{w}{(a+\alpha^2)\alpha^2}\left[e^{-\alpha x} - 1 + \alpha x - \alpha^2 C_r(x) + \alpha^3 D_r(x)\right]$	
$w\sin\beta x$	$\dfrac{w}{(\beta^2-a)\beta^2}\left[\sin\beta x - \beta x + \beta^3 D_r(x)\right]$	
$w\cos\beta x$	$\dfrac{w}{(\beta^2-a)\beta^2}\left[\cos\beta x - 1 + \beta^2 C_r(x)\right]$	
P	$0 \qquad\qquad\qquad \text{for } x < c$ $PD_r(x-c) \qquad\quad \text{for } x \geq c$	
Q	$0 \qquad\qquad\qquad \text{for } x < c$ $QC_r(x-c) \qquad\quad \text{for } x \geq c$	

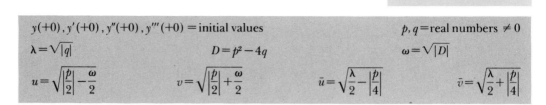

$$y^{iv} + py'' + qy = f(x)$$

$y(+0), y'(+0), y''(+0), y'''(+0) =$ initial values $\qquad p, q =$ real numbers $\neq 0$

$\lambda = \sqrt{|q|} \qquad D = p^2 - 4q \qquad \omega = \sqrt{|D|}$

$u = \sqrt{\left|\frac{p}{2}\right| - \frac{\omega}{2}} \qquad v = \sqrt{\left|\frac{p}{2}\right| + \frac{\omega}{2}} \qquad \bar{u} = \sqrt{\frac{\lambda}{2} - \left|\frac{p}{4}\right|} \qquad \bar{v} = \sqrt{\frac{\lambda}{2} + \left|\frac{p}{4}\right|}$

(1) General Solution

$$y(x) = y(+0) A_r(x) + y'(+0) B_r(x) + y''(+0) C_r(x) + y'''(+0) D_r(x) + F_r(x) \qquad \boxed{r = 1, 2, 3, 4, 5, 6}$$

where the convolution integral is

$$F_r(x) = \int_0^x f(x - \tau) D_r(\tau) \, d\tau \qquad \text{(Sec. 16.16)}$$

(2) Shape Functions

$r = 1$ $p > 0$ $D > 0$	$A_1(x) = \dfrac{q}{\omega} \left(\dfrac{\cos ux}{u^2} - \dfrac{\cos vx}{v^2} \right)$ $\qquad$ $B_1(x) = \dfrac{q}{\omega} \left(\dfrac{\sin ux}{u^3} - \dfrac{\sin vx}{v^3} \right)$ $C_1(x) = \dfrac{1}{\omega} (\cos ux - \cos vx)$ $\qquad$ $D_1(x) = \dfrac{1}{\omega} \left(\dfrac{\sin ux}{u} - \dfrac{\sin vx}{v} \right)$
$r = 2$ $p > 0$ $D = 0$	$A_2(x) = \dfrac{ux}{2} \sin ux + \cos ux$ $\qquad$ $B_2(x) = \dfrac{1}{2u} (3 \sin ux - ux \cos ux)$ $C_2(x) = \dfrac{x}{2u} \sin ux$ $\qquad$ $D_2(x) = \dfrac{1}{2u^3} (\sin ux - ux \cos ux)$
$r = 3$ $p > 0$ $D < 0$	$A_3(x) = \cosh \bar{u}x \cos \bar{v}x + \dfrac{p}{\omega} \sinh \bar{u}x \sin \bar{v}x$ $\qquad$ $B_3(x) = \dfrac{\lambda + p}{2\lambda\bar{v}} \cosh \bar{u}x \sin \bar{v}x + \dfrac{\lambda - p}{2\lambda\bar{u}} \sinh \bar{u}x \cos \bar{v}x$ $C_3(x) = \dfrac{1}{2} \dfrac{\sinh \bar{u}x}{\bar{u}} \dfrac{\sin \bar{v}x}{\bar{v}}$ $\qquad$ $D_3(x) = \dfrac{1}{2\lambda} \left(\dfrac{\cosh \bar{u}x \sin \bar{v}x}{\bar{v}} - \dfrac{\sinh \bar{u}x \cos \bar{v}x}{\bar{u}} \right)$
$r = 4$ $p < 0$ $D > 0$	$A_4(x) = \dfrac{q}{\omega} \left(\dfrac{\cosh ux}{u^2} - \dfrac{\cosh vx}{v^2} \right)$ $\qquad$ $B_4(x) = \dfrac{q}{\omega} \left(\dfrac{\sinh ux}{u^3} - \dfrac{\sinh vx}{v^3} \right)$ $C_4(x) = \dfrac{-1}{\omega} (\cosh ux + \cosh vx)$ $\qquad$ $D_4(x) = \dfrac{-1}{\omega} \left(\dfrac{\sinh ux}{u} - \dfrac{\sinh vx}{v} \right)$
$r = 5$ $p < 0$ $D = 0$	$A_5(x) = -\dfrac{ux}{2} \sinh ux + \cosh ux$ $\qquad$ $B_5(x) = \dfrac{1}{2u} (3 \sinh ux - ux \cosh ux)$ $C_5(x) = \dfrac{x}{2u} \sinh ux$ $\qquad$ $D_5(x) = \dfrac{-1}{2u^3} (\sinh ux - ux \cosh ux)$
$r = 6$ $p < 0$ $D < 0$	$A_6(x) = \cos \bar{u}x \cosh \bar{v}x - \dfrac{p}{\omega} \sin \bar{u}x \sinh \bar{v}x$ $\qquad$ $B_6(x) = \dfrac{\lambda + p}{2\lambda\bar{v}} \cos \bar{u}x \sinh \bar{v}x + \dfrac{\lambda - p}{2\lambda\bar{u}} \sin \bar{u}x \cosh \bar{v}x$ $C_6(x) = \dfrac{1}{2} \dfrac{\sin \bar{u}x}{\bar{u}} \dfrac{\sinh \bar{v}x}{\bar{v}}$ $\qquad$ $D_6(x) = \dfrac{-1}{2\lambda} \left(\dfrac{\cos \bar{u}x \sinh \bar{v}x}{\bar{v}} - \dfrac{\sin \bar{u}x \cosh \bar{v}x}{\bar{u}} \right)$

$$y^{iv} + py'' + qy = f(x)$$

The analytical expressions for $F_r(x)$ below apply for all six solutions $(r=1,2,3,4,5,6)$ (if the numerical values of p and q are used with their proper sign and the respective shape functions are applied).

$c, p, q, w, \alpha, \beta, P, Q =$ real numbers

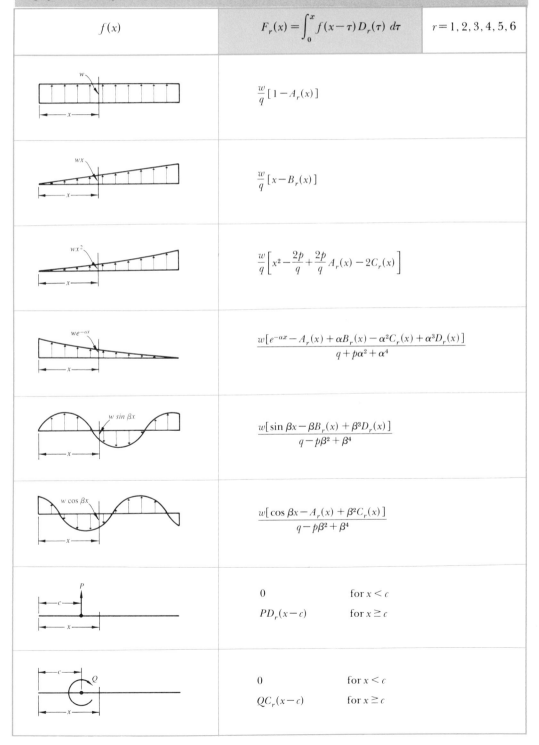

$f(x)$	$F_r(x) = \displaystyle\int_0^x f(x-\tau) D_r(\tau)\, d\tau$	$r=1,2,3,4,5,6$
w	$\dfrac{w}{q}[1 - A_r(x)]$	
wx	$\dfrac{w}{q}[x - B_r(x)]$	
wx^2	$\dfrac{w}{q}\left[x^2 - \dfrac{2p}{q} + \dfrac{2p}{q}A_r(x) - 2C_r(x)\right]$	
$we^{-\alpha x}$	$\dfrac{w[e^{-\alpha x} - A_r(x) + \alpha B_r(x) - \alpha^2 C_r(x) + \alpha^3 D_r(x)]}{q + p\alpha^2 + \alpha^4}$	
$w \sin \beta x$	$\dfrac{w[\sin \beta x - \beta B_r(x) + \beta^3 D_r(x)]}{q - p\beta^2 + \beta^4}$	
$w \cos \beta x$	$\dfrac{w[\cos \beta x - A_r(x) + \beta^2 C_r(x)]}{q - p\beta^2 + \beta^4}$	
P	0 for $x < c$ $PD_r(x-c)$ for $x \geq c$	
Q	0 for $x < c$ $QC_r(x-c)$ for $x \geq c$	

(1) Definition

The Dirac Delta function $\mathscr{D}(x - t)$ is not a function in the true sense but a definition of a distribution.

$$\mathscr{D}(x - t) = \lim_{\Delta \to 0} \begin{cases} 0 & \text{for } x < \left(t - \dfrac{\Delta}{2}\right) \\[2mm] \dfrac{1}{\Delta} & \text{for } \left(t - \dfrac{\Delta}{2}\right) < x < \left(t + \dfrac{\Delta}{2}\right) \\[2mm] 0 & \text{for } x > \left(t + \dfrac{\Delta}{2}\right) \end{cases}$$

(2) Basic Relations

$$\mathscr{D}(x) = \mathscr{D}(-x)$$

$$a\mathscr{D}(x) = \mathscr{D}\left(\frac{x}{a}\right)$$

$$f(a \pm x)\mathscr{D}(x) = f(a)\mathscr{D}(x)$$

$$\mathscr{D}(x \neq 0) = 0$$

$$x\mathscr{D}(x) = 0$$

$$f(x)\mathscr{D}(x) = f(0)\mathscr{D}(x)$$

(3) Properties

$$\int_{-\infty}^{+\infty} \mathscr{D}(x - t)\, dt = 1$$

$$\int_{-\infty}^{+\infty} f(t)\mathscr{D}(x - t)\, dt = f(x)$$

$$\int_{0}^{+\infty} \mathscr{D}(t)\, dt = 1$$

$$\int_{0}^{+\infty} f(t)\mathscr{D}(t)\, dt = f(0)$$

$$\int_{-\infty}^{+\infty} \mathscr{D}(x - t)\mathscr{D}(y - t)\, dt = \mathscr{D}(x - y)$$

(4) Derivatives

$$\frac{d\mathscr{D}(x)}{dx} = -\frac{\mathscr{D}(x)}{x}$$

$$\frac{d^n\mathscr{D}(x)}{dx^n} = (-1)^n \frac{n!\mathscr{D}(x)}{x^n}$$

(5) Integrals

$$\int_{-\infty}^{+\infty} f(t)\frac{d\mathscr{D}(x - t)}{dx}\, dt = \frac{df(x)}{dx}$$

$$\int_{-\infty}^{+\infty} f(t)\frac{d^n\mathscr{D}(x - t)}{dx^n} = (-1)^n \frac{d^n f(x)}{dx^n}$$

(6) Laplace Transforms

$$\mathscr{L}\{[\mathscr{D}(t)]\} = 1$$

$$\mathscr{L}\{[\mathscr{D}(t - a)]\} = e^{-as}$$

17
NUMERICAL METHODS

(1) Methods

Whenever the solution of an engineering problem leads to an expression, equation, or system of equations which cannot be evaluated or solved in closed form, *numerical methods* must be employed. The best-known numerical techniques, *approximate evaluation of functions, numerical solution of equations, finite differences*, and *numerical integration*, are outlined in this chapter.

(2) Errors

The *absolute error* ϵ in the result is the difference between the true result a (assumed to be known) and the approximate result $\bar{a}$. The *relative error* $\bar{\epsilon}$ is the absolute error ϵ divided by $\bar{a}$. Aside from possible outright mistakes (blunders) there are three basic types of errors in numerical calculations: *inherent errors* (due to initial data error), *truncation errors* (due to finite approximation of limiting processes), and *round-off errors* (due to use of finite number of digits).

17.02 APPROXIMATIONS BY SERIES EXPANSION

(1) Algebraic Functions

	$\epsilon_T \leqslant 0.1$ percent	$\epsilon_T \leqslant 1$ percent
$\dfrac{1}{1+x} \approx 1-x$	$-0.03 \leqslant x \leqslant +0.03$	$-0.10 \leqslant x \leqslant +0.10$
$\approx 1-x+x^2$	$-0.10 \leqslant x \leqslant +0.10$	$-0.21 \leqslant x \leqslant +0.21$
$\sqrt{1+x} \approx 1+\dfrac{x}{2}$	$-0.08 \leqslant x \leqslant +0.10$	$-0.24 \leqslant x \leqslant +0.32$
$\approx 1+\dfrac{x}{2}-\dfrac{x^2}{8}$	$-0.22 \leqslant x \leqslant +0.27$	$-0.44 \leqslant x \leqslant +0.66$
$\dfrac{1}{\sqrt{1+x}} \approx 1-\dfrac{x}{2}$	$-0.04 \leqslant x \leqslant +0.06$	$-0.15 \leqslant x \leqslant +0.17$
$\approx 1-\dfrac{x}{2}+\dfrac{3x^2}{8}$	$-0.14 \leqslant x \leqslant +0.15$	$-0.30 \leqslant x \leqslant +0.32$

(2) Transcendent Functions

	$\epsilon_T \leqslant 0.1$ percent	$\epsilon_T \leqslant 1$ percent
$e^x \approx 1+x$	$-0.04 \leqslant x \leqslant +0.04$	$-0.13 \leqslant x \leqslant +0.14$
$\approx 1+x+\dfrac{x^2}{2}$	$-0.17 \leqslant x \leqslant +0.19$	$-0.35 \leqslant x \leqslant +0.43$
$\sin x \approx x$	$\left.\begin{array}{r}-0.077 \\ -4.4°\end{array}\right\} \leqslant x \leqslant \left\{\begin{array}{l}+0.077 \\ +4.4°\end{array}\right.$	$\left.\begin{array}{r}-0.244 \\ -14.0°\end{array}\right\} \leqslant x \leqslant \left\{\begin{array}{l}+0.244 \\ +14.0°\end{array}\right.$
$\approx x-\dfrac{x^2}{6}$	$\left.\begin{array}{r}-0.578 \\ -33.1°\end{array}\right\} \leqslant x \leqslant \left\{\begin{array}{l}+0.578 \\ +33.1°\end{array}\right.$	$\left.\begin{array}{r}-1.032 \\ -59.0°\end{array}\right\} \leqslant x \leqslant \left\{\begin{array}{l}+1.032 \\ +59.0°\end{array}\right.$
$\cos x \approx 1$	$\left.\begin{array}{r}-0.045 \\ -2.6°\end{array}\right\} \leqslant x \leqslant \left\{\begin{array}{l}+0.045 \\ +2.6°\end{array}\right.$	$\left.\begin{array}{r}-0.141 \\ -8.1°\end{array}\right\} \leqslant x \leqslant \left\{\begin{array}{l}+0.141 \\ +8.1°\end{array}\right.$
$\approx 1-\dfrac{x^2}{2}$	$\left.\begin{array}{r}-0.384 \\ -22.0°\end{array}\right\} \leqslant x \leqslant \left\{\begin{array}{l}+0.384 \\ +22.0°\end{array}\right.$	$\left.\begin{array}{r}-0.650 \\ -37.2°\end{array}\right\} \leqslant x \leqslant \left\{\begin{array}{l}+0.650 \\ +37.2°\end{array}\right.$
$\tan x \approx x$	$\left.\begin{array}{r}-0.054 \\ -3.1°\end{array}\right\} \leqslant x \leqslant \left\{\begin{array}{l}+0.054 \\ +3.1°\end{array}\right.$	$\left.\begin{array}{r}-0.183 \\ -10.5°\end{array}\right\} \leqslant x \leqslant \left\{\begin{array}{l}+0.183 \\ +10.5°\end{array}\right.$
$\approx x+\dfrac{x^3}{3}$	$\left.\begin{array}{r}-0.385 \\ -22.0°\end{array}\right\} \leqslant x \leqslant \left\{\begin{array}{l}+0.385 \\ +22.0°\end{array}\right.$	$\left.\begin{array}{r}-0.533 \\ -30.5°\end{array}\right\} \leqslant x \leqslant \left\{\begin{array}{l}+0.533 \\ +30.5°\end{array}\right.$

(Ref. 17.04, p. 30)

$k, m, n =$ positive integers $P_k(x) =$ Legendre polynomial (Sec. 14.18)

$J_n(x) =$ Bessel function (Sec. 14.24) $T_k(x) =$ Chebyshev polynomial (Sec. 14.20)

(1) Bessel Series $(0 \leq x \leq 1)$

(a) Representation

$$f(x) = \sum_{k=1}^{\infty} a_k J_n(\lambda_k x) \quad \text{where } \lambda_k \text{ is the } k\text{th root of } J_n(x) = 0$$

$$a_k = \frac{2}{J_{n+1}^2(\lambda_k)} \int_0^{+1} x f(x) J_n(\lambda_k x) \, dx$$

(b) Integral

$$\int_0^{+1} x J_n(\lambda_k x) \, dx = -\frac{J_{n-1}(\lambda_k)}{\lambda_k} + \frac{n}{\lambda_k} \int_0^{+1} J_{n-1}(\lambda_k x) \, dx$$

(2) Legendre Series $(-1 \leq x \leq +1)$

(a) Representation

$$f(x) = \sum_{k=0}^{\infty} a_k P_k(x)$$

$$a_k = \frac{2k+1}{2} \int_{-1}^{+1} f(x) P_k(x) \, dx$$

(b) Integrals

$$\int_{-1}^{+1} x P_k(x) \, dx = \frac{\sqrt{2k-1}}{k!!}$$

$$\int_{-1}^{+1} x^m P_k(x) \, dx = \frac{\sqrt{2k-m}}{(m+k-1)!!}$$

(c) Powers of x

$$P_k = P_k(x)$$

$x = P_1$	$x^5 = \frac{1}{63}(27P_1 + 28P_3 + 8P_5)$
$x^2 = \frac{1}{3}(1 + 2P_2)$	$x^6 = \frac{1}{231}(33 + 110P_2 + 72P_4 + 16P_6)$
$x^3 = \frac{1}{5}(P_1 + 2P_3)$	$x^7 = \frac{1}{429}(143P_1 + 182P_3 + 88P_5 + 16P_7)$
$x^4 = \frac{1}{35}(7 + 20P_2 + 8P_4)$	$x^8 = \frac{1}{6,435}(715 + 2,600P_2 + 2,160P_4 + 832P_6 + 128P_8)$

(3) Chebyshev Series $(-1 \leq x \leq +1)$

(a) Representation

$$f(x) = \frac{a_0}{\pi\sqrt{1-x^2}} + \frac{2}{\pi\sqrt{1-x^2}} \sum_{k=1}^{\infty} a_k T_k(x)$$

$$a_k = \int_{-1}^{+1} f(x) T_k(x) \, dx$$

(b) Integrals

$$\int_{-1}^{+1} x^m T_k(x) \, dx = \begin{cases} 0 & \text{odd } (m+k) \\ 2(m)! \left[\dfrac{T_k(1)}{(m+1)!} - \dfrac{T_k'(1)}{(m+2)!} + \dfrac{T_k''(1)}{(m+3)!} - \cdots \right] & \text{even } (m+k) \end{cases}$$

(c) Powers of x

$$T_k = T_k(x)$$

$x = T_1$	$x^5 = \frac{1}{16}(10T_1 + 5T_3 + T_5)$
$x^2 = \frac{1}{2}(1 - T_2)$	$x^6 = \frac{1}{32}(10 + 15T_2 + 6T_4 + T_6)$
$x^3 = \frac{1}{4}(3T_1 + T_3)$	$x^7 = \frac{1}{64}(35T_1 + 21T_3 + 7T_5 + T_7)$
$x^4 = \frac{1}{8}(3 + 4T_2 + T_4)$	$x^8 = \frac{1}{128}(35 + 56T_2 + 28T_4 + 8T_6 + T_8)$

(1) General Properties

Every algebraic equation of the nth degree (with n real and complex roots),

$$f(x) = a_0 x^n + a_1 x^{n-1} + \cdots + a_{n-1} x + a_n = 0 \qquad a_0 \neq 0$$

can be represented as a product of n linear factors. If $x_1, x_2, \ldots, x_n$ are the roots of this equation, then

$$f(x) = (x - x_1)(x - x_2) \cdots (x - x_n) = 0$$

(2) Relations between the Roots and the Coefficients

$$x_1 + x_2 + \cdots + x_n = \sum_{i=1}^{n} x_i = -\frac{a_1}{a_0}$$

$$x_1 x_2 + x_1 x_3 + \cdots + x_{n-1} x_n = \sum_{i,j=1}^{n} x_i x_j = \frac{a_2}{a_0}$$

$$x_1 x_2 x_3 + x_1 x_2 x_4 + \cdots + x_{n-2} x_{n-1} x_n = \sum_{i,j,k=1}^{n} x_i x_j x_k = -\frac{a_3}{a_0}$$

$$\cdots \cdots \cdots \cdots \cdots \cdots \cdots \cdots \cdots \cdots \cdots \cdots$$

$$(-1)^n x_1 x_2 x_3 \cdots x_n = \frac{a_n}{a_0}$$

(3) Methods of Solution

If $n > 4$, there is no formula which gives the roots of this general equation. The following methods are useful:

(a) Roots by trial

Find a number x_1 that satisfies $f(x_1) = 0$. Then divide $f(x)$ by $x - x_1$, thus obtaining an equation of degree one less than that of the original equation. Repeat the same procedure with the reduced equation.

(b) Roots by *regula falsi* approximation

If coefficients $a_0, a_1, \ldots$ are real, introduce $x = a$ and $x = b$ so that $f(a)$ and $f(b)$ have opposite signs. Then

$$c_1 = a - f(a) \frac{b - a}{f(b) - f(a)}$$

is the first approximation of x_1. By repeating the application of this idea, real roots may be obtained to any degree of accuracy.

(c) Roots by Newton's approximation

Assume c_1 as an approximate root, calculate a better approximation by means of

$$c_2 = c_1 - \frac{f(c_1)}{f'(c_1)}$$

and repeat this process to a desired accuracy.

(d) Roots by Steinman's approximation

First locate the roots approximately, and write the equation in the form

$$a_0 x^n = a_1 x^{n-1} + a_2 x^{n-2} + \cdots + a_n$$

Then, with $x = d_1$,

$$d_2 = \frac{a_1 + 2a_2/d_1 + 3a_3/d_1{}^2 + 4a_4/d_1{}^3 + \cdots}{a_0 + a_2/d_1{}^2 + 2a_3/d_1{}^3 + 3a_4/d_1{}^4 + \cdots}$$

and repeat this process to a desired accuracy.

(1) General

The solution of a system of linear equations by the method outlined in Sec. 1.13 – 1*b* becomes impractical if $n > 4$. In such a case the employment of the numerical methods outlined in the following discussion offers a workable solution:

(2) Gauss' Elimination Method

It involves replacing the given system by combination and modification of the initial equations leading to the following *triangular system*.

$$x_1 + \alpha_{12}x_2 + \cdots + \alpha_{1,n-1}x_{n-1} + \alpha_{1,n}x_n = \beta_1$$
$$x_2 + \cdots + \alpha_{2,n-1}x_{n-1} + \alpha_{2,n}x_n = \beta_2$$
$$\cdots\cdots\cdots\cdots\cdots\cdots\cdots\cdots\cdots\cdots\cdots\cdots$$
$$x_{n-1} + \alpha_{n-1,n}x_n = \beta_{n-1}$$
$$x_n = \beta_n$$

The last equation (n) yields the value of x_n, which is then substituted in the preceding equation ($n - 1$) from which x_{n-1} is determined, etc.

(3) Gauss-Seidel Iteration Method

This method involves rearrangement and/or modification of the given system to obtain the *largest diagonal coefficients* possible. Then dividing each equation by the respective diagonal coefficient leads to the *carryover form*

$$x_j = r_{j1}x_1 + r_{j2}x_2 + \cdots + m_j + \cdots + r_{j,n-1}x_{n-1} + r_{jn}x_n$$

where $r_{j1} = - a_{j1}/a_{jj}$, $r_{j2} = - a_{j2}/a_{jj}$, . . . are the *carryover factors* and $m_j = b_j/a_{jj}$ is the *starting value*. Starting with the *trial solution*

$$x_1^{(1)} = m_1, \; x_2^{(1)} = m_2, \ldots$$

The successive approximations become

$$x_1^{(2)} = x_1^{(1)} + \sum_{j=1}^{n} r_{1j}x_j^{(1)}, \; x_2^{(2)} = x_2^{(1)} + \sum_{j=1}^{n} r_{2j}x_j^{(1)}, \ldots$$

$$x_1^{(3)} = x_1^{(1)} + x_1^{(2)} + \sum_{j=1}^{n} r_{1j}x_j^{(2)}, \; x_2^{(3)} = x_2^{(1)} + x_2^{(2)} + \sum_{j=1}^{n} r_{2j}x_j^{(2)}, \ldots$$

$$\cdots\cdots\cdots\cdots\cdots\cdots\cdots\cdots\cdots\cdots\cdots\cdots\cdots\cdots\cdots$$

Under certain conditions this procedure is rapidly convergent, but it may also converge slowly or not at all.

(4) Matrix Iterative Method

The given system is written in a *matrix carryover form* as

$$X = rX + m$$

where $X = \{x_1, x_2, \ldots, x_j, \ldots, x_{n-1}, x_n\}$
$$r = \begin{bmatrix} 0 & r_{12} & \cdots & r_{1,n-1} & r_{1n} \\ r_{21} & 0 & \cdots & r_{2,n-1} & r_{2n} \\ \cdots\cdots\cdots\cdots\cdots\cdots\cdots\cdots \end{bmatrix}$$

$$m = \{m_1, m_2, \ldots, m_j, \ldots, m_{n-1}, m_n\}$$

The successive approximations become

$$X^{(1)} = m, \; X^{(2)} = m + rm, \; X^{(3)} = m + rm + r^2m, \ldots$$

and the final solution takes the form of the series,

$$X = [1 + r + r^2 + \cdots]m$$

where the sum of the power series in the brackets equals the inverse of A defined in Sec. 1.13 – 1*c*.

(1) Forward Differences

For a given *discrete function* $y = f(x)$, a set of *equally spaced arguments* $x_n = x_0 + n \, \Delta x (n = 0, \pm 1, \pm 2, \ldots; \Delta x = h > 0)$ and a corresponding set of values $y_n = y(x_0 + n \, \Delta x)$ define the *forward differences* as follows:

$$\Delta y_n = y_{n+1} - y_n \qquad\qquad\qquad \text{First-order difference}$$

$$\Delta^2 y_n = \Delta y_{n+1} - \Delta y_n = y_{n+2} - 2y_{n+1} + y_n \qquad\qquad\qquad \text{Second-order difference}$$

$$\cdots\cdots\cdots\cdots\cdots\cdots\cdots\cdots\cdots\cdots$$

$$\Delta^k y_n = \Delta^{k-1} y_{n+1} - \Delta^{k-1} y_n = \sum_{j=0}^{k} (-1)^j \binom{k}{j} y_{n+k-j} \qquad\qquad \text{kth-order difference } k = 2, 3, \ldots$$

(2) Backward Differences

The same set of values defines the *backward differences* as follows:

$$\nabla y_n = y_n - y_{n-1} \qquad\qquad\qquad \text{First-order difference}$$

$$\nabla^2 y_n = \nabla y_n - \nabla y_{n-1} = y_n - 2y_{n-1} + y_{n-2} \qquad\qquad\qquad \text{Second-order difference}$$

$$\cdots\cdots\cdots\cdots\cdots\cdots\cdots\cdots\cdots\cdots$$

$$\nabla^k y_n = \nabla^{k-1} y_n - \nabla^{k-1} y_{n-1} = \sum_{j=0}^{k} (-1)^j \binom{k}{j} y_{n+j} \qquad\qquad \text{kth-order difference } k = 2, 3, \ldots$$

(3) Central Differences

For the same discrete function, a similar set of values y_n corresponding to $x_n = x_0 + n \, \Delta x$ defines the *central differences* as follows:

$$\delta y_n = y_{n+1/2} - y_{n-1/2} \qquad\qquad\qquad \text{First-order difference}$$

$$\delta^2 y_n = \delta(y_{n+1/2} - y_{n-1/2}) = y_{n+1} - 2y_n + y_{n-1} \qquad\qquad\qquad \text{Second-order difference}$$

$$\cdots\cdots\cdots\cdots\cdots\cdots\cdots\cdots\cdots\cdots$$

$$\delta^k y_n = \delta^{n-1}(y_{n+1/2} - y_{n-1/2}) = \sum_{j=0}^{k} (-1)^j \binom{k}{j} y_{n-j+k/2} \qquad\qquad \text{kth-order difference}$$

(4) Central Means

For the same discrete functions, the *central means* are defined as follows:

$$\mu y_n = \tfrac{1}{2}(y_{n-1/2} + y_{n+1/2}) \qquad\qquad\qquad \text{First-order mean}$$

$$\mu^2 y_n = \frac{\mu}{2}(y_{n-1/2} + y_{n+1/2}) = \tfrac{1}{4}(y_{n-1} + 2y_n + y_{n+1}) \qquad\qquad\qquad \text{Second-order mean}$$

$$\cdots\cdots\cdots\cdots\cdots\cdots\cdots\cdots\cdots\cdots$$

$$\mu^k y_n = \frac{\mu^{k-1}}{2}(y_{n-1/2} + y_{n+1/2}) = \frac{1}{2^k} \sum_{j=0}^{k} \binom{k}{j} y_{n-j+k/2} \qquad\qquad \text{kth-order mean}$$

(1) Forward Differences

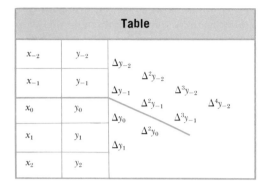

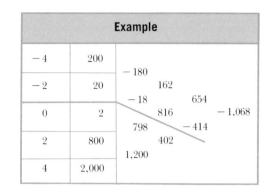

(2) Backward Differences

	Table			
x_{-2}	y_{-2}			
		∇y_{-1}		
x_{-1}	y_{-1}		$\nabla^2 y_0$	
		∇y_0		$\nabla^3 y_1$
x_0	y_0		$\nabla^2 y_1$		$\nabla^4 y_2$
		∇y_1		$\nabla^3 y_2$
x_1	y_1		$\nabla^2 y_2$	
		∇y_2		
x_2	y_2			

	Example			
-4	200			
		-180		
-2	20		162	
		-18		654
0	2		816		$-1{,}068$
		798		-414
2	800		402	
		1,200		
4	2,000			

(3) Central Differences

	Table			
x_{-2}	y_{-2}			
		$\delta y_{-3/2}$		
x_{-1}	y_{-1}		$\delta^2 y_{-1}$	
		$\delta y_{-1/2}$		$\delta^3 y_{-1/2}$
x_0	y_0		$\delta^2 y_0$		$\delta^4 y_0$
		$\delta y_{1/2}$		$\delta^3 y_{1/2}$
x_1	y_1		$\delta^2 y_1$	
		$\delta y_{3/2}$		
x_2	y_2			

	Example			
-4	200			
		-180		
-2	20		162	
		-18		654
0	2		816		$-1{,}068$
		798		-414
2	800		402	
		1,200		
4	2,000			

(4) Relationships $(k, n, p, q = 0, 1, 2, \ldots)$

$$\Delta^k y_n = \nabla^k y_{n+k} = \delta^k y_{n+k/2} \qquad \nabla^k y_n = \Delta^k y_{n-k} = \delta^k y_{n-k/2}$$

$$\delta^k y_n = \Delta^k y_{n-k/2} = \nabla^k y_{n+k/2} \qquad \Delta^p \nabla^q y_n = \nabla^q \Delta^p y_n = \Delta^{p+q} y_{n-q}$$

(1) General

The *process of interpolation* is used for finding in-between values of a tabulated function or for the development of a substitute function $\bar{y}(x)$ closely approximating a more complicated function $y(x)$ in a given interval.

(2) Lagrange's Interpolation Formula

From the tabulated $n + 1$ values given as

$$\begin{array}{|l} x_0, x_1, x_2, \ldots, x_n \\ \hline y_0, y_1, y_2, \ldots, y_n \end{array}$$

and not necessarily equally spaced, the *substitute (approximate) function* is

$$\bar{y}(x) = \frac{(x - x_0)(x - x_1) \cdots (x - x_n)}{(x_0 - x_1)(x_0 - x_2) \cdots (x_0 - x_n)} y_0 + \frac{(x - x_0)(x - x_2) \cdots (x - x_n)}{(x_1 - x_0)(x_1 - x_2) \cdots (x_1 - x_n)} y_1$$

$$+ \cdots + \frac{(x - x_0)(x - x_1) \cdots (x - x_{n-1})}{(x_n - x_0)(x_n - x_1) \cdots (x_n - x_{n-1})} y_n$$

If the number of given values is small ($n \leq 6$), this formula is a very powerful and convenient approximation model; a special case involving five equally spaced values, called *five-point formula,* is given in Sec. 17.09–7.*

(3) Newton's Interpolation Formula

From the same values the *alternate form* of the *substitute function* is

$$\bar{y}(x) = y_0 + (x - x_0) \Delta_1(x_1) + (x - x_0)(x - x_1) \Delta_2(x_2)$$

$$+ \cdots + [(x - x_0)(x - x_1) \cdots (x - x_{n-1})] \Delta_n(x_n)$$

where the *divided differences* are

$$\Delta_1(x_1) = \frac{y_1 - y_0}{x_1 - x_0} \qquad \text{First-order divided difference}$$

$$\Delta_2(x_2) = \frac{\Delta_1(x_2) - \Delta_1(x_1)}{x_2 - x_0} \qquad \text{Second-order divided difference}$$

$$\cdots \cdots \cdots \cdots \cdots \cdots \cdots \cdots \cdots$$

$$\Delta_n(x_n) = \frac{\Delta_{n-1}(x_n) - \Delta_{n-1}(x_{n-1})}{x_n - x_0} \qquad n\text{th-order divided difference}$$

Divided differences can be again computed in *tabular form* as shown below.

Table				
x_0	y_0			
x_1	y_1	$\Delta_1(x_1)$		
x_2	y_2	$\Delta_1(x_2)$	$\Delta_2(x_2)$	
x_3	y_3	$\Delta_1(x_3)$	$\Delta_2(x_3)$	$\Delta_3(x_3)$
.	.	.	.	.
.	.	.	.	.

Example				
6	1,000			
8	2,000	500		
16	6,000	500	0	
26	7,400	320	10	1
.	.	.	.	.
.	.	.	.	.

*Five-point formula is used frequently for in-table interpolation.

(1) General

If the increment $\Delta x = h$ is a fixed value, the interpolation functions may be taken in a *polynomial form*, the coefficients of which can be expressed in *differences of ascending order* (Sec. 17.06).

(2) Newton's Formula, Forward Interpolation

$$\bar{y}(x) = y_0 + \frac{u}{1!}\Delta y_0 + \frac{u(u-1)}{2!}\Delta^2 y_0 + \frac{u(u-1)(u-2)}{3!}\Delta^3 y_0$$

$$+ \cdots + \frac{u(u-1)(u-2)\cdots(u-n+1)}{n!}\Delta^n y_0 \qquad u = \frac{x-x_0}{h}$$

(3) Newton's Formula, Backward Interpolation

$$\bar{y}(x) = y_n + \frac{u}{1!}\nabla y_n + \frac{u(u+1)}{2!}\nabla^2 y_n + \frac{u(u+1)(u+2)}{3!}\nabla^3 y_n$$

$$+ \cdots + \frac{u(u+1)(u+2)\cdots(u+n-1)}{n!}\nabla^n y_n \qquad u = \frac{x-x_n}{h}$$

(4) Stirling's Interpolation Formula[1]

$$\bar{y}(x) = y_0 + u\frac{\Delta y_0 + \nabla y_0}{2} + \frac{u^2}{2!}\Delta\nabla y_0 + \frac{u(u^2-1)}{3!}\frac{\Delta^2\nabla y_0 + \Delta\nabla^2 y_0}{2}$$

$$+ \frac{u^2(u^2-1)}{4!}\Delta^2\nabla^2 y_0 + \cdots \qquad u = \frac{x-x_0}{h}$$

(5) Bessel's Interpolation Formula[1] $(v \neq 0)$

$$\bar{y}(x) = \frac{y_0 + y_1}{2} + v\,\Delta y_0 + \frac{v^2 - \frac{1}{4}}{2!}\frac{\Delta\nabla y_0 + \Delta\nabla y_1}{2} + \frac{v(v^2 - \frac{1}{4})}{3!}\Delta^2\nabla y_0$$

$$+ \frac{(v^2 - \frac{1}{4})(v^2 - \frac{9}{4})}{4!}\frac{\Delta^2\nabla^2 y_0 + \Delta^2\nabla^2 y_1}{2} + \cdots \qquad v = u - \frac{1}{2} = \frac{x-x_0}{h} - \frac{1}{2}$$

(6) Bessel's Interpolation Formula[1] $(v = 0)$

$$\bar{y}(x) = \frac{1}{2}[(y_0 + y_1) - \frac{1}{8}(\Delta\nabla y_0 + \Delta\nabla y_1) + \frac{3}{128}(\Delta^2\nabla^2 y_0 + \Delta^2\nabla^2 y_1) - \frac{5}{1.024}(\Delta^3\nabla^3 y_0 + \Delta^3\nabla^3 y_1) + \cdots]$$

(7) Lagrange's Five-point Formula

$$\bar{y}(x) = \frac{f}{24}\left(\frac{y_{-2}}{2+u} - \frac{4y_{-1}}{1+u} + \frac{6y_0}{u} + \frac{4y_1}{1-u} - \frac{y_2}{2-u}\right) \qquad u = \frac{x-x_0}{h}$$

where $f = (2+u)(1+u)u(1-u)(2-u)$

[1]For relationships of difference operators refer to Sec. 17.07-4.

(1) Concept

Whenever the closed-form integration becomes too involved or is not feasible, the numerical value of a definite integral can be found (to any degree of accuracy) by means of any of several *quadrature formulas* which express the given integral as a linear combination of a selected set of integrands. The basis of these formulas are *difference polynomials* or *orthogonal polynomials*.

(2) Trapezoidal Rule   [n **even or odd;** $h = (b-a)/n$]

$$\int_a^b y(x)\, dx \approx \int_a^b \bar{y}(x)\, dx = \frac{h}{2}(y_0 + 2y_1 + 2y_2 + \cdots + 2y_{n-2} + 2y_{n-1} + y_n) + \epsilon_T$$

Truncation error: $\epsilon_T \approx -\dfrac{n(h)^3 f''(\xi)}{12}$ $a \leqslant \xi \leqslant b$

(3) Simpson's Rule   [n **even;** $h = (b-a)/n$]

$$\int_a^b y(x)\, dx \approx \int_a^b \bar{y}(x)\, dx = \frac{h}{3}(y_0 + 4y_1 + 2y_2 + 4y_3 + \cdots + 4y_{n-3} + 2y_{n-2} + 4y_{n-1} + y_n) + \epsilon_T$$

Truncation error: $\epsilon_T \approx -\dfrac{n(h)^5 f^{iv}(\xi)}{180}$ $a \leqslant \xi \leqslant b$

(4) Weddle's Rule   [n **must be a multiple of 6;** $h = (b-a)/n$]

$$\int_a^b y(x)\, dx \approx \int_a^b \bar{y}(x)\, dx = \frac{3h}{10}[(y_0 + 5y_1 + y_2 + 6y_3 + y_4 + 5y_5 + y_6)$$

$$+ (y_6 + 5y_7 + y_8 + 6y_9 + y_{10} + 5y_{11} + y_{12})$$

$$+ \cdots + (y_{n-6} + 5y_{n-5} + y_{n-4} + 6y_{n-3} + y_{n-2} + 5y_{n-1} + y_n)] + \epsilon_T$$

Truncation error: $\epsilon_T \approx -\dfrac{n(h)^7 f^{vi}(\xi)}{140}$ $a \leqslant \xi \leqslant b$

(5) Euler's Quadrature Formula[1]   [n **even or odd;** $h = (b-a)/n$]

$$\int_a^b y(x)\, dx \approx \int_a^b \bar{y}(x)\, dx = \frac{h}{2}(y_0 + 2y_1 + 2y_2 + \cdots + 2y_{n-2} + 2y_{n-1} + y_n)$$

$$- \bar{B}_2 \frac{h^2}{2!}[y'(b) - y'(a)] - \bar{B}_4 \frac{h^4}{4!}[y'''(b) - y'''(a)] - \cdots + \epsilon_T$$

Truncation error: $\epsilon_T \approx -n\bar{B}_{2m}\dfrac{h^{2m+1}}{(2m)!} f^{(2m)}(\xi)$ $a \leqslant \xi \leqslant b$

[1]$\bar{B}_2, \bar{B}_4, \ldots, \bar{B}_{2m}$ = Bernouilli's numbers, Sec. 8.05.

(1) Gauss-Chebyshev Quadrature Formula

$$\int_a^b y(x)\, dx \approx \frac{b-a}{n}[y(X_1) + y(X_2) + \cdots + y(X_n)] + \epsilon_T$$

$$X_j = \frac{a+b}{2} + \frac{b-a}{2}x_j \qquad n = 2, 3, 4, 5, 6, 7, 9$$

Truncation error for $n = 3$: $\epsilon_T \approx \dfrac{1}{360}\left(\dfrac{b-a}{2}\right)^5 y^{iv}(X)$ $a < X < b$

(2) Gauss-Legendre Quadrature Formula

$$\int_a^b y(x)\, dx \approx (b-a)[C_1y(X_1) + C_2y(X_2) + \cdots + C_ny(X_n)] + \epsilon_T$$

$$X_j = a + (b-a)x_j \qquad n = 1, 2, 3, \ldots$$

Truncation error for $n = n$: $\epsilon_T \approx \dfrac{(n!)^4(b-a)^{2n+1}}{(2n+1)[(2n)!]^3}y^{(2n)}(X)$ $a < X < b$

(3) Table of x_j for the Gauss-Chebyshev Formula

n	x_1	x_2	x_3	x_4	x_5
2	0.577350	−0.577350			
3	0.707107	0	−0.707107		
4	0.794654	0.187592	−0.187592	−0.794654	
5	0.832497	0.374541	0	0.374541	−0.832497

(4) Table of x_j for the Gauss-Legendre Formula

n	x_1	x_2	x_3	x_4	x_5
1	0.500000				
2	0.211325	0.788675			
3	0.112702	0.500000	0.887298		
4	0.069432	0.330009	0.669991	0.930568	
5	0.046910	0.230765	0.500000	0.769235	0.953090

(5) Table of C_j for the Gauss-Legendre Formula[1]

n	C_1	C_2	C_3	C_4	C_5
1	1.000000				
2	0.500000	0.500000			
3	0.277778	0.444444	0.277778		
4	0.173927	0.326073	0.326073	0.173927	
5	0.118463	0.239314	0.284444	0.239314	0.118463

[1](Ref. 17.06, p. 286)

a, b, c, h = real numbers	m, n, r = positive integers
$u = f(x), v = g(x)$	$w = g(x+h)$

(1) Derivatives and Differences

(a) **Definitions.** If $y = f(x)$ satisfies the conditions of Sec. 7.03 and

$$y_1 = y_0 + h, \qquad y_2 = y_0 + 2h, \qquad \cdots, \qquad y_n = y_0 + nh, \qquad \cdots$$

then

$$Dy_n = \lim_{h \to 0} \frac{\Delta y_n}{h}, \qquad D^2 y_n = \lim_{h \to 0} \frac{\Delta^2 y_n}{h^2}, \qquad \cdots, \qquad D^m y_n = \lim_{h \to 0} \frac{\Delta^m y_n}{h^m}$$

where Δ = *forward difference operator* (Sec. 17.06) and D = *derivative operator*, used as

$$D(\ ) = \frac{d}{dx}(\ ), \qquad D^2(\ ) = \frac{d^2}{dx^2}(\ ), \qquad \cdots, \qquad D^m(\ ) = \frac{d^m}{dx^m}(\ )$$

(b) **Rules of differential and difference calculus bear close resemblance:**

$D[a] = 0 \qquad D[x] = 1$		$\Delta[a] = 0 \qquad \Delta[x] = h$	
$D\left[\dfrac{1}{x}\right] = -\dfrac{1}{x^2}$		$\Delta\left[\dfrac{1}{x}\right] = -\dfrac{h}{x(x+h)}$	
$D[au] = a\,Du$		$\Delta[au] = a\,\Delta u$	
$D[u+v] = Du + Dv$		$\Delta[u+v] = \Delta u + \Delta v$	
$D[uv] = v\,Du + u\,Dv$		$\Delta[uv] = v\,\Delta u + u\,\Delta v + \Delta u\,\Delta v$	
$D\left[\dfrac{u}{v}\right] = \dfrac{v\,Du - u\,Dv}{v^2}$		$\Delta\left[\dfrac{u}{v}\right] = \dfrac{v\,\Delta u - u\,\Delta v}{vw}$	

(2) Relationships of Operators

(a) D **in terms of** Δ **and** ∇ (Sec. 17.06)

$$D = \frac{1}{h}\ln(1+\Delta) = \frac{1}{h}\left(\Delta - \frac{\Delta^2}{2} + \frac{\Delta^3}{3} - \frac{\Delta^4}{4} + \frac{\Delta^5}{5} - \frac{\Delta^6}{6} + \frac{\Delta^7}{7} - \cdots\right)$$

$$D^2 = \frac{1}{h^2}\ln^2(1+\Delta) = \frac{1}{h^2}\left(\Delta^2 - \Delta^3 + \tfrac{11}{12}\Delta^4 - \tfrac{5}{6}\Delta^5 + \tfrac{137}{180}\Delta^6 - \cdots\right)$$

$$D^m = \frac{1}{h^m}\ln^m(1+\Delta) = \frac{1}{h^m}\left[\sum_{r=1}^{\infty}(-1)^r\frac{\Delta^r}{r}\right]^m = \frac{1}{h^m}\ln^m(1-\nabla) = \frac{1}{h^m}\left[\sum_{r=1}^{\infty}\frac{\nabla^r}{r}\right]^m$$

(b) D **in terms of** $\lambda = \delta/2$ (Sec. 17.06)

$$D = \frac{2}{h}\sinh^{-1}\lambda = \frac{2}{h}\left(\lambda - a\lambda^3 + b\lambda^5 - c\lambda^7 + d\lambda^9 - \cdots\right)$$

$$D^2 = \left(\frac{2}{h}\sinh^{-1}\lambda\right)^2 = \frac{4}{h^2}\left[\lambda^2 - 2a\lambda^4 + (a^2+2b)\lambda^6 - 2(ab+c)\lambda^8 + \cdots\right]$$

$$D^m = \left(\frac{2}{h}\sinh^{-1}\lambda\right)^m = \left(\frac{2}{h}\right)^m\left[\lambda + \lambda^2\sum_{r=1}^{\infty}(-1)^r\frac{(2r-1)!!}{(2r)!!}\frac{\lambda^{2r-1}}{(2r+1)}\right]^m$$

where $\qquad a = \dfrac{1}{2\cdot 3} \qquad b = \dfrac{1\cdot 3}{2\cdot 4\cdot 5} \qquad c = \dfrac{1\cdot 3\cdot 5}{2\cdot 4\cdot 6\cdot 7} \qquad d = \dfrac{1\cdot 3\cdot 5\cdot 7}{2\cdot 4\cdot 6\cdot 8\cdot 9}$

18
PROBABILITY AND STATISTICS

---

(1) Classification of Events

An *observation* (experiment, trial) may have (theoretically) a class (set) S of *possible results* (states, events) $E_1, E_2, \ldots, E_n$ permitting the following classification:

(a) **The union** $E_1 \cup E_2 \cup \cdots \cup E_n$ of S is the single event of realizing at least one of the events in S.

(b) **The intersection** $E_i \cap E_j$ of S is the joint event realizing two events E_i and E_j.

(c) **The complement** $\bar{E}$ of S is the event not included in S.

(2) Simple Probability

If E can occur (happen) in m ways (m times) out of a total of n mutually exclusive and equally likely ways, then

(a) **The probability of occurrence** (called *success*) of E is

$$P(E) = \frac{m}{n} = p$$

$$0 \leqslant P(E) \leqslant 1$$

(b) **The probability of nonoccurrence** (called *failure*) of E is

$$P(\bar{E}) = 1 - \frac{m}{n} = q$$

$$0 \leqslant P(\bar{E}) \leqslant 1$$

(3) Special Conditions

(a) **Conditional probability** of E_1, given that E_2 has occurred, or of E_2, given that E_1 has occurred, are respectively,

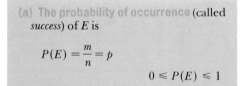

$$P\!\left(\frac{E_1}{E_2}\right) = \frac{P(E_1 \cap E_2)}{P(E_2)} \qquad\qquad P\!\left(\frac{E_2}{E_1}\right) = \frac{P(E_1 \cap E_2)}{P(E_1)}$$

where $P(E_1) \neq 0$ and $P(E_2) \neq 0$.

(b) **Statistical independence.** Two events E_1 and E_2 are statistically independent if and only if

$$P\!\left(\frac{E_1}{E_2}\right) = P(E_1) \neq 0 \qquad\qquad P\!\left(\frac{E_2}{E_1}\right) = P(E_2) \neq 0$$

(4) Probability Theorems

(a) If E_1 and E_2 are *two mutually not exclusive events*,

$$P(E_1 \cup E_2) = P(E_1) + P(E_2) - P(E_1 \cap E_2)$$

(b) If E_1 and E_2 are *two mutually exclusive events*,

$$P(E_1 \cup E_2) = P(E_1) + P(E_2)$$

(c) If E and $\bar{E}$ are *two mutually complementary events*,

$$P(E) + P(\bar{E}) = 1 = p + q$$

(d) If E_1 and E_2 are *two mutually independent events*,

$$P(E_1 \cap E_2) = P(E_1)P(E_2)$$

(e) If $E_1, E_2, \ldots, E_k, \ldots, E_n$ form a set S of mutually exclusive events, then for each pair of events E_j, E_k included in this set,

$$P\!\left(\frac{E_j}{E_k}\right) = \frac{P(E_k \cap E_j)}{P(E_k)} = \frac{P(E_j)P(E_k/E_j)}{\sum\limits_{j=1}^{n} [P(E_j)P(E_k/E_j)]}$$

where $P(E_k) \neq 0$ (Bayes theorem).

(1) Frequency Distribution

A set of numbers $x_1, x_2, \ldots, x_n$ corresponding to the events $E_1, E_2, \ldots, E_n$ which occur with *frequencies* $f_1, f_2, \ldots, f_n$, respectively, so that $f_j \geqslant 0$ and $x_1 < x_2 < \cdots < x_n$ is called the *frequency distribution*.

(2) Discrete Random Variable

If a variable X can assume a discrete set of values $x_1, x_2, \ldots, x_n$ with respect to probabilities $p_1, p_2, \ldots, p_n$, where $p_1 + p_2 + \cdots + p_n = 1$, then these values define the *discrete probability distribution*. Because x takes specific values with given probabilities, it is designated as the *discrete random variable* (change variable, stochastic variable). The probability that X takes the value x_j is

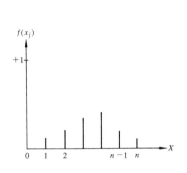

$$P(X=x_j)=f(x_j) \qquad j=1, 2, \ldots, n$$

where $f(x)$ is the *probability function of the random variable X*, with $f(x_j) \geqslant 0$, $\sum_1^n f(x_j) = 1$. The *joint probability* that X and Y take the values x_j and y_k, respectively, is

$$P(X = x_j, Y = y_j) = f(x_j, y_k) \qquad \begin{array}{l} j = 1, 2, \ldots, n \\ k = 1, 2, \ldots, r \end{array}$$

where $f(x, y)$ is the *probability function of a two-dimensional random variable*. $\left[f(x_j, y_j) \geqslant 0, \sum_1^n f(x_j) = 1, \sum_1^r f(y_k) = 1 \right]$.

(3) Continuous Random Variable

The random variable X is denoted as a continuous random variable if $f(x)$ is continuously differentiable.

$$\frac{df(x)}{dx} = \lim_{\Delta x \to 0} \frac{P(X < x \leqslant X + \Delta x)}{\Delta x} = \phi(x)$$

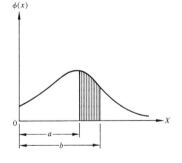

where $\phi(x)$ is the *probability density*, and the *cumulative probability distributions* are

$$P(x \leqslant X) = \int_{-\infty}^{X} \phi(x)\, dx$$

$$P(a < x \leqslant b) = \int_a^b \phi(x)\, dx = f(b) - f(a)$$

$$P(-\infty < x \leqslant \infty) = \int_{-\infty}^{\infty} \phi(x)\, dx = 1$$

A *two-dimensional random variable* is a *continuous random variable* if $f(x, y)$ is continuous for all x_j and y_k, and the *joint probability density*

$$\phi(x, y) = \frac{\partial^2 f(x, y)}{\partial x\, \partial y}$$

exists and is piecewise-continuous.

(1) Measures of Tendency and Frequency

(a) **An average** is a value which is typical for a set of numbers $X(x_1, x_2, \ldots, x_n)$. Since it tends to lie centrally within the set arranged according to magnitude, it is also a measure of central tendency.

(b) **A median** of the same set is the average of the two middle values if n is even.

(c) **A mode** is the value of x_m having a maximum frequency f_m.

(2) Means

(a) **The arithmetic mean** of a set of n numbers is

$$\overline{X} = \frac{x_1 + x_2 + \cdots + x_n}{n} = \frac{\Sigma x}{n}$$

If the numbers occur $f_1, f_2, \ldots, f_n$ times, respectively, then

$$\overline{X} = \frac{f_1 x_1 + f_2 x_2 + \cdots + f_n x_n}{f_1 + f_2 + \cdots + f_n} = \frac{\Sigma f x}{\Sigma f}$$

If the numbers are associated with a *weighting factor* $w_j \geqslant 0$, then

$$\overline{X} = \frac{w_1 x_1 + w_2 x_2 + \cdots + w_n x_n}{w_1 + w_2 + \cdots + w_n} = \frac{\Sigma w x}{\Sigma w}$$

(b) **The geometric mean** of a set of n numbers is

$$\overline{G} = \sqrt[n]{x_1 x_2 \cdots x_n}$$

If the numbers occur $f_1, f_2, \ldots, f_n$ times, respectively, then

$$\overline{G} = \sqrt[n]{x_1^{f_1} x_2^{f_2} \cdots x_n^{f_n}}$$

where $n = f_1 + f_2 + \cdots + f_n$.

(c) **The harmonic mean** of a set of n numbers is

$$\overline{H} = \frac{1}{(1/n)(1/x_1 + 1/x_2 + \cdots + 1/x_n)} = \frac{n}{\Sigma(1/x)}$$

If the numbers occur $f_1, f_2, \ldots, f_n$ times, respectively, then

$$\overline{H} = \frac{1}{(1/n)(f_1/x_1 + f_2/x_2 + \cdots + f_n/x_n)} = \frac{n}{\Sigma(f/x)}$$

where $n = f_1 + f_2 + \cdots + f_n$

(d) **The quadratic mean** of a set of n numbers is

$$\overline{Q} = \sqrt{\frac{x_1^2 + x_2^2 + \cdots + x_n^2}{n}} = \sqrt{\frac{\Sigma x^2}{n}}$$

(e) **Relations**

$$\overline{H} \leqslant \overline{G} \leqslant \overline{X}$$

(1) Dispersion

(a) The degree to which a frequency distribution of a set of numbers *tends to spread* about a point of central tendency is called the *dispersion* (spread, variance).

(b) The *range* of a set of numbers is the *difference* between the *largest* and *smallest numbers* in the set.

(2) Deviation

(a) The deviation from the arithmetic mean $\bar{X}$ of each number x_j in a set of numbers $x_1, x_2, \ldots, x_n$ is

$$D_j = x_j - \bar{X} \qquad j = 1, 2, \ldots, n$$

(b) The mean deviation of the same set is

$$\bar{D} = \frac{|x_1 + x_2 + \cdots + x_n - n\bar{X}|}{n} = \frac{\Sigma|x - \bar{X}|}{n}$$

where $|x_j - X|$ is the absolute value of D_j.
   If the numbers occur $f_1, f_2, \ldots, f_n$ times, then $n = f_1 + f_2 + \cdots + f_n$ and

$$\bar{D} = \frac{f_1|x_1 - \bar{X}| + f_2|x_2 - \bar{X}| + \cdots + f_n|x_n - \bar{X}|}{n} = \frac{\Sigma f|x - \bar{X}|}{n}$$

(c) The standard deviation of the same set is

$$\sigma = \sqrt{\frac{(x_1 - \bar{X})^2 + (x_2 - \bar{X})^2 + \cdots + (x_n - \bar{X})^2}{n}} = \sqrt{\frac{\Sigma(x - \bar{X})^2}{n}}$$

If the numbers occur $f_1, f_2, \ldots, f_n$ times, then $n = f_1 + f_2 + \cdots + f_n$ and

$$\sigma = \sqrt{\frac{f_1(x_1 - \bar{X})^2 + f_2(x_2 - \bar{X})^2 + \cdots + f_n(x_n - \bar{X})^2}{n}} = \sqrt{\frac{\Sigma f(x - \bar{X})^2}{n}}$$

(d) The variance of the same set is defined as

$$\sigma^2 = \frac{\Sigma(x - \bar{X})^2}{n} \qquad \text{or} \qquad \sigma^2 = \frac{\Sigma f(x - \bar{X})^2}{n}$$

(e) The covariance of two sets $X(x_1, x_2, \ldots, x_n)$ and $Y(y_1, y_2, \ldots, y_n)$, the arithmetic means of which are $\bar{X}$ and $\bar{Y}$, respectively, is

$$\sigma_{xy} = \frac{\Sigma(x - \bar{X})(y - \bar{Y})}{n}$$

(3) Skewness and Kurtosis

(a) The skewness of the smoothed frequency polygon is the departure of the curve from symmetry.

(b) The kurtosis of the curve is the degree of peakedness of the same curve.

(c) Moments are defined by

$$\mu_k = \frac{\Sigma f(x - \bar{X})^k}{n}$$

where k is the constant indicating the *degree of the moment*. The *first moment* is $\mu_1 = 0$, the *second moment* is $\mu_2 = \sigma^2$, the *coefficient of skewness* is $\gamma_1 = \mu_3/\sigma^3$, the *coefficient of excess* is $\gamma_2 = \mu_4/\sigma^4 - 3$, and the *kurtosis* $\beta_2 = \mu_4/\sigma^4 = \gamma_2 + 3$.

(1) Binomial Distribution

(a) Probability function

$$P(X=x) = f(x) = \frac{n!}{x!\,(n-x)!} p^x (1-p)^{n-x} \qquad \begin{array}{l} x = 1, 2, \ldots, n \\ 0 \leqslant p \leqslant 1 \end{array}$$

(b) Properties

$\mu = np = $ mean

$\sigma^2 = np(1-p) = $ variance

$\sigma = \sqrt{np(1-p)} = $ standard deviation

$\gamma_1 = \dfrac{1-2p}{\sqrt{np(1-p)}} = $ coefficient of skewness

$\gamma_2 = \dfrac{1-6p(1-p)}{np(1-p)} = $ coefficient of excess

(2) Poisson Distribution

(a) Probability function

$$P(X=x) = \frac{e^{-\lambda}\lambda^x}{x!} \qquad \begin{array}{l} x = 0, 1, 2, \ldots \\ \lambda > 0 \\ e = 2.71828\cdots \end{array}$$

(b) Properties

$\mu = \lambda = $ mean

$\sigma^2 = \lambda = $ variance

$\sigma = \sqrt{\lambda} = $ standard deviation

$\gamma_1 = \dfrac{1}{\sqrt{\lambda}} = $ coefficient of skewness

$\gamma_2 = \dfrac{1}{\lambda} = $ coefficient of excess

(3) Multinomial Distribution

A multinomial distribution is defined by

$$P(X_1 = x_1, X_2 = x_2, \ldots, X_n = x_n) = f(x_1, x_2, \ldots, x_n) = \frac{n!}{x_1! x_2! \cdots x_n!} p_1^{x_1} p_2^{x_2} \cdots p_n^{x_n}$$

where $\qquad \displaystyle\sum_{j=1}^{n} p_j = 1 \qquad \displaystyle\sum_{j=1}^{n} x_j = n \qquad p_j > 0$

(1) Normal Distribution

(a) Density function

$$\phi_N(x) = \frac{1}{\sigma\sqrt{2\pi}} e^{-(x-\mu)^2/2\sigma^2} \qquad \pi = 3.14159\cdots, e = 2.71828\cdots$$

where μ = mean, σ = standard deviation of the random variable X, and $\phi_N(x)$ is the *normal distribution density function* (gaussian function) (Sec. 18.02–3).

(b) Properties

μ = mean $\gamma_1 = 0$ = coefficient of skewness

σ^2 = variance $\gamma_2 = 0$ = coefficient of excess

σ = standard deviation

Moments about $x = \mu$:   Moments about $x = 0$:

$\mu_1 = 0$ $\nu_1 = \mu$

$\mu_2 = \sigma^2$ $\nu_2 = \mu^2 + \sigma^2$

$\mu_3 = 0$ $\nu_3 = \mu(\mu^2 + 3\sigma^2)$

$\mu_4 = 3\sigma^4$ $\nu_4 = \mu^4 + 6\mu^2\sigma^2 + 3\sigma^4$

(c) Probability function

$$P(X = x) = \int_{-\infty}^{x} \phi_N(x)\, dx = F_N(x)$$

is the *cumulative normal distribution function* (Sec. 18.02–3).

(2) Standard Normal Distribution

(a) Density function

$$\phi_N(t) = \frac{1}{\sqrt{2\pi}} e^{-t^2/2} \qquad t = \frac{x - \mu}{\sigma}$$

is the *standard normal distribution density function,* the ordinates of which are given in Sec. 18.07.

(b) Properties

$\mu = 0$ = mean

$\sigma^2 = 1$ = variance

(c) Probability function

$$P(T = t) = \int_{-\infty}^{t} \phi_N(t)\, dt = F_N(t)$$

is the *cumulative standard normal distribution function,* the ordinates of which are given in Sec. 18.08 and are areas under $\phi_N(t)$.

$$\phi_N(t) = \frac{1}{\sqrt{2\pi}}e^{-t^2/2}$$

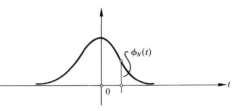

t	0	1	2	3	4	5	6	7	8	9	t
0.0	0.3989	0.3989	0.3989	0.3988	0.3986	0.3984	0.3982	0.3980	0.3977	0.3973	0.0
0.1	0.3970	0.3965	0.3961	0.3956	0.3951	0.3945	0.3939	0.3932	0.3925	0.3918	0.1
0.2	0.3910	0.3902	0.3894	0.3885	0.3876	0.3867	0.3857	0.3847	0.3836	0.3825	0.2
0.3	0.3814	0.3802	0.3790	0.3778	0.3765	0.3752	0.3739	0.3726	0.3712	0.3697	0.3
0.4	0.3683	0.3668	0.3653	0.3637	0.3621	0.3605	0.3589	0.3572	0.3555	0.3538	0.4
0.5	0.3521	0.3503	0.3485	0.3467	0.3448	0.3429	0.3410	0.3391	0.3372	0.3352	0.5
0.6	0.3332	0.3312	0.3292	0.3271	0.3251	0.3230	0.3209	0.3187	0.3166	0.3144	0.6
0.7	0.3123	0.3101	0.3079	0.3056	0.3034	0.3011	0.2989	0.2966	0.2943	0.2920	0.7
0.8	0.2897	0.2874	0.2850	0.2827	0.2803	0.2780	0.2756	0.2732	0.2709	0.2685	0.8
0.9	0.2661	0.2637	0.2613	0.2589	0.2565	0.2541	0.2516	0.2492	0.2468	0.2444	0.9
1.0	0.2420	0.2396	0.2371	0.2347	0.2323	0.2299	0.2275	0.2251	0.2227	0.2203	1.0
1.1	0.2179	0.2155	0.2131	0.2107	0.2083	0.2059	0.2036	0.2012	0.1989	0.1965	1.1
1.2	0.1942	0.1919	0.1895	0.1872	0.1849	0.1826	0.1804	0.1781	0.1758	0.1736	1.2
1.3	0.1714	0.1691	0.1669	0.1647	0.1626	0.1604	0.1582	0.1561	0.1539	0.1518	1.3
1.4	0.1497	0.1476	0.1456	0.1435	0.1415	0.1394	0.1374	0.1354	0.1334	0.1315	1.4
1.5	0.1295	0.1276	0.1257	0.1238	0.1219	0.1200	0.1182	0.1163	0.1145	0.1127	1.5
1.6	0.1109	0.1092	0.1074	0.1057	0.1040	0.1023	0.1006	0.0989	0.0973	0.0957	1.6
1.7	0.0940	0.0925	0.0909	0.0893	0.0878	0.0863	0.0848	0.0833	0.0818	0.0804	1.7
1.8	0.0790	0.0775	0.0761	0.0748	0.0734	0.0721	0.0707	0.0694	0.0681	0.0669	1.8
1.9	0.0656	0.0644	0.0632	0.0620	0.0608	0.0596	0.0584	0.0573	0.0562	0.0551	1.9
2.0	0.0540	0.0529	0.0519	0.0508	0.0498	0.0488	0.0478	0.0468	0.0459	0.0449	2.0
2.1	0.0440	0.0431	0.0422	0.0413	0.0404	0.0396	0.0387	0.0379	0.0371	0.0361	2.1
2.2	0.0355	0.0347	0.0339	0.0332	0.0325	0.0317	0.0310	0.0303	0.0297	0.0290	2.2
2.3	0.0283	0.0277	0.0270	0.0264	0.0258	0.0252	0.0246	0.0241	0.0235	0.0229	2.3
2.4	0.0224	0.0219	0.0213	0.0208	0.0203	0.0198	0.0194	0.0189	0.0184	0.0180	2.4
2.5	0.0175	0.0171	0.0167	0.0163	0.0158	0.0155	0.0151	0.0147	0.0143	0.0139	2.5
2.6	0.0136	0.0132	0.0129	0.0126	0.0122	0.0119	0.0116	0.0113	0.0110	0.0107	2.6
2.7	0.0104	0.0101	0.0099	0.0096	0.0093	0.0091	0.0088	0.0086	0.0084	0.0081	2.7
2.8	0.0079	0.0077	0.0075	0.0073	0.0071	0.0069	0.0067	0.0065	0.0063	0.0061	2.8
2.9	0.0060	0.0058	0.0056	0.0055	0.0053	0.0051	0.0050	0.0048	0.0047	0.0046	2.9
3.0	0.0044	0.0043	0.0042	0.0040	0.0039	0.0038	0.0037	0.0036	0.0035	0.0034	3.0
3.1	0.0033	0.0032	0.0031	0.0030	0.0029	0.0028	0.0027	0.0026	0.0025	0.0025	3.1
3.2	0.0024	0.0023	0.0022	0.0022	0.0021	0.0020	0.0020	0.0019	0.0018	0.0018	3.2
3.3	0.0017	0.0017	0.0016	0.0016	0.0015	0.0015	0.0014	0.0014	0.0013	0.0013	3.3
3.4	0.0012	0.0012	0.0012	0.0011	0.0011	0.0010	0.0010	0.0010	0.0009	0.0009	3.4
t	0	1	2	3	4	5	6	7	8	9	t

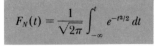

$$F_N(t) = \frac{1}{\sqrt{2\pi}} \int_{-\infty}^{t} e^{-t^2/2}\, dt$$

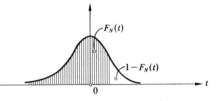

t	0	1	2	3	4	5	6	7	8	9	t
0.0	0.5000	0.5040	0.5080	0.5120	0.5160	0.5199	0.5239	0.5279	0.5319	0.5359	0.0
0.1	0.5398	0.5438	0.5478	0.5517	0.5557	0.5596	0.5636	0.5675	0.5714	0.5754	0.1
0.2	0.5793	0.5832	0.5871	0.5910	0.5948	0.5987	0.6026	0.6064	0.6103	0.6141	0.2
0.3	0.6179	0.6217	0.6255	0.6293	0.6331	0.6368	0.6406	0.6443	0.6480	0.6517	0.3
0.4	0.6554	0.6591	0.6628	0.6664	0.6700	0.6736	0.6772	0.6808	0.6844	0.6879	0.4
0.5	0.6915	0.6950	0.6985	0.7019	0.7054	0.7088	0.7123	0.7157	0.7190	0.7224	0.5
0.6	0.7258	0.7291	0.7324	0.7357	0.7389	0.7422	0.7454	0.7486	0.7517	0.7549	0.6
0.7	0.7580	0.7612	0.7642	0.7673	0.7704	0.7734	0.7764	0.7794	0.7823	0.7852	0.7
0.8	0.7881	0.7910	0.7939	0.7967	0.7996	0.8023	0.8051	0.8078	0.8106	0.8133	0.8
0.9	0.8159	0.8186	0.8212	0.8238	0.8264	0.8289	0.8315	0.8340	0.8365	0.8389	0.9
1.0	0.8413	0.8438	0.8461	0.8485	0.8508	0.8531	0.8554	0.8577	0.8599	0.8621	1.0
1.1	0.8643	0.8665	0.8686	0.8708	0.8729	0.8749	0.8770	0.8790	0.8810	0.8830	1.1
1.2	0.8849	0.8869	0.8888	0.8907	0.8925	0.8944	0.8962	0.8980	0.8997	0.9015	1.2
1.3	0.9032	0.9049	0.9066	0.9082	0.9099	0.9115	0.9131	0.9147	0.9162	0.9177	1.3
1.4	0.9192	0.9207	0.9222	0.9236	0.9251	0.9265	0.9279	0.9292	0.9306	0.9319	1.4
1.5	0.9332	0.9345	0.9357	0.9370	0.9382	0.9394	0.9406	0.9418	0.9429	0.9441	1.5
1.6	0.9452	0.9463	0.9474	0.9484	0.9495	0.9505	0.9515	0.9525	0.9535	0.9545	1.6
1.7	0.9554	0.9564	0.9573	0.9582	0.9591	0.9599	0.9608	0.9616	0.9625	0.9633	1.7
1.8	0.9641	0.9649	0.9656	0.9664	0.9671	0.9678	0.9686	0.9693	0.9699	0.9706	1.8
1.9	0.9713	0.9719	0.9726	0.9732	0.9738	0.9744	0.9750	0.9756	0.9761	0.9767	1.9
2.0	0.9773	0.9778	0.9783	0.9788	0.9793	0.9798	0.9803	0.9808	0.9812	0.9817	2.0
2.1	0.9821	0.9826	0.9830	0.9834	0.9838	0.9842	0.9846	0.9850	0.9854	0.9857	2.1
2.2	0.9861	0.9864	0.9868	0.9871	0.9875	0.9878	0.9881	0.9884	0.9887	0.9890	2.2
2.3	0.9893	0.9896	0.9898	0.9901	0.9904	0.9906	0.9909	0.9911	0.9913	0.9916	2.3
2.4	0.9918	0.9920	0.9922	0.9925	0.9927	0.9929	0.9931	0.9932	0.9934	0.9936	2.4
2.5	0.9938	0.9940	0.9941	0.9943	0.9945	0.9946	0.9948	0.9949	0.9951	0.9952	2.5
2.6	0.9953	0.9955	0.9956	0.9957	0.9959	0.9960	0.9961	0.9962	0.9963	0.9964	2.6
2.7	0.9965	0.9966	0.9967	0.9968	0.9969	0.9970	0.9971	0.9972	0.9973	0.9974	2.7
2.8	0.9974	0.9975	0.9976	0.9977	0.9977	0.9978	0.9979	0.9979	0.9980	0.9981	2.8
2.9	0.9981	0.9982	0.9982	0.9983	0.9984	0.9984	0.9985	0.9985	0.9986	0.9986	2.9
3.0	0.9987	0.9987	0.9987	0.9988	0.9988	0.9989	0.9989	0.9989	0.9990	0.9990	3.0
3.1	0.9990	0.9991	0.9991	0.9991	0.9992	0.9992	0.9992	0.9992	0.9993	0.9993	3.1
3.2	0.9993	0.9993	0.9994	0.9994	0.9994	0.9994	0.9994	0.9995	0.9995	0.9995	3.2
3.3	0.9995	0.9995	0.9996	0.9996	0.9996	0.9996	0.9996	0.9996	0.9996	0.9997	3.3
3.4	0.9997	0.9997	0.9997	0.9997	0.9997	0.9997	0.9997	0.9997	0.9997	0.9998	3.4
t	0	1	2	3	4	5	6	7	8	9	t

$$\binom{n}{0} = 1 \qquad \binom{n}{1} = n \qquad \binom{n}{n} = 1 \qquad \binom{n}{k} = \frac{n!}{k!\,(n-k)!} = \frac{n(n-1)\cdots(n-k+1)}{k!} = \binom{n}{n-k}$$

n	$\binom{n}{0}$	$\binom{n}{1}$	$\binom{n}{2}$	$\binom{n}{3}$	$\binom{n}{4}$	$\binom{n}{5}$	$\binom{n}{6}$	$\binom{n}{7}$	$\binom{n}{8}$	$\binom{n}{9}$	$\binom{n}{10}$	n
0	1											0
1	1	1										1
2	1	2	1									2
3	1	3	3	1								3
4	1	4	6	4	1							4
5	1	5	10	10	5	1						5
6	1	6	15	20	15	6	1					6
7	1	7	21	35	35	21	7	1				7
8	1	8	28	56	70	56	28	8	1			8
9	1	9	36	84	126	126	84	36	9	1		9
10	1	10	45	120	210	252	210	120	45	10	1	10
11	1	11	55	165	330	462	462	330	165	55	11	11
12	1	12	66	220	495	792	924	792	495	220	66	12
13	1	13	78	286	715	1,287	1,716	1,716	1,287	715	286	13
14	1	14	91	364	1,001	2,002	3,003	3,432	3,003	2,002	1,001	14
15	1	15	105	455	1,365	3,003	5,005	6,435	6,435	5,005	3,003	15
16	1	16	120	560	1,820	4,368	8,008	11,440	12,870	11,440	8,008	16
17	1	17	136	680	2,380	6,188	12,376	19,448	24,310	24,310	19,448	17
18	1	18	153	816	3,060	8,568	18,564	31,824	43,758	48,620	43,758	18
19	1	19	171	969	3,876	11,628	27,132	50,388	75,582	92,378	92,378	19
20	1	20	190	1,140	4,845	15,504	38,760	77,520	125,970	167,960	184,756	20
n	$\binom{n}{0}$	$\binom{n}{1}$	$\binom{n}{2}$	$\binom{n}{3}$	$\binom{n}{4}$	$\binom{n}{5}$	$\binom{n}{6}$	$\binom{n}{7}$	$\binom{n}{8}$	$\binom{n}{9}$	$\binom{n}{10}$	n

Note: For coefficients not given above use $\binom{n}{k} = \binom{n}{n-k}$;

for example, $\binom{19}{15} = \binom{19}{19-15} = \binom{19}{4} = 3{,}876.$

19
TABLES OF INDEFINITE INTEGRALS

(1) Notation

The more frequently encountered indefinite integrals and their solutions are tabulated in this chapter. Particular symbols used in the following are

$a, b, c, d, e, f = $ constants	$\alpha, \beta, \gamma = $ constant equivalents
$m, n, p, q, r = $ integers	$x = $ independent variable

$A = a + bx$	$M = a^2 + b^2x$
$B = a + bx + cx^2$	$N = a^2 - b^2x$
$C = a^2 + x^2$	$E = a^2 - x^2$
$F = x^2 - a^2$	$G = a^3 \pm x^3$
$H = a^4 + x^4$	$L = a^4 - x^4$
$P = a + bx^q$	$R = ax^q + bx^{q+r}$

In these tables, the constant of integration is omitted but implied, logarithmic expressions are for the absolute value of the respective argument, all angles are in radians, and all inverse functions represent principal values (angles).

(2) Indefinite Integrals Involving $\qquad$ $f(x^m, A^n)$ $\qquad$ $A = a + bx$ $\qquad$ $a \neq 0$

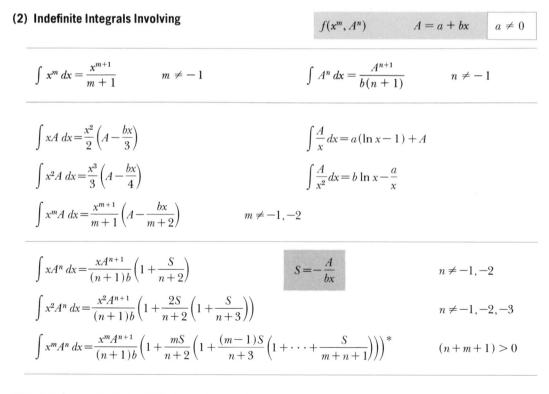

$$\int x^m \, dx = \frac{x^{m+1}}{m+1} \qquad m \neq -1 \qquad\qquad \int A^n \, dx = \frac{A^{n+1}}{b(n+1)} \qquad n \neq -1$$

$$\int xA \, dx = \frac{x^2}{2}\left(A - \frac{bx}{3}\right) \qquad\qquad \int \frac{A}{x} \, dx = a(\ln x - 1) + A$$

$$\int x^2A \, dx = \frac{x^3}{3}\left(A - \frac{bx}{4}\right) \qquad\qquad \int \frac{A}{x^2} \, dx = b \ln x - \frac{a}{x}$$

$$\int x^mA \, dx = \frac{x^{m+1}}{m+1}\left(A - \frac{bx}{m+2}\right) \qquad m \neq -1, -2$$

$$\int xA^n \, dx = \frac{xA^{n+1}}{(n+1)b}\left(1 + \frac{S}{n+2}\right) \qquad S = -\frac{A}{bx} \qquad n \neq -1, -2$$

$$\int x^2A^n \, dx = \frac{x^2A^{n+1}}{(n+1)b}\left(1 + \frac{2S}{n+2}\left(1 + \frac{S}{n+3}\right)\right) \qquad n \neq -1, -2, -3$$

$$\int x^mA^n \, dx = \frac{x^mA^{n+1}}{(n+1)b}\left(1 + \frac{mS}{n+2}\left(1 + \frac{(m-1)S}{n+3}\left(1 + \cdots + \frac{S}{m+n+1}\right)\right)\right)^{*} \qquad (n+m+1) > 0$$

*For nested sum refer to Sec. 8.11; $S = $ pocket calculator storage.

(3) Indefinite Integrals Involving*

$$f(x) = f(x^m X_p^{(r)}) \qquad m > 0, \qquad p \neq 0$$

$k, m, r =$ positive integers $\qquad \mathscr{S}_1^{(r)}, \mathscr{S}_2^{(r)}, \mathscr{S}_3^{(r)}, \ldots =$ Stirling numbers (Sec. A.05)

$p =$ signed number $\qquad X_p^{(r)} =$ factorial polynomial (Sec. 1.03)

$$X_1^{(r)} = x(x-1)(x-2)(x-3) \cdots (x-r+2)(x-r+1)$$

$$= x\mathscr{S}_1^{(r)} + x^2\mathscr{S}_2^{(r)} + x^3\mathscr{S}_3^{(r)} + \cdots + x^r\mathscr{S}_r^{(r)} = \sum_{k=1}^{r} x^k \mathscr{S}_k^{(r)}$$

$$= x(\mathscr{S}_1^{(r)} + x(\mathscr{S}_2^{(r)} + x(\mathscr{S}_3^{(r)} + \cdots + x))) = x \bigwedge_{k=1}^{r-1} [\mathscr{S}_k^{(r)} + x]^* \qquad \boxed{\mathscr{S}_r^{(r)} = 1}$$

$$X_p^{(r)} = x(x-p)(x-2p)(x-3p) \cdots (x-pr+2p)(x-pr+p)$$

$$= xp^{r-1}\mathscr{S}_1^{(r)} + x^2p^{r-2}\mathscr{S}_2^{(r)} + x^3p^{r-3}\mathscr{S}_3^{(r)} + \cdots + x^r\mathscr{S}_r^{(r)} = p^r \sum_{k=1}^{r} \left(\frac{x}{p}\right)^k \mathscr{S}_k^{(r)}$$

$$= xp^{r-1}\left(\mathscr{S}_1^{(r)} + \frac{x}{p}\left(\mathscr{S}_2^{(r)} + \frac{x}{p}\left(\mathscr{S}_3^{(r)} + \cdots + \frac{x}{p}\right)\right)\right) = xp^{r-1} \bigwedge_{k=1}^{r-1} \left[\mathscr{S}_k^{(r)} + \frac{x}{p}\right]^*$$

$$\int X_1^{(2)}\, dx = -\frac{x^2}{2}\left(1 - \frac{2x}{3}\right) \qquad\qquad \int xX_1^{(2)}\, dx = -\frac{x^3}{3}\left(1 - \frac{3x}{4}\right)$$

$$\int x^m X_1^{(2)}\, dx = -\frac{x^{m+2}}{m+2}\left(1 - \frac{(m+2)x}{m+3}\right)$$

$$\int X_1^{(r)}\, dx = \frac{x^2}{2}\left(\mathscr{S}_1^{(r)} + \frac{2x}{3}\left(\mathscr{S}_2^{(r)} + \frac{3x}{4}\left(\mathscr{S}_3^{(r)} + \cdots + \frac{rx}{r+1}\right)\right)\right)$$

$$\int xX_1^{(r)}\, dx = \frac{x^3}{3}\left(\mathscr{S}_1^{(r)} + \frac{3x}{4}\left(\mathscr{S}_2^{(r)} + \frac{4x}{5}\left(\mathscr{S}_3^{(r)} + \cdots + \frac{(r+1)x}{r+2}\right)\right)\right)$$

$$\int x^m X_1^{(r)}\, dx = \frac{x^{m+2}}{m+2}\left(\mathscr{S}_1^{(r)} + \frac{(m+2)x}{m+3}\left(\mathscr{S}_2^{(r)} + \frac{(m+3)x}{m+3}\left(\mathscr{S}_3^{(r)} + \cdots + \frac{(r+m)x}{r+m+1}\right)\right)\right)^*$$

$$\int X_p^{(2)}\, dx = -\frac{px^2}{2}\left(1 - \frac{2x}{3p}\right) \qquad\qquad \int xX_p^{(2)}\, dx = -\frac{px^3}{3}\left(1 - \frac{3x}{4p}\right)$$

$$\int x^m X_p^{(2)}\, dx = -\frac{px^{m+2}}{m+2}\left(1 - \frac{(m+2)x}{(m+3)p}\right)$$

$$\int X_p^{(r)}\, dx = \frac{p^{r-1}x^2}{2}\left(\mathscr{S}_1^{(r)} + \frac{2x}{3p}\left(\mathscr{S}_2^{(r)} + \frac{3x}{4p}\left(\mathscr{S}_3^{(r)} + \cdots + \frac{rx}{(r+1)p}\right)\right)\right)$$

$$\int xX_p^{(r)}\, dx = \frac{p^{r-1}x^3}{3}\left(\mathscr{S}_1^{(r)} + \frac{3x}{4p}\left(\mathscr{S}_2^{(r)} + \frac{4x}{5p}\left(\mathscr{S}_3^{(r)} + \cdots + \frac{(r+1)x}{(r+2)p}\right)\right)\right)$$

$$\int x^m X_p^{(r)}\, dx = \frac{p^{r-1}x^{m+2}}{m+2}\left(\mathscr{S}_1^{(r)} + \frac{(m+2)x}{(m+3)p}\left(\mathscr{S}_2^{(r)} + \frac{(m+3)x}{(m+4)p}\left(\mathscr{S}_3^{(r)} + \cdots \frac{(r+m)x}{(r+m+1)p}\right)\right)\right)^*$$

---

*For nested sum refer to Sec. 8.11.

(4) Indefinite Integrals Involving $f(x) = \dfrac{x^m}{A^n}$ $A = a + bx$ $a \neq 0$

$$\int \frac{dx}{A} = \frac{\ln A}{b}$$

$$\int \frac{dx}{A^2} = -\frac{1}{bA}$$

$$\int \frac{dx}{A^3} = -\frac{1}{2bA^2}$$

$$\int \frac{dx}{A^n} = -\frac{1}{(n-1)bA^{n-1}} \qquad n \neq 0, 1$$

$$\int \frac{x\, dx}{A} = \frac{1}{b^2}(A - a \ln A)$$

$$\int \frac{x\, dx}{A^2} = \frac{1}{b^2}\left(\frac{a}{A} + \ln A\right)$$

$$\int \frac{x\, dx}{A^3} = -\frac{2A - a}{2b^2 A^2}$$

$$\int \frac{x\, dx}{A^n} = -\frac{1}{(n-2)b^2 A^{n-1}}\left(A - \frac{n-2}{n-1}a\right) \qquad n \neq 0, 1, 2$$

$$\int \frac{x^2\, dx}{A} = \frac{1}{b^3}\left(\frac{A^2}{2} - 2aA + a^2 \ln A\right)$$

$$\int \frac{x^2\, dx}{A^2} = \frac{1}{b^3}\left(A - \frac{a^2}{A} - 2a \ln A\right)$$

$$\int \frac{x^2\, dx}{A^3} = \frac{1}{b^3}\left(\frac{2a}{A} - \frac{a^2}{2A^2} + \ln A\right)$$

$$\int \frac{x^2\, dx}{A^n} = -\frac{1}{(n-3)b^3 A^{n-1}}\left[A^2 - \frac{2(n-3)a}{n-2}A + \frac{n-3}{n-1}a^2\right] \qquad n \neq 0, 1, 2, 3$$

$$\int \frac{x^m}{A}\, dx = \frac{1}{b}\left[\left(-\frac{a}{b}\right)^m \ln A + x^m \sum_{k=0}^{m-1} \frac{1}{m-k}\left(-\frac{a}{bx}\right)^k\right]$$

$$\int \frac{x^m}{A^n}\, dx = \frac{1}{b^{m+1}} \sum_{k=0}^{m} \binom{m}{k} \frac{A^{m-n-k+1}(-a)^k}{m-n-k+1}$$

with terms $m - n - k + 1 = 0$ replaced by $\dbinom{m}{n-1}(-a)^{m-n+1} \ln A$

(5) Indefinite Integrals Involving $\qquad f(x) = \dfrac{1}{x^m A^n} \qquad A = a + bx \qquad a \neq 0$

$$\int \frac{dx}{xA} = -\frac{1}{a} \ln \frac{A}{x}$$

$$\int \frac{dx}{xA^2} = -\frac{1}{a^2} \ln \frac{A}{x} + \frac{1}{aA}$$

$$\int \frac{dx}{xA^n} = -\frac{1}{a^n} \ln \frac{A}{x} + \frac{1}{a^n} \sum_{k=1}^{n-1} \binom{n-1}{k} \frac{(-bx)^k}{kA^k} \qquad n \neq 0$$

$$\int \frac{dx}{x^2 A} = \frac{b}{a^2} \ln \frac{A}{x} - \frac{1}{ax}$$

$$\int \frac{dx}{x^2 A^2} = \frac{2b}{a^3} \ln \frac{A}{x} - \frac{1}{a^2 x} - \frac{b}{a^2 A}$$

$$\int \frac{dx}{x^2 A^3} = \frac{1}{a^4} \left(3b \ln \frac{A}{x} - \frac{A}{x} + \frac{3b^2 x}{A} - \frac{b^3 x^2}{2A^2} \right)$$

$$\int \frac{dx}{x^2 A^n} = \frac{nb}{a^{n+1}} \ln \frac{A}{x} - \frac{A}{a^{n+1}x} + \frac{A}{a^{n+1}x} \sum_{k=2}^{n} \binom{n}{k} \frac{(-bx)^k}{(k-1)A^k} \qquad n \neq 0, 1$$

$$\int \frac{dx}{x^3 A} = -\frac{b^2}{a^3} \ln \frac{A}{x} + \frac{2bA}{a^3 x} - \frac{A^2}{2a^3 x^2}$$

$$\int \frac{dx}{x^3 A^2} = -\frac{3b^2}{a^4} \ln \frac{A}{x} + \frac{3bA}{a^4 x} - \frac{A^2}{2a^4 x^2} - \frac{b^3 x}{a^4 A}$$

$$\int \frac{dx}{x^3 A^3} = -\frac{1}{a^5} \left(6b^2 \ln \frac{A}{x} - \frac{4bA}{x} + \frac{A^2}{2x^2} + \frac{4b^2 x}{A} - \frac{b^4 x^2}{2A^2} \right)$$

$$\int \frac{dx}{x^3 A^n} = -\frac{n(n+1)b^2}{2a^{n+2}} \ln \frac{A}{x} + \frac{(n+1)bA}{a^{n+2}x} - \frac{b^2 A^2}{2a^{n+2}x^2} + \frac{A^2}{a^{n+2}x^2} \sum_{k=3}^{n+1} \binom{n+1}{k} \frac{(-bx)^k}{(k-2)A^k} \qquad n \neq 0, 1, 2$$

$$\int \frac{dx}{x^m A} = \frac{1}{b} \left[\left(-\frac{b}{a} \right)^m \ln A - \frac{1}{x^m} \sum_{k=1}^{m-1} \frac{1}{m+1} \left(-\frac{a}{bx} \right)^k \right]$$

$$\int \frac{dx}{x^m A^n} = \frac{-1}{a^{m+n-1}} \sum_{k=0}^{m+n-2} \binom{m+n-2}{k} \frac{A^{m-k-1}(-b)^k}{(m-k-1)^{m-k-1}x^{m-k-1}}$$

with terms $m - k - 1 = 0$ replaced by $\dbinom{m+n-2}{n-1} (-b)^{m-1} \ln \dfrac{A}{x}$

(6) Indefinite Integrals Involving

$f(x) = f(A, D)$	$A = a + x$	$a \neq 0$
$\alpha = a - b \neq 0$	$D = b + x$	$b \neq 0$

$$\int \frac{A}{D}\,dx = \alpha \ln D + x$$

$$\int \frac{dx}{AD} = \frac{1}{\alpha} \ln \frac{D}{A}$$

$$\int \frac{x\,dx}{AD} = \frac{1}{\alpha}(a \ln A - b \ln D)$$

$$\int \frac{x^2\,dx}{AD} = x - \frac{a+b}{2} \ln AD + \frac{a^2+b^2}{2} \ln \frac{A}{D}$$

$$\int \frac{dx}{AD^2} = \frac{-1}{\alpha D} + \frac{1}{\alpha^2} \ln \frac{A}{D}$$

$$\int \frac{x\,dx}{AD^2} = \frac{b}{\alpha D} - \frac{a}{\alpha^2} \ln \frac{A}{D}$$

$$\int \frac{x^2\,dx}{AD^2} = \frac{b^2}{\alpha D} + \frac{a^2}{\alpha^2} \ln \frac{A}{D} + \ln D$$

$$\int \frac{dx}{A^2D^2} = -\frac{1}{\alpha^2}\left(\frac{1}{A} + \frac{1}{D}\right) + \frac{2}{\alpha^3} \ln \frac{A}{D}$$

$$\int \frac{x\,dx}{A^2D^2} = \frac{1}{\alpha^2}\left(\frac{a}{A} + \frac{b}{D}\right) + \frac{a+b}{\alpha^3} \ln \frac{A}{D}$$

$$\int \frac{x^2\,dx}{A^2D^2} = -\frac{1}{\alpha^2}\left(\frac{a^2}{A} + \frac{b^2}{D}\right) + \frac{2ab}{\alpha^3} \ln \frac{A}{D}$$

$$\int A^m D^n\,dx = \frac{A^{m+1}D^n}{m+n+1} + \frac{n\alpha}{m+n+1} \int A^m D^{n-1}\,dx \qquad\qquad m+n \neq -1$$

$$\int \frac{A^m}{D^n}\,dx = \frac{-A^{m+1}}{(n-1)\alpha D^{n-1}} - \frac{n-m-2}{(n-1)\alpha} \int \frac{A^m}{D^{n-1}}\,dx \qquad\qquad n \neq 1$$

$$\int \frac{dx}{A^m D^n} = \frac{-1}{(n-1)\alpha A^{m-1}D^{n-1}} - \frac{m+n-2}{(n-1)\alpha} \int \frac{dx}{A^m D^{n-1}} \qquad\qquad n \neq 1$$

$$\left.\begin{aligned} \int \frac{dx}{(c+x)AD} &= \frac{\ln A}{\alpha(a-c)} + \frac{\ln D}{\alpha(c-b)} + \frac{\ln(c+x)}{(a-c)(c-b)} \\[2mm] \int \frac{x\,dx}{(c+x)AD} &= \frac{a \ln A}{\alpha(c-a)} + \frac{b \ln D}{\alpha(b-c)} + \frac{c \ln(c+x)}{(c-a)(b-c)} \end{aligned}\right\} \qquad \begin{aligned} a-c &\neq 0 \\ b-c &\neq 0 \end{aligned}$$

(7) Indefinite Integrals Involving

$$f(x) = f(x^m, B, D) \qquad \begin{array}{l} B = a + bx + cx^2 \\ D = e + fx \end{array} \qquad a \neq 0$$

$$\omega = b + 2cx = dB/dx \qquad \gamma = 4ac - b^2 \qquad \lambda = ae^2 - bef + cf^2 \qquad m, n > 0$$

$$S = \frac{2Bc}{\gamma} \qquad \gamma \neq 0$$

$$\int B^2 \, dx = \frac{\omega \gamma^2}{60c^3} \left(1 + S(1 + \tfrac{3}{2}S) \right)$$

$$\int B^3 \, dx = \frac{\omega \gamma^3}{600c^4} \left(1 + S(1 + \tfrac{3}{2}S(1 + \tfrac{5}{3}S)) \right)$$

$$\int B^m \, dx = \left(\frac{\gamma}{c}\right)^m \frac{(m!)^2}{(2m+1)!} \frac{\omega}{2c} \overset{m}{\underset{k=1}{\Lambda}} \left[1 + \frac{(2k-1)}{k} S \right]^* \qquad \int xB^m \, dx = \frac{B^{m+1}}{2(m+1)c} - \frac{b}{2c} \int B^m \, dx$$

$$\int \frac{dx}{B} = \begin{cases} + \dfrac{2}{\sqrt{\gamma}} \tan^{-1} \dfrac{\omega}{\sqrt{\gamma}} & \gamma > 0 \\[2mm] - \dfrac{2}{\omega} & \gamma = 0 \\[2mm] - \dfrac{2}{\sqrt{-\gamma}} \tanh^{-1} \dfrac{\omega}{\sqrt{-\gamma}} & \gamma < 0 \end{cases}$$

$$\int \frac{dx}{B^n} = \frac{\omega}{(n-1)\gamma B^{n-1}} + \frac{2(2n-3)c}{(n-1)\gamma} \int \frac{dx}{B^{n-1}} \qquad \int \frac{x \, dx}{B^n} = \frac{-1}{2(n-1)cB^{n-1}} - \frac{b}{2c} \int \frac{dx}{B^n}$$

$$\int \frac{x \, dx}{B} = \frac{1}{2c} \ln B - \frac{b}{2c} \int \frac{dx}{B} \qquad \int \frac{x^2 \, dx}{B} = \frac{x}{c} - \frac{b}{2c^2} \ln B + \frac{b^2 - 2bc}{2c^2} \int \frac{dx}{B}$$

$$\int \frac{x^m \, dx}{B^n} = \frac{1}{(m - 2n + 1)c} \left[\frac{x^{m-1}}{B^{n-1}} - (m-n)b \int \frac{x^{m-1}}{B^n} \, dx - (m-1)a \int \frac{x^{m-2}}{B^n} \, dx \right]$$

$$\int \frac{dx}{xB} = \frac{1}{2a} \ln \frac{x^2}{B} - \frac{b}{2a} \int \frac{dx}{B} \qquad \int \frac{dx}{x^2 B} = \frac{b}{2a^2} \ln \frac{B}{x^2} - \frac{1}{ax} + \left(\frac{b^2}{2a^2} - \frac{c}{a} \right) \int \frac{dx}{B}$$

$$\int \frac{dx}{x^m B^n} = \frac{-1}{(m-1)a} \left[\frac{1}{x^{m-1}B^{n-1}} + (m+n-2)b \int \frac{dx}{x^{m-1}B^n} + (m+2n-3)c \int \frac{dx}{x^{m-2}B^n} \right]$$

$$\int \frac{D}{B} \, dx = \frac{f}{2c} \ln B + \frac{2ce - bf}{2c} \int \frac{dx}{B} \qquad \int \frac{D}{B^n} \, dx = \frac{f}{2(n-1)aB^{n-1}} + \frac{2ce - bf}{2c} \int \frac{dx}{B^n}$$

$$\int \frac{dx}{DB} = \frac{1}{2\lambda} \left[2e \ln D - e \ln B + (2cf - be) \int \frac{dx}{B} \right]$$

$$\int \frac{dx}{D^m B^n} = \frac{-1}{(m-1)\lambda} \left[\frac{f}{D^{m-1}B^{n-1}} + (m+n-2)(2ce - bf) \int \frac{dx}{D^{m-1}B^n} + (m+2n-3)c \int \frac{dx}{D^{m-2}B^n} \right]$$

*For nested sum refer to Sec. 8.11, S = pocket calculator storage.

(8) Indefinite Integrals Involving $\qquad f(x) = \dfrac{x^m}{C^n} \qquad C = a^2 + x^2 \qquad \begin{matrix} a^2 \neq 0 \\ b^2 \neq 0 \end{matrix}$

$$\int \frac{dx}{C} = \frac{1}{a} \tan^{-1} \frac{x}{a}$$

$$\int \frac{dx}{C^2} = \frac{1}{2a^3} \left(\frac{ax}{C} + \tan^{-1} \frac{x}{a} \right)$$

$$\int \frac{dx}{C^3} = \frac{1}{8a^5} \left(\frac{2a^3 x}{C^2} + \frac{3ax}{C} + 3 \tan^{-1} \frac{x}{a} \right)$$

$$\int \frac{dx}{C^n} = \frac{x}{2(n-1)a^2 C^{n-1}} + \frac{2n-3}{2(n-1)a^2} \int \frac{dx}{C^{n-1}} \qquad n \neq 1$$

$$\int \frac{x\,dx}{C} = \frac{1}{2} \ln C$$

$$\int \frac{x\,dx}{C^2} = \frac{-1}{2C}$$

$$\int \frac{x\,dx}{C^n} = \frac{-1}{2(n-1)C^{n-1}} \qquad n \neq 1$$

$$\int \frac{x^2\,dx}{C} = x - a \tan^{-1} \frac{x}{a}$$

$$\int \frac{x^2\,dx}{C^2} = -\frac{x}{2C} + \frac{1}{2a} \tan^{-1} \frac{x}{a}$$

$$\int \frac{x^2\,dx}{C^n} = \frac{-x}{2(n-1)C^{n-1}} + \frac{1}{2(n-1)} \int \frac{dx}{C^{n-1}} \qquad n \neq 1$$

$$\int \frac{x^3\,dx}{C} = \frac{x^2}{2} - \frac{a^2}{2} \ln C$$

$$\int \frac{x^3\,dx}{C^2} = \frac{a^2}{2C} + \frac{1}{2} \ln C$$

$$\int \frac{x^3\,dx}{C^n} = \frac{1}{2(n-2)C^{n-2}} + \frac{a^2}{2(n-1)C^{n-1}} \qquad n \neq 1,2$$

$$\int \frac{x^m\,dx}{C^n} = -\frac{x^{m-1}}{2(n-1)C^{n-1}} + \frac{m-1}{2(n-1)} \int \frac{x^{m-2}}{C^{n-1}} \qquad n \neq 1$$

$$\int \frac{e+fx}{C}\,dx = f \ln \sqrt{C} + \frac{e}{a} \tan^{-1} \frac{x}{a}$$

(9) Indefinite Integrals Involving $\qquad f(x) = \dfrac{1}{x^m C^n} \qquad C = a^2 + x^2 \qquad a^2 \neq 0$

$$\int \frac{dx}{xC} = \frac{1}{2a^2} \ln \frac{x^2}{C}$$

$$\int \frac{dx}{xC^2} = \frac{1}{2a^2 C} + \frac{1}{2a^4} \ln \frac{x^2}{C}$$

$$\int \frac{dx}{xC^3} = \frac{1}{4a^2 C^2} + \frac{1}{2a^4 C} + \frac{1}{2a^6} \ln \frac{x^2}{C}$$

$$\int \frac{dx}{xC^n} = \frac{1}{2(n-1)a^2 C^{n-1}} + \frac{1}{a^2} \int \frac{dx}{xC^{n-1}} \qquad n \neq 1$$

$$\int \frac{dx}{x^2 C} = \frac{-1}{a^2 x} - \frac{1}{a^3} \tan^{-1} \frac{x}{a}$$

$$\int \frac{dx}{x^2 C^2} = \frac{-1}{a^4 x} - \frac{x}{2a^4 C} - \frac{3}{2a^5} \tan^{-1} \frac{x}{a}$$

$$\int \frac{dx}{x^2 C^n} = \frac{-1}{a^2 x C^{n-1}} - \frac{2n-1}{a^2} \int \frac{dx}{C^n}$$

$$\int \frac{dx}{x^3 C} = \frac{-1}{2a^2 x^2} - \frac{1}{2a^4} \ln \frac{x^2}{C}$$

$$\int \frac{dx}{x^3 C^2} = \frac{-1}{2a^4 x^2} - \frac{1}{2a^4 C} - \frac{1}{a^6} \ln \frac{x^2}{C}$$

$$\int \frac{dx}{x^3 C^n} = \frac{-1}{2a^2 x^2 C^{n-1}} - \frac{n}{a^2} \int \frac{dx}{xC^n}$$

$$\int \frac{dx}{(b^2 + x^2) C} = \frac{1}{b^2 - a^2} \left(\frac{1}{a} \tan^{-1} \frac{x}{a} - \frac{1}{b} \tan^{-1} \frac{x}{b} \right)$$

$$\int \frac{x\,dx}{(b^2 + x^2) C} = \frac{1}{b^2 - a^2} \ln \sqrt{\frac{a^2 + x^2}{b^2 + x^2}}$$

$$\int \frac{x^2\,dx}{(b^2 + x^2) C} = \frac{1}{b^2 - a^2} \left(b \tan^{-1} \frac{x}{b} - a \tan^{-1} \frac{x}{a} \right)$$

$$\int \frac{dx}{x^m C^n} = \frac{-1}{(m-1)a^2 x^{m-1} C^{n-1}} - \frac{m + 2n - 3}{(m-1)a^2} \int \frac{dx}{x^{m-2} C^n} \qquad m \neq 1$$

$$\int \frac{dx}{(e + fx) C} = \frac{1}{e^2 + a^2 f^2} \left[f \ln (e + fx) - \frac{f}{2} \ln C + \frac{e}{a} \tan^{-1} \frac{x}{a} \right] \qquad e^2 \neq -a^2 f^2$$

(10) Indefinite Integrals Involving $f(x) = \dfrac{x^m}{E^n}$ $E = a^2 - x^2$ $a^2 \neq 0$

$$\int \frac{dx}{E} = \frac{1}{2a} \ln \frac{a+x}{a-x} = \begin{cases} \dfrac{1}{a} \tanh^{-1} \dfrac{x}{a} & x^2 < a^2 \\[2mm] \dfrac{1}{a} \coth^{-1} \dfrac{x}{a} & x^2 > a^2 \end{cases}$$

$$\int \frac{dx}{E^2} = \frac{1}{2a^3} \left(\frac{ax}{E} + \ln \sqrt{\frac{a+x}{a-x}} \right)$$

$$\int \frac{dx}{E^n} = \frac{x}{2(n-1)a^2 E^{n-1}} + \frac{2n-3}{2(n-1)a^2} \int \frac{dx}{E^{n-1}} \qquad n \neq 1$$

$$\int \frac{x\,dx}{E} = -\frac{1}{2} \ln E$$

$$\int \frac{x\,dx}{E^2} = \frac{1}{2E}$$

$$\int \frac{x\,dx}{E^n} = \frac{1}{2(n-1)E^{n-1}} \qquad n \neq 1$$

$$\int \frac{x^2\,dx}{E} = -x + \frac{a}{2} \ln \frac{a+x}{a-x}$$

$$\int \frac{x^2\,dx}{E^2} = \frac{x}{2E} - \frac{1}{4a} \ln \frac{a+x}{a-x}$$

$$\int \frac{x^2\,dx}{E^n} = \frac{x}{2(n-1)E^{n-1}} - \frac{1}{2(n-1)} \int \frac{dx}{E^{n-1}} \qquad n \neq 1$$

$$\int \frac{x^3\,dx}{E} = -\frac{x^2}{2} - \frac{a^2}{2} \ln E$$

$$\int \frac{x^3\,dx}{E^2} = \frac{a^2}{2E} + \frac{1}{2} \ln E$$

$$\int \frac{x^3\,dx}{E^n} = \frac{-1}{2(n-2)E^{n-2}} + \frac{a^2}{2(n-1)E^{n-1}} \qquad n \neq 1, 2$$

$$\int \frac{x^m\,dx}{E^n} = \frac{x^{m-1}}{2(n-1)E^{n-1}} - \frac{m-1}{2(n-1)} \int \frac{x^{m-2}}{E^{n-1}}\,dx \qquad n \neq 1$$

$$\int \frac{e+fx}{E} = -f \ln \sqrt{E} + \frac{e}{a} \ln \sqrt{\frac{a+x}{a-x}}$$

(11) Indefinite Integrals Involving $\qquad$ $f(x) = \dfrac{1}{x^m E^n}$ $\qquad$ $E = a^2 - x^2$ $\qquad$ $a^2 \neq 0$

$$\int \frac{dx}{xE} = \frac{1}{2a^2} \ln \frac{x^2}{E}$$

$$\int \frac{dx}{xE^2} = \frac{1}{2a^2 E} + \frac{1}{2a^4} \ln \frac{x^2}{E}$$

$$\int \frac{dx}{xE^3} = \frac{1}{4a^2 E^2} + \frac{1}{2a^4 E} + \frac{1}{2a^6} \ln \frac{x^2}{E}$$

$$\int \frac{dx}{xE^n} = \frac{1}{2(n-1)a^2 E^{n-1}} + \frac{1}{a^2} \int \frac{dx}{xE^{n-1}} \qquad n \neq 1$$

$$\int \frac{dx}{x^2 E} = \frac{-1}{a^2 x} + \frac{1}{a^3} \tanh^{-1} \frac{x}{a}$$

$$\int \frac{dx}{x^2 E^2} = \frac{-1}{a^4 x} + \frac{2}{2a^4 E} + \frac{3}{2a^5} \tanh^{-1} \frac{x}{a}$$

$$\int \frac{dx}{x^2 E^n} = \frac{-1}{a^2 x E^{n-1}} + \frac{2n-1}{a^2} \int \frac{dx}{E^n}$$

$$\int \frac{dx}{x^3 E} = \frac{-1}{2a^2 x^2} + \frac{1}{2a^4} \ln \frac{x^2}{E}$$

$$\int \frac{dx}{x^3 E^2} = \frac{-1}{2a^4 x^2} + \frac{1}{2a^4 E} + \frac{1}{a^6} \ln \frac{x^2}{E}$$

$$\int \frac{dx}{x^3 E^n} = \frac{-1}{2a^2 x^2 E^{n-1}} + \frac{n}{a^2} \int \frac{dx}{xE^n}$$

$$\int \frac{dx}{(b^2 + x^2)E} = \frac{1}{a^2 + b^2} \left(\frac{1}{b} \tan^{-1} \frac{x}{b} + \frac{1}{a} \ln \sqrt{\frac{a+x}{a-x}} \right)$$

$$\int \frac{x\, dx}{(b^2 + x^2)E} = \frac{1}{a^2 + b^2} \ln \sqrt{\frac{b^2 + x^2}{a^2 - x^2}}$$

$$\int \frac{x^2\, dx}{(b^2 + x^2)E} = \frac{1}{a^2 + b^2} \left(a \ln \sqrt{\frac{a+x}{a-x}} - b \tan^{-1} \frac{x}{b} \right)$$

$$\int \frac{dx}{x^m E^n} = \frac{-1}{(m-1)a^2 x^{m-1}} + \frac{m + 2n - 3}{(m-1)a^2} \int \frac{dx}{x^{m-2} E^n} \qquad m \neq 1$$

$$\int \frac{dx}{(e + fx)E} = \frac{-1}{e^2 - a^2 f^2} \left[f \ln (e + fx) - \frac{f}{2} \ln E - \frac{e}{a} \tanh^{-1} \frac{x}{a} \right] \qquad e^2 \neq a^2 f^2$$

(12) Indefinite Integrals Involving

$$f(x) = \frac{x^m}{G^n} \qquad G = a^3 \pm x^3 \qquad a^3 \neq 0$$

If two signs appear in a formula, the upper sign corresponds to $G = a^3 + x^3$, and the lower sign corresponds to $G = a^3 - x^3$.

$$\int \frac{dx}{G} = \frac{\pm 1}{6a^2} \ln \frac{(a \pm x)^2}{a^2 \mp ax + x^2} + \frac{1}{a^2\sqrt{3}} \tan^{-1} \frac{2x \mp a}{a\sqrt{3}}$$

$$\int \frac{dx}{G^n} = \frac{x}{3a^3(n-1)G^{n-1}} - \frac{4-3n}{3a^3(n-1)} \int \frac{dx}{G^{n-1}} \qquad n \neq 1$$

$$\int \frac{x\,dx}{G} = \frac{1}{6a} \ln \frac{a^2 \mp ax + x^2}{(a \pm x)^2} \pm \frac{1}{a\sqrt{3}} \tan^{-1} \frac{2x \mp a}{a\sqrt{3}}$$

$$\int \frac{x\,dx}{G^2} = \frac{x^2}{3a^3G} + \frac{1}{3a^3} \int \frac{x\,dx}{G}$$

$$\int \frac{x\,dx}{G^n} = \frac{x^2}{3a^3(n-1)G^{n-1}} - \frac{5-3n}{3a^3(n-1)} \int \frac{x\,dx}{G^{n-1}} \qquad n \neq 1$$

$$\int \frac{x^2\,dx}{G} = \pm \frac{\ln G}{3}$$

$$\int \frac{x^2\,dx}{G^2} = \mp \frac{1}{3G}$$

$$\int \frac{x^2\,dx}{G^n} = \frac{x^3}{3a^3(n-1)G^{n-1}} + \frac{n-2}{a^3(n-1)} \int \frac{x^2\,dx}{G^{n-1}} \qquad n \neq 1$$

$$\int \frac{x^3\,dx}{G} = \pm x \mp a^3 \int \frac{dx}{G}$$

$$\int \frac{x^3\,dx}{G^2} = \mp \frac{x}{3G} \pm \frac{1}{3} \int \frac{dx}{G}$$

$$\int \frac{x^3\,dx}{G^n} = \frac{x^4}{3a^3(n-1)G^{n-1}} + \frac{3n-7}{3a^3(n-1)} \int \frac{x^3\,dx}{G^{n-1}} \qquad n \neq 1$$

$$\int \frac{x^m\,dx}{G^n} = \frac{x^{m+1}}{3a^3(n-1)G^{n-1}} - \frac{m-3n+4}{3a^3(n-1)} \int \frac{x^m\,dx}{G^{n-1}} \qquad n \neq 1$$

$$\int \frac{a+x}{a^3+x^3} dx = \int \frac{dx}{a^2-ax+x^2} = \frac{2}{a\sqrt{3}} \tan^{-1} \frac{2x-a}{a\sqrt{3}}$$

$$\int \frac{a-x}{a^3+x^3} dx = \frac{1}{3a} \ln \frac{(a+x)^2}{a^2-ax+x^2}$$

(13) Indefinite Integrals Involving $\qquad f(x) = \dfrac{1}{x^m G^n} \qquad G = a^3 \pm x^3 \qquad \boxed{a^3 \neq 0}$

If two signs appear in a formula, the upper sign corresponds to $G = a^3 + x^3$, and the lower corresponds to $G = a^3 - x^3$.

$$\int \frac{dx}{xG} = \frac{1}{3a^3} \ln \frac{x^3}{G}$$

$$\int \frac{dx}{xG^2} = \frac{1}{3a^3 G} + \frac{1}{3a^6} \ln \frac{x^3}{G}$$

$$\int \frac{dx}{xG^3} = \frac{1}{6a^3 G^2} + \frac{1}{a^3} \int \frac{dx}{xG^2}$$

$$\int \frac{dx}{xG^n} = \frac{1}{3a^3(n-1)G^{n-1}} + \frac{1}{a^3} \int \frac{dx}{xG^{n-1}} \qquad n \neq 1$$

$$\int \frac{dx}{x^2 G} = -\frac{1}{a^3 x} \mp \frac{1}{a^3} \int \frac{dx}{G}$$

$$\int \frac{dx}{x^2 G^2} = \frac{-1}{a^6 x} \mp \frac{x^2}{3a^6 G} \mp \frac{4}{3a^6} \int \frac{x\, dx}{G}$$

$$\int \frac{dx}{x^2 G^n} = \frac{1}{3a^3(n-1)xG^{n-1}} + \frac{3n-2}{3a^3(n-1)} \int \frac{dx}{x^2 G^{n-1}} \qquad n \neq 1$$

$$\int \frac{dx}{x^3 G} = -\frac{1}{2a^3 x^2} \mp \frac{1}{a^3} \int \frac{dx}{G}$$

$$\int \frac{dx}{x^3 G^2} = -\frac{1}{2a^6 x^2} \mp \frac{x}{3a^6 G} \mp \frac{5}{3a^6} \int \frac{dx}{G}$$

$$\int \frac{dx}{x^3 G^n} = \frac{1}{3a^3(n-1)x^2 G^{n-1}} + \frac{3n-1}{3a^3(n-1)} \int \frac{dx}{x^3 G^{n-1}} \qquad n \neq 1$$

$$\int \frac{dx}{x^m G^n} = \frac{1}{3a^3(n-1)x^{m-1}G^{n-1}} + \frac{m+3n-4}{3a^3(n-1)} \int \frac{dx}{x^m G^{n-1}} \qquad n \neq 1$$

$$\int \frac{a-x}{a^3-x^3}\, dx = \int \frac{dx}{a^2+ax+x^2} = \frac{2}{a\sqrt{3}} \tan^{-1} \frac{2x+a}{a\sqrt{3}}$$

$$\int \frac{a+x}{a^3-x^3}\, dx = -\frac{1}{3a} \ln \frac{(a-x)^2}{a^2+ax+x^2}$$

(14) Indefinite Integrals Involving $\qquad f(x) = \dfrac{x^n}{H^m} \qquad f(x) = \dfrac{1}{x^m H^m} \qquad H = a^4 + x^4 \qquad a^4 \neq 0$

$$\int \frac{dx}{H} = \frac{1}{a^3\sqrt{8}}\left(\tanh^{-1}\frac{ax\sqrt{2}}{a^2+x^2} + \tan^{-1}\frac{ax\sqrt{2}}{a^2-x^2}\right)$$

$$\int \frac{dx}{H^n} = \frac{x}{4(n-1)a^4H^{n-1}} + \frac{4n-5}{4(n-1)a^4}\int \frac{dx}{H^{n-1}} \qquad\qquad n \neq 1$$

$$\int \frac{x\,dx}{H} = \frac{1}{2a^2}\tan^{-1}\left(\frac{x^2}{a^2}\right)$$

$$\int \frac{x^2\,dx}{H} = \frac{1}{a\sqrt{8}}\left(\tanh^{-1}\frac{ax\sqrt{2}}{a^2-x^2} - \tan^{-1}\frac{ax\sqrt{2}}{a^2+k^2}\right)$$

$$\int \frac{x^3\,dx}{H} = \tfrac{1}{4}\ln H$$

$$\int \frac{x^4\,dx}{H} = x - \frac{a}{\sqrt{8}}\left(\tanh^{-1}\frac{ax\sqrt{2}}{a^2+x^2} + \tan^{-1}\frac{ax\sqrt{2}}{a^2-x^2}\right)$$

$$\int \frac{x^m\,dx}{H} = \frac{x^{m-3}}{m-3} - a^4\int \frac{x^{m-4}}{H}\,dx \qquad\qquad m \neq 3$$

$$\int \frac{x^m\,dx}{H^n} = \frac{x^{m+1}}{4a^4(n-1)H^{n-1}} + \frac{4n-m-5}{4a^4(n-1)}\int \frac{x^m\,dx}{H^{n-1}} \qquad\qquad n \neq 1$$

$$\int \frac{dx}{xH} = \frac{1}{2a^4}\ln\frac{x^2}{\sqrt{H}}$$

$$\int \frac{dx}{x^2H} = -\frac{1}{a^4x} - \frac{1}{a^5\sqrt{8}}\left(\tanh^{-1}\frac{ax\sqrt{2}}{a^2-x^2} - \tan^{-1}\frac{ax\sqrt{2}}{a^2+x^2}\right)$$

$$\int \frac{dx}{x^3H} = -\frac{1}{2a^4x^2} - \frac{1}{2a^6}\tan^{-1}\frac{x^2}{a^2}$$

$$\int \frac{dx}{x^4H} = -\frac{1}{3a^4x^3} - \frac{1}{a^7\sqrt{8}}\left(\tanh^{-1}\frac{ax\sqrt{2}}{a^2+x^2} + \tan^{-1}\frac{ax\sqrt{2}}{a^2-x^2}\right)$$

$$\int \frac{dx}{x^mH} = \frac{-1}{(m-1)a^4x^{m-1}} - \frac{1}{a^4}\int \frac{dx}{x^{m-4}H} \qquad\qquad m \neq 1$$

$$\int \frac{dx}{x^mH^n} = \frac{1}{4(n-1)a^4x^{m-1}H^{n-1}} + \frac{m+4n-5}{4(n-1)a^4}\int \frac{dx}{x^mH^{n-1}} \qquad\qquad n \neq 1$$

$$= \frac{-1}{(m-1)a^4x^{m-1}H^{n-1}} - \frac{m+4n-5}{(m-1)a^4}\int \frac{dx}{x^{m-4}H^n} \qquad\qquad m \neq 1$$

(15) Indefinite Integrals Involving $\qquad f(x) = \dfrac{x^m}{L^n} \qquad f(x) = \dfrac{1}{x^m L^n} \qquad L = a^4 - x^4 \qquad a^4 \neq 0$

$$\int \frac{dx}{L} = \frac{1}{2a^3}\left(\ln\sqrt{\frac{a+x}{a-x}} + \tan^{-1}\frac{x}{a}\right)$$

$$\int \frac{dx}{L^n} = \frac{x}{4(n-1)a^4 L^{n-1}} + \frac{4n-5}{4(n-1)a^4}\int \frac{dx}{L^{n-1}} \qquad\qquad n \neq 1$$

$$\int \frac{x\,dx}{L} = \frac{1}{4a^2}\ln\frac{a^2+x^2}{a^2-x^2}$$

$$\int \frac{x^2\,dx}{L} = \frac{1}{2a}\left(\ln\sqrt{\frac{a+x}{a-x}} - \tan^{-1}\frac{x}{a}\right)$$

$$\int \frac{x^3\,dx}{L} = -\tfrac{1}{4}\ln L$$

$$\int \frac{x^4\,dx}{L} = -x + \frac{a}{2}\left(\ln\sqrt{\frac{a+x}{a-x}} + \tan^{-1}\frac{x}{a}\right)$$

$$\int \frac{x^m\,dx}{L} = -\frac{x^{m-3}}{m-3} + a^4\int \frac{x^{m-4}}{L}\,dx \qquad\qquad m \neq 3$$

$$\int \frac{x^m\,dx}{L^n} = \frac{x^{m+1}}{4a^4(n-1)L^{n-1}} + \frac{4n-m-5}{4a^4(n-1)}\int \frac{x^m\,dx}{L^{n-1}} \qquad\qquad n \neq 1$$

$$\int \frac{dx}{xL} = \frac{1}{2a^4}\ln\frac{x^2}{\sqrt{L}}$$

$$\int \frac{dx}{x^2 L} = -\frac{1}{a^4 x} + \frac{1}{2a^5}\left(\ln\sqrt{\frac{a+x}{a-x}} - \tan^{-1}\frac{x}{a}\right)$$

$$\int \frac{dx}{x^3 L} = -\frac{1}{2a^4 x^2} + \frac{1}{4a^6}\ln\frac{a^2+x^2}{a^2-x^2}$$

$$\int \frac{dx}{x^4 L} = -\frac{1}{3a^4 x^3} + \frac{1}{2a^7}\left(\ln\sqrt{\frac{a+x}{a-x}} + \tan^{-1}\frac{x}{a}\right)$$

$$\int \frac{dx}{x^m L} = \frac{-1}{(m-1)a^4 x^{m-1}} + \frac{1}{a^4}\int \frac{dx}{x^{m-4}L} \qquad\qquad m \neq 1$$

$$\int \frac{dx}{x^m L^n} = \frac{1}{4(n-1)a^4 x^{m-1}L^{n-1}} + \frac{m+4n-5}{4(n-1)a^4}\int \frac{dx}{x^m L^{n-1}} \qquad\qquad n \neq 1$$

$$= \frac{-1}{(m-1)a^4 x^{m-1}L^{n-1}} + \frac{m+4n-5}{(m-1)a^4}\int \frac{dx}{x^{m-4}L^n} \qquad\qquad m \neq 1$$

(16) Indefinite Integrals Involving

$f(x) = \sqrt[p]{x^m}/M^n, N^n, P^n$	$M = a^2 + b^2 x$	$a^2 \neq 0$
	$N = a^2 - b^2 x$	$a^2 \neq 0$
	$P = a + b\sqrt{x}$	$b \neq 0$

$$\int \sqrt{x^m}\, dx = \frac{2x\sqrt{x^m}}{m+2} \qquad m \neq -2$$

$$\int \sqrt[p]{x^m}\, dx = \frac{px\sqrt[p]{x^m}}{m+p} \qquad m+p \neq 0$$

$$\int \frac{\sqrt{x}\, dx}{M} = \frac{2\sqrt{x}}{b^2} - \frac{2a}{b^3}\tan^{-1}\frac{b\sqrt{x}}{a}$$

$$\int \frac{\sqrt{x}\, dx}{M^2} = \frac{-\sqrt{x}}{b^2 M} + \frac{1}{ab^3}\tan^{-1}\frac{b\sqrt{x}}{a}$$

$$\int \frac{\sqrt{x^3}\, dx}{M} = \frac{2\sqrt{x^3}}{3b^2} - \frac{2a^2\sqrt{x}}{b^4} + \frac{2a^3}{b^5}\tan^{-1}\frac{b\sqrt{x}}{a}$$

$$\int \frac{\sqrt{x^3}\, dx}{M^2} = \frac{2\sqrt{x^3}}{b^2 M} + \frac{3a^2\sqrt{x}}{b^4 M} - \frac{3a}{b^5}\tan^{-1}\frac{b\sqrt{x}}{a}$$

$$\int \frac{\sqrt[p]{x^m}\, dx}{M^n} = \frac{+1}{\sqrt[p]{b^{2+2m}}}\int \frac{(M - a^2)^{m/p}}{M^n}\, dM$$

$$\int \frac{\sqrt{x}\, dx}{N} = -\frac{2\sqrt{x}}{b^2} + \frac{a}{b^3}\ln\frac{a + b\sqrt{x}}{a - b\sqrt{x}}$$

$$\int \frac{\sqrt{x}\, dx}{N^2} = \frac{\sqrt{x}}{b^2 M} - \frac{1}{2ab^3}\ln\frac{a + b\sqrt{x}}{a - b\sqrt{x}}$$

$$\int \frac{\sqrt{x^3}\, dx}{N} = -\frac{2\sqrt{x^3}}{3b^2} - \frac{2a^2\sqrt{x}}{b^4} + \frac{a^3}{b^5}\ln\frac{a + b\sqrt{x}}{a - b\sqrt{x}}$$

$$\int \frac{\sqrt{x^3}\, dx}{N^2} = -\frac{2\sqrt{x^3}}{b^2 N} + \frac{3a^2\sqrt{x}}{b^4 N} - \frac{3a}{2b^5}\ln\frac{a + b\sqrt{x}}{a - b\sqrt{x}}$$

$$\int \frac{\sqrt[p]{x^m}\, dx}{N^n} = \frac{-1}{\sqrt[p]{b^{2+2m}}}\int \frac{(a^2 - N)^{m/p}}{N^n}\, dN$$

$$\int \frac{dx}{P} = \frac{2}{b^2}[P - a(1 + \ln P)]$$

$$\int \frac{dx}{P^2} = \frac{2}{Pb^2}\left(a + P\ln P\right)$$

$$\int \frac{\sqrt{x}\, dx}{P} = \frac{1}{b}\left[x - \frac{2a}{b}\left(\sqrt{x} - \frac{a}{b}\ln P\right)\right]$$

(17) Indefinite Integrals Involving

$$f(x) = \frac{M^n, N^n, P^n}{\sqrt[p]{x^m}} \qquad \begin{array}{ll} M = a^2 + b^2 x & a^2 \neq 0 \\ N = a^2 - b^2 x & a^2 \neq 0 \\ P = a + b\sqrt{x} & b \neq 0 \end{array}$$

$$\int \frac{dx}{\sqrt{x^m}} = \frac{2x}{(2-m)\sqrt{x^m}} \qquad m \neq 2$$

$$\int \frac{dx}{\sqrt[p]{x^m}} = \frac{px}{(p-m)\sqrt[p]{x^m}} \qquad m \neq p$$

$$\int \frac{dx}{M\sqrt{x}} = \frac{2}{ab} \tan^{-1} \frac{b\sqrt{x}}{a}$$

$$\int \frac{dx}{M^2\sqrt{x}} = \frac{\sqrt{x}}{a^2 M} + \frac{1}{a^3 b} \tan^{-1} \frac{b\sqrt{x}}{a}$$

$$\int \frac{dx}{M\sqrt{x^3}} = \frac{-2}{a^2\sqrt{x}} - \frac{2b}{a^3} \tan^{-1} \frac{b\sqrt{x}}{a}$$

$$\int \frac{dx}{M^2\sqrt{x^3}} = \frac{-2}{a^2 M\sqrt{x}} - \frac{3b^2\sqrt{x}}{a^4 M} - \frac{3b}{a^5} \tan^{-1} \frac{b\sqrt{x}}{a}$$

$$\int \frac{dx}{M^n\sqrt[p]{x^m}} = \frac{-p}{(m-p)a^2 x^{m/p-1} M^{n-1}} - \frac{b^2(m+pn-2p)}{a^2(m-p)} \int \frac{dx}{x^{m/p-1} M^n} \qquad m \neq p$$

$$\int \frac{dx}{N\sqrt{x}} = \frac{1}{ab} \ln \frac{a+b\sqrt{x}}{a-b\sqrt{x}}$$

$$\int \frac{dx}{N^2\sqrt{x}} = \frac{\sqrt{x}}{a^2 N} + \frac{1}{2a^3 b} \ln \frac{a+b\sqrt{x}}{a-b\sqrt{x}}$$

$$\int \frac{dx}{N\sqrt{x^3}} = \frac{-2}{a^2\sqrt{x}} + \frac{b}{a^3} \ln \frac{a+b\sqrt{x}}{a-b\sqrt{x}}$$

$$\int \frac{dx}{N^2\sqrt{x^3}} = \frac{-2}{a^2 N\sqrt{x}} + \frac{3b^2\sqrt{x}}{a^4 N} + \frac{3b}{2a^5} \ln \frac{a+b\sqrt{x}}{a-b\sqrt{x}}$$

$$\int \frac{dx}{N^n\sqrt[p]{x^m}} = \frac{-p}{(m-p)a^2 x^{m/p-1} N^{n-1}} + \frac{b^2(m+pn-2p)}{a^2(m-p)} \int \frac{dx}{x^{m/p-1} N^n} \qquad m \neq p$$

$$\int \frac{dx}{P\sqrt{x}} = \frac{2}{b} \ln P$$

$$\int \frac{dx}{Px} = \frac{1}{a} (\ln x - 2 \ln P)$$

$$\int \frac{dx}{P^2 x} = \frac{2}{Pa^2} [a + P(\ln \sqrt{x} - \ln P)]$$

(18) Indefinite Integrals Involving

$$f(x) = x^m \sqrt[p]{A^n} \qquad A = a + bx \qquad a \neq 0$$

$$S = -\frac{A}{bx}$$

$$\int \sqrt{A}\, dx = \frac{2}{3b} \sqrt{A^3}$$

$$\int \sqrt{A^n}\, dx = \frac{2}{(n+2)b} \sqrt{A^{n+2}}$$

$$\int \sqrt[p]{A^n}\, dx = \frac{p}{(n+p)b} \sqrt[p]{A^{n+p}}$$

$$\int x\sqrt{A}\, dx = \frac{2x}{3b} \sqrt{A^3} \left(1 + \frac{2S}{5}\right)$$

$$\int x\sqrt{A^n}\, dx = \frac{2x}{(n+2)b} \sqrt{A^{n+2}} \left(1 + \frac{2S}{n+4}\right)$$

$$\int x\sqrt[p]{A^n}\, dx = \frac{px}{(n+p)b} \sqrt[p]{A^{n+p}} \left(1 + \frac{pS}{n+2p}\right)$$

$$\int x^2\sqrt{A}\, dx = \frac{2x^2}{3b} \sqrt{A^3} \left(1 + \frac{4S}{5}\left(1 + \frac{2S}{7}\right)\right)$$

$$\int x^2\sqrt{A^n} = \frac{2x^2}{(n+2)b} \sqrt{A^{n+2}} \left(1 + \frac{4S}{n+4}\left(1 + \frac{2S}{n+6}\right)\right)$$

$$\int x^2\sqrt[p]{A^n}\, dx = \frac{px^2}{(n+p)b} \sqrt[p]{A^{n+p}} \left(1 + \frac{2pS}{n+2p}\left(1 + \frac{pS}{n+3p}\right)\right)$$

$$\int x^3\sqrt{A}\, dx = \frac{2x^3}{3b} \sqrt{A^3} \left(1 + \frac{6S}{5}\left(1 + \frac{4S}{7}\left(1 + \frac{2S}{9}\right)\right)\right)$$

$$\int x^3\sqrt{A^n}\, dx = \frac{2x^3}{(n+2)b} \sqrt{A^{n+2}} \left(1 + \frac{6S}{n+4}\left(1 + \frac{4S}{n+6}\left(1 + \frac{2S}{n+8}\right)\right)\right)$$

$$\int x^3\sqrt[p]{A^n}\, dx = \frac{px^3}{(n+p)b} \sqrt[p]{A^{n+p}} \left(1 + \frac{3pS}{n+2p}\left(1 + \frac{2pS}{n+3p}\left(1 + \frac{pS}{n+4p}\right)\right)\right)$$

$$\int x^m\sqrt{A}\, dx = \frac{2x^m}{3b} \sqrt{A^3} \bigwedge_{k=0}^{m-1} \left[1 + \frac{2(m-k)S}{5+2k}\right]^*$$

$$\int x^m\sqrt{A^n}\, dx = \frac{2x^m}{(n+2)b} \sqrt{A^{n+2}} \bigwedge_{k=0}^{m-1} \left[1 + \frac{2(m-k)S}{n+4+2k}\right]^*$$

$$\int x^m\sqrt[p]{A^n}\, dx = \frac{px^m}{(n+p)b} \sqrt[p]{A^{n+p}} \bigwedge_{k=0}^{m-1} \left[1 + \frac{p(m-k)S}{n+2p+kp}\right]^*$$

---

*For nested sum refer to Sec. 8.11; S = pocket calculator storage.

(19) Indefinite Integrals Involving

$$f(x) = \frac{x^m}{\sqrt[p]{A^n}} \qquad A = a + bx \qquad a \neq 0$$

$$\int \frac{dx}{\sqrt{A}} = \frac{2\sqrt{A}}{b} \qquad\qquad S = -\frac{A}{bx}$$

$$\int \frac{dx}{\sqrt{A^n}} = \frac{-2}{(n-2)b\sqrt{A^{n-2}}} \qquad n \neq 2$$

$$\int \frac{dx}{\sqrt[p]{A^n}} = \frac{p}{(p-n)b\sqrt[p]{A^{n-p}}} \qquad p \neq n$$

$$\int \frac{x\,dx}{\sqrt{A}} = \frac{2x\sqrt{A}}{b}\left(1 + \frac{2S}{3}\right)$$

$$\int \frac{x\,dx}{\sqrt{A^n}} = \frac{-2x}{(n-2)b\sqrt{A^{n-2}}}\left(1 + \frac{2S}{4-n}\right) \qquad n \neq 2, 4$$

$$\int \frac{x\,dx}{\sqrt[p]{A^n}} = \frac{px}{(p-n)b\sqrt[p]{A^{n-p}}}\left(1 + \frac{pS}{2p-n}\right) \qquad n \neq p, 2p$$

$$\int \frac{x^2\,dx}{\sqrt{A}} = \frac{2x^2\sqrt{A}}{b}\left(1 + \frac{4S}{3}\left(1 + \frac{2S}{5}\right)\right)$$

$$\int \frac{x^2\,dx}{\sqrt{A^n}} = \frac{2x^2}{(2-n)b\sqrt{A^{n-2}}}\left(1 + \frac{4S}{4-n}\left(1 + \frac{2S}{6-n}\right)\right) \qquad n \neq 2, 4, 6$$

$$\int \frac{x^2\,dx}{\sqrt[p]{A^n}} = \frac{-px^2}{(n-p)b\sqrt[p]{A^{n-p}}}\left(1 + \frac{2pS}{2p-n}\left(1 + \frac{pS}{3p-n}\right)\right) \qquad n \neq p, 2p, 3p$$

$$\int \frac{x^3\,dx}{\sqrt{A}} = \frac{2x^3\sqrt{A}}{b}\left(1 + 2S\left(1 + \frac{4S}{5}\left(1 + \frac{2S}{7}\right)\right)\right)$$

$$\int \frac{x^3\,dx}{\sqrt{A^n}} = \frac{-2x^3}{(n-2)b\sqrt{A^{n-2}}}\left(1 + \frac{6S}{4-n}\left(1 + \frac{4S}{6-n}\left(1 + \frac{2S}{8-n}\right)\right)\right) \qquad n \neq 2, 4, 6, 8$$

$$\int \frac{x^3\,dx}{\sqrt[p]{A^n}} = \frac{px^3}{(p-n)b\sqrt[p]{A^{n-p}}}\left(1 + \frac{3pS}{2p-n}\left(1 + \frac{2pS}{3p-n}\left(1 + \frac{pS}{4p-n}\right)\right)\right) \qquad n \neq p, 2p, 3p, 4p$$

$$\int \frac{x^m\,dx}{\sqrt{A}} = \frac{2x^m\sqrt{A}^{\,m-1}}{b}\overset{m-1}{\underset{k=0}{\Lambda}}\left[1 + \frac{2(m-k)S}{3+2k}\right]^*$$

$$\int \frac{x^m\,dx}{\sqrt{A^n}} = \frac{-2x^m}{(n-2)b\sqrt{A^{n-2}}}\overset{m-1}{\underset{k=0}{\Lambda}}\left[1 + \frac{2(m-k)S}{4-n+2k}\right]^* \qquad n \neq 2, 4, 6, \ldots, (m+1)2$$

$$\int \frac{x^m\,dx}{\sqrt[p]{A^n}} = \frac{px^m}{(p-n)b\sqrt[p]{A^{n-p}}}\overset{m-1}{\underset{k=0}{\Lambda}}\left[1 + \frac{p(m-k)S}{2p-n+kp}\right]^* \qquad n \neq p, 2p, \ldots, (m+1)p$$

*For nested sum refer to Sec. 8.11; S = pocket calculator storage.

(20) Indefinite Integrals Involving

$$f(x) = \frac{1}{x^m \sqrt{A^n}} \qquad f(x) = \frac{\sqrt{A^n}}{x^m} \qquad A = a + bx \qquad a \neq 0$$

$$\int \frac{dx}{x\sqrt{A}} = \begin{cases} -\dfrac{2}{\sqrt{a}} \tanh^{-1} \sqrt{\dfrac{A}{a}} & a > A > 0 \\[3mm] -\dfrac{2}{\sqrt{a}} \coth^{-1} \sqrt{\dfrac{A}{a}} & A > a > 0 \\[3mm] \dfrac{2}{\sqrt{-a}} \tan^{-1} \sqrt{\dfrac{A}{-a}} & a < 0, A > 0 \\[3mm] \dfrac{1}{\sqrt{a}} \ln \dfrac{\sqrt{A}-\sqrt{a}}{\sqrt{A}+\sqrt{a}} & a > 0, A > 0 \end{cases}$$

$$\int \frac{dx}{x\sqrt{A^n}} = \frac{2}{(n-2)a\sqrt{A^{n-2}}} + \frac{1}{a}\int \frac{dx}{x\sqrt{A^{n-2}}} \qquad n > 2$$

$$\int \frac{dx}{x^2\sqrt{A}} = -\frac{\sqrt{A}}{ax} - \frac{b}{2a}\int \frac{dx}{x\sqrt{A}}$$

$$\int \frac{dx}{x^m\sqrt{A^n}} = -\frac{1}{a(m-1)x^{m-1}\sqrt{A^{n-2}}} - \frac{b(2m+n-4)}{2a(m-1)}\int \frac{dx}{x^{m-1}\sqrt{A^n}} \qquad m \neq 1$$

$$\int \frac{\sqrt{A}}{x}\,dx = 2\sqrt{A} + a\int \frac{dx}{x\sqrt{A}}$$

$$\int \frac{\sqrt{A^n}}{x}\,dx = \frac{2\sqrt{A^n}}{n} + a\int \frac{\sqrt{A^{n-2}}}{x}\,dx$$

$$\int \frac{\sqrt{A}}{x^2}\,dx = -\frac{\sqrt{A}}{x} + \frac{b}{2}\int \frac{dx}{x\sqrt{A}}$$

$$\int \frac{\sqrt{A^n}}{x^2}\,dx = -\frac{\sqrt{A^{n+2}}}{ax} + \frac{nb}{2a}\int \frac{\sqrt{A^n}}{x}\,dx$$

$$\int \frac{\sqrt{A^n}}{x^m}\,dx = -\frac{\sqrt{A^{n+2}}}{a(m-1)x^{m-1}} - \frac{b(2m-n-4)}{2a(m-1)}\int \frac{\sqrt{A^n}}{x^{m-1}}\,dx \qquad m \neq 1$$

$$\int \sqrt{\frac{x}{A}}\,dx = \frac{\sqrt{x}}{b}A - \frac{a}{b\sqrt{b}}\ln\,(A+\sqrt{bx})$$

$$\int \sqrt{\frac{A}{x}}\,dx = \sqrt{x}\,A + \frac{a}{\sqrt{b}}\ln\,(A+\sqrt{bx})$$

$$\int \sqrt{xA}\,dx = \frac{A+bx}{4b}\sqrt{xA} - \frac{a^2}{8b\sqrt{b}}\cosh^{-1}\frac{A+bx}{a}$$

(21) Indefinite Integrals Involving

$$f(x) = f(\sqrt{A}, \sqrt{D}) \qquad A = a + bx \qquad a \neq 0$$

$$D = e + fx \qquad e \neq 0$$

$$\int \frac{\sqrt{A}}{D} \, dx = \frac{2\sqrt{A}}{f} + \frac{\beta}{f} \int \frac{dx}{D\sqrt{A}} \qquad\qquad \beta = af - be \neq 0$$

$$\int \frac{A}{\sqrt{D}} \, dx = \frac{2\sqrt{D}}{3f} \left(A - \frac{2\beta}{f} \right)$$

$$\int \sqrt{\frac{A}{D}} \, dx = \frac{\sqrt{AD}}{f} - \frac{\beta}{2f} \int \frac{dx}{\sqrt{AD}}$$

$$\int \frac{\sqrt{A}}{D^n} \, dx = \frac{-1}{(n-1)f} \left(\frac{\sqrt{A}}{D^{n-1}} - \frac{b}{2} \int \frac{dx}{D^{n-1}\sqrt{A}} \right) \qquad n \neq 1$$

$$\int \frac{A^m}{\sqrt{D}} \, dx = \frac{2}{(2m+1)f} \left(A^m \sqrt{D} - m\beta \int \frac{A^{m-1}}{\sqrt{D}} \, dx \right)$$

$$\int \sqrt{AD} \, dx = \frac{\beta + 2bD}{4bf} \sqrt{AD} - \frac{\beta^2}{8bf} \int \frac{dx}{\sqrt{AD}}$$

$$\int A^m \sqrt{D} \, dx = \frac{1}{(2m+3)f} \left(2A^{m+1}\sqrt{D} + \beta \int \frac{A^m \, dx}{\sqrt{D}} \right)$$

$$\int \frac{dx}{\sqrt{AD}} = \begin{cases} \dfrac{2}{\sqrt{-bf}} \tan^{-1} \sqrt{-\dfrac{fA}{bD}} & bf < 0 \\[3mm] \dfrac{2}{\sqrt{bf}} \ln \left(\sqrt{fA} + \sqrt{bD} \right) & bf > 0 \end{cases}$$

$$\int \frac{dx}{D\sqrt{A}} = \begin{cases} \dfrac{2}{\sqrt{-\beta f}} \tan^{-1} \sqrt{\dfrac{fA}{-\beta}} & \beta f < 0 \\[3mm] \dfrac{1}{\sqrt{\beta f}} \ln \dfrac{\sqrt{fA} - \sqrt{\beta}}{\sqrt{fA} + \sqrt{\beta}} & \beta f > 0 \end{cases}$$

$$\int \frac{dx}{D^n \sqrt{A}} = -\frac{1}{(n-1)\beta} \left[\frac{\sqrt{A}}{D^{n-1}} + \left(n - \frac{3}{2} \right) b \int \frac{dx}{D^{n-1}\sqrt{A}} \right] \qquad n \neq 1$$

$$\int \sqrt{\frac{A}{a-bx}} \, dx = -\frac{1}{b} \sqrt{(a-bx)A} + \frac{a}{b} \sin^{-1} \frac{bx}{a}$$

$$\int \sqrt{\frac{a-bx}{A}} \, dx = \frac{1}{b} \sqrt{(a-bx)A} + \frac{a}{b} \sin^{-1} \frac{bx}{a}$$

$$\int \frac{dx}{\sqrt{xA}} = \frac{1}{\sqrt{b}} \cosh^{-1} \frac{A+bx}{a}$$

(22) Indefinite Integrals Involving

$$f(x) = x^m \sqrt{B^n} \qquad f(x) = \frac{x^m}{\sqrt{B^n}} \qquad B = a + bx + cx^2 \qquad a \neq 0$$

$$\omega = b + 2cx = dB/dx \qquad \gamma = 4ac - b^2 \qquad \lambda = \gamma/8c$$

$$\int \frac{dx}{\sqrt{B}} = \begin{cases} \dfrac{1}{\sqrt{c}} \ln(\omega + 2\sqrt{cB}) & c > 0 & \gamma \neq 0 \\[2mm] \dfrac{1}{\sqrt{c}} \sinh^{-1} \dfrac{\omega}{\sqrt{\gamma}} & c > 0 & \gamma > 0 \\[2mm] \dfrac{1}{\sqrt{c}} \ln \omega & c > 0 & \gamma = 0 \\[2mm] \dfrac{1}{\sqrt{c}} \cosh^{-1} \dfrac{\omega}{\sqrt{-\gamma}} & c > 0 & \gamma < 0 \\[2mm] \dfrac{-1}{\sqrt{-c}} \sin^{-1} \dfrac{\omega}{\sqrt{-\gamma}} & c < 0 & \gamma < 0 \end{cases}$$

$$\int \frac{dx}{\sqrt{B^3}} = \frac{2\omega}{\gamma\sqrt{B}} \qquad\qquad \int \frac{dx}{\sqrt{B^5}} = \frac{2\omega}{3\gamma\sqrt{B}}\left(\frac{1}{B} + \frac{1}{\lambda}\right)$$

$$\int \frac{dx}{\sqrt{B^n}} = \frac{2\omega}{(n-2)\gamma\sqrt{B^{n-2}}} + \frac{(n-3)}{2(n-2)\lambda} \int \frac{dx}{\sqrt{B^{n-2}}} \qquad n \neq 2$$

$$\int \sqrt{B}\, dx = \frac{\omega}{4c}\sqrt{B} + \lambda \int \frac{dx}{\sqrt{B}} \qquad\qquad \int \sqrt{B^3}\, dx = \frac{\omega}{8c}\sqrt{B^3} + \frac{3\lambda}{2} \int \sqrt{B}\, dx$$

$$\int \sqrt{B^n}\, dx = \frac{\omega\sqrt{B^n}}{2(n+1)c} + \frac{2n\lambda}{n+1} \int \sqrt{B^{n-2}}\, dx \qquad n \geqslant 0$$

$$\int x\sqrt{B}\, dx = \frac{\sqrt{B^3}}{3c} - \frac{b}{2c} \int \sqrt{B}\, dx \qquad\qquad \int x\sqrt{B^3}\, dx = \frac{\sqrt{B^5}}{5c} - \frac{b}{2c} \int \sqrt{B^3}\, dx$$

$$\int x^m \sqrt{B^n}\, dx = \frac{1}{(m+n+1)c}\left[x^{m-1}\sqrt{B^{n+2}} - \frac{(2m+n)b}{2} \int x^{m-1}\sqrt{B^n}\, dx \right] \qquad n > 0$$

$$\int \frac{x\, dx}{\sqrt{B}} = \frac{\sqrt{B}}{c} - \frac{b}{2c} \int \frac{dx}{\sqrt{B}} \qquad\qquad \int \frac{x\, dx}{\sqrt{B^3}} = -\frac{2\omega}{\gamma\sqrt{B}}$$

$$\int \frac{x^m\, dx}{\sqrt{B^n}} = \frac{1}{(m-n+1)c}\left[\frac{x^{m-1}}{\sqrt{B^{n-2}}} - \frac{(2m-n)b}{2} \int \frac{x^{m-1}}{\sqrt{B^n}}\, dx - (m-1)a \int \frac{x^{m-2}}{\sqrt{B^n}}\, dx \right] \qquad n > 0$$

$$\int \frac{e+fx}{\sqrt{B}}\, dx = \frac{f}{c}\sqrt{B} + \frac{2ce - bf}{2c} \int \frac{dx}{\sqrt{B}} \qquad\qquad \int \frac{b+2cx}{\sqrt{B}}\, dx = \int \frac{dB}{\sqrt{B}} = 2\sqrt{B}$$

(23) Indefinite Integrals Involving

$$f(x) = \frac{1}{x^m \sqrt{B^n}} \qquad f(x) = \frac{\sqrt{B^n}}{x^m} \qquad B = a + bx + cx^2 \qquad a \neq 0$$

$$\omega = b + 2cx = dB/dx \qquad \gamma = 4ac - b^2 \qquad \theta = 2a + bx$$

$$\int \frac{dx}{x\sqrt{B}} = \begin{cases} \dfrac{-1}{\sqrt{a}} \ln \dfrac{\theta + 2\sqrt{aB}}{x} & a > 0 \qquad \gamma \neq 0 \\[2ex] \dfrac{-1}{\sqrt{a}} \sinh^{-1} \dfrac{\theta}{x\sqrt{\gamma}} & a > 0 \qquad \gamma > 0 \\[2ex] \dfrac{-1}{\sqrt{a}} \ln \dfrac{\theta}{x} & a > 0 \qquad \gamma = 0 \\[2ex] \dfrac{1}{\sqrt{-a}} \sin^{-1} \dfrac{\theta}{x\sqrt{-\gamma}} & a < 0 \qquad \gamma < 0 \\[2ex] \dfrac{1}{\sqrt{-a}} \tan^{-1} \dfrac{\theta}{2\sqrt{-aB}} & a < 0 \qquad \gamma \leq 0 \end{cases}$$

$$\int \frac{dx}{x^2 \sqrt{B}} = -\frac{\sqrt{B}}{ax} - \frac{b}{2a} \int \frac{dx}{x\sqrt{B}} \qquad\qquad \int \frac{dx}{x^3 \sqrt{B}} = \frac{\sqrt{B}}{a^2 x^2}\left(bx - \frac{\theta}{4}\right) + \frac{2b^2 - \gamma}{8a^2} \int \frac{dx}{x\sqrt{B}}$$

$$\int \frac{dx}{x^m \sqrt{B}} = -\frac{\sqrt{B}}{(m-1)ax^{m-1}} - \frac{b(2m-3)}{2a(m-1)} \int \frac{dx}{x^{m-1}\sqrt{B}} - \frac{c(m-2)}{a(m-1)} \int \frac{dx}{x^{m-2}\sqrt{B}} \qquad m \neq 1$$

$$\int \frac{dx}{x\sqrt{B^n}} = \frac{1}{(n-2)a\sqrt{B^{n-2}}} - \frac{b}{2a} \int \frac{dx}{\sqrt{B^n}} + \frac{1}{a} \int \frac{dx}{x\sqrt{B^{n-2}}} \qquad n \neq 2$$

$$\int \frac{\sqrt{B}}{x} dx = \sqrt{B} + \frac{b}{2} \int \frac{dx}{\sqrt{B}} + a \int \frac{dx}{x\sqrt{B}} \qquad\qquad \int \frac{\sqrt{B}}{x^2} dx = -\frac{\sqrt{B}}{x} + \frac{b}{2} \int \frac{dx}{x\sqrt{B}} + c \int \frac{dx}{\sqrt{B}}$$

$$\int \frac{\sqrt{B}}{x^m} dx = -\frac{\sqrt{B}}{(m-1)x^{m-1}} + \frac{b}{2(m-1)} \int \frac{dx}{x^{m-1}\sqrt{B}} + \frac{c}{m-1} \int \frac{dx}{x^{m-2}\sqrt{B}} \qquad m \neq 1$$

$$\int \frac{\sqrt{B^n}}{x^m} dx = -\frac{\sqrt{B^{n+2}}}{a(m-1)x^{m-1}} + \frac{b(n-2m+4)}{2a(m+1)} \int \frac{\sqrt{B^n}}{x^{m-1}} dx + \frac{c(n-m+3)}{a(m-1)} \int \frac{\sqrt{B^n}}{x^{m-2}} dx \qquad m \neq 1$$

$$\int \frac{dx}{x^m \sqrt{B^n}} = \frac{-1}{a(m-1)x^{m-1}\sqrt{B^{n-2}}} - \frac{b(n+2m-4)}{2a(m-1)} \int \frac{dx}{x^{m-1}\sqrt{B^n}}$$

$$- \frac{c(n+m-3)}{a(m-1)} \int \frac{dx}{x^{m-2}\sqrt{B^n}} \qquad m \neq 1$$

$$\int \frac{dx}{(x-f)\sqrt{B}} = -\int \frac{dz}{\sqrt{c + (b+2cf)z + (a+bf+cf^2)z^2}} \qquad z = \frac{1}{x-f}$$

(24) Indefinite Integrals Involving

$$f(x) = x^m \sqrt[p]{C^n} \qquad\qquad C = a^2 + x^2 \qquad a^2 \neq 0$$

$$\int \sqrt{C}\, dx = \tfrac{1}{2}[x\sqrt{C} + a^2 \ln(x + \sqrt{C})]$$

$$\int \sqrt{C^3}\, dx = \frac{x\sqrt{C^3}}{4} + \frac{3a^2}{4}\int \sqrt{C}\, dx$$

$$\int \sqrt{C^n}\, dx = \frac{x\sqrt{C^n}}{n+1} + \frac{na^2}{n+1}\int \sqrt{C^{n-2}}\, dx \qquad n \neq -1$$

$$\int x\sqrt{C}\, dx = \tfrac{1}{3}\sqrt{C^3}$$

$$\int x\sqrt{C^3}\, dx = \tfrac{1}{5}\sqrt{C^5}$$

$$\int x\sqrt{C^n}\, dx = \frac{1}{n+2}\sqrt{C^{n+2}} \qquad n \neq -2$$

$$\int x^2\sqrt{C}\, dx = \frac{x}{4}\sqrt{C^3} - \frac{a^2x}{8}\sqrt{C} - \frac{a^4}{8}\ln(x + \sqrt{C})$$

$$\int x^2\sqrt{C^3}\, dx = \frac{x}{6}\sqrt{C^5} - \frac{a^2x}{24}\sqrt{C^3} - \frac{a^4x}{16}\sqrt{C} - \frac{a^6}{16}\ln(x + \sqrt{C})$$

$$\int x^2\sqrt{C^n}\, dx = \frac{x\sqrt{C^{n+2}}}{n+3} - \frac{a^2}{n+3}\int \sqrt{C^n} \qquad n \neq -3$$

$$\int x^3\sqrt{C}\, dx = \frac{1}{5}\sqrt{C^5} - \frac{a^2}{3}\sqrt{C^3}$$

$$\int x^3\sqrt{C^3}\, dx = \frac{1}{7}\sqrt{C^7} - \frac{a^2}{5}\sqrt{C^5}$$

$$\int x^3\sqrt{C^n} = \left(x^2 - \frac{2a^2}{n+2}\right)\frac{\sqrt{C^{n+2}}}{n+4} \qquad n \neq -2, -4$$

$$\int x^m\sqrt[p]{C^n}\, dx = \frac{px^{m-1}\sqrt[p]{C^{n+p}}}{2n + mp + p} - \frac{a^2p(m-1)}{2n + mp + p}\int x^{m-2}\sqrt[p]{C^n}\, dx \qquad 2n \neq -p(1 + m)$$

$$\int \sqrt{ax + bx^2}\, dx = \frac{a + 2bx}{4b}\sqrt{ax + bx^2} - \frac{a^2}{8b\sqrt{b}}\cosh^{-1}\frac{a + 2bx}{a}$$

$$\int x^m\sqrt{ax + bx^2}\, dx = \frac{x^{m-1}}{b(m+2)}\sqrt{(ax + bx^2)^3} - \frac{a(2m+1)}{2b(m+2)}\int x^{m-1}\sqrt{ax + bx^2}\, dx$$

(25) Indefinite Integrals Involving $\qquad f(x) = \dfrac{x^m}{\sqrt{C^n}} \qquad C = a^2 + x^2 \qquad a^2 \neq 0$

$$\int \frac{dx}{\sqrt{C}} = \sinh^{-1}\frac{x}{a} = \ln\,(x + \sqrt{C})$$

$$\int \frac{dx}{\sqrt{C^3}} = \frac{x}{a^2\sqrt{C}}$$

$$\int \frac{dx}{\sqrt{C^n}} = \frac{x}{a^2(n-2)\sqrt{C^{n-2}}} - \frac{3-n}{a^2(n-2)}\int \frac{dx}{\sqrt{C^{n-2}}} \qquad n \neq 2$$

$$\int \frac{x\,dx}{\sqrt{C}} = \sqrt{C}$$

$$\int \frac{x\,dx}{\sqrt{C^3}} = \frac{-1}{\sqrt{C}}$$

$$\int \frac{x\,dx}{\sqrt{C^n}} = \frac{x^2}{a^2(n-2)\sqrt{C^{n-2}}} - \frac{4-n}{a^2(n-2)}\int \frac{x\,dx}{\sqrt{C^{n-2}}} \qquad n \neq 2$$

$$\int \frac{x^2\,dx}{\sqrt{C}} = \frac{x}{2}\sqrt{C} - \frac{a^2}{2}\ln\,(x + \sqrt{C})$$

$$\int \frac{x^2\,dx}{\sqrt{C^3}} = \frac{-x}{\sqrt{C}} + \ln\,(x + \sqrt{C})$$

$$\int \frac{x^2\,dx}{\sqrt{C^n}} = \frac{x^3}{a^2(n-2)\sqrt{C^{n-2}}} - \frac{5-n}{a^2(n-2)}\int \frac{x^2\,dx}{\sqrt{C^{n-2}}} \qquad n \neq 2$$

$$\int \frac{x^3\,dx}{\sqrt{C}} = \left(\frac{C}{3} - a^2\right)\sqrt{C}$$

$$\int \frac{x^3\,dx}{\sqrt{C^3}} = \frac{C + a^2}{\sqrt{C}}$$

$$\int \frac{x^3\,dx}{\sqrt{C^n}} = \frac{1}{4-n}\left(x^2 + \frac{2a^2}{n-2}\right)\frac{1}{\sqrt{C^{n-2}}} \qquad n \neq 2,4$$

$$\int \frac{x^m}{\sqrt[p]{C^n}}\,dx = \frac{p\,x^{m+1}}{2a^2(n-p)\sqrt[p]{C^{n-p}}} + \frac{2n - p(m+3)}{2a^2(n-p)}\int \frac{x^m\,dx}{\sqrt[p]{C^{n-p}}} \qquad n \neq p$$

$$\int \frac{dx}{\sqrt{ax + bx^2}} = \frac{1}{\sqrt{b}}\cosh^{-1}\frac{a + 2bx}{a}$$

$$\int \frac{x^m\,dx}{\sqrt{ax + bx^2}} = \frac{x^{m-1}}{bm}\sqrt{ax + bx^2} - \frac{a(2m-1)}{2bm}\int \frac{x^{m-1}\,dx}{\sqrt{ax + bx^2}}$$

(26) Indefinite Integrals Involving

$$f(x) = \frac{1}{x^m \sqrt[p]{C^n}} \qquad C = a^2 + x^2 \qquad a^2 \ne 0$$

$$\int \frac{dx}{x\sqrt{C}} = -\frac{1}{a} \ln \frac{a + \sqrt{C}}{x}$$

$$\int \frac{dx}{x\sqrt{C^3}} = \frac{1}{a^2\sqrt{C}} - \frac{1}{a^3} \ln \frac{a + \sqrt{C}}{x}$$

$$\int \frac{dx}{x\sqrt{C^n}} = \frac{1}{a^2(n-2)\sqrt{C^{n-2}}} + \frac{1}{a^2} \int \frac{dx}{x\sqrt{C^{n-2}}} \qquad n \ne 2$$

$$\int \frac{dx}{x^2\sqrt{C}} = -\frac{\sqrt{C}}{a^2 x}$$

$$\int \frac{dx}{x^2\sqrt{C^3}} = -\frac{\sqrt{C}}{a^4 x}\left(1 + \frac{x^2}{x}\right)$$

$$\int \frac{dx}{x^2\sqrt{C^n}} = \frac{1}{a^2(n-2)x\sqrt{C^{n-2}}} + \frac{n-1}{a^2(n-2)} \int \frac{dx}{x^2\sqrt{C^{n-2}}} \qquad n \ne 2$$

$$\int \frac{dx}{x^3\sqrt{C}} = -\frac{\sqrt{C}}{2a^2 x^2} + \frac{1}{2a^3} \ln \frac{a + \sqrt{C}}{x}$$

$$\int \frac{dx}{x^3\sqrt{C^3}} = \frac{-1}{2a^2 x^2 \sqrt{C}} - \frac{3}{2a^4\sqrt{C}} + \frac{3}{2a^5} \ln \frac{a + \sqrt{C}}{x}$$

$$\int \frac{dx}{x^3\sqrt{C^n}} = \frac{1}{a^2(n-2)x^2\sqrt{C^{n-2}}} + \frac{n}{a^2(n-2)} \int \frac{dx}{x^3\sqrt{C^{n-2}}} \qquad n \ne 2$$

$$\int \frac{dx}{x^4\sqrt{C}} = \frac{\sqrt{C}}{a^4 x}\left(1 - \frac{C}{3x^2}\right)$$

$$\int \frac{dx}{x^4\sqrt{C^3}} = \frac{x}{a^6\sqrt{C}}\left(1 + \frac{2C}{x^2} - \frac{C^2}{3x^4}\right)$$

$$\int \frac{dx}{x^4\sqrt{C^n}} = \frac{1}{a^2(n-2)x^3\sqrt{C^{n-2}}} + \frac{n+1}{a^2(n-2)} \int \frac{dx}{x^4\sqrt{C^{n-2}}} \qquad n \ne 2$$

$$\int \frac{dx}{x^m\sqrt[p]{C^n}} = \frac{p}{2a^2(n-p)x^{m-1}\sqrt[p]{C^{n-p}}} + \frac{2n+p(m-3)}{2a^2(n-p)} \int \frac{dx}{x^m\sqrt[p]{C^{n-p}}} \qquad n \ne p$$

$$\int \frac{dx}{x\sqrt{ax+bx^2}} = -\frac{2}{ax}\sqrt{ax+bx^2}$$

$$\int \frac{dx}{x^m\sqrt{ax+bx^2}} = -\frac{2\sqrt{ax+bx^2}}{a(2m-1)x^m} - \frac{2b(m-1)}{a(2m-1)} \int \frac{dx}{x^{m-1}\sqrt{ax+bx^2}}$$

(27) Indefinite Integrals Involving
$$f(x) = \frac{\sqrt[p]{C^n}}{x^m} \qquad C = a^2 + x^2 \qquad a^2 \neq 0$$

$$\int \frac{\sqrt{C}}{x}\,dx = \sqrt{C} - a\ln\frac{a + \sqrt{C}}{x}$$

$$\int \frac{\sqrt{C^3}}{x}\,dx = \frac{1}{3}\sqrt{C^3} + a^2\sqrt{C} - a^3\ln\frac{a + \sqrt{C}}{x}$$

$$\int \frac{\sqrt{C^n}}{x}\,dx = \frac{1}{n}\sqrt{C^n} + a^2\int \frac{\sqrt{C^{n-2}}}{x}\,dx$$

$$\int \frac{\sqrt{C}}{x^2}\,dx = -\frac{1}{x}\sqrt{C} + \ln\,(x + \sqrt{C})$$

$$\int \frac{\sqrt{C^3}}{x^2}\,dx = -\frac{1}{x}\sqrt{C^3} + \frac{3x}{2}\sqrt{C} + \frac{3a^2}{2}\ln\,(x + \sqrt{C})$$

$$\int \frac{\sqrt{C^n}}{x^2}\,dx = \frac{\sqrt{C^n}}{(n-1)x} + \frac{a^2 n}{n-1}\int \frac{\sqrt{C^{n-2}}}{x^2}\,dx \qquad n \neq 1$$

$$\int \frac{\sqrt{C}}{x^3}\,dx = -\frac{1}{2x^2}\sqrt{C} - \frac{1}{2a}\ln\frac{a + \sqrt{C}}{x}$$

$$\int \frac{\sqrt{C^3}}{x^3}\,dx = -\frac{1}{2x^2}\sqrt{C^3} + \frac{3}{2}\sqrt{C} - \frac{3a}{2}\ln\frac{a + \sqrt{C}}{x}$$

$$\int \frac{\sqrt{C^n}}{x^3}\,dx = \frac{\sqrt{C^n}}{(n-2)x^2} + \frac{a^2 n}{n-2}\int \frac{\sqrt{C^{n-2}}}{x^3}\,dx \qquad n \neq 2$$

$$\int \frac{\sqrt{C}}{x^4}\,dx = -\frac{\sqrt{C^3}}{3a^2 x^3}$$

$$\int \frac{\sqrt{C^3}}{x^4}\,dx = -\frac{\sqrt{C^3}}{3x^3} - \frac{\sqrt{C}}{x} + \ln\,(x + \sqrt{C})$$

$$\int \frac{\sqrt{C^n}}{x^4}\,dx = \frac{\sqrt{C^n}}{(n-3)x^3} + \frac{a^2 n}{n-3}\int \frac{\sqrt{C^{n-2}}}{x^4}\,dx \qquad n \neq 3$$

$$\int \frac{\sqrt[p]{C^n}}{x^m}\,dx = \frac{p\sqrt[p]{C^n}}{(2n - mp + p)x^{m-1}} + \frac{2a^2 n}{2n - mp + p}\int \frac{\sqrt[p]{C^{n-p}}}{x^m}\,dx \qquad 2n \neq -p(1 - m)$$

$$\int \frac{\sqrt{ax + bx^2}}{x}\,dx = \sqrt{ax + bx^2} + \frac{a}{2\sqrt{b}}\cosh^{-1}\frac{a + 2bx}{a}$$

$$\int \frac{\sqrt{ax + bx^2}}{x^m}\,dx = -\frac{2\sqrt{(ax + bx^2)^3}}{a(2m - 3)x^{m+1}} - \frac{2b(m - 3)}{a(2m - 3)}\int \frac{\sqrt{ax + bx^2}}{x^{m-1}}\,dx \qquad m \neq 1$$

(28) Indefinite Integrals Involving $\qquad$ $f(x) = x^m \sqrt[p]{E^n}$ $\qquad$ $E = a^2 - x^2$ $\qquad$ $a^2 \neq 0$

$$\int \sqrt{E}\, dx = \frac{1}{2}\left(x\sqrt{E} + a^2 \sin^{-1}\frac{x}{a}\right)$$

$$\int \sqrt{E^3}\, dx = \frac{1}{8}\left(2x\sqrt{E^3} + 3a^2 x\sqrt{E} + 3a^4 \sin^{-1}\frac{x}{a}\right)$$

$$\int \sqrt{E^n}\, dx = \frac{x\sqrt{E^n}}{n+1} + \frac{a^2 n}{n+1}\int \sqrt{E^{n-2}}\, dx \qquad n \neq -1$$

$$\int x\sqrt{E}\, dx = -\tfrac{1}{3}\sqrt{E^3}$$

$$\int x\sqrt{E^3}\, dx = -\tfrac{1}{5}\sqrt{E^5}$$

$$\int x\sqrt{E^n}\, dx = \frac{x^2\sqrt{E^n}}{n+2} + \frac{a^2 n}{n+2}\int x\sqrt{E^{n-2}}\, dx \qquad n \neq -2$$

$$\int x^2\sqrt{E}\, dx = -\frac{x}{4}\sqrt{E^3} + \frac{a^2}{8}\left(x\sqrt{E} + a^2 \sin^{-1}\frac{x}{a}\right)$$

$$\int x^2\sqrt{E^3}\, dx = -\frac{x}{6}\sqrt{E^5} + \frac{a^2 x}{24}\sqrt{E^3} + \frac{a^4}{16}\left(x\sqrt{E} + a^2 \sin^{-1}\frac{x}{a}\right)$$

$$\int x^2\sqrt{E^n}\, dx = \frac{x^3\sqrt{E^n}}{n+3} + \frac{a^2 n}{n+3}\int x^2\sqrt{E^{n-2}}\, dx \qquad n \neq -3$$

$$\int x^3\sqrt{E}\, dx = \frac{1}{5}\sqrt{E^5} - \frac{a^2}{3}\sqrt{E^3}$$

$$\int x^3\sqrt{E^3}\, dx = \frac{1}{7}\sqrt{E^7} - \frac{a^2}{5}\sqrt{E^5}$$

$$\int x^3\sqrt{E^n}\, dx = \frac{x^4\sqrt{E^n}}{n+4} + \frac{a^2 n}{n+4}\int x^3\sqrt{E^{n-2}}\, dx \qquad n \neq -4$$

$$\int x^m\sqrt[p]{E^n}\, dx = \frac{p x^{m+1}\sqrt[p]{E^n}}{2n + mp + p} + \frac{2a^2 n}{2n + mp + p}\int x^m\sqrt[p]{E^{n-p}}\, dx \qquad 2n \neq -p(1+m)$$

$$\int \sqrt{ax - bx^2}\, dx = \frac{2bx - a}{4b}\sqrt{ax - bx^2} + \frac{a^2}{4b\sqrt{b}}\sin^{-1}\sqrt{\frac{bx}{a}}$$

$$\int x^m\sqrt{ax - bx^2}\, dx = -\frac{x^{m-1}}{b(m+2)}\sqrt{(ax - bx^2)^3} + \frac{a(2m+1)}{2b(m+2)}\int x^{m-1}\sqrt{ax - bx^2}\, dx$$

(29) Indefinite Integrals Involving

$$f(x) = \frac{x^m}{\sqrt[p]{E^n}} \qquad E = a^2 - x^2 \qquad a^2 \neq 0$$

$$\int \frac{dx}{\sqrt{E}} = \sin^{-1}\frac{x}{a}$$

$$\int \frac{dx}{\sqrt{E^3}} = \frac{x}{a^2\sqrt{E}}$$

$$\int \frac{dx}{\sqrt{E^n}} = \frac{x}{a^2(n-2)\sqrt{E^{n-2}}} + \frac{n-3}{a^2(n-2)} \int \frac{dx}{\sqrt{E^{n-2}}} \qquad n \neq 2$$

$$\int \frac{x\,dx}{\sqrt{E}} = -\sqrt{E}$$

$$\int \frac{x\,dx}{\sqrt{E^3}} = \frac{1}{\sqrt{E}}$$

$$\int \frac{x\,dx}{\sqrt{E^n}} = \frac{x^2}{a^2(n-2)\sqrt{E^{n-2}}} + \frac{n-4}{a^2(n-2)} \int \frac{x\,dx}{\sqrt{E^{n-2}}} \qquad n \neq 2$$

$$\int \frac{x^2\,dx}{\sqrt{E}} = -\frac{x}{2}\sqrt{E} + \frac{a^2}{2}\sin^{-1}\frac{x}{a}$$

$$\int \frac{x^2\,dx}{\sqrt{E^3}} = \frac{x}{\sqrt{E}} - \sin^{-1}\frac{x}{a}$$

$$\int \frac{x^2\,dx}{\sqrt{E^n}} = \frac{x^3}{a^2(n-2)\sqrt{E^{n-2}}} + \frac{n-5}{a^2(n-2)} \int \frac{x^2\,dx}{\sqrt{E^{n-2}}} \qquad n \neq 2$$

$$\int \frac{x^3\,dx}{\sqrt{E}} = \left(\frac{E}{3} - a^2\right)\sqrt{E}$$

$$\int \frac{x^3\,dx}{\sqrt{E^3}} = \frac{E + a^2}{\sqrt{E}}$$

$$\int \frac{x^3\,dx}{\sqrt{E^n}} = \frac{x^4}{a^2(n-2)\sqrt{E^{n-2}}} + \frac{n-6}{a^2(n-2)} \int \frac{x^3\,dx}{\sqrt{E^{n-2}}} \qquad n \neq 2$$

$$\int \frac{x^m\,dx}{\sqrt[p]{E^n}} = \frac{px^{m+1}}{2a^2(n-p)\sqrt[p]{E^{n-p}}} + \frac{2n - p(m+3)}{2a^2(n-p)} \int \frac{x^m\,dx}{\sqrt[p]{E^{n-p}}} \qquad n \neq p$$

$$\int \frac{dx}{\sqrt{ax - bx^2}} = \frac{2}{\sqrt{b}}\sin^{-1}\sqrt{\frac{bx}{a}}$$

$$\int \frac{x^m\,dx}{\sqrt{ax - bx^2}} = -\frac{x^{m-1}}{bm}\sqrt{ax - bx^2} + \frac{(2m-1)a}{2\,bm} \int \frac{x^{m-1}\,dx}{\sqrt{ax - bx^2}}$$

(30) Indefinite Integrals Involving $\qquad f(x) = \dfrac{1}{x^m\sqrt[p]{E^n}} \qquad E = a^2 - x^2 \qquad a^2 \neq 0$

$$\int \frac{dx}{x\sqrt{E}} = -\frac{1}{a}\ln\frac{a+\sqrt{E}}{x}$$

$$\int \frac{dx}{x\sqrt{E^3}} = \frac{1}{a^2\sqrt{E}} - \frac{1}{a^3}\ln\frac{a+\sqrt{E}}{x}$$

$$\int \frac{dx}{x\sqrt{E^n}} = \frac{1}{a^2(n-2)\sqrt{E^{n-2}}} + \frac{n-2}{a^2(n-2)}\int \frac{dx}{x\sqrt{E^{n-2}}} \qquad n \neq 2$$

$$\int \frac{dx}{x^2\sqrt{E}} = -\frac{\sqrt{E}}{a^2 x}$$

$$\int \frac{dx}{x^2\sqrt{E^3}} = -\frac{\sqrt{E}}{a^4 x}\left(1 - \frac{x^2}{E}\right)$$

$$\int \frac{dx}{x^2\sqrt{E^n}} = \frac{1}{a^2(n-2)x\sqrt{E^{n-2}}} + \frac{n-1}{a^2(n-2)}\int \frac{dx}{x^2\sqrt{E^{n-2}}} \qquad n \neq 2$$

$$\int \frac{dx}{x^3\sqrt{E}} = -\frac{\sqrt{E}}{2a^2 x^2} - \frac{1}{2a^3}\ln\frac{a+\sqrt{E}}{x}$$

$$\int \frac{dx}{x^3\sqrt{E^3}} = \frac{-1}{2a^2 x^2\sqrt{E}} + \frac{3}{2a^4\sqrt{E}} - \frac{3}{2a^5}\ln\frac{a+\sqrt{E}}{x}$$

$$\int \frac{dx}{x^3\sqrt{E^n}} = \frac{1}{a^2(n-2)x^2\sqrt{E^{n-2}}} + \frac{n}{a^2(n-2)}\int \frac{dx}{x^3\sqrt{E^{n-2}}} \qquad n \neq 2$$

$$\int \frac{dx}{x^4\sqrt{E}} = -\frac{\sqrt{E}}{a^4 x}\left(1 + \frac{E}{3x^2}\right)$$

$$\int \frac{dx}{x^4\sqrt{E^3}} = \frac{1}{a^6\sqrt{E}}\left(x - \frac{2E}{x} - \frac{E^2}{3x^3}\right)$$

$$\int \frac{dx}{x^4\sqrt{E^n}} = \frac{1}{a^2(n-2)x^3\sqrt{E^{n-2}}} + \frac{n+1}{a^2(n-2)}\int \frac{dx}{x^4\sqrt{E^{n-2}}} \qquad n \neq 2$$

$$\int \frac{dx}{x^m\sqrt[p]{E^n}} = \frac{p}{2a^2(n-p)x^{m-1}\sqrt[p]{E^{n-p}}} + \frac{2n+p(m-3)}{2a^2(n-p)}\int \frac{dx}{x^m\sqrt[p]{E^{n-p}}} \qquad n \neq p$$

$$\int \frac{dx}{x\sqrt{ax-bx^2}} = -\frac{2}{ax}\sqrt{ax-bx^2}$$

$$\int \frac{dx}{x^m\sqrt{ax-bx^2}} = -\frac{2\sqrt{ax-bx^2}}{a(2m-1)x^m} + \frac{2b(m-1)}{a(2m-1)}\int \frac{dx}{x^{m-1}\sqrt{ax-bx^2}}$$

(31) Indefinite Integrals Involving $\quad f(x) = \dfrac{\sqrt[p]{E^n}}{x^m} \qquad E = a^2 - x^2 \qquad a^2 \neq 0$

$$\int \frac{\sqrt{E}}{x}\,dx = \sqrt{E} - a\ln\frac{a + \sqrt{E}}{x}$$

$$\int \frac{\sqrt{E^3}}{x}\,dx = \frac{1}{3}\sqrt{E^3} + a^2\sqrt{E} - a^3\ln\frac{a + \sqrt{E}}{x}$$

$$\int \frac{\sqrt{E^n}}{x}\,dx = \frac{1}{n}\sqrt{E^n} + a^2\int \frac{\sqrt{E^{n-2}}}{x}\,dx$$

$$\int \frac{\sqrt{E}}{x^2}\,dx = -\frac{1}{x}\sqrt{E} - \sin^{-1}\frac{x}{a}$$

$$\int \frac{\sqrt{E^3}}{x^2}\,dx = -\frac{1}{x}\sqrt{E^3} - \frac{3x}{2}\sqrt{E} - \frac{3a^2}{2}\sin^{-1}\frac{x}{a}$$

$$\int \frac{\sqrt{E^n}}{x^2}\,dx = \frac{\sqrt{E^n}}{(n-1)x} + \frac{a^2 n}{n-1}\int \frac{\sqrt{E^{n-2}}}{x^2}\,dx \qquad n \neq 1$$

$$\int \frac{\sqrt{E}}{x^3}\,dx = -\frac{1}{2x^2}\sqrt{E} + \frac{1}{2a}\ln\frac{a + \sqrt{E}}{x}$$

$$\int \frac{\sqrt{E^3}}{x^3}\,dx = -\frac{1}{2x^2}\sqrt{E^3} - \frac{3}{2}\sqrt{E} + \frac{3a}{2}\ln\frac{a + \sqrt{E}}{x}$$

$$\int \frac{\sqrt{E^n}}{x^3}\,dx = \frac{\sqrt{E^n}}{(n-2)x^2} + \frac{a^2 n}{n-2}\int \frac{\sqrt{E^{n-2}}}{x^3}\,dx \qquad n \neq 2$$

$$\int \frac{\sqrt{E}}{x^4}\,dx = -\frac{1}{3a^2 x^3}\sqrt{E^3}$$

$$\int \frac{\sqrt{E^3}}{x^4}\,dx = -\frac{1}{3x^3}\sqrt{E^3} + \frac{1}{x}\sqrt{E} + \sin^{-1}\frac{x}{a}$$

$$\int \frac{\sqrt{E^n}}{x^4}\,dx = \frac{\sqrt{E^n}}{(n-3)x^3} + \frac{a^2 n}{n-3}\int \frac{\sqrt{E^{n-2}}}{x^4}\,dx \qquad n \neq 3$$

$$\int \frac{\sqrt[p]{E^n}}{x^m}\,dx = \frac{p\sqrt[p]{E^n}}{(2n - mp + p)x^{m-1}} + \frac{2a^2 n}{2n - mp + p}\int \frac{\sqrt[p]{E^{n-p}}}{x^m}\,dx \qquad 2n \neq -p(1 - m)$$

$$\int \frac{\sqrt{ax - bx^2}}{x}\,dx = \sqrt{ax - bx^2} + \frac{a}{\sqrt{b}}\sin^{-1}\sqrt{\frac{bx}{a}}$$

$$\int \frac{\sqrt{ax - bx^2}}{x^m}\,dx = \frac{2\sqrt{(ax - bx^2)^3}}{a(3 - 2m)x^m} + \frac{2b(3 - m)}{a(3 - 2m)}\int \frac{\sqrt{ax - bx^2}}{x^{m-1}}\,dx$$

(32) Indefinite Integrals Involving $\qquad$ $f(x) = x^m \sqrt[p]{F^n}$ $\qquad$ $F = x^2 - a^2$ $\qquad$ $a^2 \neq 0$

$$\int \sqrt{F}\, dx = \frac{x}{2}\sqrt{F} - \frac{a^2}{2}\ln\left(x + \sqrt{F}\right)$$

$$\int \sqrt{F^3}\, dx = \frac{x\sqrt{F^3}}{4} - \frac{3a^2}{4}\int \sqrt{F}\, dx$$

$$\int \sqrt{F^n}\, dx = \frac{x\sqrt{F^n}}{n+1} - \frac{a^2 n}{n+1}\int \sqrt{F^{n-2}}\, dx \qquad n \neq -1$$

$$\int x\sqrt{F}\, dx = \frac{1}{3}\sqrt{F^3}$$

$$\int x\sqrt{F^3}\, dx = \frac{1}{5}\sqrt{F^5}$$

$$\int x\sqrt{F^n}\, dx = \frac{x^2\sqrt{F^n}}{n+2} - \frac{a^2 n}{n+2}\int x\sqrt{F^{n-2}}\, dx \qquad n \neq -2$$

$$\int x^2\sqrt{F}\, dx = \frac{x}{4}\sqrt{F^3} + \frac{a^2 x}{8}\sqrt{F} - \frac{a^4}{8}\ln\left(x + \sqrt{F}\right)$$

$$\int x^2\sqrt{F^3}\, dx = \frac{x}{6}\sqrt{F^5} + \frac{a^2 x}{24}\sqrt{F^3} - \frac{a^4 x}{16}\sqrt{F} + \frac{a^6}{16}\ln\left(x + \sqrt{F}\right)$$

$$\int x^2\sqrt{F^n}\, dx = \frac{x^3\sqrt{F^n}}{n+3} - \frac{a^2 n}{n+3}\int x^2\sqrt{F^{n-2}}\, dx \qquad n \neq -3$$

$$\int x^3\sqrt{F}\, dx = \frac{1}{5}\sqrt{F^5} + \frac{a^2}{3}\sqrt{F^3}$$

$$\int x^3\sqrt{F^3}\, dx = \frac{1}{7}\sqrt{F^7} + \frac{a^2}{5}\sqrt{F^5}$$

$$\int x^3\sqrt{F^n}\, dx = \frac{x^4\sqrt{E^n}}{n+4} - \frac{a^2 n}{n+4}\int x^3\sqrt{F^{n-2}}\, dx \qquad n \neq -4$$

$$\int x^m\sqrt[p]{F^n}\, dx = \frac{px^{m+1}\sqrt[p]{F^n}}{2n + mp + p} - \frac{2a^2 n}{2n + mp + p}\int x^m\sqrt[p]{F^{n-p}}\, dx \qquad 2n \neq -p(m+1)$$

$$\int \sqrt{ax^2 - bx}\, dx = \frac{2ax - b}{4a}\sqrt{ax^2 - bx} - \frac{b^2}{8a\sqrt{a}}\ln\left(2ax - b + 2\sqrt{a}\sqrt{ax^2 - bx}\right)$$

$$\int x^m\sqrt{ax^2 - bx}\, dx = \frac{x^{m-1}}{a(m+2)}\sqrt{(ax^2 - bx)^3} + \frac{b(2m+1)}{2a(m+1)}\int x^{m-1}\sqrt{ax^2 - bx}\, dx$$

(33) Indefinite Integrals Involving

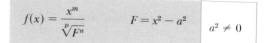

$$f(x) = \frac{x^m}{\sqrt[p]{F^n}} \qquad F = x^2 - a^2 \qquad a^2 \neq 0$$

$$\int \frac{dx}{\sqrt{F}} = \ln\left(x + \sqrt{F}\right) = \cosh^{-1}\frac{x}{a}$$

$$\int \frac{dx}{\sqrt{F^3}} = -\frac{x}{a^2\sqrt{F}}$$

$$\int \frac{dx}{\sqrt{F^n}} = -\frac{x}{a^2(n-2)\sqrt{F^{n-2}}} - \frac{n-3}{a^2(n-2)}\int \frac{dx}{\sqrt{F^{n-2}}} \qquad n \neq 2$$

$$\int \frac{x\,dx}{\sqrt{F}} = \sqrt{F}$$

$$\int \frac{x\,dx}{\sqrt{F^3}} = \frac{-1}{\sqrt{F}}$$

$$\int \frac{x\,dx}{\sqrt{F^n}} = -\frac{x^2}{a^2(n-2)\sqrt{F^{n-2}}} - \frac{n-4}{a^2(n-2)}\int \frac{x\,dx}{\sqrt{F^{n-2}}} \qquad n \neq 2$$

$$\int \frac{x^2\,dx}{\sqrt{F}} = \frac{x}{2}\sqrt{F} + \frac{a^2}{2}\ln\left(x + \sqrt{F}\right)$$

$$\int \frac{x^2\,dx}{\sqrt{F^3}} = -\frac{x}{\sqrt{F}} + \ln\left(x + \sqrt{F}\right)$$

$$\int \frac{x^2\,dx}{\sqrt{F^n}} = -\frac{x^3}{a^2(n-2)\sqrt{F^{n-2}}} - \frac{n-5}{a^2(n-2)}\int \frac{x^2\,dx}{\sqrt{F^{n-2}}} \qquad n \neq 2$$

$$\int \frac{x^3\,dx}{\sqrt{F}} = \left(\frac{F}{3} + a^2\right)\sqrt{F}$$

$$\int \frac{x^3\,dx}{\sqrt{F^3}} = \frac{F - a^2}{\sqrt{F}}$$

$$\int \frac{x^3\,dx}{\sqrt{F^n}} = -\frac{x^4}{a^2(n-2)\sqrt{F^{n-2}}} - \frac{n-6}{a^2(n-2)}\int \frac{x^3\,dx}{\sqrt{F^{n-2}}} \qquad n \neq 2$$

$$\int \frac{x^m\,dx}{\sqrt[p]{F^n}} = -\frac{px^{m+1}}{2a^2(n-p)\sqrt[p]{F^{n-p}}} - \frac{2n - p(m+3)}{2a^2(n-p)}\int \frac{x^m\,dx}{\sqrt[p]{F^{n-p}}} \qquad n \neq p$$

$$\int \frac{dx}{\sqrt{ax^2 - bx}} = \frac{2}{\sqrt{a}}\ln\left(\sqrt{ax} + \sqrt{ax - b}\right)$$

$$\int \frac{x^m\,dx}{\sqrt{ax^2 - bx}} = \frac{x^{m-1}}{am}\sqrt{ax^2 - bx} + \frac{b(2m-1)}{2am}\int \frac{x^{m-1}\,dx}{\sqrt{ax^2 - bx}}$$

(34) Indefinite Integrals Involving $\quad f(x) = \dfrac{1}{x^m\sqrt[p]{F^n}} \qquad F = x^2 - a^2 \qquad a^2 \neq 0$

$$\int \frac{dx}{x\sqrt{F}} = \frac{1}{a}\cos^{-1}\frac{a}{x}$$

$$\int \frac{dx}{x\sqrt{F^3}} = \frac{-1}{a^2\sqrt{F}} - \frac{1}{a^3}\cos^{-1}\frac{a}{x}$$

$$\int \frac{dx}{x\sqrt{F^n}} = -\frac{1}{a^2(n-2)\sqrt{F^{n-2}}} - \frac{1}{a^2}\int\frac{dx}{x\sqrt{F^{n-2}}} \qquad n \neq 2$$

$$\int \frac{dx}{x^2\sqrt{F}} = \frac{\sqrt{F}}{a^2 x}$$

$$\int \frac{dx}{x^2\sqrt{F^3}} = -\frac{\sqrt{F}}{a^4 x}\left(1 + \frac{x^2}{F}\right)$$

$$\int \frac{dx}{x^2\sqrt{F^n}} = -\frac{1}{a^2(n-2)x\sqrt{F^{n-2}}} - \frac{n-1}{a^2(n-2)}\int\frac{dx}{x^2\sqrt{F^{n-2}}} \qquad n \neq 2$$

$$\int \frac{dx}{x^3\sqrt{F}} = \frac{\sqrt{F}}{2a^2 x^2} + \frac{1}{2a^3}\cos^{-1}\frac{a}{x}$$

$$\int \frac{dx}{x^3\sqrt{F^3}} = \frac{1}{2a^2 x^2\sqrt{F}} - \frac{3}{2a^4\sqrt{F}} - \frac{3}{2a^5}\cos^{-1}\frac{a}{x}$$

$$\int \frac{dx}{x^3\sqrt{F^n}} = -\frac{1}{a^2(n-2)x^2\sqrt{F^{n-2}}} - \frac{n}{a^2(n-2)}\int\frac{dx}{x^3\sqrt{F^{n-2}}} \qquad n \neq 2$$

$$\int \frac{dx}{x^4\sqrt{F}} = \frac{\sqrt{F}}{a^4 x}\left(1 - \frac{F^2}{3x^2}\right)$$

$$\int \frac{dx}{x^4\sqrt{F^3}} = -\frac{x}{a^6\sqrt{F}}\left(1 + \frac{2F}{x^2} - \frac{F^2}{3x^4}\right)$$

$$\int \frac{dx}{x^4\sqrt{F^n}} = \frac{1}{a^2(n-2)x^3\sqrt{F^{n-2}}} - \frac{n+1}{a^2(n-2)}\int\frac{dx}{x^4\sqrt{F^{n-2}}} \qquad n \neq 2$$

$$\int \frac{dx}{x^m\sqrt[p]{F^n}} = \frac{p}{2a^2(n-p)x^{m-1}\sqrt[p]{F^{n-p}}} - \frac{2n+p(m-3)}{2a^2(n-p)}\int\frac{dx}{x^m\sqrt[p]{F^{n-p}}} \qquad n \neq p$$

$$\int \frac{dx}{x\sqrt{ax^2-bx}} = \frac{2}{bx}\sqrt{ax^2-bx}$$

$$\int \frac{dx}{x^m\sqrt{ax^2-bx}} = \frac{2\sqrt{ax^2-bx}}{b(2m-1)x^m} + \frac{2a(m-1)}{b(2m-1)}\int\frac{dx}{x^{m-1}\sqrt{ax^2-bx}}$$

(35) Indefinite Integrals Involving

$$f(x) = \frac{\sqrt[p]{F^n}}{x^m} \qquad F = x^2 - a^2 \qquad a^2 \neq 0$$

$$\int \frac{\sqrt{F}}{x} \, dx = \sqrt{F} - a \cos^{-1} \frac{a}{x}$$

$$\int \frac{\sqrt{F^3}}{x} \, dx = \frac{1}{3}\sqrt{F^3} - a^2\sqrt{F} + a^3 \cos^{-1} \frac{a}{x}$$

$$\int \frac{\sqrt{F^n}}{x} \, dx = \frac{1}{n}\sqrt{F^n} - a^2 \int \frac{\sqrt{F^{n-2}}}{x} \, dx$$

$$\int \frac{\sqrt{F}}{x^2} \, dx = -\frac{1}{x}\sqrt{F} + \ln(x + \sqrt{F})$$

$$\int \frac{\sqrt{F^3}}{x^2} \, dx = -\frac{1}{x}\sqrt{F^3} + \frac{3x}{2}\sqrt{F} - \frac{3a^2}{2}\ln(x + \sqrt{F})$$

$$\int \frac{\sqrt{F^n}}{x^2} \, dx = \frac{\sqrt{F^n}}{(n-1)x} - \frac{a^2 n}{n-1} \int \frac{\sqrt{F^{n-2}}}{x^2} \, dx \qquad n \neq 1$$

$$\int \frac{\sqrt{F}}{x^3} \, dx = \frac{-1}{2x^2}\sqrt{F} + \frac{1}{2a}\cos^{-1}\frac{a}{x}$$

$$\int \frac{\sqrt{F^3}}{x^3} \, dx = -\frac{1}{2x^2}\sqrt{F^3} + \frac{3}{2}\sqrt{F} - \frac{3a}{2}\cos^{-1}\frac{a}{x}$$

$$\int \frac{\sqrt{F^n}}{x^3} \, dx = \frac{\sqrt{F^n}}{(n-2)x^2} - \frac{a^2 n}{n-2} \int \frac{\sqrt{F^{n-2}}}{x^3} \, dx \qquad n \neq 2$$

$$\int \frac{\sqrt{F}}{x^4} \, dx = \frac{1}{3a^2 x^3}\sqrt{F^3}$$

$$\int \frac{\sqrt{F^3}}{x^4} \, dx = -\frac{1}{3x^3}\sqrt{F^3} - \frac{1}{x}\sqrt{F} + \ln(x + \sqrt{F})$$

$$\int \frac{\sqrt{F^n}}{x^4} \, dx = \frac{\sqrt{F^n}}{(n-3)x^3} - \frac{a^2 n}{n-3} \int \frac{\sqrt{F^{n-2}}}{x^4} \, dx \qquad n \neq 3$$

$$\int \frac{\sqrt[p]{F^n}}{x^m} \, dx = \frac{p\sqrt[p]{F^n}}{(2n - mp + p)x^{m-1}} - \frac{2a^2 n}{2n - mp + p} \int \frac{\sqrt[p]{F^{n-p}}}{x^m} \, dx \qquad 2n \neq -p(1-m)$$

$$\int \frac{\sqrt{ax^2 - bx}}{x} \, dx = \sqrt{ax^2 - bx} - \frac{b}{\sqrt{a}} \ln(\sqrt{ax} + \sqrt{ax - b})$$

$$\int \frac{\sqrt{ax^2 - bx}}{x^m} \, dx = \frac{2\sqrt{(ax^2 - bx)^3}}{b(2m - 3)x^{m-1}} + \frac{2a(m-3)}{b(2m-3)} \int \frac{\sqrt{ax^2 - bx}}{x^{m-1}} \, dx$$

(36) Indefinite Integrals Involving $f(x) = x^m \sqrt[p]{P^n}$ $\lambda = \dfrac{n}{p}$ $P = a + bx^q$ $\begin{aligned} a &\neq 0 \\ b &\neq 0 \end{aligned}$

$$\int P^\lambda \, dx = \frac{1}{1 + q\lambda}\left(xP^\lambda + aq\lambda \int P^{\lambda-1} dx\right)$$

$$\int \frac{dx}{P^\lambda} = \frac{-1}{aq(1 - \lambda)}\left[\frac{x}{P^{\lambda-1}} - (1 + q - q\lambda) \int \frac{dx}{P^{\lambda-1}}\right]$$

$$\int x^m P^\lambda \, dx = \frac{1}{1 + m + q\lambda}\left(x^{m+1}P^\lambda + aq\lambda \int x^m P^{\lambda-1} dx\right)$$

$$= \frac{-1}{aq(1 + \lambda)}\left[x^{m+1}P^{\lambda+1} - (1 + m + q + q\lambda) \int x^m P^{\lambda+1} dx\right]$$

$$= \frac{1}{a(1 + m)}\left[x^{m+1}P^{\lambda+1} - b(1 + m + q + q\lambda) \int x^{m+q}P^\lambda dx\right]$$

$$= \frac{1}{b(1 + m + q^\lambda)}\left[x^{m-q+1}P^{\lambda+1} - a(1 + m - q) \int x^{m-q}P^\lambda dx\right]$$

(37) Indefinite Integrals Involving $f(x) = \dfrac{x^{p-1}}{x^{2m+1} \pm a^{2m+1}}$ $a \neq 0$

$$\int \frac{x^{p-1} \, dx}{x^{2m+1} + a^{2m+1}} = \frac{2(-1)^{p-1}}{(2m + 1)a^{2m-p+1}} \sum_{k=1}^{m} \sin\frac{2kp\pi}{2m + 1}\tan^{-1}\left\{\frac{x + a\cos\left[2k\pi/(2m + 1)\right]}{a\sin\left[2k\pi/(2m + 1)\right]}\right\}$$

$$- \frac{(-1)^{p-1}}{(2m + 1)a^{2m-p+1}} \sum_{k=1}^{m} \cos\frac{2kp\pi}{2m + 1}\ln\left(x^2 + 2ax\cos\frac{2k\pi}{2m + 1} + a^2\right)$$

$$+ \frac{(-1)^{p-1}\ln(x + a)}{(2m + 1)a^{2m-p+1}} \qquad 0 < p \leqslant 2m + 1$$

$$\int \frac{x^{p-1} \, dx}{x^{2m+1} - a^{2m+1}} = \frac{-2}{(2m + 1)a^{2m-p+1}} \sum_{k=1}^{m} \sin\frac{2kp\pi}{2m + 1}\tan^{-1}\left\{\frac{x - a\cos\left[2k\pi/(2m + 1)\right]}{a\sin\left[2k\pi/(2m + 1)\right]}\right\}$$

$$+ \frac{1}{(2m + 1)a^{2m-p+1}} \sum_{k=1}^{m} \cos\frac{2kp\pi}{2m + 1}\ln\left(x^2 - 2ax\cos\frac{2k\pi}{2m + 1} + a^2\right)$$

$$+ \frac{\ln(x - a)}{(2m + 1)a^{2m-p+1}} \qquad 0 < p \leqslant 2m + 1$$

(Ref. 19.08, p. 75).

(38) Indefinite Integrals Involving $\boxed{f(x) = x^m \sqrt[p]{R^n} \qquad \lambda = n/p \qquad R = ax^q + bx^{q+r} \qquad \begin{array}{l} a \neq 0 \\ b \neq 0 \end{array}}$

$$\int R^\lambda \, dx = \frac{\lambda}{1 + \lambda(q + r)} \left(xR^\lambda + ar\lambda \int x^q R^{\lambda-1} \, dx \right)$$

$$\int \frac{dx}{R^\lambda} = \frac{-1}{ar(1 + \lambda)} \left[xR^{\lambda+1} - (1 + r + q\lambda + r\lambda) \int \frac{x^q \, dx}{R^{\lambda-1}} \right]$$

$$\int x^m R^\lambda \, dx = \frac{1}{1 + m + q\lambda + r\lambda} \left(x^{m+1} R^\lambda + ar\lambda \int x^{m+q} R^{\lambda-1} \, dx \right)$$

$$= \frac{-1}{ar(1 + \lambda)} \left[x^{m+1} R^{\lambda+1} - (1 + r + q\lambda + r\lambda) \int x^{m-q} R^{\lambda+1} \, dx \right]$$

(39) Indefinite Integrals Involving $\boxed{f(x) = \dfrac{x^m}{(x - a_1)(x - a_2) \cdots (x - a_k)} \qquad a_i \neq a_j \neq 0}$

$$\int \frac{x^m \, dx}{(x - a_1)(x - a_2) \cdots (x - a_k)} = \frac{a_1{}^m \ln (x - a_1)}{(a_1 - a_2)(a_1 - a_3) \cdots (a_1 - a_k)}$$

$$+ \frac{a_2{}^m \ln (x - a_2)}{(a_2 - a_1)(a_2 - a_3) \cdots (a_2 - a_k)} + \cdots + \frac{a_k{}^m \ln (x - a_k)}{(a_k - a_1)(a_k - a_2) \cdots (a_k - a_{k-1})}$$

(40) Indefinite Integrals Involving $\boxed{f(x) = \dfrac{x^{p-1}}{x^{2m} \pm a^{2m}} \qquad a \neq 0}$

$$\int \frac{x^{p-1} \, dx}{x^{2m} + a^{2m}} = \frac{1}{ma^{2m-p}} \sum_{k=1}^{m} \sin \frac{(2k - 1)p\pi}{2m} \tan^{-1} \left\{ \frac{x + a \cos [(2k - 1)\pi/2m]}{a \sin [(2k - 1)\pi/2m]} \right\}$$

$$- \frac{1}{2ma^{2m-p}} \sum_{k=1}^{m} \cos \frac{(2k - 1)p\pi}{2m} \ln \left[x^2 + 2ax \cos \frac{(2k - 1)\pi}{2m} + a^2 \right] \qquad 0 < p \leq 2m$$

$$\int \frac{x^{p-1} \, dx}{x^{2m} - a^{2m}} = \frac{1}{2ma^{2m-p}} \sum_{k=1}^{m-1} \cos \frac{kp\pi}{m} \ln \left(x^2 - 2ax \cos \frac{k\pi}{m} + a^2 \right)$$

$$- \frac{1}{na^{2m-p}} \sum_{k=1}^{m-1} \sin \frac{kp\pi}{m} \tan^{-1} \frac{x - a \cos (k\pi/m)}{a \sin (k\pi/m)} + \frac{1}{2ma^{2m-p}} [\ln (x - a) + (-1)^p \ln (x + a)]$$

$$0 < p \leq 2m$$

(Ref. 19.08, p. 75)

(41) Indefinite Integrals Involving $\qquad$ $f(x) = \sin^n A$ $\qquad$ $f(x) = \dfrac{1}{\sin^n A}$ $\qquad$ $A = bx$

$$\int \sin A \, dx = -\frac{\cos A}{b}$$

$$\int \sin^2 A \, dx = -\frac{\sin 2A}{4b} + \frac{x}{2}$$

$$\int \sin^3 A \, dx = -\frac{\cos A}{b} + \frac{\cos^3 A}{3b}$$

$$\int \sin^4 A \, dx = -\frac{\sin 2A}{4b} + \frac{\sin 4A}{32b} + \frac{3x}{8}$$

$$\int \sin^5 A \, dx = -\frac{5 \cos A}{8b} + \frac{5 \cos 3A}{48b} - \frac{\cos 5A}{80b}$$

$$\int \sin^6 A \, dx = -\frac{15 \sin 2A}{64b} + \frac{3 \sin 4A}{64b} - \frac{\sin 6A}{192b} + \frac{5x}{16}$$

$$\int \sin^7 A \, dx = -\frac{35 \cos A}{64b} + \frac{7 \cos 3A}{64b} - \frac{7 \cos 5A}{320b} + \frac{\cos 7A}{448b}$$

$$\int \sin^n A \, dx = -\frac{\cos A \, \sin^{n-1} A}{b \quad n} + \frac{n-1}{n} \int \sin^{n-2} A \, dx \qquad n > 0$$

$$\int \frac{dx}{\sin A} = \frac{1}{b} \ln \tan \frac{A}{2}$$

$$\int \frac{dx}{\sin^2 A} = -\frac{\cot A}{b}$$

$$\int \frac{dx}{\sin^3 A} = -\frac{\cos A}{2b \sin^2 A} + \frac{1}{2b} \ln \tan \frac{A}{2}$$

$$\int \frac{dx}{\sin^4 A} = -\frac{\cot A}{b} - \frac{\cot^3 A}{3b}$$

$$\int \frac{dx}{\sin^5 A} = -\frac{\cos A}{4b \sin^4 A} - \frac{3 \cos A}{8b \sin^2 A} - \frac{3}{8b} \ln \tan \frac{A}{2}$$

$$\int \frac{dx}{\sin^6 A} = -\frac{\cot A}{b} - \frac{2 \cot^3 A}{3b} - \frac{\cot^5 A}{5b}$$

$$\int \frac{dx}{\sin^7 A} = -\frac{\cos A}{6b \sin^6 A} - \frac{5 \cos A}{24b \sin^4 A} - \frac{5 \cos A}{16b \sin^2 A} + \frac{5}{16b} \ln \tan \frac{A}{2}$$

$$\int \frac{dx}{\sin^n A} = -\frac{\cos A}{b(n-1) \sin^{n-1} A} + \frac{n-2}{n-1} \int \frac{dx}{\sin^{n-2} A} \qquad n > 1$$

(42) Indefinite Integrals Involving

$f(x) = \cos^n A$	$f(x) = \dfrac{1}{\cos^n A}$	$A = bx$

$$\int \cos A \, dx = \frac{\sin A}{b}$$

$$\int \cos^2 A \, dx = \frac{\sin 2A}{4b} + \frac{x}{2}$$

$$\int \cos^3 A \, dx = \frac{\sin A}{b} - \frac{\sin^3 A}{3b}$$

$$\int \cos^4 A \, dx = \frac{\sin 2A}{4b} + \frac{\sin 4A}{32b} + \frac{3x}{8}$$

$$\int \cos^5 A \, dx = \frac{5 \sin A}{8b} + \frac{5 \sin 3A}{48b} + \frac{\sin 5A}{80b}$$

$$\int \cos^6 A \, dx = \frac{15 \sin 2A}{64b} + \frac{3 \sin 4A}{64b} + \frac{\sin 6A}{192b} + \frac{5x}{16}$$

$$\int \cos^7 A \, dx = \frac{35 \sin A}{64b} + \frac{7 \sin 3A}{64b} + \frac{7 \sin 5A}{320b} + \frac{\sin 7A}{448b}$$

$$\int \cos^n A \, dx = \frac{\sin A \, \cos^{n-1} A}{b} \cdot \frac{1}{n} + \frac{n-1}{n} \int \cos n^{n-2}A \, dx \qquad n > 0$$

$$\int \frac{dx}{\cos A} = \frac{1}{b} \ln \tan \left(\frac{\pi}{4} + \frac{A}{2} \right)$$

$$\int \frac{dx}{\cos^2 A} = \frac{1}{b} \tan A$$

$$\int \frac{dx}{\cos^3 A} = \frac{\sin A}{2b \cos^2 A} + \frac{1}{2b} \ln \tan \left(\frac{\pi}{4} + \frac{A}{2} \right)$$

$$\int \frac{dx}{\cos^4 A} = \frac{\tan A}{b} + \frac{\tan^3 A}{3b}$$

$$\int \frac{dx}{\cos^5 A} = \frac{\sin A}{4b \cos^4 A} + \frac{3 \sin A}{8 \cos^2 A} + \frac{3}{8b} \ln \tan \left(\frac{\pi}{4} + \frac{A}{2} \right)$$

$$\int \frac{dx}{\cos^6 A} = \frac{\tan A}{b} + \frac{2 \tan^3 A}{3b} + \frac{\tan^5 A}{5b}$$

$$\int \frac{dx}{\cos^7 A} = \frac{\sin A}{6b \cos^6 A} + \frac{5 \sin A}{24b \cos^4 A} + \frac{5 \sin A}{16b \cos^2 A} + \frac{5}{16b} \tan \left(\frac{\pi}{4} + \frac{A}{2} \right)$$

$$\int \frac{dx}{\cos^n A} = \frac{\sin A}{b(n-1) \cos^{n-1} A} + \frac{n-2}{n-1} \int \frac{dx}{\cos^{n-2} A} \qquad n > 1$$

(43) Indefinite Integrals Involving $\qquad$ $f(x) = \sin^m A \cos^n A$ $\quad$ $A = bx$

$$\int \sin A \cos A \, dx = -\frac{\cos^2 A}{2b}$$

$$\int \sin A \cos^n A \, dx = -\frac{\cos^{n+1} A}{(n+1)b} \qquad n \neq -1 \qquad\qquad \int \sin^m A \cos A \, dx = \frac{\sin^{m+1} A}{(m+1)b} \qquad m \neq -1$$

$$\int \sin^2 A \cos A \, dx = \frac{\sin^3 A}{3b}$$

$$\int \sin^2 A \cos^2 A \, dx = \frac{x}{8} - \frac{\sin 4A}{32b}$$

$$\int \sin^2 A \cos^3 A \, dx = \frac{\sin^3 A \cos^2 A}{5b} + 2\frac{\sin^3 A}{15b}$$

$$\int \sin^2 A \cos^4 A \, dx = \frac{x}{16} + \frac{\sin 2A}{64b} - \frac{\sin 4A}{64b} - \frac{\sin 6A}{192b}$$

$$\int \sin^2 A \cos^n A \, dx = -\frac{\sin A \cos^{n+1} A}{(n+2)b} + \int \frac{\cos^n A \, dx}{n+2} \qquad n \neq -2$$

$$\int \sin^3 A \cos A \, dx = \frac{\sin^4 A}{4b}$$

$$\int \sin^3 A \cos^2 A \, dx = -\frac{\cos^3 A}{3b} + \frac{\cos^5 A}{5b}$$

$$\int \sin^3 A \cos^3 A \, dx = -\frac{3\cos 2A}{64b} + \frac{\cos 6A}{192b}$$

$$\int \sin^3 A \cos^n A \, dx = -\frac{\cos^{n+1} A}{(n+1)b} + \frac{\cos^{n+3} A}{(n+3)b} \qquad n \neq -1, -3$$

$$\int \sin^4 A \cos A \, dx = \frac{\sin^5 A}{5b}$$

$$\int \sin^4 A \cos^2 A \, dx = \frac{1}{192b}(\sin 6A - 3\sin 4A - 3\sin 2A + 12A)$$

$$\int \sin^4 A \cos^3 A \, dx = \frac{\sin^5 A}{5b} - \frac{\sin^7 A}{7b}$$

$$\int \sin^m A \cos^n A \, dx = \frac{\sin^{m+1} A \cos^{n-1} A}{b(m+n)} + \frac{n-1}{m+n}\int \sin^m A \cos^{n-2} A \, dx$$

$$= -\frac{\sin^{m-1} A \cos^{n+1} A}{b(m+n)} + \frac{m-1}{m+n}\int \sin^{m-2} A \cos^n A \, dx \qquad \begin{array}{l} m > 0 \\ n > 0 \end{array}$$

(44) Indefinite Integrals Involving
$$f(x) = \frac{1}{\sin^m A \cos^n A} \qquad A = bx$$

$$\int \frac{dx}{\sin A \cos A} = \frac{1}{b} \ln(\tan A)$$

$$\int \frac{dx}{\sin A \cos^2 A} = \frac{1}{b} \ln\left(\tan\frac{A}{2}\right) + \frac{1}{b \cos A}$$

$$\int \frac{dx}{\sin A \cos^n A} = \frac{1}{b(n-1)\cos^{n-1} A} + \int \frac{dx}{\sin A \cos^{n-2} A} \qquad n \neq 1$$

$$\int \frac{dx}{\sin^2 A \cos A} = -\frac{1}{b \sin A} + \frac{1}{b} \ln\left[\tan\left(\frac{\pi}{4}+\frac{A}{2}\right)\right]$$

$$\int \frac{dx}{\sin^2 A \cos^2 A} = -\frac{2}{b} \cot 2A$$

$$\int \frac{dx}{\sin^2 A \cos^3 A} = \frac{1}{2b \sin A}\left(\frac{1}{\cos^2 A} - 3\right) + \frac{3}{2b} \ln\left[\tan\left(\frac{\pi}{4}+\frac{A}{2}\right)\right]$$

$$\int \frac{dx}{\sin^2 A \cos^4 A} = \frac{1}{3b \sin A \cos^3 A} - \frac{8}{3b \tan 2A}$$

$$\int \frac{dx}{\sin^2 A \cos^n A} = \frac{1 - n \cos^2 A}{b(n-1)\sin A \cos^{n-1} A} + \frac{n(n-2)}{n-1} \int \frac{dx}{\cos^{n-2} A} \qquad n \neq 1$$

$$\int \frac{dx}{\sin^3 A \cos A} = \frac{-1}{2b \sin^2 A} + \frac{1}{b} \ln(\tan A)$$

$$\int \frac{dx}{\sin^3 A \cos^2 A} = \frac{1}{b \cos A} - \frac{\cos A}{2b \sin^2 A} + \frac{3}{2b} \ln\left(\tan\frac{A}{2}\right)$$

$$\int \frac{dx}{\sin^3 A \cos^3 A} = -\frac{2 \cos 2A}{b \sin^2 2A} + \frac{2}{b} \ln(\tan A)$$

$$\int \frac{dx}{\sin^3 A \cos^4 A} = \frac{2}{b \cos A} + \frac{1}{3b \cos^3 A} - \frac{\cos A}{2b \sin^2 A} + \frac{5}{2b} \ln\left(\tan\frac{A}{2}\right)$$

$$\int \frac{dx}{\sin^m A \cos A} = \frac{-1}{b(m-1)\sin^{m-1} A} + \int \frac{dx}{\sin^{m-2} A \cos A} \qquad m \neq 1$$

$$\int \frac{dx}{\sin^m A \cos^n A} = \begin{cases} \dfrac{-1}{b(m-1)\sin^{m-1} A \cos^{n-1} A} + \dfrac{m+n-2}{m-1} \displaystyle\int \dfrac{dx}{\sin^{m-2} A \cos^n A} & m \neq 1 \\[4mm] \dfrac{1}{b(n-1)\sin^{m-1} A \cos^{n-1} A} + \dfrac{m+n-2}{n-1} \displaystyle\int \dfrac{dx}{\sin^m A \cos^{n-2} A} & n \neq 1 \end{cases}$$

(45) Indefinite Integrals Involving

$$f(x) = \frac{\sin^m A}{\cos^n A} \qquad A = bx$$

$$\int \frac{\sin A \, dx}{\cos A} = -\frac{\ln(\cos A)}{b}$$

$$\int \frac{\sin A \, dx}{\cos^n A} = \frac{1}{b(n-1)\cos^{n-1} A} \qquad n \neq 1$$

$$\int \frac{\sin^2 A \, dx}{\cos A} = -\frac{\sin A}{b} + \frac{1}{b} \ln\left[\tan\left(\frac{\pi}{4} + \frac{A}{2}\right)\right]$$

$$\int \frac{\sin^m A \, dx}{\cos^{m+2} A} = \frac{\tan^{m+1} A}{b(m+1)} \qquad m \neq -1$$

$$\int \frac{\sin^2 A \, dx}{\cos^2 A} = -x + \frac{1}{b}\tan A$$

$$\int \frac{\sin^2 A \, dx}{\cos^3 A} = \frac{\sin A}{2b\cos^2 A} - \frac{1}{2b}\ln\left[\tan\left(\frac{\pi}{4} + \frac{A}{2}\right)\right]$$

$$\int \frac{\sin^2 A \, dx}{\cos^4 A} = \frac{\tan^3 A}{3b}$$

$$\int \frac{\sin^2 A \, dx}{\cos^5 A \, dx} = \frac{\sin A}{4b\cos^4 A} - \frac{\sin A}{8b\cos^2 A} - \frac{1}{8b}\ln\left[\tan\left(\frac{\pi}{4} + \frac{A}{2}\right)\right]$$

$$\int \frac{\sin^2 A \, dx}{\cos^n A} = \frac{\sin A}{b(n-1)\cos^{n-1} A} - \frac{1}{n-1}\int \frac{dx}{\cos^{n-2} A} \qquad n \neq 1$$

$$\int \frac{\sin^3 A \, dx}{\cos A} = -\frac{\sin^2 A}{2b} - \frac{1}{b}\ln(\cos A)$$

$$\int \frac{\sin^3 A \, dx}{\cos^2 A} = \frac{\cos A}{b} + \frac{1}{b\cos A}$$

$$\int \frac{\sin^3 A \, dx}{\cos^3 A} = \frac{\tan^2 A}{2b} + \frac{1}{b}\ln(\cos A)$$

$$\int \frac{\sin^3 A \, dx}{\cos^4 A} = \frac{1}{3b\cos^3 A} - \frac{1}{b\cos A}$$

$$\int \frac{\sin^3 A \, dx}{\cos^5 A} = \frac{1}{4b\cos^4 A} - \frac{1}{2b\cos^2 A}$$

$$\int \frac{\sin^3 A \, dx}{\cos^n A} = \frac{1}{b(n-1)\cos^{n-1} A} - \frac{1}{b(n-3)\cos^{n-3} A} \qquad n \neq 1, 3$$

$$\int \frac{\sin^m A \, dx}{\cos A} = -\frac{\sin^{m-1} A}{b(m-1)} + \int \frac{\sin^{m-2} A \, dx}{\cos A} \qquad m \neq 1$$

$$\int \frac{\sin^m A \, dx}{\cos^n A} = \begin{cases} -\dfrac{\sin^{m-1} A}{b(m-n)\cos^{n-1} A} + \dfrac{m-1}{m-n}\displaystyle\int \frac{\sin^{m-2} A \, dx}{\cos^n A} & m \neq n \\[4mm] \dfrac{\sin^{m-1} A}{b(n-1)\cos^{n-1} A} - \dfrac{m-1}{n-1}\displaystyle\int \frac{\sin^{m-2} A \, dx}{\cos^{n-2} A} & n \neq 1 \end{cases}$$

(46) Indefinite Integrals Involving

$$f(x) = \frac{\cos^n A}{\sin^m A} \qquad A = bx$$

$$\int \frac{\cos A \, dx}{\sin A} = \frac{\ln(\sin A)}{b}$$

$$\int \frac{\cos A \, dx}{\sin^m A} = \frac{-1}{b(m-1)\sin^{m-1} A} \qquad m \neq 1$$

$$\int \frac{\cos^2 A \, dx}{\sin A} = \frac{\cos A}{b} + \frac{1}{b}\ln\left(\tan\frac{A}{2}\right)$$

$$\int \frac{\cos^n A \, dx}{\sin^{n+2} A} = -\frac{\cot^{n+1} A}{b(n+1)} \qquad n \neq -1$$

$$\int \frac{\cos^2 A \, dx}{\sin^2 A} = -x - \frac{1}{b}\cot A$$

$$\int \frac{\cos^2 A \, dx}{\sin^3 A} = -\frac{\cos A}{2b\sin^2 A} - \frac{1}{2b}\ln\left(\tan\frac{A}{2}\right)$$

$$\int \frac{\cos^2 A \, dx}{\sin^4 A} = -\frac{\cot^3 A}{3b}$$

$$\int \frac{\cos^2 A \, dx}{\sin^5 A} = -\frac{\cos A}{4b\sin^4 A} + \frac{\cos A}{8b\sin^2 A} - \frac{1}{8b}\ln\left(\tan\frac{A}{2}\right)$$

$$\int \frac{\cos^2 A \, dx}{\sin^m A} = \frac{-\cos A}{b(m-1)\sin^{m-1} A} - \frac{1}{m-1}\int \frac{dx}{\sin^{m-2} A} \qquad m \neq 1$$

$$\int \frac{\cos^3 A \, dx}{\sin A} = \frac{\cos^2 A}{2b} + \frac{1}{b}\ln(\sin A)$$

$$\int \frac{\cos^3 A \, dx}{\sin^2 A} = -\frac{\sin A}{b} - \frac{1}{b\sin A}$$

$$\int \frac{\cos^3 A \, dx}{\sin^3 A} = -\frac{\cot^2 A}{2b} - \frac{1}{b}\ln(\sin A)$$

$$\int \frac{\cos^3 A \, dx}{\sin^4 A} = -\frac{1}{3b\sin^3 A} + \frac{1}{b\sin A}$$

$$\int \frac{\cos^3 A \, dx}{\sin^5 A} = -\frac{1}{4b\sin^4 A} + \frac{1}{2b\sin^2 A}$$

$$\int \frac{\cos^3 A \, dx}{\sin^m A} = \frac{-1}{b(m-1)\sin^{m-1} A} + \frac{1}{b(m-3)\sin^{m-3} A} \qquad m \neq 1, 3$$

$$\int \frac{\cos^n A \, dx}{\sin A} = \frac{\cos^{n-1} A}{b(n-1)} + \int \frac{\cos^{n-2} A \, dx}{\sin A} \qquad n \neq 1$$

$$\int \frac{\cos^n A \, dx}{\sin^m A} = \begin{cases} -\dfrac{\cos^{n-1} A}{b(m-n)\sin^{m-1} A} - \dfrac{n-1}{m-n}\displaystyle\int \frac{\cos^{n-2} A \, dx}{\sin^m A} & m \neq n \\[4mm] -\dfrac{\cos^{n-1} A}{b(m-1)\sin^{m-1} A} - \dfrac{n-1}{m-1}\displaystyle\int \frac{\cos^{n-2} A \, dx}{\sin^{m-2} A} & m \neq 1 \end{cases}$$

$$f(x) = x^m \sin^n A \qquad A = bx$$

$$D_1 = \int \sin^n A\, dx \qquad\qquad D_2 = \int D_1\, dx \qquad\qquad D_k = \int D_{k-1}\, dx$$

$$\int x \sin A\, dx = -\frac{x \cos A}{b} + \frac{\sin A}{b^2}$$

$$\int x \sin^2 A\, dx = \frac{x^2}{4} - \frac{x \sin 2A}{4b} - \frac{\cos 2A}{8b^2}$$

$$\int x \sin^3 A\, dx = \frac{x \cos 3A}{12b} - \frac{\sin 3A}{36b^2} + \frac{3x \cos A}{4b} + \frac{3 \sin A}{4b^2}$$

$$\int x \sin^n A\, dx = xD_1 - D_2$$

$$\int x^2 \sin A\, dx = -\frac{x^2 \cos A}{b} + \frac{2x \sin A}{b^2} + \frac{2 \cos A}{b^3}$$

$$\int x^2 \sin^2 A\, dx = \frac{1}{2} \int x^2 (1 - \cos 2A)\, dx$$

$$\int x^2 \sin^3 A\, dx = \frac{1}{4} \int x^2 (3 \sin A - \sin 3A)\, dx$$

$$\int x^2 \sin^n A\, dx = x^2 D_1 - 2xD_2 + 2D_3$$

$$\int x^3 \sin A\, dx = -\frac{x^3 \cos A}{b} + \frac{3x^2 \sin A}{b^2} + \frac{6x \cos A}{b^3} - \frac{6 \sin A}{b^4}$$

$$\int x^3 \sin^2 A\, dx = \frac{1}{2} \int x^3 (1 - \cos 2A)\, dx$$

$$\int x^3 \sin^3 A\, dx = \frac{1}{4} \int x^3 (3\sin A - \sin 3A)\, dx$$

$$\int x^3 \sin^n A\, dx = x^3 D_1 - 3x^2 D_2 + 6xD_3 - 6D_4$$

$$\int x^m \sin A\, dx = \frac{m! \sin A}{b} \left[\frac{x^{m-1}}{(m-1)!b} - \frac{x^{m-3}}{(m-3)!b^3} + \frac{x^{m-5}}{(m-5)!b^5} - \cdots \right]$$

$$-\frac{m! \cos A}{b} \left[\frac{x^m}{m!} - \frac{x^{m-2}}{(m-2)!b^2} + \frac{x^{m-4}}{(m-4)!b^4} - \cdots \right]$$

Series terminates with the term involving x^{m-m}

$$\int x^m \sin^n A\, dx = x^m D_1 - mx^{m-1} D_2 + m(m-1)x^{m-2} D_3 - \cdots$$

(48) Indefinite Integrals Involving $\qquad$ $f(x) = x^m \cos^n A$ $\quad$ $A = bx$

$$D_1 = \int \cos^n A \, dx \qquad D_2 = \int D_1 \, dx \qquad D_k = \int D_{k-1} \, dx$$

$$\int x \cos A \, dx = \frac{x \sin A}{b} + \frac{\cos A}{b^2}$$

$$\int x \cos^2 A \, dx = \frac{x^2}{4} + \frac{x \sin 2A}{4b} + \frac{\cos 2A}{8b^2}$$

$$\int x \cos^3 A \, dx = \frac{x \sin 3A}{12b} + \frac{\cos 3A}{36b^2} + \frac{3x \sin A}{4b} + \frac{3 \cos A}{4b^2}$$

$$\int x \cos^n A \, dx = x D_1 - D_2$$

$$\int x^2 \cos A \, dx = \frac{x^2 \sin A}{b} + \frac{2x \cos A}{b^2} - \frac{2 \sin A}{b^3}$$

$$\int x^2 \cos^2 A \, dx = \frac{1}{2} \int x^2 (1 + \cos 2A) \, dx$$

$$\int x^2 \cos^3 A \, dx = \frac{1}{4} \int x^2 (3 \cos A + \cos 3A) \, dx$$

$$\int x^2 \cos^n A \, dx = x^2 D_1 - 2x D_2 + D_3$$

$$\int x^3 \cos A \, dx = \frac{x^3 \sin A}{b} + \frac{3x^2 \cos A}{b^2} - \frac{6x \sin A}{b^3} - \frac{6 \cos A}{b^4}$$

$$\int x^3 \cos^2 A \, dx = \frac{1}{2} \int x^3 (1 + \cos 2A) \, dx$$

$$\int x^3 \cos^3 A \, dx = \frac{1}{4} \int x^3 (3 \cos A + \cos 3A) \, dx$$

$$\int x^3 \cos^n A \, dx = x^3 D_1 - 3x^2 D_2 + 6x D_3 - 6D_4$$

$$\int x^m \cos A \, dx = \frac{m! \sin A}{b} \left[\frac{x^m}{m!} - \frac{x^{m-2}}{(m-2)! b^2} + \frac{x^{m-4}}{(m-4)! b^4} - \cdots \right]$$
$$+ \frac{m! \cos A}{b} \left[\frac{x^{m-1}}{(m-1)! b} - \frac{x^{m-3}}{(m-3)! b^3} + \frac{x^{m-5}}{(m-5)! b^5} - \cdots \right]$$

Series terminates with the term involving x^{m-m}

$$\int x^m \cos^n A \, dx = x^m D_1 - mx^{m-1} D_2 + m(m-1) x^{m-2} D_3 - \cdots$$

(49) Indefinite Integrals Involving

$$f(x) = \frac{x^m}{\sin^n A} \qquad f(x) = \frac{\sin^n A}{x^m} \qquad A = bx$$

$\text{Si}(A) = $ sine-integral function (Sec. 13.01-1) $\qquad$ $B_k = $ auxiliary Bernoulli number (Sec. 8.05-1)

$\text{Ci}(A) = $ cosine-integral function (Sec. 13.01-1) $\qquad$ $E_k = $ auxiliary Euler number (Sec. 8.06-1)

$b_0 = 1, \qquad b_k = \dfrac{1}{(2k)!} B_k$ (Sec. 8.15-2) $\qquad\qquad d_0 = 1, \qquad d_k = \dfrac{4^k - 2}{(2k)!} B_k$ (Sec. 8.15-2)

$$\int \frac{\sin A}{x} dx = \text{Si}(A)$$

$$\int \frac{\sin A}{x^2} dx = b\, \text{Ci}(A) - \frac{\sin A}{x}$$

$$\int \frac{\sin A}{x^3} dx = -\frac{\sin A}{2x^2} - \frac{b \cos A}{2x} - \frac{b^2}{2} \text{Si}(A)$$

$$\int \frac{\sin A}{x^m} dx = -\frac{\sin A}{(m-1)x^{m-1}} - \frac{b \cos A}{(m-1)(m-2)x^{m-2}} - \frac{b^2}{(m-1)(m-2)} \int \frac{\sin A}{x^{m-2}} dx \qquad m > 2$$

$$\int \frac{x\, dx}{\sin A} = \frac{A}{b^2} \sum_{k=0}^{\infty} d_k \frac{A^{2k}}{2k+1} = \frac{A}{b^2} \bigwedge_{k=1}^{\infty} \left[1 + \frac{(2k-1) d_{k+1}}{(2k+1) d_{k-1}} A^2 \right]^*$$

$$\int \frac{x\, dx}{\sin^2 A} = -\frac{A}{b^2} \cot A + \frac{1}{b^2} \ln \sin A$$

$$\int \frac{x\, dx}{\sin^3 A} = -\frac{A \cos A}{2b^2 \sin^2 A} - \frac{1}{2b^2 \sin A} + \frac{1}{2} \int \frac{x\, dx}{\sin A}$$

$$\int \frac{x\, dx}{\sin^n A} = -\frac{A \cos A}{(n-1) b^2 \sin^{n-1} A} - \frac{1}{(n-1)(n-2) b^2 \sin^{n-2} A} + \frac{n-2}{n-1} \int \frac{x\, dx}{\sin^{n-2} A} \qquad n > 2$$

$$\int \frac{x^2\, dx}{\sin A} = \frac{A^2}{2b^3} \sum_{k=0}^{\infty} d_k \frac{A^{2k}}{k+1} = \frac{A^2}{2b^3} \bigwedge_{k=1}^{\infty} \left[1 + \frac{k d_k}{(k+1) d_{k-1}} A^2 \right]^*$$

$$\int \frac{x^2\, dx}{\sin^2 A} = \frac{2A}{b^3} \left[1 - \frac{A}{2} \cot A - \sum_{k=1}^{\infty} b_k \frac{(2A)^{2k}}{2k+1} \right]$$

$$= \frac{2A}{b^3} \left\{ -\bigwedge_{k=1}^{\infty} \left[1 + \frac{(2k-1) b_k}{(2k+1) b_{k-1}} (2A)^2 \right]^* - \frac{A}{2} \cot A + 2 \right\}$$

$$\int \frac{x^m\, dx}{\sin A} = \frac{A^m}{b^{m+1}} \sum_{k=0}^{\infty} d_k \frac{A^{2k}}{m+2k} = \frac{A^m}{mb^{m+1}} \bigwedge_{k=1}^{\infty} \left[1 + \frac{(m+2k-2) d_k}{(m+2k) d_{k-1}} A^2 \right]^*$$

$$\int \frac{\sin^{2n-1} A}{x^m} dx = \left(-\frac{1}{4} \right)^{n-1} \int \left\{ \sum_{k=0}^{n-1} (-1)^k \binom{2n-1}{k} \sin\left[(2n-2k-1) A \right] \right\} \frac{dx}{x^m}$$

$$\int \frac{\sin^{2n} A}{x^m} dx = 2 \left(-\frac{1}{4} \right)^n \int \left\{ \sum_{k=0}^{n-1} (-1)^k \binom{2n}{k} \cos\left[(2n-2k) A \right] \right\} \frac{dx}{x^m} - \frac{\binom{2n}{n}}{(m-1) 4^n x^{m-1}}$$

---

*For nested sum refer to Sec. 8.11.

(50) Indefinite Integrals Involving $\quad f(x) = \dfrac{x^m}{\cos^n A} \qquad f(x) = \dfrac{\cos^n A}{x^m} \qquad A = bx$

$\text{Si}(A) = \text{sine-integral function (Sec. 13.01-1)} \qquad B_k = \text{auxiliary Bernoulli number (Sec. 8.05-1)}$

$\text{Ci}(A) = \text{cosine-integral function (Sec. 13.01-1)} \qquad E_k = \text{auxiliary Euler number (Sec. 8.06-1)}$

$a_0 = 1, \qquad a_k = \dfrac{4^k - 1}{(2k)!} B_k \text{ (Sec. 8.15-2)} \qquad c_0 = 1, \qquad c_k = \dfrac{1}{(2k)!} E_k \text{ (Sec. 8.15-2)}$

$$\int \frac{\cos A}{x}\, dx = \text{Ci}(A) \qquad\qquad \int \frac{\cos A}{x^2}\, dx = -b\,\text{Si}(A) - \frac{\cos A}{x}$$

$$\int \frac{\cos A}{x^3}\, dx = -\frac{\cos A}{2x^2} + \frac{b \sin A}{2x} - \frac{b^2}{2}\,\text{Ci}(A)$$

$$\int \frac{\cos A}{x^m}\, dx = -\frac{\cos A}{(m-1)x^{m-1}} + \frac{b \sin A}{(m-1)(m-2)x^{m-2}} - \frac{b^2}{(m-1)(m-2)} \int \frac{\cos A}{x^{m-2}}\, dx \qquad m > 2$$

$$\int \frac{x\, dx}{\cos A} = \frac{A^2}{2b^2} \sum_{k=0}^{\infty} c_k \frac{A^{2k}}{k+1} = \frac{A^2}{2b^2} \bigwedge_{k=1}^{\infty}\left[1 + \frac{k c_k}{(k+1)c_{k-1}} A^2\right]^*$$

$$\int \frac{x\, dx}{\cos^2 A} = \frac{A}{b^2}\tan A + \frac{1}{b^2}\ln\cos A$$

$$\int \frac{x\, dx}{\cos^3 A} = \frac{A \sin A}{2b^2 \cos^2 A} - \frac{1}{2b^2 \cos A} + \frac{1}{2}\int \frac{x\, dx}{\cos A}$$

$$\int \frac{x\, dx}{\cos^n A} = \frac{A \sin A}{(n-1)b^2 \cos^{n-1} A} - \frac{1}{(n-1)(n-2)b^2 \cos^{n-2} A} + \frac{n-2}{n-1}\int \frac{x\, dx}{\cos^{n-2} A} \qquad n > 2$$

$$\int \frac{x^2\, dx}{\cos A} = \frac{A^3}{b^3} \sum_{k=0}^{\infty} c_k \frac{A^{2k}}{2k+3} = \frac{A^3}{3b^3} \bigwedge_{k=1}^{\infty}\left[1 + \frac{(2k+1)c_k}{(2k+3)c_{k-1}} A^2\right]^*$$

$$\int \frac{x^2\, dx}{\cos^2 A} = \frac{2A}{b^3}\left[\frac{A}{2}\tan A - \sum_{k=1}^{\infty} \frac{a_k}{2k+1}(2A)^{2k}\right]$$

$$= \frac{2A}{b^3}\left\{-\bigwedge_{k=1}^{\infty}\left[1 + \frac{(2k-1)a_k}{(2k+1)a_{k-1}}(2A)^2\right]^* + \frac{A}{2}\tan A + 1\right\}$$

$$\int \frac{x^m\, dx}{\cos A} = \frac{A^{m+1}}{b^{m+1}} \sum_{k=0}^{\infty} c_k \frac{A^{2k}}{m+2k+1} = \frac{A^{m+1}}{(m+1)b^{m+1}} \bigwedge_{k=1}^{\infty}\left[1 + \frac{(m+2k-1)c_k}{(m+2k+1)c_{k-1}} A^2\right]^*$$

$$\int \frac{\cos^{2n-1} A}{x^m}\, dx = \left(\frac{1}{4}\right)^{n-1}\int\left\{\sum_{k=0}^{n-1}\binom{2n-1}{k}\cos\left[(2n-2k-1)A\right]\right\}\frac{dx}{x^m}$$

$$\int \frac{\cos^{2n} A}{x^m}\, dx = 2\left(\frac{1}{4}\right)^{n}\int\left\{\sum_{k=0}^{n-1}\binom{2n}{k}\cos\left[(2n-2k)A\right]\right\}\frac{dx}{x^m} - \frac{\binom{2n}{n}}{(m-1)4^n x^{m-1}}$$

*For nested sum refer to Sec. 8.11.

(51) Indefinite Integrals Involving $\qquad\qquad f(x) = f(\sin X_k, \ldots, \cos X_k, \ldots)$

$\alpha_1 = a+b$	$\alpha_2 = a-b$	$\beta_1 = a+2b$	$\beta_2 = -a+2b$
$\lambda_1 = a+b+c$	$\lambda_2 = -a+b+c$	$\lambda_3 = a-b+c$	$\lambda_4 = a+b-c$
$A_1 = ax+b$	$A_2 = cx+d$	$B_1 = A_1 + A_2$	$B_2 = A_1 - A_2$
$C_1 = \omega x + a$	$C_2 = \omega x + b$	$D_1 = C_1 + C_2$	$D_2 = C_1 - C_2$

$$\int \sin ax \sin bx \, dx = -\frac{\sin \alpha_1 x}{2\alpha_1} + \frac{\sin \alpha_2 x}{2\alpha_2} \qquad\qquad a \neq b$$

$$\int \sin ax \sin^2 bx \, dx = +\frac{\cos \beta_1 x}{4\beta_1} - \frac{\cos \beta_2 x}{4\beta_2} - \frac{\cos ax}{2a} \qquad\qquad a \neq 2b$$

$$\int \sin ax \sin bx \sin cx \, dx = +\frac{\cos \lambda_1 x}{4\lambda_1} - \frac{\cos \lambda_2 x}{4\lambda_2} - \frac{\cos \lambda_3 x}{4\lambda_3} - \frac{\cos \lambda_4 x}{4\lambda_4} \qquad\qquad \lambda_k \neq 0$$

$$\int \sin ax \cos bx \, dx = -\frac{\cos \alpha_1 x}{2\alpha_1} - \frac{\cos \alpha_2 x}{2\alpha_2} \qquad\qquad a \neq b$$

$$\int \sin ax \cos^2 bx \, dx = -\frac{\cos \beta_1 x}{4\beta_1} + \frac{\cos \beta_2 x}{4\beta_2} - \frac{\cos ax}{2a} \qquad\qquad a \neq 2b$$

$$\int \sin ax \cos bx \cos cx \, dx = -\frac{\cos \lambda_1 x}{4\lambda_1} + \frac{\cos \lambda_2 x}{4\lambda_2} - \frac{\cos \lambda_3 x}{4\lambda_3} - \frac{\cos \lambda_4 x}{4\lambda_4} \qquad\qquad \lambda_k \neq 0$$

$$\int x \sin ax \sin bx \, dx = -x\left(\frac{\sin \alpha_1 x}{2\alpha_1} - \frac{\sin \alpha_2 x}{2\alpha_2}\right) - \left(\frac{\cos \alpha_1 x}{2\alpha_1^2} - \frac{\cos \alpha_2 x}{2\alpha_2^2}\right)$$

$$\int x \sin ax \cos bx \, dx = -x\left(\frac{\cos \alpha_1 x}{2\alpha_1} + \frac{\cos \alpha_2 x}{2\alpha_2}\right) + \left(\frac{\sin \alpha_1 x}{2\alpha_1^2} + \frac{\sin \alpha_2 x}{2\alpha_2^2}\right) \qquad\qquad a \neq b$$

$$\int A_1 \sin A_2 \, dx = -\frac{A_1}{c}\cos A_2 + \frac{a}{c^2}\sin A_2 \qquad\qquad a, c \neq 0$$

$$\int \sin A_1 \sin A_2 \, dx = -\frac{\sin B_1}{2(a+c)} + \frac{\sin B_2}{2(a-c)} \qquad\qquad a \neq c$$

$$\int \sin A_1 \cos A_2 \, dx = -\frac{\cos B_1}{2(a+c)} - \frac{\cos B_2}{2(a-c)} \qquad\qquad a \neq c$$

$$\int C_1 \sin C_2 \, dx = -\frac{C_1}{\omega}\cos C_2 + \frac{1}{\omega}\sin C_2$$

$$\int \sin C_1 \sin C_2 \, dx = -\frac{\sin D_1}{4\omega} + \frac{x}{2}\cos D_2 \qquad\qquad \omega \neq 0$$

$$\int \sin C_1 \cos C_2 \, dx = -\frac{\cos D_1}{4\omega} + \frac{x}{2}\sin D_2$$

(52) Indefinite Integrals Involving $\qquad f(x) = f(\cos X_k, \ldots, \sin X_k, \ldots)$

$\alpha_1 = a+b$	$\alpha_2 = a-b$	$\beta_1 = a+2b$	$\beta_2 = 2b-a$
$\lambda_1 = a+b+c$	$\lambda_2 = -a+b+c$	$\lambda_3 = a-b+c$	$\lambda_4 = a+b-c$
$A_1 = ax+b$	$A_2 = cx+d$	$B_1 = A_1+A_2$	$B_2 = A_1-A_2$
$C_1 = \omega x + a$	$C_2 = \omega x + b$	$D_1 = C_1 + C_2$	$D_2 = C_1 - C_2$

$$\int \cos ax \cos bx \, dx = +\frac{\sin \alpha_1 x}{2\alpha_1} + \frac{\sin \alpha_2 x}{2\alpha_2} \qquad a \neq b$$

$$\int \cos ax \cos^2 bx \, dx = +\frac{\sin \beta_1 x}{4\beta_1} + \frac{\sin \beta_2 x}{4\beta_2} + \frac{\sin ax}{2a} \qquad a \neq 2b$$

$$\int \cos ax \cos bx \cos cx \, dx = +\frac{\sin \lambda_1 x}{4\lambda_1} + \frac{\sin \lambda_2 x}{4\lambda_2} + \frac{\sin \lambda_3 x}{4\lambda_3} + \frac{\sin \lambda_4 x}{4\lambda_4} \qquad \lambda_k \neq 0$$

$$\int \cos ax \sin bx \, dx = -\frac{\cos \alpha_1 x}{2\alpha_1} + \frac{\cos \alpha_2 x}{2\alpha_2} \qquad a \neq b$$

$$\int \cos ax \sin^2 bx \, dx = -\frac{\sin \beta_1 x}{4\beta_1} - \frac{\sin \beta_2 x}{4\beta_2} + \frac{\sin ax}{2a} \qquad a \neq 2b$$

$$\int \cos ax \sin bx \sin cx \, dx = -\frac{\sin \lambda_1 x}{4\lambda_1} - \frac{\sin \lambda_2 x}{4\lambda_2} + \frac{\sin \lambda_3 x}{4\lambda_3} + \frac{\sin \lambda_4 x}{4\lambda_4} \qquad \lambda_k \neq 0$$

$$\int x \cos ax \cos bx \, dx = +x\left(\frac{\sin \alpha_1 x}{2\alpha_1} + \frac{\sin \alpha_2 x}{2\alpha_2}\right) + \left(\frac{\cos \alpha_1 x}{2\alpha_1^2} + \frac{\cos \alpha_2 x}{2\alpha_2^2}\right)$$

$$\int x \cos ax \sin bx \, dx = -x\left(\frac{\cos \alpha_1 x}{2\alpha_1} - \frac{\cos \alpha_2 x}{2\alpha_2}\right) + \left(\frac{\sin \alpha_1 x}{2\alpha_1^2} - \frac{\sin \alpha_2 x}{2\alpha_2^2}\right)$$

$\qquad a \neq b$

$$\int A_1 \cos A_2 \, dx = +\frac{A_1}{c} \sin A_2 + \frac{a}{c^2} \cos A_2 \qquad a, c \neq 0$$

$$\int \cos A_1 \cos A_2 \, dx = +\frac{\sin B_1}{2(a+c)} + \frac{\sin B_2}{2(a-c)} \qquad a \neq c$$

$$\int \cos A_1 \sin A_2 \, dx = -\frac{\cos B_1}{2(a+c)} + \frac{\cos B_2}{2(a-c)} \qquad a \neq c$$

$$\int C_1 \cos C_2 \, dx = +\frac{C_1}{\omega} \sin C_2 + \frac{1}{\omega} \cos C_2$$

$$\int \cos C_1 \cos C_2 \, dx = +\frac{\sin D_1}{4\omega} + \frac{x}{2} \cos D_2 \qquad \omega \neq 0$$

$$\int \cos C_1 \sin C_2 \, dx = -\frac{\cos D_1}{4\omega} - \frac{x}{2} \sin D_2$$

(53) Indefinite Integrals Involving

$$f(x) = f(1 \pm \sin A) \qquad A = bx$$

$$\int \frac{dx}{1+\sin A} = -\frac{1}{b}\tan\left(\frac{\pi}{4}-\frac{A}{2}\right)$$

$$\int \frac{x\,dx}{1+\sin A} = -\frac{x}{b}\tan\left(\frac{\pi}{4}-\frac{A}{2}\right) + \frac{2}{b^2}\ln\left[\cos\left(\frac{\pi}{4}-\frac{A}{2}\right)\right]$$

$$\int \frac{\sin A\,dx}{1+\sin A} = \frac{1}{b}\tan\left(\frac{\pi}{4}-\frac{A}{2}\right) + x$$

$$\int \frac{\cos A\,dx}{1+\sin A} = \frac{1}{b}\ln\left(1+\sin A\right)$$

$$\int \frac{dx}{\sin A\,(1+\sin A)} = \frac{1}{b}\tan\left(\frac{\pi}{4}-\frac{A}{2}\right) + \frac{1}{b}\ln\left(\tan\frac{A}{2}\right)$$

$$\int \frac{dx}{\cos A(1+\sin A)} = \frac{-1}{2b(1+\sin A)} + \frac{1}{2b}\ln\left[\tan\left(\frac{\pi}{4}+\frac{A}{2}\right)\right]$$

$$\int \frac{dx}{1-\sin A} = \frac{1}{b}\tan\left(\frac{\pi}{4}+\frac{A}{2}\right)$$

$$\int \frac{x\,dx}{1-\sin A} = \frac{x}{b}\cot\left(\frac{\pi}{4}-\frac{A}{2}\right) + \frac{2}{b^2}\ln\left[\sin\left(\frac{\pi}{4}-\frac{A}{2}\right)\right]$$

$$\int \frac{\sin A\,dx}{1-\sin A} = \frac{1}{b}\tan\left(\frac{\pi}{4}+\frac{A}{2}\right) - x$$

$$\int \frac{\cos A\,dx}{1-\sin A} = -\frac{1}{b}\ln\left(1-\sin A\right)$$

$$\int \frac{dx}{\sin A\,(1-\sin A)} = \frac{1}{b}\tan\left(\frac{\pi}{4}+\frac{A}{2}\right) + \frac{1}{b}\ln\left(\tan\frac{A}{2}\right)$$

$$\int \frac{dx}{\cos A(1-\sin A)} = \frac{1}{2b(1-\sin A)} + \frac{1}{2b}\left[\tan\left(\frac{\pi}{4}+\frac{A}{2}\right)\right]$$

$$\int \frac{dx}{(1+\sin A)^2} = \frac{-1}{2b}\tan\left(\frac{\pi}{4}-\frac{A}{2}\right) - \frac{1}{6b}\tan^3\left(\frac{\pi}{4}-\frac{A}{2}\right)$$

$$\int \frac{\sin A\,dx}{(1+\sin A)^2} = \frac{-1}{2b}\tan\left(\frac{\pi}{4}-\frac{A}{2}\right) + \frac{1}{6b}\tan^3\left(\frac{\pi}{4}-\frac{A}{2}\right)$$

$$\int \frac{dx}{(1-\sin A)^2} = \frac{1}{2b}\cot\left(\frac{\pi}{4}-\frac{A}{2}\right) + \frac{1}{6b}\cot^3\left(\frac{\pi}{4}-\frac{A}{2}\right)$$

$$\int \frac{\sin A\,dx}{(1-\sin A)^2} = -\frac{1}{2b}\cot\left(\frac{\pi}{4}-\frac{A}{2}\right) + \frac{1}{6b}\cot^3\left(\frac{\pi}{4}-\frac{A}{2}\right)$$

(54) Indefinite Integrals Involving

$$f(x) = f(1 \pm \cos A) \qquad A = bx$$

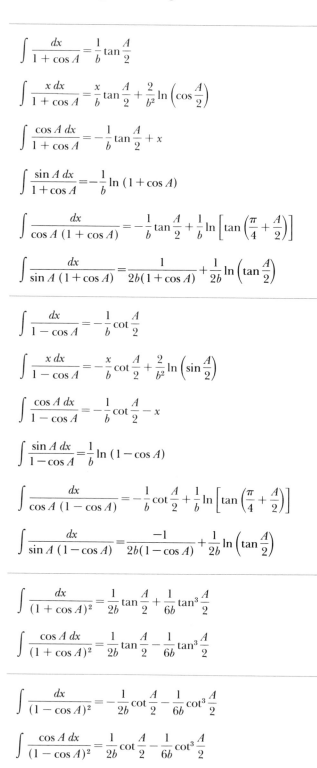

$$\int \frac{dx}{1+\cos A} = \frac{1}{b}\tan\frac{A}{2}$$

$$\int \frac{x\,dx}{1+\cos A} = \frac{x}{b}\tan\frac{A}{2} + \frac{2}{b^2}\ln\left(\cos\frac{A}{2}\right)$$

$$\int \frac{\cos A\,dx}{1+\cos A} = -\frac{1}{b}\tan\frac{A}{2} + x$$

$$\int \frac{\sin A\,dx}{1+\cos A} = -\frac{1}{b}\ln\left(1+\cos A\right)$$

$$\int \frac{dx}{\cos A\,(1+\cos A)} = -\frac{1}{b}\tan\frac{A}{2} + \frac{1}{b}\ln\left[\tan\left(\frac{\pi}{4}+\frac{A}{2}\right)\right]$$

$$\int \frac{dx}{\sin A\,(1+\cos A)} = \frac{1}{2b(1+\cos A)} + \frac{1}{2b}\ln\left(\tan\frac{A}{2}\right)$$

$$\int \frac{dx}{1-\cos A} = -\frac{1}{b}\cot\frac{A}{2}$$

$$\int \frac{x\,dx}{1-\cos A} = -\frac{x}{b}\cot\frac{A}{2} + \frac{2}{b^2}\ln\left(\sin\frac{A}{2}\right)$$

$$\int \frac{\cos A\,dx}{1-\cos A} = -\frac{1}{b}\cot\frac{A}{2} - x$$

$$\int \frac{\sin A\,dx}{1-\cos A} = \frac{1}{b}\ln\left(1-\cos A\right)$$

$$\int \frac{dx}{\cos A\,(1-\cos A)} = -\frac{1}{b}\cot\frac{A}{2} + \frac{1}{b}\ln\left[\tan\left(\frac{\pi}{4}+\frac{A}{2}\right)\right]$$

$$\int \frac{dx}{\sin A\,(1-\cos A)} = \frac{-1}{2b(1-\cos A)} + \frac{1}{2b}\ln\left(\tan\frac{A}{2}\right)$$

$$\int \frac{dx}{(1+\cos A)^2} = \frac{1}{2b}\tan\frac{A}{2} + \frac{1}{6b}\tan^3\frac{A}{2}$$

$$\int \frac{\cos A\,dx}{(1+\cos A)^2} = \frac{1}{2b}\tan\frac{A}{2} - \frac{1}{6b}\tan^3\frac{A}{2}$$

$$\int \frac{dx}{(1-\cos A)^2} = -\frac{1}{2b}\cot\frac{A}{2} - \frac{1}{6b}\cot^3\frac{A}{2}$$

$$\int \frac{\cos A\,dx}{(1-\cos A)^2} = \frac{1}{2b}\cot\frac{A}{2} - \frac{1}{6b}\cot^3\frac{A}{2}$$

(55) Indefinite Integrals Involving*

$$f(x) = f(1 \pm a \sin A) \qquad A = bx$$

$$\int \frac{dx}{1 + a \sin A} = \begin{cases} \dfrac{2}{b\sqrt{1 - a^2}} \tan^{-1} \dfrac{a + \tan (A/2)}{\sqrt{1 - a^2}} & a^2 < 1 \\[3mm] \dfrac{1}{b\sqrt{1 - a^2}} \sin^{-1} \dfrac{a + \sin A}{1 + a \sin A} & a^2 < 1 \\[3mm] \dfrac{1}{b\sqrt{a^2 - 1}} \ln \dfrac{\tan (A/2) + a - \sqrt{a^2 - 1}}{\tan (A/2) + a + \sqrt{a^2 - 1}} & a^2 > 1 \end{cases}$$

$$\int \frac{dx}{(1 + a \sin A)^2} = \frac{a \cos A}{b(1 - a^2)(1 + a \sin A)} + \frac{1}{1 - a^2} \int \frac{dx}{1 + a \sin A}$$

$$\int \frac{dx}{(1 + a \sin A)^n} = \frac{a \cos A}{b(n - 1)(1 - a^2)(1 + a \sin A)^{n-1}} + \frac{(2n - 3)}{(n - 1)(1 - a^2)} \int \frac{dx}{(1 + a \sin A)^{n-1}}$$

$$- \frac{(n - 2)}{(n - 1)(1 - a^2)} \int \frac{dx}{(1 + a \sin A)^{n-2}} \qquad \begin{matrix} a^2 \neq 1 \\ n \neq 1 \end{matrix}$$

$$\int \frac{\sin A \, dx}{1 + a \sin A} = \frac{x}{a} - \frac{1}{a} \int \frac{dx}{1 + a \sin A}$$

$$\int \frac{\sin A \, dx}{(1 + a \sin A)^2} = \frac{\cos A}{b(a^2 - 1)(1 + a \sin A)} + \frac{a}{a^2 - 1} \int \frac{dx}{1 + a \sin A} \qquad a^2 \neq 1$$

$$\int \frac{\cos A \, dx}{1 \pm a \sin A} = \pm \frac{\ln (1 \pm a \sin A)}{ab}$$

$$\int \frac{\cos A \, dx}{(1 \pm a \sin A)^2} = \mp \frac{1}{ab(1 \pm a \sin A)}$$

$$\int \frac{\cos A \, dx}{(1 \pm a \sin A)^n} = \mp \frac{1}{ab(n - 1)(1 \pm a \sin A)^{n-1}} \qquad n \neq 1$$

$$\int \frac{dx}{(1 + a \sin A) \sin A} = \frac{1}{b} \ln \left(\tan \frac{A}{2} \right) - a \int \frac{dx}{1 + a \sin A}$$

$$\int \frac{dx}{(1 \pm \sin A) \cos A} = \frac{\mp 1}{2b(1 \pm \sin A)} + \frac{1}{2b} \ln \left[\tan \left(\frac{\pi}{4} + \frac{A}{2} \right) \right]$$

$$\int \frac{\sin A \, dx}{(1 \pm \sin A) \cos A} = \frac{1}{2b(1 \pm \sin A)} \pm \frac{1}{2b} \ln \left[\tan \left(\frac{\pi}{4} + \frac{A}{2} \right) \right]$$

$$\int \frac{\cos A \, dx}{(1 \pm \sin A) \sin A} = -\frac{1}{b} \ln \frac{1 \pm \sin A}{\sin A}$$

$$\int \frac{1 + c \sin A}{1 + a \sin A} \, dx = \frac{cx}{ab} + \frac{a - c}{ab} \int \frac{dx}{1 + a \sin A} \qquad c \neq 0$$

*If $a = 1$ refer to Sec. 19.53.

(56) Indefinite Integrals Involving* $\boxed{f(x) = f(1 \pm a \cos A)} \quad \boxed{A = bx}$

$$\int \frac{dx}{1 + a \cos A} = \begin{cases} \dfrac{2}{b\sqrt{1-a^2}} \tan^{-1}\left(\sqrt{\dfrac{1-a}{1+a}}\,\tan\dfrac{A}{2}\right) & a^2 < 1 \\[3ex] \dfrac{1}{b\sqrt{1-a^2}} \cos^{-1}\dfrac{a + \cos A}{1 + a \cos A} & a^2 < 1 \\[3ex] \dfrac{1}{b\sqrt{a^2-1}} \ln \dfrac{(\sqrt{a+1}) + (\sqrt{a-1})\tan(A/2)}{(\sqrt{a+1}) - (\sqrt{a-1})\tan(A/2)} & a^2 > 1 \end{cases}$$

$$\int \frac{dx}{(1 + a \cos A)^2} = \frac{-a \sin A}{b(1 - a^2)(1 + a \cos A)} + \frac{1}{1 - a^2}\int \frac{dx}{1 + a \cos A}$$

$$\int \frac{dx}{(1 + a \cos A)^n} = -\frac{a \sin A}{b(n-1)(1-a^2)(1 + a \cos A)^{n-1}} + \frac{2n-3}{(n-1)(1-a^2)}\int \frac{dx}{(1 + a \cos A)^{n-1}}$$
$$-\frac{n-2}{(n-1)(1-a^2)}\int \frac{dx}{(1 + a \cos A)^{n-2}} \qquad \begin{array}{l} a^2 \neq 1 \\ n \neq 1 \end{array}$$

$$\int \frac{\cos A\, dx}{1 + a \cos A} = \frac{x}{a} - \frac{1}{a}\int \frac{dx}{1 + a \cos A}$$

$$\int \frac{\cos A\, dx}{(1 + a \cos A)^2} = \frac{-\sin A}{b(a^2-1)(1 + a \cos A)} + \frac{a}{a^2-1}\int \frac{dx}{1 + a \cos A} \qquad a^2 \neq 1$$

$$\int \frac{\sin A\, dx}{1 \pm a \cos A} = \mp \frac{\ln(1 \pm a \cos A)}{ab}$$

$$\int \frac{\sin A\, dx}{(1 \pm a \cos A)^2} = \pm \frac{1}{ab(1 \pm a \cos A)}$$

$$\int \frac{\sin A\, dx}{(1 \pm a \cos A)^n} = \pm \frac{1}{ab(n-1)(1 \pm a \cos A)^{n-1}} \qquad n \neq 1$$

$$\int \frac{dx}{(1 + a \cos A)\cos A} = \frac{1}{b}\ln\left[\tan\left(\frac{A}{2} + \frac{\pi}{4}\right)\right] - a \int \frac{dx}{1 + a \cos A}$$

$$\int \frac{dx}{(1 \pm \cos A)\sin A} = \frac{\pm 1}{2b(1 \pm \cos A)} + \frac{1}{2b}\ln\left(\tan\frac{A}{2}\right)$$

$$\int \frac{\cos A\, dx}{(1 \pm \cos A)\sin A} = \frac{1}{2b(1 \pm \cos A)} \pm \frac{1}{2b}\ln\left(\tan\frac{A}{2}\right)$$

$$\int \frac{\sin A\, dx}{(1 \pm \cos A)\cos A} = \frac{1}{b}\ln\frac{1 \pm \cos A}{\cos A}$$

$$\int \frac{1 + c \cos A}{1 + a \cos A}\, dx = \frac{cx}{ab} + \frac{a-c}{ab}\int \frac{dx}{1 + a \cos A} \qquad c \neq 0$$

*If $a = 1$ refer to Sec. 19.54.

(57) Indefinite Integrals Involving $f(x) = f(1 \pm a \sin^2 A)$ $A = bx$

$\alpha = \sqrt{1+a}$ $\beta = \sqrt{1-a}$ $\gamma = \sqrt{a-1}$

$$\int \frac{dx}{1+\sin^2 A} = \frac{1}{b\sqrt{2}} \tan^{-1}(\sqrt{2} \tan A)$$

$$\int \frac{dx}{1+a\sin^2 A} = \frac{1}{\alpha b} \tan^{-1}(\alpha \tan A) \qquad\qquad a>0$$

$$\int \frac{dx}{(1+a\sin^2 A)^2} = \frac{a \sin 2A}{4\alpha^2 b(1+a\sin^2 A)} + \frac{2+a}{2\alpha^3 b} \tan^{-1}(\alpha \tan A) \qquad\qquad a>0$$

$$\int \frac{dx}{1-\sin^2 A} = \frac{1}{b} \tan A$$

$$\int \frac{dx}{1-a\sin^2 A} = \begin{cases} \dfrac{1}{\beta b} \tan^{-1}(\beta \tan A) & 0<a<1 \\[2mm] \dfrac{1}{2\gamma b} \ln \dfrac{\gamma \tan A + 1}{\gamma \tan A - 1} & a>1 \end{cases}$$

$$\int \frac{dx}{(1-a\sin^2 A)^2} = \frac{-a \sin 2A}{4b\beta^2(1-a\sin^2 A)} + \frac{2-a}{2b\beta^2} \begin{cases} \dfrac{1}{\beta} \tan^{-1}(\beta \tan A) & 0<a<1 \\[2mm] \dfrac{1}{2\gamma} \ln \dfrac{\gamma \tan A + 1}{\gamma \tan A - 1} & a>1 \end{cases}$$

$$\int \frac{\sin^2 A \, dx}{1+a\sin^2 A} = \frac{x}{a} - \frac{1}{\alpha a b} \tan^{-1}(\alpha \tan A) \qquad\qquad a>0$$

$$\int \frac{\cos^2 A \, dx}{1+a\sin^2 A} = \frac{\alpha}{ab} \tan^{-1}(\alpha \tan A) - \frac{x}{a} \qquad\qquad a>0$$

$$\int \frac{\sin A \cos A \, dx}{1 \pm a\sin^2 A} = \pm \frac{1}{ab} \ln\sqrt{1 \pm a\sin^2 A} \qquad\qquad a>0$$

$$\int \frac{\sin^2 A \, dx}{1-a\sin^2 A} = \begin{cases} \dfrac{1}{ab\beta} \tan^{-1}(\beta \tan A) - \dfrac{x}{a} & 0<a<1 \\[2mm] \dfrac{1}{2ab\gamma} \ln \left(\dfrac{\gamma \tan A + 1}{\gamma \tan A - 1}\right) - \dfrac{x}{a} & a>0 \end{cases}$$

$$\int \frac{\cos^2 A \, dx}{1-a\sin^2 A} = \begin{cases} -\dfrac{\beta}{ab} \tan^{-1}(\beta \tan A) + \dfrac{x}{a} & 0<a<1 \\[2mm] \dfrac{\gamma}{2ab} \ln \left(\dfrac{\gamma \tan A + 1}{\gamma \tan A - 1}\right) + \dfrac{x}{a} & a>0 \end{cases}$$

(58) Indefinite Integrals Involving

$$f(x) = f(1 \pm a \cos^2 A) \qquad A = bx$$

$$\alpha = \sqrt{1+a} \qquad\qquad \beta = \sqrt{1-a} \qquad\qquad \gamma = \sqrt{a-1}$$

$$\int \frac{dx}{1+\cos^2 A} = \frac{1}{b\sqrt{2}} \tan^{-1} \frac{\tan A}{\sqrt{2}}$$

$$\int \frac{dx}{1+a\cos^2 A} = \frac{1}{\alpha b} \tan^{-1} \frac{\tan A}{\alpha} \qquad a > 0$$

$$\int \frac{dx}{(1+a\cos^2 A)^2} = \frac{-a\sin 2A}{4\alpha^2 b(1+a\cos^2 A)} + \frac{2+a}{2\alpha^3 b} \tan^{-1} \frac{\tan A}{\alpha} \qquad a > 0$$

$$\int \frac{dx}{1-\cos^2 A} = -\frac{1}{b}\cot A$$

$$\int \frac{dx}{1-a\cos^2 A} = \begin{cases} \dfrac{1}{\beta b} \tan^{-1} \dfrac{\tan A}{\beta} & 0 < a < 1 \\[2ex] \dfrac{1}{2\gamma b} \ln \dfrac{\tan A - \gamma}{\tan A + \gamma} & a > 1 \end{cases}$$

$$\int \frac{dx}{(1-a\cos^2 A)^2} = \frac{a\sin 2A}{4b\beta^2(1-a\cos^2 A)} + \frac{2-a}{2b\beta^2} \begin{cases} \dfrac{1}{\beta} \tan^{-1} \dfrac{\tan A}{\beta} & 0 < a < 1 \\[2ex] \dfrac{1}{2\gamma} \ln \dfrac{\tan A - \gamma}{\tan A + \gamma} & a > 1 \end{cases}$$

$$\int \frac{\cos^2 A \, dx}{1+a\cos^2 A} = \frac{x}{a} - \frac{1}{\alpha ab} \tan^{-1} \frac{\tan A}{\alpha} \qquad a > 0$$

$$\int \frac{\sin^2 A \, dx}{1+a\cos^2 A} = \frac{\alpha}{ab} \tan^{-1} \left(\frac{\tan A}{\alpha}\right) - \frac{x}{a} \qquad a > 0$$

$$\int \frac{\sin A \cos A \, dx}{1 \pm a \cos^2 A} = \mp \frac{1}{ab} \ln \sqrt{1 \pm a \cos^2 A} \qquad a > 0$$

$$\int \frac{\cos^2 A \, dx}{1-a\cos^2 A} = \begin{cases} \dfrac{1}{ab\beta} \tan^{-1} \left(\dfrac{\tan A}{\beta}\right) - \dfrac{x}{a} & 0 < a < 1 \\[2ex] \dfrac{1}{2ab\gamma} \ln \left(\dfrac{\tan A - \gamma}{\tan A + \gamma}\right) - \dfrac{x}{a} & a > 0 \end{cases}$$

$$\int \frac{\sin^2 A \, dx}{1-a\cos^2 A} = \begin{cases} -\dfrac{\beta}{ab} \tan^{-1} \left(\dfrac{\tan A}{\beta}\right) + \dfrac{x}{a} & 0 < a < 1 \\[2ex] \dfrac{\gamma}{2ab} \ln \left(\dfrac{\tan A - \gamma}{\tan A + \gamma}\right) + \dfrac{x}{a} & a > 0 \end{cases}$$

(59) Indefinite Integrals Involving*

$$f(x) = f(p \sin A + q \cos A) \qquad A = bx$$

$$r = \sqrt{p^2 + q^2} \neq 0 \qquad s = \sqrt{a^2 - r^2} \qquad \alpha = \tan^{-1}\frac{q}{p} \qquad \beta = A + \alpha \qquad \lambda = \sin A \pm \cos A$$

$$\int \frac{dx}{\sin A \pm \cos A} = \frac{1}{b\sqrt{2}} \ln \tan \left(\frac{A}{2} \pm \frac{\pi}{8}\right) \qquad \int \frac{dx}{(\sin A \pm \cos A)^2} = \frac{1}{2b} \tan \left(A \mp \frac{\pi}{4}\right)$$

$$\int \frac{\sin A \, dx}{\sin A \pm \cos A} = \frac{1}{2b}(A \mp \ln \lambda) \qquad \int \frac{\cos A \, dx}{\sin A \pm \cos A} = \frac{1}{2b}(\ln \lambda \pm A)$$

$$\int \frac{dx}{p \sin A + q \cos A} = \frac{1}{br} \ln \left(\tan \frac{\beta}{2}\right)$$

$$\int \frac{dx}{(p \sin A + q \cos A)^2} = -\frac{1}{br} \cot \beta = \frac{1}{br} \frac{p \sin A - q \cos A}{p \sin A + q \cos A}$$

$$\int \frac{dx}{(p \sin A + q \cos A)^n} = \frac{1}{r^n} \int \frac{d\beta}{\sin^n \beta} = -\frac{\cos \beta}{br^n(n-1) \sin^{n-1} \beta} + \frac{n-2}{(n-1)r^n} \int \frac{d\beta}{\sin^{n-2} \beta}$$

$$\int \frac{\sin A \, dx}{p \sin A + q \cos A} = \frac{1}{br^2}[pA - q \ln (p \sin A + q \cos A)]$$

$$\int \frac{\cos A \, dx}{p \sin A + q \cos A} = \frac{1}{br^2}[q \ln (p \sin A + q \cos A) + pA]$$

$$\int \frac{dx}{a + p \sin A + q \cos A} = \begin{cases} \dfrac{2}{bs} \tan^{-1}\left[\dfrac{(a-q) \tan (A/2) + p}{s}\right] & a^2 > r^2 \\[3mm] \dfrac{1}{ab}\left[\dfrac{a - (p+q) \cos A - (p-q) \sin A}{a - (p-q) \cos A + (p+q) \sin A}\right] & a^2 = r^2 \\[3mm] \dfrac{1}{bsi} \ln \left[\dfrac{(a-q) \tan (A/2) + p - si}{(a-q) \tan (A/2) + p + si}\right] & a^2 < r^2 \end{cases}$$

$$\int \frac{(p + q \sin A) \, dx}{\sin A (1 \pm \cos A)} = \frac{p}{2b}\left[\ln \left(\tan \frac{A}{2}\right) \pm \frac{1}{1 \pm \cos A}\right] + q \int \frac{dx}{1 \pm \cos A}$$

$$\int \frac{(p + q \sin A) \, dx}{\cos A (1 \pm \cos A)} = \frac{p}{b}\left[\ln \tan \left(\frac{\pi}{4} + \frac{A}{2}\right)\right] + \frac{q}{b} \ln \left(\frac{1 \pm \cos A}{\cos A}\right) - p \int \frac{dx}{1 \pm \cos A}$$

$$\int \frac{(p + q \cos A) \, dx}{\sin A (1 \pm \sin A)} = \frac{p}{b} \ln \left(\tan \frac{A}{2}\right) - \frac{q}{b} \ln \left(\frac{1 \pm \sin A}{\sin A}\right) - p \int \frac{dx}{1 \pm \sin A}$$

$$\int \frac{(p + q \cos A) \, dx}{\cos A (1 \pm \sin A)} = \frac{p}{2b}\left[\ln \tan \left(\frac{\pi}{4} + \frac{A}{2}\right) \mp \frac{1}{1 \pm \sin A}\right] + q \int \frac{dx}{1 \pm \sin A}$$

---

*$p, q =$ signed integers or fractions.

(60) Indefinite Integrals Involving*

$$f(x) = f(p^2 \sin^2 A \pm q^2 \cos^2 A) \qquad A = bx$$

$\alpha = p^2 \pm q^2$	$\beta = p^2 \mp q^2$	$\gamma = 4p^2 r^2 - q^4$

$$\int (p^2 \sin^2 A \pm q^2 \cos^2 A)\ dx = \frac{\alpha A}{2b} - \frac{\beta \sin 2A}{4b}$$

$$\int (p^2 \sin^2 A \pm q^2 \cos^2 A) \sin A \cos A\ dx = \frac{1}{4b} (p^2 \sin^4 A \mp q^2 \cos^4 A)$$

$$\int (p^2 \sin^2 A \pm q^2 \cos^2 A)^m \sin A \cos A\ dx = \frac{1}{2(m+1)\beta b} (p^2 \sin^2 A \pm q^2 \cos^2 A)^{m+1} \qquad m > 1 \qquad \beta \neq 0$$

$$\int \frac{dx}{p^2 \sin^2 A + q^2 \cos^2 A} = \frac{1}{bpq} \tan^{-1}\left(\frac{p}{q} \tan A\right)$$

$$\int \frac{dx}{p^2 \sin^2 A - q^2 \cos^2 A} = \frac{1}{bpq} \ln \sqrt{\frac{p \tan A - q}{p \tan A + q}}$$

$$\int \frac{\sin A \cos A\ dx}{p^2 \sin^2 A \pm q^2 \cos^2 A} = \frac{1}{2\beta b} \ln (p^2 \sin^2 A \pm q^2 \cos^2 A) \qquad\qquad \beta \neq 0$$

$$\int \frac{\sin A \cos A\ dx}{\sqrt{p^2 \sin^2 A \pm q^2 \cos^2 A}} = \frac{1}{\beta b} \sqrt{p^2 \sin^2 A \pm q^2 \cos^2 A}$$

$$\left.\begin{array}{l} \displaystyle\int \frac{dx}{(p^2 \sin^2 A \pm q^2 \cos^2 A)^2} = \frac{1}{2b(pq)^3} (\alpha z \pm \beta \sin z \cos z) \\[2em] \displaystyle\int \frac{dx}{(p^2 \sin^2 A \pm q^2 \cos^2 A)^n} = \frac{1}{b(pq)^{2n-1}} \int (p^2 \sin^2 z \pm q^2 \cos^2 z)^{n-1}\ dz \end{array}\right\} \qquad z = \tan^{-1}\left(\frac{p}{q} \tan A\right)$$

$$\int \frac{dx}{p^2 \sin^2 A + q^2 \sin A \cos A + r^2 \cos^2 A} = \begin{cases} \dfrac{2}{b\sqrt{\gamma}} \tan^{-1}\left(\dfrac{2p^2 \tan A + q^2}{\sqrt{\gamma}}\right) & \gamma > 0 \\[1.5em] -\dfrac{2}{b(2p^2 \tan A + q^2)} & \gamma = 0 \\[1.5em] \dfrac{1}{b\sqrt{-\gamma}} \ln \left(\dfrac{2p^2 \tan A + q^2 - \sqrt{-\gamma}}{2p^2 \tan A + q^2 + \sqrt{-\gamma}}\right) & \gamma < 0 \end{cases}$$

$$\int \frac{dx}{\sin^2 A (p^2 \pm q^2 \cos^2 A)} = \frac{1}{\alpha b} \left(\int \frac{\pm q^2\ dA}{p^2 \pm q^2 \cos^2 A} - \cot A\right) \qquad\qquad \alpha \neq 0$$

$$\int \frac{dx}{\cos^2 A (p^2 \pm q^2 \sin^2 A)} = \frac{1}{\alpha b} \left(\int \frac{\pm q^2\ dA}{p^2 \pm q^2 \sin^2 A} + \tan A\right)$$

---

*p, q = positive integers or fractions.

(61) Indefinite Integrals Involving $\boxed{f(x) = f(\sqrt{p \pm q \sin A})}$ $\boxed{A = bx}$

$E(k, \phi) =$ incomplete elliptic integral of the second kind (Sec. 13.05)

$F(k, \phi) =$ incomplete elliptic integral of the first kind (Sec. 13.05)

$G(k, \phi) = \tan \phi \sqrt{1 - k^2 \sin \phi}$, $r = \sqrt{p^2 + q^2}$, $\lambda = p/q$, $a > r$, $p > 0$, $q > 0$

$$\int \sqrt{1 + \sin A} \, dx = -\frac{2\sqrt{2}}{b} \cos\left(\frac{\pi}{4} + \frac{A}{2}\right)$$

$$\int \sqrt{1 - \sin A} \, dx = \frac{2\sqrt{2}}{b} \sin\left(\frac{\pi}{4} + \frac{A}{2}\right)$$

$$\int \sqrt{p + q \sin A} \, dx = -\frac{2}{b} \sqrt{p + q} \, E\left(\sqrt{\frac{2}{1 + \lambda}}, \frac{\cos^{-1}(\sin A)}{2}\right)$$

$$\int \sqrt{p - q \sin A} = -\frac{2}{b} \sqrt{p + q} \left[E\left(\sqrt{\frac{2}{1 + \lambda}}, \sin^{-1}\sqrt{\frac{\lambda - \sin A}{1 - \sin A}}\right) - G\left(\sqrt{\frac{2}{1 + \lambda}}, \sin^{-1}\sqrt{\frac{\lambda - \sin A}{1 - \sin A}}\right) \right]$$

$$\int \frac{dx}{\sqrt{\sin 2A}} = \frac{\sqrt{2}}{b} F\left(\frac{1}{\sqrt{2}}, \sin^{-1}\sqrt{\frac{2 \sin A}{1 + \sin A + \cos A}}\right)$$

$$\int \frac{dx}{\sqrt{1 + \sin A}} = \frac{\sqrt{2}}{b} \ln\left[\tan\left(\frac{A}{4} + \frac{\pi}{8}\right)\right]$$

$$\int \frac{dx}{\sqrt{1 - \sin A}} = \frac{\sqrt{2}}{b} \ln\left[\tan\left(\frac{A}{4} - \frac{\pi}{8}\right)\right]$$

$$\int \frac{dx}{\sqrt{p + q \sin A}} = \frac{-2}{b\sqrt{p + q}} F\left(\sqrt{\frac{2}{1 + \lambda}}, \sin^{-1}\sqrt{\frac{1 - \sin A}{2}}\right)$$

$$\int \frac{dx}{\sqrt{p - q \sin A}} = \sqrt{\frac{2}{qb^2}} F\left(\sqrt{\lambda}, \sin^{-1}\sqrt{\frac{1 - \sin A}{\lambda - \sin A}}\right)$$

$$\int \frac{\cos A \, dx}{\sqrt{p \pm q \sin A}} = \frac{\pm 1}{2bq} \sqrt{p \pm q \sin A}$$

$$\int \frac{\sin A \, dx}{\sqrt{p + q \sin A}} = \frac{2p}{b\sqrt{p + q}} F\left(\sqrt{\frac{2}{1 + \lambda}}, \sin^{-1}\sqrt{\frac{1 - \sin A}{2}}\right) - \frac{2\sqrt{p + q}}{qb} E\left(\sqrt{\frac{2}{1 + \lambda}}, \sin^{-1}\sqrt{\frac{1 - \sin A}{2}}\right)$$

$$\int \frac{\sin A \, dx}{\sqrt{p - q \sin A}} = -\sqrt{\frac{2}{qb^2}} F\left(\sqrt{\lambda}, \sin^{-1}\sqrt{\frac{1 - \sin A}{\lambda - \sin A}}\right) - \sqrt{\frac{q}{2b^2}} E\left(\sqrt{\lambda}, \sin^{-1}\sqrt{\frac{1 - \sin A}{\lambda - \sin A}}\right)$$

$$\int \sqrt{a + p \sin A + q \cos A} \, dx = -\frac{2}{b} \sqrt{a + r} \, E\left(\sqrt{\frac{2r}{a + r}}, \sin^{-1}\sqrt{\frac{r - p \sin A - q \cos A}{2r}}\right)$$

(62) Indefinite Integrals Involving $\boxed{f(x) = f(\sqrt{p \pm q \cos A})}$ $\boxed{A = bx}$

$E(k, \phi) = $ incomplete elliptic integral of the second kind (Sec. 13.05)

$F(k, \phi) = $ incomplete elliptic integral of the first kind (Sec. 13.05)

$G(k, \phi) = \tan \phi \sqrt{1 - k^2 \sin \phi}, r = \sqrt{p^2 + q^2}, \lambda = p/q, a > r, p > 0, q > 0$

$$\int \sqrt{1 + \cos A}\ dx = \frac{2\sqrt{2}}{b} \sin \frac{A}{2}$$

$$\int \sqrt{1 - \cos A}\ dx = -\frac{2\sqrt{2}}{b} \cos \frac{A}{2}$$

$$\int \sqrt{p + q \cos A}\ dx = \frac{2}{b} \sqrt{p + q}\ E\left(\sqrt{\frac{2}{1 + \lambda}}, \frac{A}{2}\right)$$

$$\int \sqrt{p - q \cos A}\ dx = \frac{2}{b} \sqrt{p + q}\ E\left(\sqrt{\frac{2}{1 + \lambda}}, \sin^{-1}\sqrt{\frac{(p + q)(1 - \cos A)}{2(p - q \cos A)}}\right) - \frac{2q}{b} \frac{\sin A}{\sqrt{p - q \cos A}}$$

$$\int \frac{dx}{\sqrt{\cos 2A}} = \frac{1}{b\sqrt{2}} F\left(\frac{1}{\sqrt{2}}, \sin^{-1}\left(\sqrt{2} \sin A\right)\right)$$

$$\int \frac{dx}{\sqrt{1 + \cos A}} = \frac{\sqrt{2}}{b} \ln\left(\tan \frac{\pi + A}{4}\right)$$

$$\int \frac{dx}{\sqrt{1 - \cos A}} = \frac{\sqrt{2}}{b} \ln\left(\tan \frac{A}{4}\right)$$

$$\int \frac{dx}{\sqrt{p + q \cos A}} = \frac{2}{b\sqrt{p + q}} F\left(\sqrt{\frac{2}{1 + \lambda}}, \frac{A}{2}\right)$$

$$\int \frac{dx}{\sqrt{p - q \cos A}} = \frac{2}{b\sqrt{p + q}} F\left(\sqrt{\frac{2}{1 + \lambda}}, \sin^{-1}\sqrt{\frac{(p + q)(1 - \cos A)}{2(p - q \cos A)}}\right)$$

$$\int \frac{\sin A\ dx}{\sqrt{p \pm q \cos A}} = \frac{\mp 1}{2bq} \sqrt{p \pm q \cos A}$$

$$\int \frac{\cos A\ dx}{\sqrt{p + q \cos A}} = \frac{2\sqrt{p + q}}{bq} E\left(\sqrt{\frac{2}{1 + \lambda}}, \frac{A}{2}\right) - \frac{2\lambda}{b\sqrt{p + q}} F\left(\sqrt{\frac{2}{1 + \lambda}}, \frac{A}{2}\right)$$

$$\int \frac{\cos A\ dx}{\sqrt{p - q \cos A}} = \frac{2}{bq\sqrt{p + q}} E\left(\sqrt{\frac{2}{1 + \lambda}}, \frac{A}{2}\right) - \frac{2\sqrt{p + q}}{b\lambda} F\left(\sqrt{\frac{2}{1 + \lambda}}, \frac{A}{2}\right)$$

$$\int \frac{dx}{\sqrt{a + p \sin A + q \cos A}} = \frac{2}{b\sqrt{a + r}} F\left(\sqrt{\frac{2r}{a + r}}, \sin^{-1}\sqrt{\frac{r - p \sin A - q \cos A}{2r}}\right)$$

(63) Indefinite Integrals Involving

$$f(x) = f(\sqrt{1 \pm a^2 \sin^2 A}) \qquad A = bx$$

$E(k, \phi) =$ incomplete elliptic integral of the second kind (Sec. 13.05)

$F(k, \phi) =$ incomplete elliptic integral of the first kind (Sec. 13.05)

$$\alpha = \sqrt{\frac{1}{1+a^2}} \qquad \beta = \sqrt{\frac{1}{1-a^2}} \qquad \lambda = \sqrt{\frac{a^2}{1+a^2}}$$

$$\int \sqrt{1+a^2 \sin^2 A}\, dx = -\frac{1}{\alpha b} E\left(\lambda, \frac{\pi}{2}-A\right)$$

$$\int \sin A \sqrt{1+a^2 \sin^2 A}\, dx = -\frac{\cos A}{2b}\sqrt{1+a^2 \sin^2 A} - \frac{\sin^{-1}(\lambda \cos A)}{2ab\alpha^2}$$

$$\int \cos A \sqrt{1+a^2 \sin^2 A}\, dx = \frac{\sin A}{2b}\sqrt{1+a^2 \sin^2 A} + \frac{\ln (a \sin A + \sqrt{1+a^2 \sin^2 A})}{2ab}$$

$$\int \sqrt{1-a^2 \sin^2 A}\, dx = \frac{1}{b} E(a, A)$$

$$\int \sin A \sqrt{1-a^2 \sin^2 A}\, dx = -\frac{\cos A}{2b}\sqrt{1-a^2 \sin^2 A} - \frac{\ln (a \cos A + \sqrt{1-a^2 \sin^2 A})}{2\,ab\beta^2}$$

$$\int \cos A \sqrt{1-a^2 \sin^2 A}\, dx = \frac{\sin A}{2b}\sqrt{1-a^2 \sin^2 A} + \frac{\sin^{-1}(a \sin A)}{2ab}$$

$$\int \frac{dx}{\sqrt{1+a^2 \sin^2 A}} = -\frac{\alpha}{b} F\left(\lambda, \frac{\pi}{2}-A\right)$$

$$\int \frac{\sin A\, dx}{\sqrt{1+a^2 \sin^2 A}} = -\frac{1}{ab} \sin^{-1}(\lambda \cos A)$$

$$\int \frac{\cos A\, dx}{\sqrt{1+a^2 \sin^2 A}} = \frac{1}{ab} \ln (a \sin A + \sqrt{1+a^2 \sin^2 A})$$

$$\int \frac{dx}{\sqrt{1-a^2 \sin^2 A}} = \frac{1}{b} F(a, A)$$

$$\int \frac{\sin A\, dx}{\sqrt{1-a^2 \sin^2 A}} = -\frac{1}{ab} \ln (a \cos A + \sqrt{1-a^2 \sin^2 A})$$

$$\int \frac{\cos A\, dx}{\sqrt{1-a^2 \sin^2 A}} = \frac{1}{ab} \sin^{-1}(a \sin A)$$

$$\int \frac{dx}{\sqrt{p^2 \cos^2 A + q^2 \sin^2 A}} = -\frac{1}{bp} F\left(\sqrt{1-\frac{q^2}{p^2}}, A\right) \qquad q < p$$

(64) Indefinite Integrals Involving

$$f(x) = f(\sqrt{1 \pm a^2 \cos^2 A}) \qquad A = bx$$

$E(k, \phi) =$ incomplete elliptic integral of the second kind (Sec. 13.05)

$F(k, \phi) =$ incomplete elliptic integral of the first kind (Sec. 13.05)

$$\alpha = \sqrt{\frac{1}{1 + a^2}} \qquad \beta = \sqrt{\frac{1}{1 - a^2}} \qquad \lambda = \sqrt{\frac{a^2}{1 + a^2}} \qquad \kappa = \sqrt{\frac{a^2}{1 - a^2}}$$

$$\int \sqrt{1 + a^2 \cos^2 A}\, dx = \frac{1}{b\alpha} E(\lambda, A)$$

$$\int \sin A \sqrt{1 + a^2 \cos^2 A}\, dx = -\frac{\cos A}{2\alpha b} \sqrt{1 - \lambda^2 \sin^2 A} - \frac{\ln(\lambda \cos A + \sqrt{1 - \lambda^2 \sin^2 a})}{2ab}$$

$$\int \cos A \sqrt{1 + a^2 \cos^2 A}\, dx = \frac{\sin A}{2\alpha b} \sqrt{1 - \lambda^2 \sin^2 A} + \frac{\sin^{-1}(\lambda \sin A)}{2ab\alpha^2}$$

$$\int \sqrt{1 - a^2 \cos^2 A}\, dx = -\frac{1}{b\beta} E\left(\kappa, \frac{\pi}{2} - A\right)$$

$$\int \sin A \sqrt{1 - a^2 \cos^2 A}\, dx = -\frac{\cos A}{2b\beta} \sqrt{1 + \kappa^2 \sin^2 A} - \frac{\sin^{-1}(a \cos A)}{2ab}$$

$$\int \cos A \sqrt{1 - a^2 \cos^2 A}\, dx = \frac{\sin A}{2b\beta} \sqrt{1 + \kappa^2 \sin^2 A} + \frac{\ln(\kappa \sin A + \sqrt{1 + \kappa^2 \sin^2 A})}{2ab\beta^2}$$

$$\int \frac{dx}{\sqrt{1 + a^2 \cos^2 A}} = \frac{\alpha}{b} F(\lambda, A)$$

$$\int \frac{\sin A\, dx}{\sqrt{1 + a^2 \cos^2 A}} = -\frac{1}{ab} \ln(\lambda \cos A + \sqrt{1 - \lambda^2 \sin^2 A})$$

$$\int \frac{\cos A\, dx}{\sqrt{1 + a^2 \cos^2 A}} = \frac{1}{ab} \sin^{-1}(\lambda \sin A)$$

$$\int \frac{dx}{\sqrt{1 - a^2 \cos^2 A}} = -\frac{\beta}{b} F\left(\kappa, \frac{\pi}{2} - A\right)$$

$$\int \frac{\sin A\, dx}{\sqrt{1 - a^2 \cos^2 A}} = -\frac{1}{ab} \sin^{-1}(a \cos A)$$

$$\int \frac{\cos A\, dx}{\sqrt{1 - a^2 \cos^2 A}} = \frac{1}{ab} \ln(\kappa \sin A + \sqrt{1 + \kappa^2 \sin^2 A})$$

$$\int \frac{dx}{\sqrt{p^2 \cos^2 A - q^2 \sin^2 A}} = \frac{1}{b\sqrt{p^2 + q^2}} F\left(\sqrt{\frac{p^2}{p^2 + q^2}}, \cos^{-1} \frac{\sqrt{p^2 \cos^2 A + q^2 \sin^2 A}}{p}\right)$$

(65) Indefinite Integrals Involving $\boxed{f(x) = f(x, \tan A)}$ $\boxed{A = bx}$

$$\int \tan A \, dx = -\frac{\ln (\cos A)}{b}$$

$$\int \tan^2 A \, dx = \frac{\tan A}{b} - x$$

$$\int \tan^3 A \, dx = \frac{\tan^2 A}{2b} + \frac{\ln (\cos A)}{b}$$

$$\int \tan^4 A \, dx = \frac{\tan^3 A}{3b} - \frac{\tan A}{b} + x$$

$$\int \tan^n A \, dx = \frac{\tan^{n-1} A}{(n-1)b} - \int \tan^{n-2} A \, dx \qquad n > 1$$

$$\int \frac{dx}{\tan A} = \frac{\ln (\sin A)}{b}$$

$$\int \frac{dx}{\cos^2 A \tan A} = \frac{\ln (\tan A)}{b}$$

$$\int \frac{dx}{\tan^2 A} = -\frac{\cot A}{b} - x$$

$$\int \frac{dx}{\tan^3 A} = -\frac{\cot^2 A}{2b} - \frac{\ln (\sin A)}{b}$$

$$\int \frac{dx}{\tan^n A} = \frac{-1}{(n-1)b \tan^{n-1} A} - \int \frac{dx}{\tan^{n-2} A} \qquad n > 1$$

$$\int x \tan A \, dx = \frac{A}{b^2} \sum_{k=1}^{\infty} a_k \frac{(2A)^{2k}}{2k+1} = \frac{A^3}{3b^2} \bigwedge_{k=1}^{\infty} \left[1 + \frac{(2k+1)a_{k+1}}{(2k+3)a_k} (2A)^2 \right]^*$$

$$\int \frac{\tan A \, dx}{x} = \frac{1}{A} \sum_{k=1}^{\infty} a_k \frac{(2A)^{2k}}{2k-1} = A \bigwedge_{k=1}^{\infty} \left[1 + \frac{(2k-1)a_{k+1}}{(2k+1)a_k} (2A)^2 \right]^*$$

$$\int \frac{dx}{p + q \tan A} = \frac{1}{b(p^2 + q^2)} [pA + q \ln (q \sin A + p \cos A)]$$

$$\int \frac{\tan A \, dx}{p + q \tan A} = \frac{1}{b(p^2 + q^2)} [qA - p \ln (q \sin A + p \cos A)]$$

$$\int \frac{dx}{\sqrt{p + q \tan^2 A}} = \frac{1}{b\sqrt{p-q}} \sin^{-1} \left(\sqrt{\frac{p-q}{p}} \sin A \right) \qquad p > q$$

$$\int \frac{\tan A \, dx}{\sqrt{p + q \tan^2 A}} = \frac{-1}{b\sqrt{p-q}} \ln (\sqrt{p-q} \cos A + \sqrt{p \cos^2 A + q \sin^2 A}) \qquad p > q$$

$$\int \frac{\sin A \, dx}{\sqrt{p + q \tan^2 A}} = \frac{1}{b(q-p)} \sqrt{p \cos^2 A + q \sin^2 A}$$

*a_k = numerical factor (Sec. 8.15–2); for nested sum refer to Sec. 8.11.

(66) Indefinite Integrals Involving

$$f(x) = f(x, \cot A) \qquad A = bx$$

$$\int \cot A \, dx = \frac{\ln (\sin A)}{b}$$

$$\int \cot^2 A \, dx = -\frac{\cot A}{b} - x$$

$$\int \cot^3 A \, dx = -\frac{\cot^2 A}{2b} - \frac{\ln (\sin A)}{b}$$

$$\int \cot^4 A \, dx = -\frac{\cot^3 A}{3b} + \frac{\cot A}{b} + x$$

$$\int \cot^n A \, dx = -\frac{\cot^{n-1} A}{(n-1)b} - \int \cot^{n-2} A \, dx \qquad n > 1$$

$$\int \frac{dx}{\cot A} = -\frac{\ln (\cos A)}{b}$$

$$\int \frac{dx}{\sin^2 A \cot A} = -\frac{\ln (\cot A)}{b}$$

$$\int \frac{dx}{\cot^2 A} = \frac{\tan A}{b} - x$$

$$\int \frac{dx}{\cot^3 A} = \frac{\tan^2 A}{2b} + \frac{\ln (\cos A)}{b}$$

$$\int \frac{dx}{\cot^n A} = \frac{1}{(n-1)b \cot^{n-1} A} - \int \frac{dx}{\cot^{n-2} A} \qquad n > 1$$

$$\int x \cot A \, dx = \frac{A}{b^2} \left[1 - \sum_{k=1}^{\infty} b_k \frac{(2A)^{2k}}{2k+1} \right] = \frac{A}{b^2} \left\{ 1 - \frac{A^2}{9} \bigwedge_{k=1}^{\infty} \left[1 + \frac{(2k+1)b_{k+1}}{(2k+3)b_k} (2A)^2 \right] \right\}^*$$

$$\int \frac{\cot A}{x} \, dx = -\frac{1}{A} \left[1 + \sum_{k=1}^{\infty} b_k \frac{(2A)^{2k}}{2k-1} \right] = -\frac{1}{A} \left\{ 1 + \frac{A^2}{3} \bigwedge_{k=1}^{\infty} \left[1 + \frac{(2k-1)b_{k+1}}{(2k+1)b_k} (2A)^2 \right] \right\}^*$$

$$\int \frac{dx}{p + q \cot A} = \frac{1}{b(p^2 + q^2)} \left[pA - q \ln (p \sin A + q \cos A) \right]$$

$$\int \frac{\cot A \, dx}{p + q \cot A} = \frac{1}{b(p^2 + q^2)} \left[qA + p \ln (p \sin A + q \cos A) \right]$$

$$\int \frac{dx}{\sqrt{p + q \cot^2 A}} = \frac{1}{b\sqrt{p-q}} \cos^{-1} \left(\sqrt{\frac{p-q}{p}} \cos A \right) \qquad p > q$$

$$\int \frac{\cot A \, dx}{\sqrt{p + q \cot^2 A}} = \frac{1}{b\sqrt{p-q}} \ln (\sqrt{p-q} \sin A + \sqrt{p \sin^2 A + q \cos^2 A}) \qquad p > q$$

$$\int \frac{\cos A \, dx}{\sqrt{p + q \cot^2 A}} = \frac{1}{b(p-q)} \sqrt{p \sin^2 A + q \cos^2 A}$$

---

*b_k = numerical factor (Sec. 8.15–2); for nested sum refer to Sec. 8.11.

(67) Indefinite Integrals Involving $\qquad$ $\boxed{f(x) = f(x, \sin^{-1} B) \quad\Big|\quad B = \dfrac{x}{b}}$

$$\int \sin^{-1} B \, dx = x \sin^{-1} B + \sqrt{b^2 - x^2}$$

$$\int x \sin^{-1} B \, dx = \left(\frac{x^2}{2} - \frac{b^2}{4}\right) \sin^{-1} B + \frac{x\sqrt{b^2 - x^2}}{4}$$

$$\int x^2 \sin^{-1} B \, dx = \frac{x^3}{3} \sin^{-1} B + \frac{(2b^2 + x^2)\sqrt{b^2 - x^2}}{9}$$

$$\int x^m \sin^{-1} B \, dx = \frac{x^{m+1}}{m+1} \sin^{-1} B - \frac{1}{m+1} \int \frac{x^{m+1}}{\sqrt{b^2 - x^2}} \, dx \qquad m \neq -1$$

$$\int \frac{\sin^{-1} B \, dx}{x} = B + \frac{B^3}{(2)(3)(3)} + \frac{(1)(3)B^5}{(2)(4)(5)(5)} + \frac{(1)(3)(5)B^7}{(2)(4)(6)(7)(7)} + \cdots$$

$$\int \frac{\sin^{-1} B \, dx}{x^2} = -\frac{\sin^{-1} B}{x} - \frac{1}{b} \ln \frac{b + \sqrt{b^2 - x^2}}{x}$$

$$\int \frac{\sin^{-1} B \, dx}{x^m} = -\frac{\sin^{-1} B}{(m-1)x^{m-1}} + \frac{1}{m-1} \int \frac{dx}{x^{m-1}\sqrt{b^2 - x^2}} \qquad m \neq 1$$

$$\int (\sin^{-1} B)^2 \, dx = x(\sin^{-1} B)^2 - 2x + 2\sqrt{b^2 - x^2} \sin^{-1} B$$

(68) Indefinite Integrals Involving $\qquad$ $\boxed{f(x) = f(x, \tan^{-1} B) \quad\Big|\quad B = \dfrac{x}{b}}$

$$\int \tan^{-1} B \, dx = x \tan^{-1} B - b \ln \sqrt{b^2 + x^2}$$

$$\int x \tan^{-1} B \, dx = \frac{b^2 + x^2}{2} \tan^{-1} B - \frac{bx}{2}$$

$$\int x^m \tan^{-1} B \, dx = \frac{x^{m+1}}{m+1} \tan^{-1} B - \frac{b}{m+1} \int \frac{x^{m+1} \, dx}{b^2 + x^2} \qquad m \neq -1$$

$$\int \frac{\tan^{-1} B \, dx}{x} = B - \frac{B^3}{3^2} + \frac{B^5}{5^2} - \frac{B^7}{7^2} + \cdots$$

$$\int \frac{\tan^{-1} B \, dx}{x^2} = -\frac{1}{b}\left(\frac{\tan^{-1} B}{B} + \ln \frac{\sqrt{1 + B^2}}{B}\right)$$

$$\int \frac{\tan^{-1} B \, dx}{x^m} = -\frac{\tan^{-1} B}{(m-1)x^{m-1}} + \frac{b}{m-1} \int \frac{dx}{(b^2 + x^2)x^{m-1}} \qquad m \neq 1$$

(69) Indefinite Integrals Involving $f(x) = f(x, \cos^{-1} B)$ $B = \dfrac{x}{b}$

$$\int \cos^{-1} B \, dx = x \cos^{-1} B - \sqrt{b^2 - x^2}$$

$$\int x \cos^{-1} B \, dx = \left(\frac{x^2}{2} - \frac{b^2}{4}\right) \cos^{-1} B - \frac{x\sqrt{b^2 - x^2}}{4}$$

$$\int x^2 \cos^{-1} B \, dx = \frac{x^3}{3} \cos^{-1} B - \frac{(2b^2 + x^2)\sqrt{b^2 - x^2}}{9}$$

$$\int x^m \cos^{-1} B \, dx = \frac{x^{m+1}}{m+1} \cos^{-1} B + \frac{1}{m+1} \int \frac{x^{m+1}}{\sqrt{b^2 - x^2}} \, dx \qquad m \neq -1$$

$$\int \frac{\cos^{-1} B \, dx}{x} = \frac{\pi}{2} \ln x - B - \frac{B^3}{(2)(3)(3)} - \frac{(1)(3)B^5}{(2)(4)(5)(5)} - \frac{(1)(3)(5)B^7}{(2)(4)(6)(7)(7)} - \cdots$$

$$\int \frac{\cos^{-1} B \, dx}{x^2} = -\frac{\cos^{-1} B}{x} + \frac{1}{b} \ln \frac{b + \sqrt{b^2 - x^2}}{x}$$

$$\int \frac{\cos^{-1} B \, dx}{x^m} = -\frac{\cos^{-1} B}{(m-1)x^{m-1}} - \frac{1}{m-1} \int \frac{dx}{x^{m-1}\sqrt{b^2 - x^2}} \qquad m \neq 1$$

$$\int (\cos^{-1} B)^2 \, dx = x(\cos^{-1} B)^2 - 2x - 2\sqrt{b^2 - x^2} \cos^{-1} B$$

(70) Indefinite Integrals Involving $f(x) = f(x, \cot^{-1} B)$ $B = \dfrac{x}{b}$

$$\int \cot^{-1} B \, dx = x \cot^{-1} B + b \ln \sqrt{b^2 + x^2}$$

$$\int x \cot^{-1} B \, dx = \frac{b^2 + x^2}{2} \cot^{-1} B + \frac{bx}{2}$$

$$\int x^m \cot^{-1} B \, dx = \frac{x^{m+1}}{m+1} \cot^{-1} B + \frac{b}{m+1} \int \frac{x^{m+1}}{b^2 + x^2} \, dx \qquad m \neq -1$$

$$\int \frac{\cot^{-1} B \, dx}{x} = \frac{\pi}{2} \ln x - B + \frac{B^3}{3^2} - \frac{B^5}{5^2} + \frac{B^7}{7^2} - \cdots$$

$$\int \frac{\cot^{-1} B \, dx}{x^2} = -\frac{1}{b}\left(\frac{\cot^{-1} B}{B} - \ln \frac{\sqrt{1 + B^2}}{B}\right)$$

$$\int \frac{\cot^{-1} B \, dx}{x^m} = -\frac{\cot^{-1} B}{(m-1)x^{m-1}} - \frac{b}{m-1} \int \frac{dx}{(b^2 + x^2)x^{m-1}} \qquad m \neq 1$$

(71) Indefinite Integrals Involving

$$f(x) = f(x^m, e^{\pm A}) \qquad A = bx$$

$$\int e^A\, dx = \frac{e^A}{b}$$

$$\int x e^A\, dx = \frac{(A-1)e^A}{b^2}$$

$$\int x^2 e^A\, dx = \frac{x^3 e^A}{A}\left(1 - \frac{2}{A}\left(1 - \frac{1}{A}\right)\right)$$

$$\int x^3 e^A\, dx = \frac{x^4 e^A}{A}\left(1 - \frac{3}{A}\left(1 - \frac{2}{A}\left(-\frac{1}{A}\right)\right)\right)$$

$$\int x^m e^A\, dx = \frac{x^{m+1} e^A}{A}\left(1 - \frac{m}{A}\left(1 - \frac{m-1}{A}\left(1 - \frac{m-2}{A}\left(1 - \cdots - \frac{1}{A}\right)\right)\right)\right)^*$$

$$\int \frac{e^A}{x}\, dx = \ln x + \frac{A}{1\cdot 1}\left(1 + \frac{A}{2\cdot 2}\left(1 + \frac{2A}{3\cdot 3}\left(1 + \frac{3A}{4\cdot 4}(1 + \cdots)\right)\right)\right)^*$$

$$\int \frac{e^A}{x^2}\, dx = -\frac{e^A}{x} + b\int \frac{e^A}{x}\, dx$$

$$\int \frac{e^A}{x^3}\, dx = -\frac{e^A}{2x^2}(1 + A) + \frac{b^2}{2}\int \frac{e^A}{x}\, dx$$

$$\int \frac{e^A}{x^m}\, dx = -\frac{e^A}{(m-1)x^{m-1}}\left(1 + \frac{A}{m-2}\left(1 + \frac{A}{m-3}\left(1 + \frac{A}{m-4}(1 + \cdots + A)\right)\right)\right)^* + \frac{b^{m-1}}{(m-1)!}\int \frac{e^A}{x}\, dx$$

$$\int \frac{dx}{p + qe^A} = \frac{1}{bp}\left[A - \ln(p + qe^A)\right]$$

$$\int \frac{e^A\, dx}{p + qe^A} = \frac{1}{bp}\ln(p + qe^A)$$

$$\int \frac{dx}{pe^A + qe^{-A}} = \begin{cases} \dfrac{1}{b\sqrt{pq}}\tan^{-1} e^A\sqrt{\dfrac{p}{q}} & pq > 0 \\[3mm] \dfrac{1}{2b\sqrt{-pq}}\ln\dfrac{q + e^A\sqrt{-pq}}{q - e^A\sqrt{-pq}} & pq < 0 \end{cases}$$

$$\int \frac{e^A\, dx}{\sqrt{x}} = 2\sqrt{x}\left(1 + \frac{A}{1\cdot 3}\left(1 + \frac{3A}{2\cdot 5}\left(1 + \frac{5A}{3\cdot 7}(1 + \cdots)\right)\right)\right)^*$$

$$\int \frac{dx}{\sqrt{p + qe^A}} = \begin{cases} \dfrac{1}{b\sqrt{p}}\ln\dfrac{\sqrt{p + qe^A} - \sqrt{p}}{\sqrt{p + qe^A} + \sqrt{p}} & p > 0 \\[3mm] \dfrac{1}{b\sqrt{-p}}\tan^{-1}\dfrac{\sqrt{p + qe^A}}{\sqrt{-p}} & p < 0 \end{cases}$$

$$\int f(x) e^A\, dx = \frac{e^A}{b}\left(f(x) - \frac{1}{b}\left(f'(x) - \frac{1}{b}\left(f''(x) - \frac{1}{b}\left(f'''(x) - \cdots\right)\right)\right)\right)^*$$

*For nested sum refer to Sec. 8.11.

(72) Indefinite Integrals Involving $\quad\quad f(x)=f(x^m, a^A)\quad\quad A=bx \quad\quad B=bx\ln a$

$$\int a^A\,dx=\frac{a^A}{b\ln a}\qquad\qquad\qquad \int xa^A\,dx=\frac{(B-1)\,a^A}{(b\ln a)^2}$$

$$\int x^2a^A\,dx=\frac{x^3a^A}{B}\left(1-\frac{2}{B}\left(1-\frac{1}{B}\right)\right)$$

$$\int x^3a^A\,dx=\frac{x^4a^A}{B}\left(1-\frac{3}{B}\left(1-\frac{2}{B}\left(1-\frac{1}{B}\right)\right)\right)$$

$$\int x^ma^A\,dx=\frac{x^{m+1}a^A}{B}\left(1-\frac{m}{B}\left(1-\frac{m-1}{B}\left(1-\frac{m-2}{B}\left(1-\cdots-\frac{1}{B}\right)\right)\right)\right)^*$$

$$\int\frac{a^A}{x}\,dx=\ln x+\frac{B}{1\cdot 1}\left(1+\frac{B}{2\cdot 2}\left(1+\frac{2B}{3\cdot 3}\left(1+\frac{3B}{4\cdot 4}\left(1+\cdots\right)\right)\right)\right)^*$$

$$\int\frac{a^A}{x^2}\,dx=-\frac{a^A}{x}+(b\ln a)\int\frac{a^A}{x}\,dx$$

$$\int\frac{a^A}{x^3}\,dx=-\frac{a^A}{2x^2}\,(1+B)+\frac{(b\ln a)^2}{2}\int\frac{a^A}{x}\,dx$$

$$\int\frac{a^A}{x^m}\,dx=-\frac{a^A}{(m-1)x^{m-1}}\left(1+\frac{B}{m-2}\left(1+\frac{B}{m-3}\left(1+\cdots+B\right)\right)\right)^*+\frac{(b\ln a)^{m-1}}{(m-1)!}\int\frac{a^A}{x}\,dx$$

$$\int\frac{dx}{p+qa^A}=\frac{x}{p}-\frac{x}{pB}\ln\,(p+qa^A)\qquad\qquad \int\frac{a^A\,dx}{p+qa^A}=\frac{x}{pB}\ln\,(p+qa^A)$$

$$\int\frac{dx}{pa^A+qa^{-A}}=\begin{cases}\dfrac{x}{B\sqrt{pq}}\tan^{-1}\left(a^A\sqrt{\dfrac{p}{q}}\right) & pq>0 \\[3ex] \dfrac{x}{2B\sqrt{-pq}}\ln\dfrac{q+a^A\sqrt{-pq}}{q-a^A\sqrt{-pq}} & pq<0\end{cases}$$

$$\int\frac{a^A\,dx}{\sqrt{x}}=2\sqrt{x}\left(1+\frac{B}{1\cdot 3}\left(1+\frac{3B}{2\cdot 5}\left(1+\frac{5B}{3\cdot 7}\left(1+\cdots\right)\right)\right)\right)^*$$

$$\int\frac{dx}{\sqrt{p+qa^A}}=\begin{cases}\dfrac{x}{B\sqrt{p}}\ln\dfrac{\sqrt{p+qa^A}-\sqrt{p}}{\sqrt{p+qa^A}+\sqrt{p}} & p>0 \\[3ex] \dfrac{x}{B\sqrt{-p}}\tan^{-1}\dfrac{\sqrt{p+qa^A}}{\sqrt{-p}} & p<0\end{cases}$$

$$\int f(x)a^A\,dx=\frac{a^A}{b\ln a}\left(f(x)-\frac{1}{b\ln a}\left(f'(x)-\frac{1}{b\ln a}\left(f''(x)-\frac{1}{b\ln a}\left(f'''(x)-\cdots\right)\right)\right)\right)^*$$

*For nested sum refer to Sec. 8.11.

(73) Indefinite Integrals Involving $f(x) = f(x^m, e^A, \sin B, \cos B)$ $A = \alpha x + a, B = \beta x + b$

$|\alpha| > 0, |\beta| > 0$ $|a| \geq 0, |b| \geq 0$ $\omega = \sqrt{\alpha^2 + \beta^2}$ $\phi = \tan^{-1}\left(-\dfrac{\beta}{\alpha}\right)$

$$\int e^{\alpha x} \sin \beta x\, dx = \frac{e^{\alpha x}}{\omega} \sin (\beta x + \phi) \qquad\qquad \int e^{\alpha x} \cos \beta x\, dx = \frac{e^{\alpha x}}{\omega} \cos (\beta x + \phi)$$

$$\int e^{\alpha x} \sin^2 \beta x\, dx = \frac{e^{\alpha x}}{2\alpha}\left[1 - \frac{\alpha}{\alpha^2 + 4\beta^2}(\alpha \cos 2\beta x + 2\beta \sin 2\beta x)\right]$$

$$\int e^{\alpha x} \sin^{2n} \beta x\, dx = \frac{e^{\alpha x}}{\alpha 2^{2n}}\left\{\frac{(2n)!}{(n!)^2} + 2\alpha \sum_{k=1}^{n} \frac{(-1)^k (2n)!\,[\alpha \cos (2k\beta x) + 2k\beta \sin (2k\beta x)]}{(\alpha^2 + k^2\beta^2)(n+k)!(n-k)!}\right\}$$

$$\int e^{\alpha x} \sin^{2n+1} \beta x\, dx = \frac{e^{\alpha x}}{2^{2n}} \sum_{k=0}^{n} \frac{(-1)^k (2n+1)!\,[\alpha \sin (2k+1)\beta x - (2k+1)\beta \cos (2k+1)\beta x]}{[\alpha^2 + (2k+1)^2\beta^2](n+k+1)!(n-k)!}$$

$$\int x e^A \sin B\, dx = \frac{x e^A}{\omega}\left[\sin (B + \phi) - \frac{\sin (B + 2\phi)}{\omega x}\right]$$

$$\int x^2 e^A \sin B\, dx = \frac{x^2 e^A}{\omega}\left[\sin (B + \phi) - \frac{2 \sin (B + 2\phi)}{\omega x} + \frac{2 \sin (B + 3\phi)}{(\omega x)^2}\right]$$

$$\int x^m e^A \sin B\, dx = \frac{x^m e^A}{\omega} \sin (B + \phi) \bigwedge_{k=1}^{m}\left[1 - \frac{(m+1-k) \sin (B + \phi + k\phi)}{\omega x \sin (B + k\phi)}\right]^*$$

$$\int e^{\alpha x} \cos^2 \beta x\, dx = \frac{e^{\alpha x}}{2\alpha}\left[1 + \frac{\alpha}{\alpha^2 + 4\beta^2}(\alpha \cos 2\beta x + 2\beta \sin 2\beta x)\right]$$

$$\int e^{\alpha x} \cos^{2n} \beta x\, dx = \frac{e^{\alpha x}}{\alpha 2^{2n}}\left\{\frac{(2n)!}{(n!)^2} + 2\alpha \sum_{k=1}^{n} \frac{(2n)!\,[\alpha \cos (2k\beta x) + 2k\beta \sin (2k\beta x)]}{(\alpha^2 + 4k^2\beta^2)(n+k)!(n-k)!}\right\}$$

$$\int e^{\alpha x} \cos^{2n+1} \beta x\, dx = \frac{e^{\alpha x}}{2^{2n}} \sum_{k=0}^{n} \frac{(2n+1)!\,[\alpha \cos (2k+1)\beta x + (2k+1)\beta \sin (2k+1)\beta x]}{[\alpha^2 + (2k+1)^2\beta^2](n+k+1)!(n-k)!}$$

$$\int x e^A \cos B\, dx = \frac{x e^A}{\omega}\left[\cos (B + \phi) - \frac{\cos (B + 2\phi)}{\omega x}\right]$$

$$\int x^2 e^A \cos B\, dx = \frac{x^2 e^A}{\omega}\left[\cos (B + \phi) - \frac{2 \cos (B + 2\phi)}{\omega x} + \frac{2 \cos (B + 3\phi)}{(\omega x)^2}\right]$$

$$\int x^m e^A \cos B\, dx = \frac{x^m e^A}{\omega} \cos (B + \phi) \bigwedge_{k=1}^{m}\left[1 - \frac{(m+1-k) \cos (B + \phi + k\phi)}{\omega x \cos (B + k\phi)}\right]^*$$

$$\int \frac{e^{\alpha x}}{\sin \beta x}\, dx = \int \frac{1}{\sin \beta x}\left[\sum_{k=0}^{\infty} \frac{(\alpha x)^k}{k!}\right] dx\dagger \qquad\qquad \int \frac{e^{\alpha x}}{\cos \beta x}\, dx = \int \frac{1}{\cos \beta x}\left[\sum_{k=0}^{\infty} \frac{(\alpha x)^k}{k!}\right] dx\dagger$$

*For nested sum refer to Sec. 8.11.
†No closed form available; both functions must be expressed in terms of their respective series and the product
(Sec. 8.14–1) integrated term by term.

(74) Indefinite Integrals Involving $\boxed{f(x)=f(e^A,\sin B,\cos B,\sin C,\cos C)}$ $\boxed{A=\alpha x, B=\beta x, C=\gamma x}$

$$\omega_1=\sqrt{\alpha^2+(\beta+\gamma)^2} \qquad \omega_2=\sqrt{\alpha^2+(\beta-\gamma)^2} \qquad \phi_1=\tan^{-1}\frac{\alpha}{\beta+\gamma} \qquad \phi_2=\tan^{-1}\frac{\alpha}{\beta-\gamma}$$

$$\int e^A \sin B \sin C\, dx = -\frac{e^A}{2}\left[\frac{\sin(B+C+\phi_1)}{\omega_1}-\frac{\sin(B-C+\phi_2)}{\omega_2}\right]$$

$$\int e^A \sin B \cos C\, dx = -\frac{e^A}{2}\left[\frac{\cos(B+C+\phi_1)}{\omega_1}+\frac{\cos(B-C+\phi_2)}{\omega_2}\right]$$

$$\int e^A \cos B \sin C\, dx = -\frac{e^A}{2}\left[\frac{\cos(B+C+\phi_1)}{\omega_1}-\frac{\cos(B-C+\phi_2)}{\omega_2}\right]$$

$$\int e^A \cos B \cos C\, dx = +\frac{e^A}{2}\left[\frac{\sin(B+C+\phi_1)}{\omega_1}+\frac{\sin(B-C+\phi_2)}{\omega_2}\right]$$

$$\phi_1=\tan^{-1}\frac{\alpha}{\gamma} \qquad \phi_{2,3}=\tan^{-1}\frac{\alpha}{2\beta\pm\gamma} \qquad \phi_4=\tan^{-1}\frac{\alpha}{\beta} \qquad \phi_{5,6}=\tan^{-1}\frac{\alpha}{\beta\pm2\gamma}$$

$$\int e^A \sin^2 B \cos C\, dx = +\frac{e^A}{4}\left[\frac{2\sin(C+\phi_1)}{\sqrt{\alpha^2+\gamma^2}}-\frac{\sin(2B+C+\phi_2)}{\sqrt{\alpha^2+(2\beta+\gamma)^2}}-\frac{\sin(2B-C+\phi_3)}{\sqrt{\alpha^2+(2\beta-\gamma)^2}}\right]$$

$$\int e^A \sin B \cos^2 C\, dx = -\frac{e^A}{4}\left[\frac{2\cos(B+\phi_4)}{\sqrt{\alpha^2+\beta^2}}+\frac{\cos(B+2C+\phi_5)}{\sqrt{\alpha^2+(\beta+2\gamma)^2}}+\frac{\cos(B-2C+\phi_6)}{\sqrt{\alpha^2+(\beta-2\gamma)^2}}\right]$$

$$B+b=\beta x+b \qquad \omega=\sqrt{\alpha^2+4\beta^2} \qquad \phi=\tan^{-1}\frac{\alpha}{2\beta}$$

$$\int e^A \sin B \sin(B+b) = +\frac{e^A}{2}\left[\frac{\cos b}{\alpha}-\frac{\sin(2B+b+\phi)}{\omega}\right]$$

$$\int e^A \sin B \cos(B+b) = -\frac{e^A}{2}\left[\frac{\sin b}{\alpha}+\frac{\cos(2B+b+\phi)}{\omega}\right]$$

$$\int e^A \cos B \sin(B+b) = +\frac{e^A}{2}\left[\frac{\sin b}{\alpha}-\frac{\cos(2B+b+\phi)}{\omega}\right]$$

$$\int e^A \cos B \cos(B+b) = +\frac{e^A}{2}\left[\frac{\cos b}{\alpha}+\frac{\sin(2B+b+\phi)}{\omega}\right]$$

(75) Indefinite Integrals Involving $\boxed{f(x)=f(e^A,\tan x,\cot x)}$ $\boxed{A=\alpha x}$

$$\int e^A \tan^n x\, dx = +\frac{e^A}{n-1}\tan^{n-1}x-\frac{\alpha}{n-1}\int e^A \tan^{n-1}x\, dx-\int e^A \tan^{n-2}x\, dx$$

$$\int e^A \cot^n x\, dx = -\frac{e^A}{n-1}\cot^{n-1}x+\frac{\alpha}{n-1}\int e^A \cot^{n-1}x\, dx-\int e^A \cot^{n-2}x\, dx$$

(76) Indefinite Integrals Involving $f(x) = f(x^m, \sinh A)$ $\boxed{A = bx}$

$$D_1 = \int_0^{} \sinh^n A \; dx \qquad D_2 = \int D_1 \; dx \qquad D_k = \int D_{k-1} \; dx$$

$$\int \sinh A \; dx = \frac{\cosh A}{b}$$

$$\int \sinh^2 A \; dx = \frac{\sinh A \cosh A}{2b} - \frac{x}{2}$$

$$\int \sinh^n A \; dx = \frac{\sinh^{n-1} A \cosh A}{bn} - \frac{n-1}{n} \int \sinh^{n-2} A \; dx$$

$$\int x \sinh A \; dx = \frac{x \cosh A}{b} - \frac{\sinh A}{b^2}$$

$$\int x^2 \sinh A \; dx = \left(\frac{x^2}{b} + \frac{2}{b^3}\right) \cosh A - \frac{2x}{b^2} \sinh A$$

$$\int x^m \sinh A \; dx = \frac{x^m \cosh A}{b} - \frac{m}{b} \int x^{m-1} \cosh A \; dx$$

$$\int x^m \sinh^n A \; dx = x^m D_1 - m x^{m-1} D_2 + m(m-1) x^{m-2} D_3 - \cdots (-1)^m m! D_{m+1}$$

$$\int \frac{\sinh A \; dx}{x} = A + \frac{A^3}{(3)(3!)} + \frac{A^5}{(5)(5!)} + \cdots$$

$$\int \frac{\sinh A \; dx}{x^2} = -\frac{\sinh A}{x} + b\left(\ln x + \frac{A^2}{(2)(2!)} + \frac{A^4}{(4)(4!)} + \cdots\right)$$

$$\int \frac{\sinh A \; dx}{x^m} = \frac{-\sinh A}{(m-1)x^{m-1}} + \frac{b}{m-1} \int \frac{\cosh A}{x^{m-1}} dx \qquad m \neq 1$$

$$\int \frac{dx}{\sinh A} = \frac{\ln[\tanh(A/2)]}{b}$$

$$\int \frac{dx}{\sinh^2 A} = -\frac{\coth A}{b}$$

$$\int \frac{dx}{\sinh^n A} = \frac{-\cosh A}{b(n-1)\sinh^{n-2} A} - \frac{n-2}{n-1} \int \frac{dx}{\sinh^{n-2} A} \qquad n \neq 1$$

$$\int \frac{x \; dx}{\sinh^n A} = \frac{-x \cosh A}{b(n-1)\sinh^{n-1} A} - \frac{1}{b^2(n-1)(n-2)\sinh^{n-2} A} - \frac{n-2}{n-1} \int \frac{x \; dx}{\sinh^{n-2} A} \qquad n \neq 1, 2$$

(77) Indefinite Integrals Involving $\quad\boxed{f(x) = f(x^m, \cosh A)}\quad\boxed{A = bx}$

$$D_1 = \int \cosh^n A \, dx \qquad D_2 = \int D_1 \, dx \qquad D_k = \int D_{k-1} \, dx$$

$$\int \cosh A \, dx = \frac{\sinh A}{b}$$

$$\int \cosh^2 A \, dx = \frac{\sinh A \cosh A}{2b} + \frac{x}{2}$$

$$\int \cosh^n A \, dx = \frac{\cosh^{n-1} A \sinh A}{bn} + \frac{n-1}{n} \int \cosh^{n-2} A \, dx$$

$$\int x \cosh A \, dx = \frac{x \sinh A}{b} - \frac{\cosh A}{b^2}$$

$$\int x^2 \cosh A \, dx = \left(\frac{x^2}{b} + \frac{2}{b^3}\right) \sinh A - \frac{2x}{b^2} \cosh A$$

$$\int x^m \cosh A \, dx = \frac{x^m \sinh A}{b} - \frac{m}{b} \int x^{m-1} \sinh A \, dx$$

$$\int x^m \cosh^n A \, dx = x^m D_1 - mx^{m-1} D_2 + m(m-1)x^{m-1} D_3 - \cdots (-1)^m m! D_{m+1}$$

$$\int \frac{\cosh A \, dx}{x} = \ln x + \frac{A^2}{(2)(2!)} + \frac{A^4}{(4)(4!)} + \cdots$$

$$\int \frac{\cosh A \, dx}{x^2} = -\frac{\cosh A}{x} + b\left[A + \frac{A^3}{(3)(3!)} + \frac{A^5}{(5)(5!)} + \cdots\right]$$

$$\int \frac{\cosh A \, dx}{x^m} = \frac{-\cosh A}{(m-1)x^{m-1}} + \frac{b}{m-1} \int \frac{\sinh A}{x^{m-1}} dx \qquad m \neq 1$$

$$\int \frac{dx}{\cosh A} = \frac{\tan^{-1}(\sinh A)}{b}$$

$$\int \frac{dx}{\cosh^2 A} = \frac{\tanh A}{b}$$

$$\int \frac{dx}{\cosh^n A} = \frac{\sinh A}{b(n-1)\cosh^{n-1} A} + \frac{n-2}{n-1} \int \frac{dx}{\cosh^{n-2} A} \qquad n \neq 1$$

$$\int \frac{x \, dx}{\cosh^n A} = \frac{x \sinh A}{b(n-1)\cosh^{n-1} A} + \frac{1}{b^2(n-1)(n-2)\cosh^{n-2} A} + \frac{n-2}{n-1} \int \frac{x \, dx}{\cosh^{n-2} A} \qquad n \neq 1, 2$$

(78) Indefinite Integrals Involving $\qquad$ $f(x) = f(\sinh A, \cosh A)$ $\quad$ $A = bx$

$$\int \sinh A \cosh A \, dx = \frac{\sinh^2 A}{2b}$$

$$\int \sinh^2 A \cosh^2 A \, dx = \frac{\sinh 4A}{32b} - \frac{x}{8}$$

$$\int \sinh^n A \cosh A \, dx = \frac{\sinh^{n+1} A}{(n+1)b} \qquad n \neq -1$$

$$\int \sinh A \cosh^n A \, dx = \frac{\cosh^{n+1} A}{(n+1)b} \qquad n \neq -1$$

$$\int \frac{dx}{\sinh A \cosh A} = \frac{\ln (\tanh A)}{b}$$

$$\int \frac{dx}{\sinh^2 A \cosh A} = -\frac{\tan^{-1} (\sinh A) + \operatorname{csch} A}{b}$$

$$\int \frac{dx}{\sinh A \cosh^2 A} = \frac{\ln [\tanh (A/2)] + \operatorname{sech} A}{b}$$

$$\int \frac{dx}{\sinh^2 A \cosh^2 A} = -\frac{2 \coth 2A}{b}$$

$$\int \frac{\sinh^2 A}{\cosh A} \, dx = \frac{\sinh A - \tan^{-1} (\sinh A)}{b}$$

$$\int \frac{\cosh^2 A}{\sinh A} \, dx = \frac{\cosh A + \ln [\tanh (A/2)]}{b}$$

$$\int \frac{\sinh A}{\cosh^n A} \, dx = -\frac{1}{(n-1)b \cosh^{n-1} A} \qquad n \neq 1$$

$$\int \frac{\cosh A}{\sinh^n A} \, dx = -\frac{1}{(n-1)b \sinh^{n-1} A} \qquad n \neq 1$$

$$\int \frac{\sinh A}{\cosh A \pm 1} \, dx = \frac{1}{b} \ln (\cosh A \pm 1) \qquad \int \frac{\cosh A}{1 \pm \sinh A} \, dx = \pm \frac{1}{b} \ln (1 \pm \sinh A)$$

$$\int e^{ax} \sinh A \, dx = \frac{(a \sinh A - b \cosh A) e^{ax}}{a^2 - b^2}$$
$$\qquad a^2 \neq b^2$$
$$\int e^{ax} \cosh A \, dx = \frac{(a \cosh A - b \sinh A) e^{ax}}{a^2 - b^2}$$

(79) Indefinite Integrals Involving $\qquad$ $f(x) = f(\sinh \alpha x, \cosh \alpha x, \sinh \beta x, \cosh \beta x)$

$$\left.\begin{aligned}
\int \sinh \alpha x \sinh \beta x &= \frac{\sinh (\alpha + \beta)x}{2(\alpha + \beta)} - \frac{\sinh (\alpha - \beta)x}{2(\alpha - \beta)} \\[2mm]
\int \sinh \alpha x \cosh \beta x &= \frac{\cosh (\alpha + \beta)x}{2(\alpha + \beta)} + \frac{\cosh (\alpha - \beta)x}{2(\alpha - \beta)} \\[2mm]
\int \cosh \alpha x \cosh \beta x &= \frac{\sinh (\alpha + \beta)x}{2(\alpha + \beta)} + \frac{\sinh (\alpha - \beta)x}{2(\alpha - \beta)}
\end{aligned}\right\} \alpha^2 \neq \beta^2$$

(80) Indefinite Integrals Involving $\qquad$ $f(x) = f(\sinh \alpha x, \cosh \alpha x, \sin \beta x, \cos \beta x)$

$$\int \sinh \alpha x \sin \beta x = \frac{\alpha \cosh \alpha x \sin \beta x - \beta \sinh \alpha x \cos \beta x}{\alpha^2 + \beta^2}$$

$$\int \sinh \alpha x \cos \beta x = \frac{\alpha \cosh \alpha x \cos \beta x + \beta \sinh \alpha x \sin \beta x}{\alpha^2 + \beta^2}$$

$$\int \cosh \alpha x \sin \beta x = \frac{\alpha \sinh \alpha x \sin \beta x - \beta \cosh \alpha x \cos \beta x}{\alpha^2 + \beta^2}$$

$$\int \cosh \alpha x \cos \beta x = \frac{\alpha \sinh \alpha x \cos \beta x + \beta \cosh \alpha x \sin \beta x}{\alpha^2 + \beta^2}$$

(81) Indefinite Integrals Involving $\qquad$ $f(x) = f(\tanh A, \coth A)$ $\quad\boxed{A = bx}$

$$\int \tanh A \, dx = \frac{\ln (\cosh A)}{b} \qquad\qquad \int \tanh^2 A \, dx = x - \frac{\tanh A}{b}$$

$$\int \tanh^n A \, dx = -\frac{\tanh^{n-1} A}{b(n-1)} + \int \tanh^{n-2} A \, dx \qquad n \neq 1$$

$$\int \coth A \, dx = \frac{\ln (\sinh A)}{b} \qquad\qquad \int \coth^2 A \, dx = x - \frac{\coth A}{b}$$

$$\int \coth^n A \, dx = -\frac{\coth^{n-1} A}{b(n-1)} + \int \coth^{n-2} A \, dx \qquad n \neq 1$$

$$\int x^m \tanh x \, dx = -\sum_{k=1}^{\infty} \frac{(-4)^k a_k x^{m+2k}}{m+2k} \qquad\qquad \int x^m \coth x \, dx = \frac{x^m}{m} - \sum_{k=1}^{\infty} \frac{(-4)^k b_k x^{m+2k}}{m+2k}$$

$a_k, b_k =$ numerical factors (Sec. 8.15–2)

(82) Indefinite Integrals Involving

$$f(x) = f(x, \sinh^{-1} B) \qquad B = \frac{x}{b}$$

$$\int \sinh^{-1} B \, dx = x \sinh^{-1} B - \sqrt{x^2 + b^2}$$

$$\int x \sinh^{-1} B \, dx = \left(\frac{x^2}{2} + \frac{b^2}{4}\right) \sin^{-1} B - \frac{x\sqrt{x^2 + b^2}}{4}$$

$$\int x^2 \sinh^{-1} B \, dx = \frac{x^3}{3} \sinh^{-1} B + \frac{(2b^2 - x^2)\sqrt{x^2 + b^2}}{9}$$

$$\int x^m \sinh^{-1} B \, dx = \frac{x^{m+1}}{m+1} \sinh^{-1} B - \frac{1}{m+1} \int \frac{x^{m+1}}{\sqrt{x^2 + b^2}} \, dx \qquad m \neq -1$$

$$\int \frac{\sinh^{-1} B \, dx}{x} = B - \frac{B^3}{(2)(3)(3)} + \frac{(1)(3)B^5}{(2)(4)(5)(5)} - \frac{(1)(3)(5)B^7}{(2)(4)(6)(7)(7)} + \cdots \qquad x^2 < b^2$$

$$\int \frac{\sinh^{-1} B \, dx}{x^2} = -\frac{\sinh^{-1} B}{x} - \frac{1}{b} \ln \frac{b + \sqrt{x^2 + b^2}}{x}$$

(83) Indefinite Integrals Involving

$$f(x) = f(x, \tanh^{-1} B) \qquad B = \frac{x}{b}$$

$$\int \tanh^{-1} B \, dx = x \tanh^{-1} B + b \ln \sqrt{b^2 - x^2}$$

$$\int x \tanh^{-1} B \, dx = \frac{x^2 - b^2}{2} \tanh^{-1} B + \frac{bx}{2}$$

$$\int x^m \tanh^{-1} B \, dx = \frac{x^{m+1}}{m+1} \tanh^{-1} B - \frac{b}{m+1} \int \frac{x^{m+1}}{b^2 - x^2} \, dx \qquad m \neq 1$$

$$\int \frac{\tanh^{-1} B \, dx}{x} = B + \frac{B^3}{3^2} + \frac{B^5}{5^2} + \frac{B^7}{7^2} + \cdots$$

$$\int \frac{\tanh^{-1} B \, dx}{x^2} = -\frac{1}{b}\left(\frac{\tanh^{-1} B}{B} + \ln \frac{\sqrt{1 - B^2}}{B}\right)$$

$$\int \frac{\tanh^{-1} B \, dx}{x^3} = -\frac{1}{2x^2} \left[B - (B^2 - 1) \tanh^{-1} B\right]$$

$$\int \frac{\tanh^{-1} B \, dx}{x^m} = -\frac{\tanh^{-1} B}{(m-1)x^{m-1}} + \frac{b}{m-1} \int \frac{dx}{(b^2 - x^2)x^{m-1}}$$

(84) Indefinite Integrals Involving $\qquad$ $f(x) = f(x, \cosh^{-1} B)$ $\quad B = \dfrac{x}{b}$

$$\int \cosh^{-1} B \, dx = x \cosh^{-1} B \mp \sqrt{x^2 - b^2}$$

$$\int x \cosh^{-1} B \, dx = \left(\frac{x^2}{2} - \frac{b^2}{2}\right) \cosh^{-1} B \mp \frac{x\sqrt{x^2 - b^2}}{4}$$

$$\int x^2 \cosh^{-1} B \, dx = \frac{x^3}{3} \cosh^{-1} B \mp \frac{(2b^2 + x^2)\sqrt{x^2 - b^2}}{9}$$

$$\int x^m \cosh^{-1} B \, dx = \frac{x^{m+1}}{m+1} \cosh^{-1} B \mp \frac{1}{m+1} \int \frac{x^{m+1} \, dx}{\sqrt{x^2 - b^2}} \qquad m \neq -1$$

$\left. \begin{array}{l} \\[5.5em] \end{array} \right\}$ $\begin{array}{l} -\text{if } \cosh^{-1} B > 0 \\ +\text{if } \cosh^{-1} B < 0 \end{array}$

$$\int \frac{\cosh^{-1} B \, dx}{x} = \mp \left[\frac{1}{2}(\ln 2B)^2 + \frac{B^2}{(2)(2)(2)} + \frac{(1)(3)B^4}{(2)(4)(4)(4)} \right.$$
$$\left. + \frac{(1)(3)(5)B^6}{(2)(4)(6)(6)(6)} + \cdots \right]$$

$$\int \frac{\cosh^{-1} B \, dx}{x^2} = -\frac{\cosh^{-1} B}{x} \mp \frac{1}{b} \ln \frac{b + \sqrt{x^2 + b^2}}{x}$$

$\left. \begin{array}{l} \\[3em] \end{array} \right\}$ $\begin{array}{l} -\text{if } \cosh^{-1} B < 0 \\ +\text{if } \cosh^{-1} B > 0 \end{array}$

(85) Indefinite Integrals Involving $\qquad$ $f(x) = f(x, \coth^{-1} B)$ $\quad B = \dfrac{x}{b}$

$$\int \coth^{-1} B \, dx = x \coth^{-1} B + b \ln \sqrt{x^2 - b^2}$$

$$\int x \coth^{-1} B \, dx = \frac{x^2 - b^2}{2} \coth^{-1} B + \frac{bx}{2}$$

$$\int x^m \coth^{-1} B \, dx = \frac{x^{m+1}}{m+1} \coth^{-1} B + \frac{b}{m+1} \int \frac{x^{m+1}}{x^2 - b^2} \, dx \qquad m \neq -1$$

$$\int \frac{\coth^{-1} B \, dx}{x} = -B - \frac{B^3}{3^2} - \frac{B^5}{5^2} - \frac{B^7}{7^2} - \cdots$$

$$\int \frac{\coth^{-1} B \, dx}{x^2} = -\frac{1}{b}\left(\frac{\coth^{-1} B}{B} + \ln \frac{\sqrt{B^2 - 1}}{B}\right)$$

$$\int \frac{\coth^{-1} B \, dx}{x^3} = -\frac{1}{2x^2}[B - (B^2 - 1)\coth^{-1} B]$$

$$\int \frac{\coth^{-1} B \, dx}{x^m} = -\frac{\coth^{-1} B}{(m-1)x^{m-1}} + \frac{b}{m-1} \int \frac{dx}{(b^2 - x^2)x^{m-1}}$$

(86) Indefinite Integrals Involving $f(x) = f(x, \ln A)$ $A = bx$

$$\int \ln A \, dx = x(\ln A - 1)$$

$$\int x \ln A \, dx = \left(\frac{x}{2}\right)^2 [\ln (A^2) - 1]$$

$$\int x^m \ln A \, dx = \frac{x^{m+1}}{(m+1)^2}(\ln A^{m+1} - 1) \qquad\qquad m \neq -1$$

$$\int (\ln A)^n \, dx = x(\ln A)^n - n \int (\ln A)^{n-1} \, dx \qquad\qquad n \neq -1$$

$$\int x^m (\ln A)^n \, dx = \frac{x^{m+1}(\ln A)^n - n \int x^m (\ln A)^{n-1} \, dx}{m+1} \qquad\qquad m, n \neq -1$$

$$\int \frac{\ln A \, dx}{x} = \frac{\ln^2 A}{2}$$

$$\int \frac{(\ln A)^n \, dx}{x} = \frac{(\ln A)^{n+1}}{n+1} \qquad\qquad n \neq -1$$

$$\int \frac{\ln A \, dx}{x^m} = -\frac{1 + (m-1)\ln A}{(m-1)^2 x^{m-1}} \qquad\qquad m \neq 1$$

$$\int \frac{(\ln A)^n \, dx}{x^m} = -\frac{(\ln A)^n}{(m-1)x^{m-1}} + \frac{n}{m-1}\int \frac{(\ln A)^{n-1}}{x^m} \, dx \qquad m \neq 1$$

$$\int \frac{dx}{\ln A} = \frac{1}{b}\left[\ln (\ln A) + \ln A + \frac{(\ln A)^2}{(2)(2!)} + \frac{(\ln A)^3}{(3)(3!)} + \cdots\right]$$

$$\int \frac{x^m \, dx}{\ln A} = \frac{1}{b^{m+1}}\left[\ln (\ln A) + (m+1)\ln A + \frac{(m+1)^2(\ln A)^2}{(2)(2!)} + \frac{(m+1)^3(\ln A)^3}{(3)(3!)} + \cdots\right] \qquad m > 0$$

$$\int \sin (\ln A) \, dx = \frac{x}{2}[\sin (\ln A) - \cos (\ln A)]$$

$$\int \cos (\ln A) \, dx = \frac{x}{2}[\sin (\ln A) + \cos (\ln A)]$$

$$\int \ln (\sin A) \, dx = x\left[\ln A - 1 - \sum_{k=1}^{\infty} \frac{b_k (2A)^{2k}}{2k(2k+1)}\right] \qquad\qquad |A| < \pi$$

$$\int \ln (\cos A) \, dx = -x \sum_{k=1}^{\infty} \frac{a_k (2A)^{2k}}{2k(2k+1)} \qquad\qquad |A| < \frac{\pi}{2}$$

$a_k, b_k =$ numerical factors (Sec. 8.15–2)

20
TABLES OF
DEFINITE
INTEGRALS

(1) Definite Integrals

The more frequently encountered definite integrals of elementary functions are tabulated in this chapter. Particular symbols used in the following are:

$a, b, c = $ constants	$\alpha, \beta, \gamma, \lambda = $ constant equivalents
$k, m, n, p, q, r = $ integers	$x = $ independent variable

In these tables, logarithmic expressions are for the absolute value of the respective argument, all angles are in radians, and all inverse functions represent principal values (angles). Special theorems and rules useful in the evaluation of the definite integrals are given in Secs. (2) through (8).

(2) Improper Integrals

If either or both of the *limits* of a definite integral are *infinitely large* or if the *integrand becomes infinite* in the interval of integration, the integral is called an *improper integral*.

$$\int_a^{+\infty} f(x)\,dx = \lim_{b \to +\infty} \int_a^b f(x)\,dx \qquad\qquad \int_{-\infty}^b f(x)\,dx = \lim_{a \to -\infty} \int_a^b f(x)\,dx$$

$$\int_{-\infty}^{+\infty} f(x)\,dx = \lim_{\substack{a \to -\infty \\ b \to +\infty}} \int_a^b f(x)\,dx$$

If $\lim_{x \to c} f(x) = \infty$, then

$$\int_a^b f(x)\,dx = \int_a^c f(x)\,dx + \int_c^b f(x)\,dx = \lim_{\epsilon_1 \to 0} \int_a^{c-\epsilon_1} f(x)\,dx + \lim_{\epsilon_2 \to 0} \int_{c+\epsilon_2}^b f(x)\,dx \qquad \begin{array}{c} \epsilon_1 > 0 \\ \epsilon_2 > 0 \end{array}$$

(3) First Mean-Value Theorem

If $f(x)$ and $g(x)$ are continuous in $[a, b]$, and $g(x)$ is integrable over this interval and *does not change sign* anywhere in it, then there exists at least one point c in (a, b) such that

$$\int_a^b f(x)g(x)\,dx = f(c) \int_a^b g(x)\,dx \qquad\qquad \text{and for } g(x) = 1, \qquad\qquad \int_a^b f(x)\,dx = (b-a)f(c)$$

(4) Second Mean-Value Theorem

If $f(x)$ and $g(x)$ are continuous in $[a, b]$, $f(x)$ is a *positive, monotonic, decreasing function* in this interval, and $g(x)$ is integrable in it, then there exists at least one point c in (a, b) such that

$$\int_a^b f(x)g(x)\,dx = f(a) \int_a^c g(x)\,dx$$

If the same conditions are valid but $f(x)$ is a *positive, monotonic, increasing function*, then there exists at least one point c in (a, b) such that

$$\int_a^b f(x)g(x)\,dx = f(b) \int_c^b g(x)\,dx$$

In general,

$$\int_a^b f(x)g(x)\,dx = f(a) \int_a^c g(x)\,dx + f(b) \int_c^b g(x)\,dx$$

where $f(x)$ is *monotonic increasing or decreasing* and *is not necessarily always positive* in (a, b).

(5) Even and Odd Functions

If $f(x)$ and $g(x)$ are integrable in $[-a, a]$, $f(x)$ is an *even function* such that $f(-x) = f(x)$, and $g(x)$ is an odd function such that $g(-x) = -g(x)$, then

$$\int_{-a}^{a} f(x) \, dx = 2 \int_{0}^{a} f(x) \, dx$$

$$\int_{-a}^{a} g(x) \, dx = 0$$

(6) Periodic and Antiperiodic Functions

If $n =$ integer, $T =$ positive number (period), $f(x)$ and $g(x)$ are integrable in $[0, l]$, $f(x)$ is a periodic function such that $f(nT + x) = f(x)$, and $g(x)$ is an antiperiodic function such that $g[(2n-1)T+x] = -g(x)$, $g(2nT+x) = g(x)$, then with

$$l = a = nT + t: \qquad \int_{0}^{a} f(x) \, dx = nF(T) + F(t) - (n+1)F(0)$$

$$l = b = (2n-1)T + t: \qquad \int_{0}^{b} g(x) \, dx = G(T) - G(t)$$

$$l = c = 2nT + t: \qquad \int_{0}^{c} g(x) \, dx = G(t) - G(0)$$

where $dF(x)/dx = f(x)$, $dG(x)/dx = g(x)$, and $t < T$.

(7) Trigonometric Identities

If $f(\sin x)$ and $f(\cos x)$ are, respectively and exclusively, functions of $\sin x$ and $\cos x$, then

$$\int_{0}^{\pi/2} f(\sin x) \, dx = \int_{0}^{\pi/2} f(\cos x) \, dx = \frac{1}{2} \int_{0}^{\pi} f(\sin x) \, dx$$

$$\int_{0}^{\lambda \pi/2} f\left(\sin \frac{x}{\lambda}\right) dx = \int_{0}^{\lambda \pi/2} f\left(\cos \frac{x}{\lambda}\right) dx = \frac{\lambda}{2} \int_{0}^{\pi} f(\sin x) \, dx$$

where $\lambda =$ integer or fraction.

(8) Change in Limits and Variables

If $f(x)$ is integrable in $[a, b]$ and $x = g(t)$ and its derivative $dg(t)/dt$ are continuous in $\alpha \le t \le \beta$, then with $a = g(\alpha)$, $b = g(\beta)$,

$$\int_{x=a}^{x=b} f(x) \, dx = \int_{t=\alpha}^{t=\beta} f(g(t)) \frac{dg(t)}{dt} \, dt$$

Particular cases of this transformation are:

$$\int_{0}^{b} f(x) \, dx = \int_{0}^{b} f(b-x) \, dx \qquad \int_{a}^{b} f(x) \, dx = \int_{a}^{b} f(a+b-x) \, dx$$

$$\int_{a}^{b} f(x) \, dx = \frac{1}{\lambda} \int_{\lambda a}^{\lambda b} f\left(\frac{x}{\lambda}\right) dx \qquad \int_{a}^{b} f(x) \, dx = \int_{a\pm\lambda}^{b\pm\lambda} f(x \mp \lambda) \, dx$$

$$\int_{a}^{b} f(x) \, dx = \int_{0}^{b-a} f(x+a) \, dx = (b-a) \int_{0}^{1} f((b-a)x + a) \, dx$$

(9) Definite Integrals Involving

$$f(x) = f(a^p \pm x^p) \qquad [a, b]$$

m, n, p = positive integers $\neq 0$
a, b, c, α, β = real numbers $\neq 0$

$\Gamma(\ )$ = gamma function (Sec. 13.03)
$B(\ )$ = beta function (Sec. 13.03)

$$\int_0^a x^m (a-x)^n \, dx = \frac{m! \, n! \, a^{m+n+1}}{(m+n+1)!}$$

$$\int_a^b (x-a)^m (b-x)^n \, dx = \frac{m! \, n! \, (b-a)^{m+n+1}}{(m+n+1)!}$$

$$\int_0^a x^\alpha (a-x)^\beta \, dx = \frac{\Gamma(\alpha+1)\Gamma(\beta+1)}{\Gamma(\alpha+\beta+2)} \, a^{\alpha+\beta+1} = B(\alpha+1, \beta+1) \, a^{\alpha+\beta+1}$$

$$\int_a^b (x-a)^\alpha (b-x)^\beta \, dx = \frac{\Gamma(\alpha+1)\Gamma(\beta+1)}{\Gamma(\alpha+\beta+2)} \, (b-a)^{\alpha+\beta+1} = B(\alpha+1, \beta+1) \, (b-a)^{\alpha+\beta+1}$$

$$\int_0^a x^m (a^n - x^n)^p \, dx = \frac{p! \, n^p \, a^{np+m+1}}{(m+1)(m+1+p)(m+1+2p) \cdots (m+1+np)}$$

$$\int_0^a x^\alpha (a^n - x^n)^\beta \, dx = \frac{\Gamma\left(\dfrac{\alpha+1}{n}\right) \Gamma(\beta+1)}{n \, \Gamma\left(\dfrac{\alpha+1}{n} + \beta + 1\right)} \, a^{\alpha+n\beta+1} = \frac{1}{n} \, B\left(\frac{\alpha+1}{n}, \beta+1\right) a^{\alpha+n\beta+1}$$

$$\int_a^b \frac{1}{x-c} \left(\frac{x-a}{b-x}\right)^\alpha dx = \frac{\pi}{\sin \alpha\pi} \left[1 - \left(\frac{c-a}{b-c}\right)^\alpha \cos \alpha\pi\right] \qquad a < c < b, |\alpha| < 1$$

$$\int_0^a \frac{x^m}{a+x} \, dx = (-a)^m \left[\ln 2 + \sum_{k=1}^m (-1)^k \frac{1}{k}\right]$$

$$\int_0^a \frac{x^m}{a^n + x^n} \, dx = a^{m-n+1} \left[\sum_{k=0}^\infty (-1)^k \frac{1}{m+1+kn}\right]$$

$$\int_0^a \frac{x^b}{(a-x)^b} \, dx = \frac{\pi ab}{\sin b\pi} \qquad |b| < 1$$

$$\int_0^a \frac{x^b}{(a-x)^{b+1}} \, dx = \frac{\pi}{\sin b\pi} \qquad 0 < b < 1$$

$$\int_0^1 \frac{x^\alpha - x^{-\alpha}}{x \pm 1} \, dx = \begin{cases} \dfrac{1}{\alpha} - \dfrac{\pi}{\sin \alpha\pi} & \text{for } + \\[2ex] \dfrac{1}{\alpha} - \dfrac{\pi}{\tan \alpha\pi} & \text{for } - \end{cases} \qquad |\alpha| < 1$$

$$\int_0^1 \frac{x^\alpha - x^{-\alpha}}{x^2 \pm 1} \, dx = \begin{cases} \dfrac{1}{\alpha} - \dfrac{\pi}{2 \sin (\alpha\pi/2)} & \text{for } + \\[2ex] \dfrac{1}{\alpha} - \dfrac{\pi}{2 \tan (\alpha\pi/2)} & \text{for } - \end{cases} \qquad |\alpha| < 1$$

$$\int_0^a \frac{(x/a)^{m-1} + (x/a)^{n-1}}{(a+x)^{m+n}} \, dx = \frac{a(m-1)!(n-1)!}{a^{m+n}(m+n-1)!}$$

$$\int_0^a \frac{(x/a)^{\alpha-1} + (x/a)^{\beta-1}}{(a+x)^{\alpha+\beta}} \, dx = \frac{B(\alpha, \beta)}{a^{\alpha+\beta-1}}$$

$$\int_0^a \frac{dx}{a^2 + ax + x^2} = \frac{\pi}{3a\sqrt{3}}$$

$$\int_0^a \frac{dx}{a^2 - ax + x^2} = \frac{2\pi}{3a\sqrt{3}}$$

(10) Definite Integrals Involving

$$f(x) = f(\sqrt{a^p \pm x^p}) \qquad [0, a]$$

$m, n, p =$ positive integers $\neq 0$ $\qquad\qquad$ $\Gamma(\ ) =$ gamma function (Sec. 13.03)
$a, \alpha, \beta =$ real numbers $\neq 0$ $\qquad\qquad$ $B(\ ) =$ beta function (Sec. 13.03)

$$\int_0^a \sqrt{a^2 + x^2}\, dx = \frac{a^2}{2}\left[\sqrt{2} + \ln\left(\sqrt{2} + 1\right)\right]$$

$$\int_0^a \sqrt{a^2 - x^2}\, dx = \frac{\pi a^2}{4}$$

$$\int_0^a x\sqrt{a^2 + x^2}\, dx = \frac{a^3}{3}\left(\sqrt{8} - 1\right)$$

$$\int_0^a x\sqrt{a^2 - x^2}\, dx = \frac{a^3}{3}$$

$$\int_0^a x^{2m+1}\sqrt{a^2 - x^2}\, dx = \frac{(2m)(2m-2)\cdots 6\cdot 4\cdot 2}{(2m+1)(2m-1)\cdots 5\cdot 3\cdot 1}\frac{a^{2m+3}}{2m+3} = \frac{(2m)!!}{(2m+3)!!}a^{2m+3}$$

$$\int_0^a x^{2m}\sqrt{a^2 - x^2}\, dx = \frac{(2m-1)(2m-3)\cdots 5\cdot 3\cdot 1}{(2m)(2m-2)\cdots 6\cdot 4\cdot 2}\frac{\pi a^{2m+2}}{(2m+2)2} = \frac{(2m-1)!!}{(2m+2)!!}\frac{\pi a^{2m+2}}{2}$$

$$\int_0^a x^\alpha\sqrt{a^2 - x^2}\, dx = \frac{\sqrt{\pi}}{2}\frac{\Gamma\left(\dfrac{\alpha+1}{2}\right)}{\Gamma\left(\dfrac{\alpha+2}{2}\right)}\frac{a^{\alpha+2}}{\alpha+2}$$

$$\int_0^a x^\alpha\sqrt[p]{a^n - x^n}\, dx = a^{\alpha+1}\frac{\sqrt[p]{a^n}}{n}B\left(\frac{\alpha+1}{n}, \frac{p+1}{p}\right)$$

$$\int_0^a \frac{x^m\, dx}{\sqrt{a-x}} = \frac{2m(2m-2)\cdots 6\cdot 4\cdot 2}{(2m+1)(2m-1)\cdots 5\cdot 3\cdot 1}\frac{2a^{m+1}}{\sqrt{a}} = \frac{(2m)!!}{(2m+1)!!}\frac{2a^{m+1}}{\sqrt{a}}$$

$$\int_0^a \frac{x\, dx}{\sqrt{a-x}} = \frac{4a^2}{3\sqrt{a}}$$

$$\int_0^a \frac{x^2\, dx}{\sqrt{a-x}} = \frac{16a^3}{15\sqrt{a}}$$

$$\int_0^a \frac{dx}{\sqrt{a^2 + x^2}} = \ln\left(\sqrt{2} + 1\right)$$

$$\int_0^a \frac{dx}{\sqrt{a^2 - x^2}} = \frac{\pi}{2}$$

$$\int_0^a \frac{x\, dx}{\sqrt{a^2 + x^2}} = \left(\sqrt{2} - 1\right)a$$

$$\int_0^a \frac{x\, dx}{\sqrt{a^2 - x^2}} = a$$

$$\int_0^a \frac{x^{2m+1}\, dx}{\sqrt{a^2 - x^2}} = \frac{2m(2m-2)\cdots 6\cdot 4\cdot 2}{(2m+1)(2m-1)\cdots 5\cdot 3\cdot 1}a^{2m+1} = \frac{(2m)!!\,a^{2m+1}}{(2m+1)!!}$$

$$\int_0^a \frac{x^{2m}\, dx}{\sqrt{a^2 - x^2}} = \frac{(2m-1)(2m-3)\cdots 5\cdot 3\cdot 1}{(2m)(2m-2)\cdots 6\cdot 4\cdot 2}\frac{\pi a^{2m}}{2} = \frac{(2m-1)!!}{(2m)!!}\frac{\pi a^{2m}}{2}$$

$$\int_0^a \frac{dx}{\sqrt{a^3 - x^3}} = \frac{1.403\,160\cdots}{\sqrt{a}}$$

$$\int_0^a \frac{dx}{\sqrt{a^4 - x^4}} = \frac{5.244\,115\cdots}{a}$$

$$\int_0^a \frac{dx}{\sqrt{a^n - x^n}} = \frac{a}{n}\sqrt{\frac{\pi}{a^n}}\frac{\Gamma(1/n)}{\Gamma(1/n + \frac{1}{2})}$$

$$\int_0^a \frac{dx}{\sqrt[p]{a^n - x^n}} = \frac{a}{n\sqrt[p]{a^n}}B\left(\frac{p-1}{p}, \frac{1}{n}\right)$$

$$\int_0^a \frac{x^m\, dx}{\sqrt{a^n - x^n}} = \frac{a^{m+1}}{n}\sqrt{\frac{\pi}{a^n}}\frac{\Gamma\left(\dfrac{m+1}{n}\right)}{\Gamma\left(\dfrac{m+1}{n} + \dfrac{1}{2}\right)}$$

$$\int_0^a \frac{x^m\, dx}{\sqrt[p]{a^n - x^n}} = \frac{a^{m+1}}{n\sqrt[p]{a^n}}B\left(\frac{p-1}{p}, \frac{m+1}{n}\right)$$

(11) Definite Integrals Involving

$$f(x) = f(a^p + x^p) \quad\quad [0, \infty]$$

m, n, p = positive integers $\neq 0$
a, b, λ = positive real numbers $\neq 0$
m, n as arguments in beta function may also be fractions.

( )!! = double factorial (Sec. 1.03)
B( ) = beta function (Sec. 13.03)

$$\int_0^\infty \frac{x^\lambda \, dx}{a+x} = \frac{\pi a^\lambda}{\sin(\lambda+1)\pi} \quad\quad 0 < \lambda < 1$$

$$\int_0^\infty \frac{x^{-\lambda} \, dx}{a+x} = \frac{\pi a^{-\lambda}}{\sin \lambda \pi} \quad\quad 0 < \lambda < 1$$

$$\int_0^\infty \frac{dx}{a^2 + x^2} = \frac{\pi}{2a}$$

$$\int_0^\infty \frac{x \, dx}{a^2 + x^2} = \infty$$

$$\int_0^\infty \frac{dx}{a^3 + x^3} = \frac{2\pi}{3a^2 \sqrt{3}}$$

$$\int_0^\infty \frac{x \, x}{a^3 + x^3} = \frac{2\pi}{3a\sqrt{3}}$$

$$\int_0^\infty \frac{dx}{a^4 + x^4} = \frac{\pi}{2a^3 \sqrt{2}}$$

$$\int_0^\infty \frac{x \, dx}{a^4 + x^4} = \frac{\pi}{4a^2}$$

$$\int_0^\infty \frac{dx}{a^n + x^n} = \frac{a\pi}{na^n \sin(\pi/n)}$$

$$\int_0^\infty \frac{x \, dx}{a^n + x^n} = \frac{\pi}{na^{n-2} \sin(2\pi/n)}$$

$$\int_0^\infty \frac{x^m \, dx}{a^n + x^n} = \frac{\pi a^{m+1}}{na^n \sin[(m+1)\pi/n]}$$

$$\int_0^\infty \frac{x \, dx}{a^n + x^n} = \frac{\pi}{na^{n-3} \sin(3\pi/n)}$$

$$\int_0^\infty \frac{dx}{(a+bx)^2} = \frac{1}{ab}$$

$$\int_0^\infty \frac{x \, dx}{(a+bx)^3} = \frac{1}{2ab^2}$$

$$\int_0^\infty \frac{dx}{(a+bx)^n} = \frac{B(1, n-1)}{a^{n-1}b}$$

$$\int_0^\infty \frac{x^m \, dx}{(a+bx)^n} = \frac{B(m+1, n-m-1)}{a^{n-m-1}b^{m+1}}$$

$$\int_0^\infty \frac{x^{m-1} \, dx}{(1+ax)^n} = \frac{B(m, n-m)}{a^m} \quad\quad |a| < \pi$$

$$n > m > 0$$

$$\int_0^\infty \frac{dx}{(a^2+x^2)(b^2+x^2)} = \frac{\pi}{2ab(a+b)}$$

$$\int_0^\infty \frac{dx}{(a^2+x^2)(a^n+x^n)} = \frac{\pi}{4a^{n+1}}$$

$$\int_0^\infty \frac{dx}{(a^2+x^2)^n} = \frac{(2n-3)(2n-5)\cdots 5 \cdot 3 \cdot 1}{(2n-2)(2n-4)\cdots 6 \cdot 4 \cdot 2} \frac{\pi}{2a^{2n-1}} = \frac{(2n-3)!!}{(2n-2)!!} \frac{\pi}{2a^{2n-1}}$$

$$\int_0^\infty \frac{x^{2m} \, dx}{(a+bx^2)^n} = \frac{(2m-1)!!(2n-2m-3)!!\pi}{(2n-2)!!2a^{n-m-1}b^m\sqrt{ab}} \quad\quad n > (m+1)$$

$$\int_0^\infty \frac{x^{2m+1} \, dx}{(a+bx^2)^n} = \frac{m!(n-m-2)!}{(n-1)!2a^{n-m-1}b^{m+1}} \quad\quad n > (m+1) \geq 1$$

$$\int_0^\infty \frac{dx}{a + 2bx + cx^2} = \frac{1}{\sqrt{ac-b^2}} \cot^{-1} \frac{b}{\sqrt{ac-b^2}} \quad\quad (ac-b^2) > 0$$

(12) Definite Integrals Involving $\qquad$ $f(x) = f(x^m, e^{-ax}, e^{-ax^2})$ $\qquad$ $[0, \infty]$

Ei( ) = exponential integral function (Sec. 13.01) $\qquad$ Γ( ) = gamma function (Sec. 13.03) -
erf( ) = error function (Sec. 13.01) $\qquad$ Z( ) = zeta function (Sec. A.06)
For definitions of $a, b, m, n,$ and ( )!! see opposite page.

$$\int_0^\infty e^{-ax}\,dx = \frac{1}{a}$$

$$\int_0^\infty xe^{-x}\,dx = 1$$

$$\int_0^\infty x^m e^{-x}\,dx = \Gamma(m+1) = m!$$

$$\int_0^\infty x^m e^{-ax}\,dx = \frac{\Gamma(m+1)}{a^{m+1}} = \frac{m!}{a^{m+1}}$$

$$\int_0^\infty \frac{e^{-ax}\,dx}{x} = \infty$$

$$\int_0^\infty \frac{e^{-ax}\,dx}{b+x} = -e^{ab}\,\text{Ei}(ab)$$

$$\int_0^\infty \frac{e^{-ax}\,dx}{\sqrt{x}} = \sqrt{\frac{\pi}{a}}$$

$$\int_0^\infty \frac{e^{-ax}\,dx}{\sqrt{b+x}} = \sqrt{\frac{\pi}{a}}\,e^{ab}[1-\text{erf}(\sqrt{ab})]$$

$$\int_0^\infty \frac{dx}{e^{ax}+1} = \frac{\ln 2}{a}$$

$$\int_0^\infty \frac{dx}{e^{ax}-1} = \infty$$

$$\int_0^\infty \frac{x\,dx}{e^{ax}+1} = \frac{\pi^2}{12a^2}$$

$$\int_0^\infty \frac{x\,dx}{e^{ax}-1} = \frac{\pi^2}{6a^2}$$

$$\int_0^\infty \frac{x^2\,dx}{e^{ax}+1} = \frac{3}{2a^3}\,Z(3)$$

$$\int_0^\infty \frac{x^2\,dx}{e^{ax}-1} = \frac{2}{a^3}\,Z(3)$$

$$\int_0^\infty \frac{x^b\,dx}{e^{ax}+1} = \frac{\Gamma(b+1)}{a^{b+1}}\sum_{k=0}^\infty \frac{1}{(2k+1)^{b+1}} = \frac{\Gamma(b+1)}{a^{b+1}}\left(1-\frac{1}{2^b}\right)Z(b+1)$$

$$\int_0^\infty \frac{x^b\,dx}{e^{ax}-1} = \frac{\Gamma(b+1)}{a^{b+1}}\sum_{k=1}^\infty \frac{1}{k^{b+1}} = \frac{\Gamma(b+1)}{a^{b+1}}\,Z(b+1)$$

$$\int_0^\infty e^{-x^2}\,dx = \frac{\sqrt{\pi}}{2}$$

$$\int_0^\infty xe^{-x^2}\,dx = \frac{1}{2}$$

$$\int_0^\infty e^{-ax^2}\,dx = \frac{1}{2}\sqrt{\frac{\pi}{a}}$$

$$\int_0^\infty xe^{-ax^2}\,dx = \frac{1}{2a}$$

$$\int_0^\infty x^{2m+1}e^{-ax^2}\,dx = \frac{m!}{2a^{m+1}}$$

$$\int_0^\infty x^b e^{-ax^2}\,dx = \frac{\Gamma\left(\dfrac{b+1}{2}\right)}{2\sqrt{a^{b+1}}}$$

$$\int_0^\infty x^{2m}e^{-ax^2}\,dx = \frac{(2m-1)(2m-3)\cdots 5\cdot 3\cdot 1}{2(2a)^m}\sqrt{\frac{\pi}{a}} = \frac{(2m-1)!!}{2(2a)^m}\sqrt{\frac{\pi}{a}}$$

$$\int_0^\infty \frac{e^{-ax}-e^{-bx}}{x}\,dx = \ln\frac{b}{a}$$

$$\int_0^\infty \frac{e^{-ax^2}-e^{-bx^2}}{x}\,dx = \ln\sqrt{\frac{b}{a}}$$

(13) Definite Integrals Involving

$$f(x) = f(\sin^m x, \cos^m x, \sin^n x, \cos^n x) \qquad \left[0, \frac{\pi}{2}\right]$$

m, n = positive integers $\neq 0$
$( )!!$ = double factorial (Sec. 1.03)

$\Gamma( )$ = gamma function (Sec. 13.03)
$B( )$ = beta function (Sec. 13.03)

$$\int_0^{\pi/2} \sin x \, dx = \int_0^{\pi/2} \cos x \, dx = 1$$

$$\int_0^{\pi/2} \sin^2 x \, dx = \int_0^{\pi/2} \cos^2 x \, dx = \frac{\pi}{4}$$

$$\int_0^{\pi/2} \sin^3 x \, dx = \int_0^{\pi/2} \cos^3 x \, dx = \frac{2}{3}$$

$$\int_0^{\pi/2} \sin^4 x \, dx = \int_0^{\pi/2} \cos^4 x \, dx = \frac{3\pi}{16}$$

$$\int_0^{\pi/2} \sin^{2m+1} x \, dx = \int_0^{\pi/2} \cos^{2m+1} x \, dx = \frac{(2m)(2m-2)\cdots 6\cdot 4\cdot 2}{(2m+1)(2m-1)\cdots 5\cdot 3\cdot 1} = \frac{(2m)!!}{(2m+1)!!}$$

$$\int_0^{\pi/2} \sin^{2m} x \, dx = \int_0^{\pi/2} \cos^{2m} x \, dx = \frac{(2m-1)(2m-3)\cdots 5\cdot 3\cdot 1}{(2m)(2m-2)\cdots 6\cdot 4\cdot 2}\frac{\pi}{2} = \frac{(2m-1)!!}{(2m)!!}\frac{\pi}{2}$$

$$\int_0^{\pi/2} \sin x \cos x \, dx = \frac{1}{2}$$

$$\int_0^{\pi/2} \sin^2 x \cos^2 x \, dx = \frac{\pi}{16}$$

$$\int_0^{\pi/2} \sin^3 x \cos^3 x \, dx = \frac{1}{12}$$

$$\int_0^{\pi/2} \sin^4 x \cos^4 x \, dx = \frac{3\pi}{256}$$

$$\int_0^{\pi/2} \sin^5 x \cos^5 x \, dx = \frac{1}{60}$$

$$\int_0^{\pi/2} \cos^6 x \sin^6 x \, dx = \frac{15\pi}{6144}$$

$$\int_0^{\pi/2} \sin^{2m+1} x \cos^{2n+1} x \, dx = \frac{m!\,n!}{(m+n+1)!}\frac{1}{2} = \frac{\Gamma(m+1)\Gamma(n+1)}{2\Gamma(m+n+2)} = \frac{1}{2}B(m+1, n+1)$$

$$\int_0^{\pi/2} \sin^{2m} x \cos^{2n} x \, dx = \frac{(2m-1)!!(2n-1)!!}{(2m+2n)!!}\frac{\pi}{2} = \frac{\Gamma(m+\frac{1}{2})\Gamma(n+\frac{1}{2})}{2\Gamma(m+n+1)} = \frac{1}{2}B(m+\frac{1}{2}, n+\frac{1}{2})$$

$$\int_0^{\pi/2} \sin x \cos^2 x \, dx = \frac{1}{3}$$

$$\int_0^{\pi/2} \cos x \sin^2 x \, dx = \frac{1}{3}$$

$$\int_0^{\pi/2} \sin x \cos^3 x \, dx = \frac{1}{4}$$

$$\int_0^{\pi/2} \cos x \sin^3 x \, dx = \frac{1}{4}$$

$$\int_0^{\pi/2} \sin x \cos^n x \, dx = \frac{1}{n+1}$$

$$\int_0^{\pi/2} \cos x \sin^n x \, dx = \frac{1}{n+1}$$

$$\int_0^{\pi/2} \sin^{2m+1} x \cos^{2n} x \, dx = \frac{(2m)!!(2n-1)!!}{(2m+2n+1)!!} = \frac{\Gamma^\bullet(m+1)\Gamma(n+\frac{1}{2})}{2\Gamma(m+n+\frac{3}{2})} = \frac{1}{2}B(m+1, n+\frac{1}{2})$$

$$\int_0^{\pi/2} \sin^{2m} x \cos^{2n+1} x \, dx = \frac{(2n)!!(2m-1)!!}{(2m+2n+1)!!} = \frac{\Gamma(n+1)\Gamma(m+\frac{1}{2})}{2\Gamma(m+n+\frac{3}{2})} = \frac{1}{2}B(m+\frac{1}{2}, n+1)$$

(14) Definite Integrals Involving $\qquad f(x) = f(a \sin x \pm b \cos x)$ $\quad \boxed{\left[0, \dfrac{\pi}{2}\right]}$

$a, b =$ positive real numbers $\neq 0$	$G = \overline{Z}(2) = 0.915\,965\,594$ (Sec. A.06)
$k =$ modulus ($\lvert k \rvert < 1$) (Sec. 13.05–1)	$K, E =$ complete elliptic integrals (Sec. 13.05–2)

$$\int_0^{\pi/2} \frac{dx}{1+\sin x} = \int_0^{\pi/2} \frac{dx}{1+\cos x} = 1 \qquad\qquad \int_0^{\pi/2} \frac{\sin x\, dx}{1+\sin x} = \int_0^{\pi/2} \frac{\cos x\, dx}{1+\cos x} = \frac{\pi}{2} - 1$$

$$\int_0^{\pi/2} \frac{x\, dx}{1+\sin x} = \ln 2 \qquad\qquad\qquad\qquad \int_0^{\pi/2} \frac{x\, dx}{1+\cos x} = \frac{\pi}{2} - \ln 2$$

$$\int_0^{\pi/2} \frac{x \cos x\, dx}{1+\sin x} = \pi \ln 2 - 4G \qquad\qquad \int_0^{\pi/2} \frac{x \sin x\, dx}{1+\cos x} = -\frac{\pi}{2} \ln 2 + 2G$$

$$\int_0^{\pi/2} \frac{dx}{\sin x \pm \cos x} = \mp \frac{1}{\sqrt{2}} \ln\left(\tan \frac{\pi}{8}\right) \qquad\qquad \int_0^{\pi/2} \frac{dx}{(\sin x \pm \cos x)^2} = \pm 1$$

$$\int_0^{\pi/2} \frac{dx}{1 \pm a \sin x} = \int_0^{\pi/2} \frac{dx}{1 \pm a \cos x} = \frac{\pi \mp 2 \sin^{-1} a}{2\sqrt{1-a^2}} \qquad\qquad 0 < a < 1$$

$$\int_0^{\pi/2} \frac{dx}{(1 \pm a \sin x)^2} = \int_0^{\pi/2} \frac{dx}{(1 \pm a \cos x)^2} = \frac{\pi \mp 2 \sin^{-1} a}{2\sqrt{(1-a^2)^3}} \mp \frac{a}{1-a^2} \qquad \begin{array}{c} 0 < \sin^{-1} a < \dfrac{\pi}{2} \end{array}$$

$$\int_0^{\pi/2} \frac{dx}{(a \sin x + b \cos x)^2} = \frac{1}{ab} \qquad\qquad \int_0^{\pi/2} \frac{x\, dx}{(a \sin x + b \cos x)^2} = \frac{ab}{a^2+b^2} \frac{\pi}{2} - \frac{\ln ab}{a^2+b^2}$$

$$\int_0^{\pi/2} \frac{dx}{1 \pm a^2 \sin^2 x} = \int_0^{\pi/2} \frac{dx}{1 \pm a^2 \cos^2 x} = \frac{\pi}{2\sqrt{1 \pm a^2}}$$

$$\int_0^{\pi/2} \frac{dx}{(1 \pm a^2 \sin^2 x)^2} = \int_0^{\pi/2} \frac{d}{(1 \pm a^2 \cos^2 x)^2} = \frac{\pi(2 \pm a^2)}{4\sqrt{(1 \pm a^2)^3}} \qquad\qquad 0 < a^2 < 1$$

$$\int_0^{\pi/2} \frac{dx}{a^2 \sin^2 x + b^2 \cos^2 x} = \frac{\pi}{2ab} \qquad\qquad \int_0^{\pi/2} \frac{dx}{(a^2 \sin^2 x + b^2 \cos^2 x)^2} = \frac{2\pi(a^2+b^2)}{(2ab)^3}$$

$$\int_0^{\pi/2} \frac{\cos^2 x\, dx}{a^2 \sin^2 x + b^2 \cos^2 x} = \frac{\pi}{2b(a+b)} \qquad\qquad \int_0^{\pi/2} \frac{\cos^2 x\, dx}{(a^2 \sin^2 x + b^2 \cos^2 x)^2} = \frac{\pi}{4ab^3}$$

$$\int_0^{\pi/2} \frac{\sin^2 x\, dx}{a^2 \sin^2 x + b^2 \cos^2 x} = \frac{\pi}{2a(a+b)} \qquad\qquad \int_0^{\pi/2} \frac{\sin^2 x\, dx}{(a^2 \sin^2 x + b^2 \cos^2 x)^2} = \frac{\pi}{4a^3b}$$

$$\int_0^{\pi/2} \frac{\sin x\, dx}{\sqrt{1 - k^2 \sin^2 x}} = \frac{1}{2k} \ln \frac{1+k}{1-k} \qquad\qquad \int_0^{\pi/2} \frac{\cos x\, dx}{\sqrt{1 - k^2 \sin^2 x}} = \frac{1}{k} \sin^{-1} k$$

$$\int_0^{\pi/2} \frac{\sin^2 x\, dx}{\sqrt{1 - k^2 \sin^2 x}} = \frac{1}{k^2}(K - E) \qquad\qquad \int_0^{\pi/2} \frac{\cos^2 x\, dx}{\sqrt{1 - k^2 \sin^2 x}} = \frac{1}{k^2}[E - (1 - k^2)K]$$

(15) Definite Integrals Involving $f(x) = f(\sin^p x, \cos^p x, \sin qx, \cos qx)$ $[0, \pi], [-a, a]$

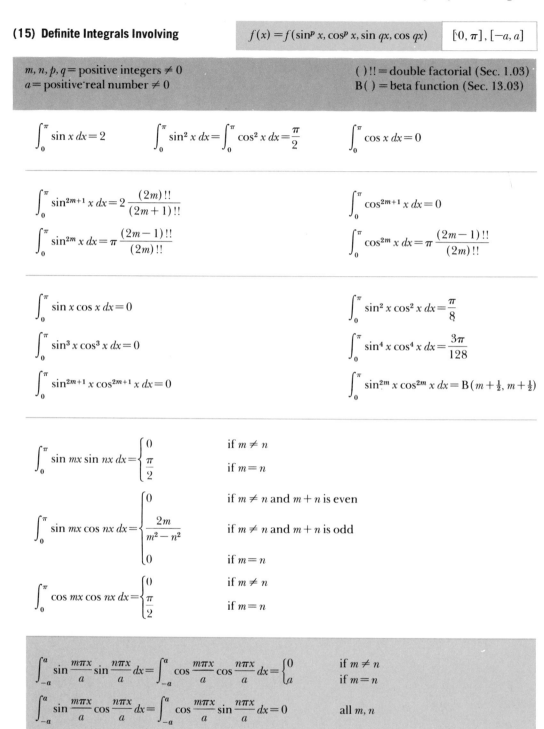

m, n, p, q = positive integers $\neq 0$ ( )!! = double factorial (Sec. 1.03)
a = positive real number $\neq 0$ B( ) = beta function (Sec. 13.03)

$$\int_0^\pi \sin x \, dx = 2 \qquad \int_0^\pi \sin^2 x \, dx = \int_0^\pi \cos^2 x \, dx = \frac{\pi}{2} \qquad \int_0^\pi \cos x \, dx = 0$$

$$\int_0^\pi \sin^{2m+1} x \, dx = 2\frac{(2m)!!}{(2m+1)!!} \qquad\qquad \int_0^\pi \cos^{2m+1} x \, dx = 0$$

$$\int_0^\pi \sin^{2m} x \, dx = \pi \frac{(2m-1)!!}{(2m)!!} \qquad\qquad \int_0^\pi \cos^{2m} x \, dx = \pi \frac{(2m-1)!!}{(2m)!!}$$

$$\int_0^\pi \sin x \cos x \, dx = 0 \qquad\qquad\qquad \int_0^\pi \sin^2 x \cos^2 x \, dx = \frac{\pi}{8}$$

$$\int_0^\pi \sin^3 x \cos^3 x \, dx = 0 \qquad\qquad\qquad \int_0^\pi \sin^4 x \cos^4 x \, dx = \frac{3\pi}{128}$$

$$\int_0^\pi \sin^{2m+1} x \cos^{2m+1} x \, dx = 0 \qquad\qquad \int_0^\pi \sin^{2m} x \cos^{2m} x \, dx = B\left(m + \tfrac{1}{2}, m + \tfrac{1}{2}\right)$$

$$\int_0^\pi \sin mx \sin nx \, dx = \begin{cases} 0 & \text{if } m \neq n \\ \dfrac{\pi}{2} & \text{if } m = n \end{cases}$$

$$\int_0^\pi \sin mx \cos nx \, dx = \begin{cases} 0 & \text{if } m \neq n \text{ and } m + n \text{ is even} \\ \dfrac{2m}{m^2 - n^2} & \text{if } m \neq n \text{ and } m + n \text{ is odd} \\ 0 & \text{if } m = n \end{cases}$$

$$\int_0^\pi \cos mx \cos nx \, dx = \begin{cases} 0 & \text{if } m \neq n \\ \dfrac{\pi}{2} & \text{if } m = n \end{cases}$$

$$\int_{-a}^a \sin \frac{m\pi x}{a} \sin \frac{n\pi x}{a} \, dx = \int_{-a}^a \cos \frac{m\pi x}{a} \cos \frac{n\pi x}{a} \, dx = \begin{cases} 0 & \text{if } m \neq n \\ a & \text{if } m = n \end{cases}$$

$$\int_{-a}^a \sin \frac{m\pi x}{a} \cos \frac{n\pi x}{a} \, dx = \int_{-a}^a \cos \frac{m\pi x}{a} \sin \frac{n\pi x}{a} \, dx = 0 \qquad \text{all } m, n$$

(16) Definite Integrals Involving $\qquad$ $f(x) = f(x, \sin^p x, \cos^p x)$ $\quad [0, \pi]$

| $m, n, p = $ positive integers $\neq 0$ | $\alpha = \sqrt{1 + a^2}$ | $1 < \alpha^2 < 2$ |
| $a = $ positive real number $\neq 0$ | $\beta = \sqrt{1 - a^2}$ | $0 < \beta^2 < 1$ |

$$\int_0^\pi x \sin x \, dx = \pi$$

$$\int_0^\pi x \cos x \, dx = -2$$

$$\int_0^\pi x \sin^2 x \, dx = \frac{\pi^2}{4}$$

$$\int_0^\pi x \cos^2 x \, dx = \frac{\pi^2}{4}$$

$$\int_0^\pi x \sin^{2n+1} x \, dx = \frac{(2n)!!}{(2n+1)!!} \pi$$

$$\int_0^\pi x \cos^{2n+1} x \, dx = -\frac{2}{4^n} \sum_{k=0}^n \binom{2n+1}{k} \frac{1}{(2n-2k-1)^2}$$

$$\int_0^\pi x \sin^{2n} x \, dx = \frac{(2n-1)!!}{(2n)!!} \frac{\pi^2}{2}$$

$$\int_0^\pi x \cos^{2n} x \, dx = \frac{(2n-1)!!}{(2n)!!} \frac{\pi^2}{2}$$

$$\int_0^\pi \frac{dx}{1 \pm a \sin x} = \frac{\pi \mp 2 \sin^{-1} a}{\beta}$$

$$\int_0^\pi \frac{dx}{1 \pm a \cos x} = \frac{\pi}{\beta}$$

$$\int_0^\pi \frac{dx}{(1 \pm a \sin x)^2} = \frac{\pi \mp 2 \sin^{-1} a}{\beta^3} + \frac{2a}{\beta^2}$$

$$\int_0^\pi \frac{dx}{(1 \pm a \cos x)^2} = \frac{\pi}{\beta^3}$$

$$\int_0^\pi (\alpha^2 - 2a \cos x)^n \, dx = \pi \left[1 + \binom{n}{1}^2 a^2 + \binom{n}{2}^2 a^4 + \cdots + \binom{n}{n}^2 a^{2n} \right]$$

$$\int_0^\pi \frac{dx}{\alpha^2 \pm 2a \cos x} = \frac{\pi}{|\beta^2|}$$

$$\int_0^\pi \frac{dx}{\alpha^2 \pm 2a \sin x} = \frac{\pi \mp 2 \sin^{-1} (2a/\alpha^2)}{|\beta^2|}$$

$$\int_0^\pi \frac{dx}{(\alpha^2 - 2a \cos x)^n} = \frac{\pi}{\beta^{2n}} \sum_{k=0}^{n-1} \binom{2k}{k} \binom{n+k-1}{n-k-1} \left(\frac{a}{\beta}\right)^{2k}$$

$$\int_0^\pi \frac{\sin x \, dx}{\alpha^2 - 2a \cos x} = \frac{2}{a} \tanh^{-1} a$$

$$\int_0^\pi \frac{\cos x \, dx}{\alpha^2 - 2a \cos x} = \frac{\pi a}{\beta^2}$$

$$\int_0^\pi \frac{\sin^2 x \, dx}{\alpha^2 - 2a \cos x} = \frac{\pi}{2}$$

$$\int_0^\pi \frac{\cos^2 x \, dx}{\alpha^2 - 2a \cos x} = \frac{\pi}{2} \left(\frac{\alpha}{\beta}\right)^2$$

$$\int_0^\pi \frac{\sin x \sin mx \, dx}{\alpha^2 - 2a \cos x} = \frac{\pi a^m}{2a}$$

$$\int_0^\pi \frac{\cos x \cos mx \, dx}{\alpha^2 - 2a \cos x} = \frac{\pi a^m}{2a} \left(\frac{\alpha}{\beta}\right)^2$$

$$\int_0^\pi \frac{\cos mx \, dx}{(\alpha^2 - 2a \cos x)^n} = \left(\frac{a}{\beta^2}\right)^{2n-1} a^{m-1} \pi \sum_{k=0}^{n-1} \binom{m+n-1}{k} \binom{2n-k-2}{n-1} \left(\frac{\beta}{a}\right)^{2k}$$

$$\int_0^\pi \frac{dx}{a^2 \sin^2 x + b^2 \cos^2 x} = \frac{\pi^2}{2ab}$$

$$\int_0^\pi \frac{x \sin x \cos x \, dx}{a^2 \sin^2 x - b^2 \cos^2 x} = \frac{\pi}{b^2 - a^2} \ln \frac{a+b}{2b}$$

(17) Definite Integrals Involving

$$f(x) = f(\sin nx) \qquad [0, \pi]$$

$n =$ positive integer $\neq 0$ $\qquad\qquad a, b =$ real numbers $\neq 0$ $\qquad\qquad \lambda = (-1)^n$

$$\alpha_1 = a + b \qquad\qquad \beta_1 = (b+n)\pi \qquad\qquad \omega_1 = 2\sqrt{a^2 + (b+n)^2} \qquad\qquad \phi_1 = \tan^{-1}\frac{a}{b+n}$$

$$\alpha_2 = a - b \qquad\qquad \beta_2 = (b-n)\pi \qquad\qquad \omega_2 = 2\sqrt{a^2 + (b-n)^2} \qquad\qquad \phi_2 = \tan^{-1}\frac{a}{b-n}$$

$$\int_0^\pi \sin nx\, dx = \frac{1-\lambda}{n}$$

$$\int_0^\pi x \sin nx\, dx = -\frac{\lambda\pi}{n}$$

$$\int_0^\pi x^2 \sin nx\, dx = \frac{2(\lambda-1)}{n^3} - \frac{\lambda\pi^2}{n}$$

$$\int_0^\pi x^3 \sin nx\, dx = \frac{6\lambda\pi}{n^3} - \frac{\lambda\pi^3}{n}$$

$$\int_0^\pi e^{ax} \sin nx\, dx = \frac{n(1 - \lambda e^{a\pi})}{a^2 + n^2}$$

$$\int_0^\pi a^{bx} \sin nx\, dx = \frac{n(1 - \lambda a^{b\pi})}{(b \ln a)^2 + n^2}$$

$$\int_0^\pi x e^{ax} \sin nx\, dx = -\frac{n\lambda\pi e^{a\pi}}{a^2 + n^2} + \frac{2an}{(a^2 + n^2)^2}(\lambda e^{a\pi} - 1)$$

$$\int_0^\pi \sin ax \sin nx\, dx = \frac{n\lambda \sin a\pi}{a^2 - n^2}$$

$$\int_0^\pi \sinh ax \sin nx\, dx = -\frac{n\lambda \sinh a\pi}{a^2 + n^2}$$

$$\int_0^\pi \cos ax \sin nx\, dx = \frac{n(\lambda \cos a\pi - 1)}{a^2 - n^2}$$

$$\int_0^\pi \cosh ax \sin nx\, dx = -\frac{n(\lambda \cosh a\pi - 1)}{a^2 + n^2}$$

$$\int_0^\pi x \sin bx \sin nx\, dx = -\frac{\pi^2}{2}\left[(\lambda \sin b\pi)\left(\frac{1}{\beta_1} - \frac{1}{\beta_2}\right) + (\lambda \cos b\pi - 1)\left(\frac{1}{\beta_1^2} - \frac{1}{\beta_2^2}\right)\right]$$

$$\int_0^\pi x \cos bx \sin nx\, dx = -\frac{\pi^2}{2}\left[(\lambda \cos b\pi)\left(\frac{1}{\beta_1} - \frac{1}{\beta_2}\right) - (\lambda \sin b\pi)\left(\frac{1}{\beta_1^2} - \frac{1}{\beta_2^2}\right)\right]$$

$$\int_0^\pi e^{ax} \sin bx \sin nx\, dx = -e^{a\pi}\left[\frac{\sin(\phi_1 + \beta_1)}{\omega_1} - \frac{\sin(\phi_2 + \beta_2)}{\omega_2}\right] + \frac{\sin \phi_1}{\omega_1} - \frac{\sin \phi_2}{\omega_2}$$

$$\int_0^\pi e^{ax} \cos bx \sin nx\, dx = -e^{a\pi}\left[\frac{\cos(\phi_1 + \beta_1)}{\omega_1} - \frac{\cos(\phi_2 + \beta_2)}{\omega_2}\right] + \frac{\cos \phi_1}{\omega_1} - \frac{\cos \phi_2}{\omega_2}$$

$$\int_0^\pi \sin ax \sin bx \sin nx\, dx = +\frac{n(1 - \lambda \cos \alpha_1 \pi)}{2(\alpha_1^2 - n^2)} - \frac{n(1 - \lambda \cos \alpha_2 \pi)}{2(\alpha_2^2 - n^2)}$$

$$\int_0^\pi \cos ax \sin bx \sin nx\, dx = +\frac{n\lambda \sin \alpha_1 \pi}{2(\alpha_1^2 - n^2)} - \frac{n\lambda \sin \alpha_2 \pi}{2(\alpha_2^2 - n^2)}$$

$$\int_0^\pi \cos ax \cos bx \sin nx\, dx = +\frac{n(\lambda \cos \alpha_1 \pi - 1)}{2(\alpha_1^2 - n^2)} + \frac{n(\lambda \cos \alpha_2 \pi - 1)}{2(\alpha_2^2 - n^2)}$$

(18) Definite Integrals Involving

$$f(x) = f(\cos nx) \qquad [0, \pi]$$

n = positive integer $\neq 0$		a, b = real numbers $\neq 0$	$\lambda = (-1)^n$
$\alpha_1 = a + b$	$\beta_1 = (b+n)\pi$	$\omega_1 = 2\sqrt{a^2 + (b+n)^2}$	$\phi_1 = \tan^{-1}\dfrac{a}{b+n}$
$\alpha_2 = a - b$	$\beta_2 = (b-n)\pi$	$\omega_2 = 2\sqrt{a^2 + (b-n)^2}$	$\phi_2 = \tan^{-1}\dfrac{a}{b-n}$

$$\int_0^\pi \cos nx \, dx = 0 \qquad\qquad \int_0^\pi x \cos nx \, dx = \frac{\lambda - 1}{n^2}$$

$$\int_0^\pi x^2 \cos nx \, dx = \frac{2\lambda\pi}{n^2} \qquad\qquad \int_0^\pi x^3 \cos nx \, dx = \frac{3\lambda\pi^2}{n^2} - \frac{6(\lambda - 1)}{nx}$$

$$\int_0^\pi e^{ax} \cos nx \, dx = \frac{a(\lambda e^{a\pi} - 1)}{a^2 + n^2} \qquad\qquad \int_0^\pi a^{bx} \cos nx \, dx = \frac{(b \ln a)(\lambda a^{b\pi} - 1)}{(b \ln a)^2 + n^2}$$

$$\int_0^\pi xe^{ax} \cos nx \, dx = +\frac{n\lambda\pi e^{a\pi}}{a^2 + n^2} + \frac{a^2 - n^2}{(a^2 + n^2)^2}(1 - \lambda e^{a\pi})$$

$$\int_0^\pi \sin ax \cos nx \, dx = \frac{a(1 - \lambda \cos a\pi)}{a^2 - n^2} \qquad\qquad \int_0^\pi \sinh ax \cos nx \, dx = -\frac{a(1 - \lambda \cosh a\pi)}{a^2 + n^2}$$

$$\int_0^\pi \cos ax \cos nx \, dx = \frac{a\lambda \sin a\pi}{a^2 - n^2} \qquad\qquad \int_0^\pi \cosh ax \cos nx \, dx = +\frac{a\lambda \sinh a\pi}{a^2 + n^2}$$

$$\int_0^\pi x \sin bx \cos nx \, dx = -\frac{\pi^2}{2}\left[(\lambda \cos b\pi)\left(\frac{1}{\beta_1} + \frac{1}{\beta_2}\right) - (\lambda \sin b\pi)\left(\frac{1}{\beta_1^2} + \frac{1}{\beta_2^2}\right)\right]$$

$$\int_0^\pi x \cos bx \cos nx \, dx = +\frac{\pi^2}{2}\left[(\lambda \sin b\pi)\left(\frac{1}{\beta_1} + \frac{1}{\beta_2}\right) - (\lambda \cos b\pi - 1)\left(\frac{1}{\beta_1^2} + \frac{1}{\beta_2^2}\right)\right]$$

$$\int_0^\pi e^{ax} \sin bx \cos nx \, dx = -e^{a\pi}\left[\frac{\cos(\phi_1 + \beta_1)}{\omega_1} + \frac{\cos(\phi_2 + \beta_2)}{\omega_2}\right] + \frac{\cos\phi_1}{\omega_1} + \frac{\cos\phi_2}{\omega_2}$$

$$\int_0^\pi e^{ax} \cos bx \cos nx \, dx = +e^{a\pi}\left[\frac{\sin(\phi_1 + \beta_1)}{\omega_1} + \frac{\sin(\phi_2 + \beta_2)}{\omega_2}\right] - \frac{\sin\phi_1}{\omega_1} - \frac{\sin\phi_2}{\omega_2}$$

$$\int_0^\pi \sin ax \sin bx \cos nx \, dx = -\frac{\lambda\alpha_1 \sin \alpha_1\pi}{2(\alpha_1^2 - n^2)} + \frac{\lambda\alpha_2 \sin \alpha_2\pi}{2(\alpha_2^2 - n^2)}$$

$$\int_0^\pi \sin ax \cos bx \cos nx \, dx = +\frac{\alpha_1(1 - \lambda \cos \alpha_1\pi)}{2(\alpha_1^2 - n^2)} + \frac{\alpha_2(1 - \lambda \cos \alpha_2\pi)}{2(\alpha_2^2 - n^2)}$$

$$\int_0^\pi \cos ax \cos bx \cos nx \, dx = +\frac{\lambda\alpha_1 \sin \alpha_1\pi}{2(\alpha_1^2 - n^2)} + \frac{\lambda\alpha_2 \sin \alpha_2\pi}{2(\alpha_2^2 - n^2)}$$

(19) Definite Integrals Involving $f(x) = f(\sin nx)$ $[-\pi, +\pi]$

$n =$ positive integer $\neq 0$	$a, b =$ real numbers $\neq 0$	$\lambda = (-1)^n$
$\alpha_1 = a + b$ $\beta_1 = (b + n)\pi$	$\omega_1 = 2\sqrt{a^2 + (b + n)^2}$	$\phi_1 = \tan^{-1}\dfrac{a}{b + n}$
$\alpha_2 = a - b$ $\beta_2 = (b - n)\pi$	$\omega_2 = 2\sqrt{a^2 + (b - n)^2}$	$\phi_2 = \tan^{-1}\dfrac{a}{b - n}$

$$\int_{-\pi}^{+\pi} \sin nx\, dx = 0$$

$$\int_{-\pi}^{+\pi} x \sin nx\, dx = -\frac{2\lambda\pi}{n}$$

$$\int_{-\pi}^{+\pi} x^2 \sin nx\, dx = 0$$

$$\int_{-\pi}^{+\pi} x^3 \sin nx\, dx = \frac{12\lambda\pi}{n^3} - \frac{2\lambda\pi^3}{n}$$

$$\int_{-\pi}^{+\pi} e^{ax} \sin nx\, dx = -\frac{2n\lambda \sinh a\pi}{a^2 + n^2}$$

$$\int_{-\pi}^{+\pi} a^{bx} \sin nx\, dx = \frac{2n\lambda \sinh (\pi b \ln a)}{(b \ln a)^2 + n^2}$$

$$\int_{-\pi}^{+\pi} x e^{ax} \sin nx\, dx = -\frac{2n\lambda \sinh a\pi}{(a^2 + n^2)^2}\left[\pi(a^2 + n^2) - 2a\right]$$

$$\int_{-\pi}^{+\pi} \sin ax \sin nx\, dx = \frac{2n\lambda \sin a\pi}{a^2 - n^2}$$

$$\int_{-\pi}^{+\pi} \sinh ax \sin nx\, dx = -\frac{2n\lambda \sinh a\pi}{a^2 + n^2}$$

$$\int_{-\pi}^{+\pi} \cos ax \sin nx\, dx = 0$$

$$\int_{-\pi}^{+\pi} \cosh ax \sin nx\, dx = 0$$

$$\int_{-\pi}^{+\pi} x \sin bx \sin nx\, dx = 0$$

$$\int_{-\pi}^{+\pi} x \cos bx \sin nx\, dx = -\pi^2\left[(\lambda \cos b\pi)\left(\frac{1}{\beta_1} - \frac{1}{\beta_2}\right) - (\lambda \sin b\pi)\left(\frac{1}{\beta_1^2} - \frac{1}{\beta_2^2}\right)\right]$$

$$\int_{-\pi}^{+\pi} e^{ax} \sin bx \sin nx\, dx = -e^{a\pi}\left[\frac{\sin(\phi_1 + \beta_1)}{\omega_1} - \frac{\sin(\phi_2 + \beta_2)}{\omega_2}\right] + e^{-a\pi}\left[\frac{\sin(\phi_1 - \beta_1)}{\omega_1} - \frac{\sin(\phi_2 - \beta_2)}{\omega_2}\right]$$

$$\int_{-\pi}^{+\pi} e^{ax} \cos bx \sin nx\, dx = -e^{a\pi}\left[\frac{\cos(\phi_1 + \beta_1)}{\omega_1} - \frac{\cos(\phi_2 + \beta_2)}{\omega_2}\right] + e^{-a\pi}\left[\frac{\cos(\phi_1 - \beta_1)}{\omega_1} - \frac{\cos(\phi_2 - \beta_2)}{\omega_2}\right]$$

$$\int_{-\pi}^{+\pi} \sin ax \sin bx \sin nx\, dx = 0$$

$$\int_{-\pi}^{+\pi} \cos ax \sin bx \sin nx\, dx = +\frac{n\lambda \sin \alpha_1\pi}{\alpha_1^2 - n^2} - \frac{n\lambda \sin \alpha_2\pi}{\alpha_2^2 - n^2}$$

$$\int_{-\pi}^{+\pi} \cos ax \cos bx \sin nx\, dx = 0$$

(20) Definite Integrals Involving $\quad\boxed{f(x)=f(\cos nx)}\quad\boxed{[-\pi,+\pi]}$

$n=$ positive integer $\neq 0$	$a,b=$ real numbers $\neq 0$	$\lambda=(-1)^n$
$\alpha_1=a+b$	$\beta_1=(b+n)\pi \qquad \omega_1=2\sqrt{a^2+(b+n)^2}$	$\phi_1=\tan^{-1}\dfrac{a}{b+n}$
$\alpha_2=a-b$	$\beta_2=(b-n)\pi \qquad \omega_2=2\sqrt{a^2+(b-n)^2}$	$\phi_2=\tan^{-1}\dfrac{a}{b-n}$

$$\int_{-\pi}^{+\pi}\cos nx\,dx=0 \qquad\qquad \int_{-\pi}^{+\pi}x\cos nx\,dx=0$$

$$\int_{-\pi}^{+\pi}x^2\cos nx\,dx=\frac{4\lambda\pi}{n^2} \qquad\qquad \int_{-\pi}^{+\pi}x^3\cos nx\,dx=0$$

$$\int_{-\pi}^{+\pi}e^{ax}\cos nx\,dx=\frac{2a\lambda\sinh a\pi}{a^2+n^2} \qquad\qquad \int_{-\pi}^{+\pi}a^{bx}\cos nx\,dx=\frac{2(b\ln a)\lambda\sinh(\pi b\ln a)}{(b\ln a)^2+n^2}$$

$$\int_{-\pi}^{+\pi}xe^{ax}\cos nx\,dx=\frac{2\lambda\sinh a\pi}{(a^2+n^2)^2}\left[n\pi(a^2+n^2)-a^2+n^2\right]$$

$$\int_{-\pi}^{+\pi}\sin ax\cos nx\,dx=0 \qquad\qquad \int_{-\pi}^{+\pi}\sinh ax\cos nx\,dx=0$$

$$\int_{-\pi}^{+\pi}\cos ax\cos nx\,dx=\frac{2a\lambda\sin a\pi}{a^2-n^2} \qquad\qquad \int_{-\pi}^{+\pi}\cosh ax\cos nx\,dx=\frac{2a\lambda\sinh a\pi}{a^2+b^2}$$

$$\int_{-\pi}^{+\pi}x\sin ax\cos nx\,dx=-\pi^2\left[(\lambda\cos a\pi-1)\left(\frac{1}{\beta_1}+\frac{1}{\beta_2}\right)-(\lambda\sin a\pi)\left(\frac{1}{\beta_1^{\,2}}+\frac{1}{\beta_2^{\,2}}\right)\right]$$

$$\int_{-\pi}^{+\pi}x\cos ax\cos nx\,dx=0$$

$$\int_{-\pi}^{+\pi}e^{ax}\sin bx\cos nx\,dx=-e^{a\pi}\left[\frac{\cos(\phi_1+\beta_1)}{\omega_1}+\frac{\cos(\phi_2+\beta_2)}{\omega_2}\right]+e^{-a\pi}\left[\frac{\cos(\phi_1-\beta_1)}{\omega_1}+\frac{\cos(\phi_2-\beta_2)}{\omega_2}\right]$$

$$\int_{-\pi}^{+\pi}e^{ax}\cos bx\cos nx\,dx=+e^{a\pi}\left[\frac{\sin(\phi_1+\beta_1)}{\omega_1}+\frac{\sin(\phi_2+\beta_2)}{\omega_2}\right]-e^{-a\pi}\left[\frac{\sin(\phi_1-\beta_1)}{\omega_1}+\frac{\sin(\phi_2-\beta_2)}{\omega_2}\right]$$

$$\int_{-\pi}^{+\pi}\sin ax\sin bx\cos nx\,dx=-\frac{\lambda\alpha_1\sin\alpha_1\pi}{\alpha_1^{\,2}-n^2}+\frac{\lambda\alpha_2\sin\alpha_2\pi}{\alpha_2^{\,2}-n^2}$$

$$\int_{-\pi}^{+\pi}\sin ax\cos bx\cos nx\,dx=0$$

$$\int_{-\pi}^{+\pi}\cos ax\cos bx\cos nx\,dx=+\frac{\lambda\alpha_1\sin\alpha_1\pi}{\alpha_1^{\,2}-n^2}+\frac{\lambda\alpha_2\sin\alpha_2\pi}{\alpha_2^{\,2}-n^2}$$

(21) Definite Integrals Involving

$$f(x)=f\left(\frac{1}{x^m}\right)\sin^p ax, \cos^p bx) \qquad [0,\infty]$$

$m, n, p =$ positive integers $\neq 0$ $(\ )!! =$ double factorial (Sec. 1.03)

$a, b, c =$ positive real numbers $\neq 0$ $\Gamma(\ ) =$ gamma function (Sec. 13.03)

$$\int_0^\infty \frac{\sin(\pm ax)}{x}\,dx = \pm\frac{\pi}{2}$$

$$\int_0^\infty \frac{\sin^{2m} ax}{x^2}\,dx = \frac{(2m-3)!!}{(2m-2)!!}\frac{a\pi}{2} \qquad m>1$$

$$\int_0^\infty \frac{\sin^2 ax}{x^2}\,dx = \frac{\pi a}{2}$$

$$\int_0^\infty \frac{\sin^4 ax}{x^2}\,dx = \frac{\pi a}{4}$$

$$\int_0^\infty \frac{\sin^4 ax}{x^3}\,dx = a^2 \ln 2$$

$$\int_0^\infty \frac{\tan(\pm ax)}{x}\,dx = \pm\frac{\pi}{2}$$

$$\int_0^\infty \frac{\sin^{2m+1} ax}{x}\,dx = \frac{(2m-1)!!}{(2m)!!}\frac{\pi}{2} \qquad m>0$$

$$\int_0^\infty \frac{\sin^{2m} ax}{x}\,dx = \int_0^\infty \frac{\cos^{2m} ax}{x}\,dx = \infty$$

$$\int_0^\infty \frac{\sin^3 ax}{x}\,dx = \frac{\pi}{4}$$

$$\int_0^\infty \frac{\sin^3 ax}{x^2}\,dx = \frac{3}{4}a \ln 3$$

$$\int_0^\infty \frac{\sin ax}{\sqrt{x}}\,dx = \sqrt{\frac{\pi}{2a}}$$

$$\int_0^\infty \frac{\sin ax}{\sqrt[b]{x}}\,dx = \frac{\pi\sqrt[b]{a}}{2a\Gamma(1/b)\sin(\pi/2b)}$$

$$\int_0^\infty \frac{\cos ax}{\sqrt{x}}\,dx = \sqrt{\frac{\pi}{2a}}$$

$$\int_0^\infty \frac{\cos ax}{\sqrt[b]{x}}\,dx = \frac{\pi\sqrt[b]{a}}{2a\Gamma(1/b)\cos(\pi/2b)}$$

$$\int_0^\infty \frac{\sin ax \sin bx}{x}\,dx = \ln\sqrt{\frac{a+b}{a-b}}$$

$$\int_0^\infty \frac{\cos ax \cos bx}{x}\,dx = \infty$$

$$\int_0^\infty \frac{\sin ax \cos bx}{x}\,dx = \begin{cases} 0 & \text{if } b>a\geq 0 \\ \pi/2 & \text{if } a>b\geq 0 \\ \pi/4 & \text{if } a=b>0 \end{cases}$$

$$\int_0^\infty \frac{\sin^2 ax}{b^2+x^2}\,dx = \frac{\pi}{4b}(1-e^{-2ab})$$

$$\int_0^\infty \frac{\sin ax \sin bx}{c^2+x^2}\,dx = \frac{\pi}{2c}e^{-ac}\sinh bc \qquad a\geq b$$

$$\int_0^\infty \frac{\cos^2 ax}{b^2+x^2}\,dx = \frac{\pi}{4b}(1+e^{-2ab})$$

$$\int_0^\infty \frac{\cos ax \cos bx}{c^2+x^2}\,dx = \frac{\pi}{2c}e^{-ac}\cosh bc \qquad a\geq b$$

$$\int_0^\infty \sin(x^{m+1})\,dx = \Gamma\left(1+\frac{1}{m+1}\right)\sin\frac{\pi}{2(m+1)}$$

$$\int_0^\infty \cos(x^{m+1})\,dx = \Gamma\left(1+\frac{1}{m+1}\right)\cos\frac{\pi}{2(m+1)}$$

(22) Definite Integrals Involving

$$f(x)=f(e^{-ax}, \sin^p ax, \cos^p bx) \quad [0, \infty]$$

For definition of $a, b, c, m, n, p,$ and $\Gamma(\ )$ see opposite page.

$$\alpha=a^2+(b+c)^2 \qquad\qquad \beta=a^2+(b-c)^2 \qquad\qquad \gamma=a^2+b^2-c^2$$

$$\int_0^\infty e^{-ax} \sin bx \, dx = \frac{b}{a^2+b^2}$$

$$\int_0^\infty e^{-ax} \cos bx \, dx = \frac{a}{a^2+b^2}$$

$$\int_0^\infty xe^{-ax} \sin bx \, dx = \frac{2ab}{(a^2+b^2)^2}$$

$$\int_0^\infty xe^{-ax} \cos bx \, dx = \frac{a^2-b^2}{(a^2+b^2)^2}$$

$$\int_0^\infty x^a e^{-bx} \sin cx \, dx = \frac{\Gamma(a+1) \sin \omega}{\sqrt{(b^2+c^2)^{a+1}}}$$

$$\omega = (a+1) \tan^{-1}\frac{c}{b}$$

$$\int_0^\infty x^a e^{-bx} \cos cx \, dx = \frac{\Gamma(a+1) \cos \omega}{\sqrt{(b^2+c^2)^{a+1}}}$$

$$\int_0^\infty e^{-ax} \sin (bx+c) \, dx = \frac{1}{a^2+b^2} (a \sin c + b \cos c)$$

$$\int_0^\infty e^{-ax} \cos (bx+c) \, dx = \frac{1}{a^2+b^2} (a \cos c - b \sin c)$$

$$\int_0^\infty e^{-ax} \sin^2 bx \, dx = \frac{2b^2}{a(a^2+4b^2)}$$

$$\int_0^\infty e^{-ax} \cos^2 bx \, dx = \frac{a^2+2b^2}{a(a^2+4b^2)}$$

$$\int_0^\infty \frac{e^{-ax}}{x} \sin bx \, dx = \tan^{-1}\frac{b}{a}$$

$$\int_0^\infty \frac{e^{-ax}}{x} \cos bx \, dx = \infty$$

$$\int_0^\infty \frac{e^{-ax}}{x} \sin^2 bx \, dx = \ln \sqrt[4]{\frac{a^2+4b^2}{a^2}}$$

$$\int_0^\infty \frac{e^{-ax}}{x^m} \cos^n bx \, dx = \infty$$

$$\int_0^\infty \frac{e^{-ax}}{x^2} \sin^2 bx \, dx = b \tan^{-1}\frac{2b}{a} - a \ln \sqrt[4]{\frac{a^2+4b^2}{a^2}}$$

$$\int_0^\infty \frac{e^{-ax}-e^{-bx}}{x} \sin cx \, dx = \tan^{-1}\frac{c(b-a)}{ab+c^2}$$

$$\int_0^\infty \frac{e^{-ax}-e^{-bx}}{x} \cos cx \, dx = \ln \sqrt{\frac{b^2+c^2}{a^2+c^2}}$$

$$\int_0^\infty \frac{e^{-ax}-e^{-bx}}{x^2} \sin cx \, dx = -a \tan^{-1}\frac{c}{a} + b \tan^{-1}\frac{c}{b} + c \ln \sqrt{\frac{b^2+c^2}{a^2+c^2}}$$

$$\int_0^\infty e^{-ax} \sin bx \sin cx \, dx = \frac{2abc}{\alpha\beta}$$

$$\int_0^\infty e^{-ax} \cos bx \cos cx \, dx = \frac{a(\alpha+\beta)}{2\alpha\beta}$$

$$\int_0^\infty e^{-ax} \sin bx \cos cx \, dx = \frac{b\gamma}{\alpha\beta}$$

$$\int_0^\infty \frac{e^{-ax}}{x} \sin bx \sin cx \, dx = \frac{1}{4} \ln \frac{\alpha}{\beta}$$

(23) Definite Integrals Involving $\quad f(x) = f\left(\dfrac{1}{x}, \sqrt{1 \pm a^2 \sin^2 x}, \sqrt{1 \pm a^2 \cos^2 x}\right)\quad$ $[0, \infty]$

Complete elliptic integral of the first kind (Sec. 13.05–2):

$$K(a) = F\left(a, \frac{\pi}{2}\right) = \int_0^{\pi/2} \frac{d\phi}{\sqrt{1 - a^2 \sin^2 \phi}} = \int_0^{\pi/2} \frac{d\phi}{\sqrt{b^2 + a^2 \cos^2 \phi}}$$

Complete elliptic integral of the second kind (Sec. 13.05–2):

$$E(a) = E\left(a, \frac{\pi}{2}\right) = \int_0^{\pi/2} \sqrt{1 - a^2 \sin^2 \phi}\, d\phi = \int_0^{\pi/2} \sqrt{b^2 + a^2 \cos^2 \phi}\, d\phi$$

Modulus: $k = a, 0 < a < 1$ $\qquad\qquad\qquad\qquad$ Complementary modulus: $k' = b = \sqrt{1 - a^2}$

$$\int_0^\infty \frac{\sin x}{x} \sqrt{1 - a^2 \sin^2 x}\, dx = E(a) \qquad\qquad \int_0^\infty \frac{\sin x}{x} \sqrt{1 - a^2 \cos^2 x}\, dx = E(a)$$

$$\int_0^\infty \frac{\sin x\, dx}{x\sqrt{1 - a^2 \sin^2 x}} = K(a) \qquad\qquad \int_0^\infty \frac{\sin x\, dx}{x\sqrt{1 - a^2 \cos^2 x}} = K(a)$$

$$\int_0^\infty \frac{\sin x \cos x\, dx}{x\sqrt{1 - a^2 \sin^2 x}} = \frac{E(a) - b^2 K(a)}{a^2} \qquad\qquad \int_0^\infty \frac{\sin x \cos x\, dx}{x\sqrt{1 - a^2 \cos^2 x}} = \frac{K(a) - E(a)}{a^2}$$

$$\int_0^\infty \frac{\sin x \cos^2 x\, dx}{x\sqrt{1 - a^2 \sin^2 x}} = \frac{E(a) - b^2 K(a)}{a^2} \qquad\qquad \int_0^\infty \frac{\sin x \cos^2 x\, dx}{x\sqrt{1 - a^2 \cos^2 x}} = \frac{K(a) - E(a)}{a^2}$$

$$\int_0^\infty \frac{\sin x\, dx}{x\sqrt{1 + \sin^2 x}} = \sqrt{\tfrac{1}{2}}\, K(\sqrt{\tfrac{1}{2}}) \qquad\qquad \int_0^\infty \frac{\sin x\, dx}{x\sqrt{1 + \cos^2 x}} = \sqrt{\tfrac{1}{2}}\, K(\sqrt{\tfrac{1}{2}})$$

$$\int_0^\infty \frac{\sin x \cos x\, dx}{x\sqrt{1 + \sin^2 x}} = \sqrt{2}\,[K(\sqrt{\tfrac{1}{2}}) - E(\sqrt{\tfrac{1}{2}})] \qquad\qquad \int_0^\infty \frac{\sin x \cos x\, dx}{x\sqrt{1 + \cos^2 x}} = \sqrt{2}\,[E(\sqrt{\tfrac{1}{2}}) - \tfrac{1}{2}K(\sqrt{\tfrac{1}{2}})]$$

$$\int_0^\infty \frac{\sin x \cos^2 x\, dx}{x\sqrt{1 + \sin^2 x}} = \sqrt{2}\,[K(\sqrt{\tfrac{1}{2}}) - E(\sqrt{\tfrac{1}{2}})] \qquad\qquad \int_0^\infty \frac{\sin x \cos^2 x\, dx}{x\sqrt{1 + \cos^2 x}} = \sqrt{2}\,[E(\sqrt{\tfrac{1}{2}}) - \tfrac{1}{2}K(\sqrt{\tfrac{1}{2}})$$

$$\int_0^\infty \frac{\sin^3 x\, dx}{x\sqrt{1 + \sin^2 x}} = \sqrt{\tfrac{1}{2}}\,[2E(\sqrt{\tfrac{1}{2}}) - K(\sqrt{\tfrac{1}{2}})] \qquad\qquad \int_0^\infty \frac{\sin^3 x\, dx}{x\sqrt{1 + \cos^2 x}} = \sqrt{\tfrac{1}{2}}\,[K(\sqrt{\tfrac{1}{2}}) - E(\sqrt{\tfrac{1}{2}})]$$

$$\int_0^\infty \frac{\tan x}{x} \sqrt{1 - a^2 \sin^2 x}\, dx = E(a) \qquad\qquad \int_0^\infty \frac{\tan x}{x} \sqrt{1 - a^2 \cos^2 x}\, dx = E(a)$$

$$\int_0^\infty \frac{\tan x\, dx}{x\sqrt{1 - a^2 \sin^2 x}} = K(a) \qquad\qquad \int_0^\infty \frac{\tan x\, dx}{x\sqrt{1 - a^2 \cos^2 x}} = K(a)$$

$$\int_0^\infty \frac{\tan x \sin^2 x\, dx}{x\sqrt{1 - a^2 \sin^2 x}} = \frac{K(a) - E(a)}{a^2} \qquad\qquad \int_0^\infty \frac{\tan x \sin^2 x\, dx}{x\sqrt{1 - a^2 \cos^2 x}} = \frac{E(a) - b^2 K(a)}{a^2}$$

$$\int_0^\infty \frac{\tan x\, dx}{x\sqrt{1 + \sin^2 x}} = \sqrt{\tfrac{1}{2}}\, K(\sqrt{\tfrac{1}{2}}) \qquad\qquad \int_0^\infty \frac{\tan x\, dx}{x\sqrt{1 + \cos^2 x}} = \sqrt{\tfrac{1}{2}}\, K(\sqrt{\tfrac{1}{2}})$$

(24) Definite Integrals Involving

$f(x) = f(\ln x)$ [0, 1]

$m, n = $ positive integers $\neq 0$

$a, b = $ positive real numbers $\neq 0$

Exponential integral functions (Sec. 13.01):

$C = 0.577\,215\,665 = $ Euler's constant (Sec. 13.01)

$\Gamma(\ ) = $ gamma function (Sec. 13.03)

$$\mathrm{Ei}(x) = C + \ln x + \sum_{k=1}^{\infty} \frac{(-x)^k}{k(k)!}$$

$$\bar{\mathrm{Ei}}(x) = C + \ln x + \sum_{k=1}^{\infty} \frac{x^k}{k(k)!}$$

$$\int_0^1 \ln \frac{1}{x}\, dx = 1$$

$$\int_0^1 \left(\ln \frac{1}{x}\right)^a dx = \Gamma(a+1)$$

$$\int_0^1 x \ln \frac{1}{x}\, dx = \frac{1}{4}$$

$$\int_0^1 x^a \left(\ln \frac{1}{x}\right)^b dx = \frac{\Gamma(b+1)}{(a+1)^{b+1}}$$

$$\int_0^1 \sqrt{\ln \frac{1}{x}}\, dx = \frac{\sqrt{\pi}}{2}$$

$$\int_0^1 \frac{dx}{\sqrt{\ln\,(1/x)}} = \sqrt{\pi}$$

$$\int_0^1 (\ln x)^m\, dx = (-1)^m m!$$

$$\int_0^1 x^m (\ln x)^n\, dx = \frac{(-1)^n n!}{(m+1)^{n+1}}$$

$$\int_0^1 \ln\,(1+x)\, dx = 2 \ln 2 - 1$$

$$\int_0^1 \ln\,(1-x)\, dx = -1$$

$$\int_0^1 x \ln\,(1+x)\, dx = \tfrac{1}{4}$$

$$\int_0^1 x \ln\,(1-x)\, dx = -\tfrac{3}{4}$$

$$\int_0^1 \frac{\ln x}{1+x}\, dx = -\frac{\pi^2}{12}$$

$$\int_0^1 \frac{\ln x}{1-x}\, dx = -\frac{\pi^2}{6}$$

$$\int_0^1 \frac{\ln\,(1+x^a)}{x}\, dx = +\frac{\pi^2}{12a}$$

$$\int_0^1 \frac{\ln\,(1-x^a)}{x}\, dx = -\frac{\pi^2}{6a}$$

$$\int_0^1 \frac{dx}{a+\ln x} = +e^{-a}\,\bar{\mathrm{Ei}}(a)$$

$$\int_0^1 \frac{dx}{a-\ln x} = -e^a\,\mathrm{Ei}(a)$$

$$\int_0^1 \frac{dx}{(a+\ln x)^2} = e^{-a}\,\bar{\mathrm{Ei}}(a) - \frac{1}{a}$$

$$\int_0^1 \frac{dx}{(a-\ln x)^2} = e^a\,\mathrm{Ei}(a) + \frac{1}{a}$$

$$\int_0^1 \frac{dx}{(a+\ln x)^m} = \frac{1}{(m-1)!} \left[\frac{1}{e^a}\,\bar{\mathrm{Ei}}(a) - \frac{1}{a^m} \sum_{k=1}^{m-1} (m-k-1)!\,a^k\right]$$

$$\int_0^1 \frac{dx}{(a-\ln x)^m} = \frac{(-1)^m}{(m-1)!}\left[e^a\,\mathrm{Ei}(a) - \frac{1}{a^m} \sum_{k=1}^{m-1} (m-k-1)!\,(-a)^k\right]$$

$$\int_0^1 \ln x[\ln\,(1+x)]\, dx = 2 - \ln 4 - \frac{\pi^2}{12}$$

$$\int_0^1 \ln x[\ln\,(1-x)]\, dx = 2 - \frac{\pi^2}{6}$$

(25) Definite Integrals Involving $f(x) = f[\ln(\sin x), \ln(\cos x), \ldots]$ $\left[0, \dfrac{\pi}{k}\right]$

$m, k = \text{positive integers} \neq 0$ $L = \dfrac{\pi \ln 2}{8}$ $M = \dfrac{\bar{Z}(2)}{2} = 0.457\,982\,797$

$\bar{Z}(m) = \displaystyle\sum_{k=1}^{\infty} \dfrac{(-1)^{k+1}}{(2k-1)^m} = \text{complementary zeta function (Sec. A.06)}$

$\displaystyle\int_0^{\pi/4} \ln(\sin x)\, dx = -2L - M$ $\displaystyle\int_0^{\pi/4} \ln(\cos x)\, dx = -2L + M$

$\displaystyle\int_0^{\pi/4} \ln(\sin x + \cos x)\, dx = -(L - M)$ $\displaystyle\int_0^{\pi/4} \ln(\cos x - \sin x)\, dx = -(L + M)$

$\displaystyle\int_0^{\pi/4} \ln(\tan x)\, dx = -2M$ $\displaystyle\int_0^{\pi/4} \ln(\cot x)\, dx = +2M$

$\displaystyle\int_0^{\pi/4} \ln(1 + \tan x)\, dx = L$ $\displaystyle\int_0^{\pi/4} \ln(\cot x + 1)\, dx = L + 2M$

$\displaystyle\int_0^{\pi/4} \ln(1 - \tan x)\, dx = L - 2M$ $\displaystyle\int_0^{\pi/4} \ln(\cot x - 1)\, dx = L$

$\displaystyle\int_0^{\pi/4} \ln(\tan x + \cot x)\, dx = 4L$ $\displaystyle\int_0^{\pi/4} \ln(\cot x - \tan x)\, dx = 2L$

$\displaystyle\int_0^{\pi/4} [\ln(\tan x)]^2\, dx = \dfrac{\pi^3}{16}$ $\displaystyle\int_0^{\pi/4} [\ln(\tan x)]^m\, dx = m!(-1)^m \bar{Z}(m+1)$

$\displaystyle\int_0^{\pi/2} \ln(\sin x)\, dx = \int_0^{\pi/2} \ln(\cos x)\, dx = -4L$

$\displaystyle\int_0^{\pi/2} [\ln(\sin x)]^2\, dx = \int_0^{\pi/2} [\ln(\cos x)]^2\, dx = \dfrac{\pi}{2}\left[(\ln 2)^2 + \dfrac{\pi^2}{12}\right]$

$\displaystyle\int_0^{\pi/2} [\ln(\sin x)] \sin x\, dx = \int_0^{\pi/2} [\ln(\cos x)] \cos x\, dx = \ln 2 - 1$

$\displaystyle\int_0^{\pi/2} [\ln(\sin x)] \cos x\, dx = \int_0^{\pi/2} [\ln(\cos x)] \sin x\, dx = -1$

$\displaystyle\int_0^{\pi/2} \ln(1 \pm \sin x)\, dx = \int_0^{\pi/2} \ln(1 \pm \cos x)\, dx = -4(L \mp M)$

$\displaystyle\int_0^{\pi} \ln(\sin x)\, dx = -8L$ $\displaystyle\int_0^{\pi} x \ln(\sin x)\, dx = -4\pi L$

$\displaystyle\int_0^{\pi} \ln(1 \pm \sin x)\, dx = -8(L \mp M)$ $\displaystyle\int_0^{\pi} \ln(1 \pm \cos x)\, dx = -8L$

(26) Definite Integrals Involving

$$f(x) = f[\ln(x, e^{-ax}, \sin bx, \cos cx)] \qquad [0, \infty]$$

$a, b, c, \lambda = $ positive real numbers $\neq 0$

For definition of L and M see opposite page

$C = 0.577\ 215\ 665 = $ Euler's constant (Sec. 13.01)

$$\int_0^\infty \frac{\ln x}{1+x^2}\,dx = 0$$

$$\int_0^\infty \frac{\ln x}{1-x^2}\,dx = -\frac{\pi^2}{4}$$

$$\int_0^\infty \frac{\ln x}{a^2+x^2}\,dx = \frac{\pi}{2a}\ln a$$

$$\int_0^\infty \frac{\ln x\,dx}{(a+x)(b+x)} = \frac{(\ln a)^2 - (\ln b)^2}{2(a-b)} \qquad a \neq b$$

$$\int_0^\infty \frac{\ln x}{a^2+b^2x^2}\,dx = \frac{\pi}{2ab}\ln\frac{a}{b}$$

$$\int_0^\infty \frac{\ln x}{a^2-b^2x^2}\,dx = -\frac{\pi^2}{4ab}$$

$$\int_0^\infty \frac{\ln(x+1)}{1+x^2}\,dx = 2(L+M)$$

$$\int_0^\infty \frac{\ln(x-1)}{1+x^2}\,dx = L$$

$$\int_0^\infty \ln\frac{a^2+x^2}{b^2+x^2}\,dx = (a-b)\pi$$

$$\int_0^\infty \ln\frac{a^2-x^2}{b^2-x^2}\,dx = (a+b)\pi$$

$$\int_0^\infty \frac{\ln(a^2+b^2x^2)}{c^2+x^2}\,dx = \frac{\pi}{c}\ln(ac+b)$$

$$\int_0^\infty \frac{\ln(a^2+b^2x^2)}{c^2-x^2}\,dx = -\frac{\pi}{c}\tan^{-1}\frac{bc}{a}$$

$$\int_0^\infty \frac{\ln(\sin ax)}{b^2+x^2}\,dx = \frac{\pi}{2b}\ln\frac{\sinh ab}{e^{ab}}$$

$$\int_0^\infty \frac{\ln(\sin ax)}{b^2-x^2}\,dx = -\frac{\pi^2}{4b}+\frac{a\pi}{2}$$

$$\int_0^\infty \frac{\ln(\cos ax)}{b^2+x^2}\,dx = \frac{\pi}{2b}\ln\frac{\cosh ab}{e^{ab}}$$

$$\int_0^\infty \frac{\ln(\cos x)}{b^2-x^2}\,dx = \frac{a\pi}{2}$$

$$\int_0^\infty \ln(1+e^{-x}) = \frac{\pi^2}{12}$$

$$\int_0^\infty \ln(1-e^{-x})\,dx = -\frac{\pi^2}{6}$$

$$\int_0^\infty e^{-ax}\ln x\,dx = \frac{-1}{a}(C+\ln a)$$

$$\int_0^\infty \frac{e^{-ax}}{\ln x}\,dx = 0$$

$$\int_0^\infty e^{-ax}\ln\frac{1}{x}\,dx = \frac{1}{a}(C+\ln a)$$

$$\int_0^\infty xe^{-ax}\ln\frac{1}{x}\,dx = \frac{2}{a^2}(C+\ln a-1)$$

$$\int_0^\infty e^{-ax}(\ln x)\sin bx\,dx = \frac{b}{a^2+b^2}\left[\ln\sqrt{a^2+b^2}+\frac{a}{b}\tan^{-1}\frac{b}{a}-C\right]$$

$$\int_0^\infty e^{-ax}(\ln x)\cos bx\,dx = \frac{-a}{a^2+b^2}\left[\ln\sqrt{a^2+b^2}+\frac{b}{a}\tan^{-1}\frac{b}{a}+C\right]$$

$$\int_0^\infty \frac{\ln(a^2\sin^2\lambda x+b^2\cos^2\lambda x)\,dx}{c^2+x^2} = \frac{\pi}{c}[\ln(a\sinh\lambda c+b\cosh\lambda c)-\lambda c]$$

(27) Definite Integrals Involving

$$f(x) = f(\sinh x, \cosh x) \qquad [0, \infty]$$

$m =$ positive integer $\neq 0$ $a, b, c =$ positive real numbers $\neq 0$

$$\alpha = \frac{a\pi}{2c} \qquad A = \cosh 2\alpha + \cosh 2\beta \qquad \mathrm{Z}(\ ) = \text{zeta function (Sec. A.06)}$$

$$\beta = \frac{b\pi}{2c} \qquad B = \cos 2\alpha + \cosh 2\beta \qquad \bar{\mathrm{Z}}(\ ) = \text{complementary zeta function (Sec. A.06)}$$

$$\int_0^\infty \frac{dx}{\sinh ax}\, dx = \infty$$

$$\int_0^\infty \frac{dx}{\cosh ax} = \frac{\pi}{2a}$$

$$\int_0^\infty \frac{x\, dx}{\sinh ax}\, dx = \left(\frac{\pi}{2a}\right)^2$$

$$\int_0^\infty \frac{x\, dx}{\cosh ax} = \frac{1.831\,329\,803\cdots}{a^2}$$

$$\int_0^\infty \frac{x^m}{\sinh bx}\, dx = \frac{2^{m+1}-1}{a(2a)^m}\, m!\,\mathrm{Z}(m+1)$$

$$\int_0^\infty \frac{x^m}{\cosh ax}\, dx = \frac{2^{m+1}}{a(2a)^m}\, m!\,\bar{\mathrm{Z}}(m+1)$$

$$\int_0^\infty \frac{dx}{a + \sinh bx} = \frac{1}{b\sqrt{1+a^2}} \ln \frac{1+a+\sqrt{1+a^2}}{1+a-\sqrt{1+a^2}} \qquad (1+a) > \sqrt{1+a^2}$$

$$\int_0^\infty \frac{dx}{a + \cosh bx} = \frac{1}{b\sqrt{a^2-1}} \ln \frac{1+a+\sqrt{a^2-1}}{1+a-\sqrt{a^2-1}} \qquad a^2 > 1$$

$$\int_0^\infty \frac{dx}{a \sinh cx + b \cosh cx} = \frac{1}{c\sqrt{a^2-b^2}} \ln \frac{a+b+\sqrt{a^2-b^2}}{a+b-\sqrt{a^2-b^2}} \qquad a^2 > b^2$$

$$\int_0^\infty \frac{\sin ax}{\sinh cx}\, dx = \frac{\alpha}{a} \tanh \alpha$$

$$\int_0^\infty \frac{\cos ax}{\cosh cx}\, dx = \frac{\alpha}{a} \operatorname{sech} \alpha$$

$$\int_0^\infty \frac{\sinh ax}{\sinh cx}\, dx = \frac{\alpha}{a} \tan \alpha$$

$$\int_0^\infty \frac{\cosh ax}{\cosh cx}\, dx = \frac{\alpha}{a} \sec \alpha$$

$$\int_0^\infty \frac{x \sin ax}{\cosh cx}\, dx = \left(\frac{\alpha}{a}\right)^2 \frac{\tanh \alpha}{\cosh \alpha}$$

$$\int_0^\infty \frac{x \cos ax}{\sinh cx}\, dx = \left(\frac{\alpha}{a}\right)^2 \operatorname{sech}^2 \alpha$$

$$\int_0^\infty \frac{\sin ax \sin bx}{\cosh cx}\, dx = \frac{\pi}{cA} \sinh \alpha \sinh \beta$$

$$\int_0^\infty \frac{\sinh ax \sin bx}{\cosh cx}\, dx = \frac{\pi}{cB} \sin \alpha \sinh \beta$$

$$\int_0^\infty \frac{\sin ax \cos bx}{\sinh cx}\, dx = \frac{\pi}{2cA} \sinh 2\alpha$$

$$\int_0^\infty \frac{\sinh ax \cos bx}{\sinh cx}\, dx = \frac{\pi}{2cB} \sin 2\alpha$$

$$\int_0^\infty \frac{\cos ax \cos bx}{\cosh cx}\, dx = \frac{\pi}{cA} \cosh \alpha \cosh \beta$$

$$\int_0^\infty \frac{\cosh ax \cos bx}{\cosh cx}\, dx = \frac{\pi}{cB} \cos \alpha \cosh \beta$$

$$\int_0^\infty e^{-ax} \sinh bx\, dx = \frac{b}{a^2 - b^2}$$

$$\int_0^\infty e^{-ax} \cosh bx\, dx = \frac{a}{a^2 - b^2}$$

Appendix A
NUMERICAL TABLES

| $n! = n$ factorial (Sec. 1.03 – 1) | $n!! = n$ double factorial (Sec. 1.03 – 7) | $n = 1 - 100$ |

n	$n!$		$n!!$		n	$n!$		$n!!$	
1	1.000 000 000	(00)	1.000 000 000	(00)	51	1.551 118 753	(066)	2.980 227 914	(33)
2	2.000 000 000	(00)	2.000 000 000	(00)	52	8.065 817 517	(067)	2.706 443 182	(34)
3	6.000 000 000	(00)	3.000 000 000	(00)	53	4.274 883 284	(069)	1.579 520 794	(35)
4	2.400 000 000	(01)	8.000 000 000	(00)	54	2.308 436 973	(071)	1.461 479 318	(36)
5	1.200 000 000	(02)	1.500 000 000	(01)	55	1.269 640 335	(073)	8.687 364 368	(36)
6	7.200 000 000	(02)	4.800 000 000	(01)	56	7.109 985 878	(074)	8.184 284 181	(37)
7	5.040 000 000	(03)	1.050 000 000	(02)	57	4.052 691 950	(076)	4.951 797 690	(38)
8	4.032 000 000	(04)	3.840 000 000	(02)	58	2.350 561 331	(078)	4.746 884 825	(39)
9	3.628 800 000	(05)	9.450 000 000	(02)	59	1.386 831 185	(080)	2.921 560 637	(40)
10	3.628 800 000	(06)	3.840 000 000	(03)	60	8.320 987 112	(081)	2.848 130 895	(41)
11	3.991 680 000	(07)	1.039 500 000	(04)	61	5.075 802 139	(083)	1.782 151 989	(42)
12	4.790 016 000	(08)	4.608 000 000	(04)	62	3.146 997 326	(085)	1.765 841 155	(43)
13	6.227 020 800	(09)	1.351 350 000	(05)	63	1.982 608 315	(087)	1.122 755 753	(44)
14	8.717 829 120	(10)	6.451 200 000	(05)	64	1.268 869 322	(089)	1.130 138 339	(45)
15	1.307 574 368	(12)	2.027 025 000	(06)	65	8.247 650 562	(090)	7.297 912 393	(46)
16	2.092 278 989	(13)	1.032 192 000	(07)	66	5.443 449 391	(092)	7.458 913 039	(46)
17	3.556 874 281	(14)	3.445 942 500	(07)	67	3.647 111 092	(094)	4.889 601 304	(47)
18	6.402 373 706	(15)	1.857 945 600	(08)	68	2.480 035 542	(096)	5.072 060 866	(48)
19	1.216 451 004	(17)	6.547 290 750	(08)	69	1.711 224 524	(098)	3.373 824 899	(49)
20	2.432 902 008	(18)	3.715 891 200	(09)	70	1.197 857 167	(100)	3.550 442 606	(50)
21	5.109 094 217	(19)	1.374 931 058	(10)	71	8.504 785 855	(101)	2.395 415 679	(51)
22	1.124 000 728	(21)	8.174 960 640	(10)	72	6.123 445 837	(103)	2.556 318 677	(52)
23	2.585 201 674	(22)	3.162 341 432	(11)	73	4.470 115 461	(105)	1.748 653 445	(53)
24	6.204 484 017	(23)	1.961 990 554	(12)	74	3.307 885 441	(107)	1.891 675 821	(54)
25	1.551 121 004	(25)	7.905 853 581	(12)	75	2.480 914 081	(109)	1.311 490 084	(55)
26	4.032 914 611	(26)	5.101 175 439	(13)	76	1.885 494 702	(111)	1.437 673 624	(56)
27	1.088 886 945	(28)	2.134 580 467	(14)	77	1.451 830 920	(113)	1.009 847 365	(57)
28	3.048 883 446	(29)	1.428 329 123	(15)	78	1.132 428 118	(115)	1.121 385 427	(58)
29	8.841 761 994	(30)	6.190 283 354	(15)	79	8.946 182 130	(116)	7.977 794 181	(58)
30	2.652 528 598	(32)	4.284 987 369	(16)	80	7.156 945 704	(118)	8.971 083 412	(59)
31	8.222 838 654	(33)	1.918 987 840	(17)	81	5.797 126 020	(120)	6.462 013 287	(60)
32	2.631 308 369	(35)	1.371 195 858	(18)	82	4.753 643 337	(122)	7.356 228 389	(61)
33	8.683 317 619	(36)	6.332 659 871	(18)	83	3.945 523 970	(124)	5.363 471 028	(62)
34	2.952 327 990	(38)	4.662 066 258	(19)	84	3.314 240 134	(126)	6.179 282 254	(63)
35	1.033 314 797	(40)	2.216 430 955	(20)	85	2.817 104 114	(128)	4.558 950 374	(64)
36	3.719 933 268	(41)	1.678 343 853	(21)	86	2.422 709 638	(130)	5.314 182 739	(65)
37	1.376 375 309	(43)	8.200 794 533	(21)	87	2.107 757 298	(132)	3.966 286 825	(66)
38	5.230 226 175	(44)	6.377 706 640	(22)	88	1.854 826 422	(134)	4.676 480 810	(67)
39	2.039 788 208	(46)	3.198 309 868	(23)	89	1.650 795 516	(136)	3.529 995 274	(68)
40	8.159 152 832	(47)	2.551 082 656	(24)	90	1.485 715 964	(138)	4.208 832 729	(69)
41	3.345 252 661	(49)	1.311 307 046	(25)	91	1.352 001 528	(140)	3.212 295 700	(70)
42	1.405 006 118	(51)	1.071 454 716	(26)	92	1.243 841 405	(142)	3.872 126 111	(71)
43	5.041 526 306	(52)	5.638 620 297	(26)	93	1.156 772 507	(144)	3.987 435 001	(72)
44	2.658 271 575	(54)	4.714 400 749	(27)	94	1.087 366 157	(146)	3.639 798 544	(73)
45	1.196 222 209	(56)	2.537 379 134	(28)	95	1.032 997 849	(148)	2.838 063 251	(74)
46	5.502 622 160	(57)	2.168 624 344	(29)	96	9.916 779 348	(149)	3.494 206 602	(75)
47	2.586 232 415	(59)	1.192 568 193	(30)	97	9.619 275 968	(151)	2.752 921 353	(76)
48	1.241 391 553	(61)	1.040 939 685	(31)	98	9.426 890 448	(153)	3.424 322 470	(77)
49	6.082 818 640	(62)	5.843 585 145	(31)	99	9.332 621 544	(155)	2.725 393 140	(78)
50	3.041 409 320	(64)	5.204 698 426	(32)	100	9.332 621 544	(157)	3.424 322 470	(79)

[1] $\Gamma(n+1) = n! =$ gamma function (Sec. 13.03 – 1) $\qquad$ $\Pi(n) = n! =$ pi function (Sec. 13.03 – 2)

$\Gamma(u+1) = u!$ (Sec. 13.03 − 1)		$\Pi(u) = u!$ (Sec. 13.03 − 2)				$u = 0.005 - 1.000$	
u	$u!$	u	$u!$	u	$u!$	u	$u!$
0.005	0.997 138 535	0.255	0.905 385 766	0.505	0.886 398 974	0.755	0.920 209 222
0.010	0.994 325 851	0.260	0.904 397 118	0.510	0.886 591 685	0.760	0.921 374 885
0.015	0.991 561 289	0.265	0.903 426 295	0.515	0.886 804 980	0.765	0.922 559 518
0.020	0.988 844 203	0.270	0.902 503 065	0.520	0.887 038 783	0.770	0.923 763 128
0.025	0.986 173 963	0.275	0.901 597 199	0.525	0.887 293 023	0.775	0.924 985 721
0.030	0.983 549 951	0.280	0.900 718 477	0.530	0.887 567 838	0.780	0.926 227 306
0.035	0.980 971 561	0.285	0.899 866 677	0.535	0.887 862 529	0.785	0.927 487 893
0.040	0.978 438 201	0.290	0.899 041 586	0.540	0.888 177 659	0.790	0.928 767 490
0.045	0.975 949 292	0.295	0.898 242 995	0.545	0.888 512 953	0.795	0.930 066 112
0.050	0.973 504 266	0.300	0.897 470 696	0.550	0.888 868 348	0.800	0.931 383 771
0.055	0.971 102 566	0.305	0.896 724 490	0.555	0.889 243 783	0.805	0.932 720 481
0.060	0.968 743 650	0.310	0.896 004 177	0.560	0.889 638 199	0.810	0.934 076 259
0.065	0.966 426 982	0.315	0.895 309 564	0.565	0.890 054 539	0.815	0.935 451 120
0.070	0.964 152 043	0.320	0.894 640 463	0.570	0.890 489 746	0.820	0.936 845 083
0.075	0.961 918 319	0.325	0.893 996 687	0.575	0.890 944 769	0.825	0.938 258 168
0.080	0.959 725 311	0.330	0.893 378 054	0.580	0.891 419 554	0.830	0.939 690 395
0.085	0.957 572 527	0.335	0.892 784 385	0.585	0.891 914 052	0.835	0.941 141 786
0.090	0.955 459 488	0.340	0.892 215 507	0.590	0.892 428 214	0.840	0.942 612 363
0.095	0.953 385 723	0.345	0.891 671 249	0.595	0.892 961 995	0.845	0.944 102 152
0.100	0.951 350 770	0.350	0.891 151 442	0.600	0.893 515 349	0.850	0.945 611 176
0.105	0.949 541 178	0.355	0.890 655 924	0.605	0.894 098 234	0.855	0.947 129 464
0.110	0.947 955 504	0.360	0.890 184 532	0.610	0.894 680 609	0.860	0.948 687 042
0.115	0.945 474 315	0.365	0.889 737 112	0.615	0.895 292 433	0.865	0.950 253 939
0.120	0.943 590 186	0.370	0.889 313 807	0.620	0.895 923 669	0.870	0.951 840 186
0.125	0.941 742 700	0.375	0.888 913 569	0.625	0.896 574 280	0.875	0.953 445 813
0.130	0.939 931 450	0.380	0.888 537 149	0.630	0.897 244 233	0.880	0.955 070 853
0.135	0.939 138 036	0.385	0.888 184 104	0.635	0.897 933 493	0.885	0.956 715 340
0.140	0.936 416 066	0.390	0.887 854 292	0.640	0.898 642 030	0.890	0.958 379 308
0.145	0.934 711 134	0.395	0.887 547 575	0.645	0.899 369 814	0.895	0.960 062 793
0.150	0.933 040 931	0.400	0.887 263 818	0.650	0.900 116 816	0.900	0.961 765 832
0.155	0.931 405 022	0.405	0.887 002 888	0.655	0.900 983 010	0.905	0.963 488 463
0.160	0.929 803 967	0.410	0.886 764 658	0.660	0.901 668 371	0.910	0.965 230 726
0.165	0.928 234 712	0.415	0.886 548 999	0.665	0.902 472 875	0.915	0.966 992 661
0.170	0.926 699 611	0.420	0.886 355 790	0.670	0.903 296 500	0.920	0.968 774 309
0.175	0.925 197 423	0.425	0.886 184 908	0.675	0.904 139 224	0.925	0.970 575 713
0.180	0.923 727 814	0.430	0.886 036 236	0.680	0.905 001 030	0.930	0.972 396 918
0.185	0.922 290 459	0.435	0.885 909 659	0.685	0.905 881 900	0.935	0.974 237 967
0.190	0.920 885 037	0.440	0.885 805 064	0.690	0.906 781 816	0.940	0.976 098 908
0.195	0.919 511 234	0.445	0.885 722 340	0.695	0.907 700 765	0.945	0.977 979 786
0.200	0.918 168 742	0.450	0.885 661 380	0.700	0.908 638 733	0.950	0.979 880 651
0.205	0.916 857 261	0.455	0.885 622 080	0.705	0.909 595 708	0.955	0.981 801 552
0.210	0.915 576 493	0.460	0.885 604 336	0.710	0.910 571 680	0.960	0.983 742 540
0.215	0.914 326 150	0.465	0.885 608 050	0.715	0.911 566 639	0.965	0.985 703 666
0.220	0.913 105 948	0.470	0.885 633 122	0.720	0.912 580 578	0.970	0.987 684 984
0.225	0.911 915 607	0.475	0.885 679 458	0.725	0.913 613 490	0.975	0.989 686 546
0.230	0.910 754 856	0.480	0.885 746 965	0.730	0.914 665 373	0.980	0.991 708 408
0.235	0.909 623 427	0.485	0.885 835 552	0.735	0.915 736 217	0.985	0.993 750 627
0.240	0.908 521 058	0.490	0.885 945 132	0.740	0.916 826 025	0.990	0.995 813 260
0.245	0.907 447 492	0.495	0.886 075 617	0.745	0.917 934 795	0.995	0.997 896 364
0.250	0.906 402 477	0.500	0.886 226 926	0.750	0.919 062 527	1.000	1.000 000 000

(1) Polynomials $\bar{B}_m(x) = \sum\limits_{k=0}^{m} b_k x^k$ (Sec. 8.05−2)

b_k \ $\bar{B}_m(x)$	b_0	b_1	b_2	b_3	b_4	b_5	b_6	b_7	b_8	b_9	b_{10}
$\bar{B}_0(x)$	1										
$\bar{B}_1(x)$	$-\frac{1}{2}$	1									
$\bar{B}_2(x)$	$\frac{1}{6}$	-1	1								
$\bar{B}_3(x)$	0	$\frac{1}{2}$	$-\frac{3}{2}$	1							
$\bar{B}_4(x)$	$-\frac{1}{30}$	0	1	-2	1						
$\bar{B}_5(x)$	0	$-\frac{1}{6}$	0	$\frac{5}{3}$	$-\frac{5}{2}$	1					
$\bar{B}_6(x)$	$\frac{1}{42}$	0	$-\frac{1}{2}$	0	$\frac{5}{2}$	-3	1				
$\bar{B}_7(x)$	0	$\frac{1}{6}$	0	$-\frac{7}{6}$	0	$\frac{7}{2}$	$-\frac{7}{2}$	1			
$\bar{B}_8(x)$	$-\frac{1}{30}$	0	$\frac{2}{3}$	0	$-\frac{7}{3}$	0	$\frac{14}{3}$	-4	1		
$\bar{B}_9(x)$	0	$-\frac{3}{10}$	0	2	0	$-\frac{21}{5}$	0	6	$-\frac{9}{2}$	1	
$\bar{B}_{10}(x)$	$\frac{5}{66}$	0	$-\frac{3}{2}$	0	5	0	-7	0	$\frac{15}{2}$	-5	1

(2) Numbers B_m **and** $\bar{B}_m$ (Sec. 8.05−1)

B_m	$\bar{B}_m$	Fractions	Decimals
	$\bar{B}_0$	$1:1$	$1.000\,000\,000\,000$ $(+00)$
	$-\bar{B}_1$	$1:2$	$5.000\,000\,000\,000$ (-01)
B_1	$\bar{B}_2$	$1:6$	$1.666\,666\,666\,667$ (-01)
B_2	$-\bar{B}_4$	$1:30$	$3.333\,333\,333\,333$ (-02)
B_3	$\bar{B}_6$	$1:42$	$2.380\,952\,380\,952$ (-02)
B_4	$-\bar{B}_8$	$1:30$	$3.333\,333\,333\,333$ (-02)
B_5	$\bar{B}_{10}$	$5:66$	$7.575\,757\,575\,758$ (-02)
B_6	$-\bar{B}_{12}$	$691:2\,730$	$2.531\,135\,531\,136$ (-01)
B_7	$\bar{B}_{14}$	$7:6$	$1.166\,666\,666\,667$ $(+00)$
B_8	$-\bar{B}_{16}$	$3\,617:510$	$7.092\,156\,862\,745$ $(+00)$
B_9	$\bar{B}_{18}$	$43\,867:798$	$5.497\,117\,794\,486$ $(+01)$
B_{10}	$-\bar{B}_{20}$	$174\,611:330$	$5.291\,242\,424\,242$ $(+02)$
$\bar{B}_3 = \bar{B}_5 = \bar{B}_7 = \cdots = 0$			

(1) Polynomials　$\bar{E}_m(x) = \sum_{k=0}^{m} e_k x^k$　　(Sec. 8.06−2)

$\bar{E}_m(x)$ \ e_k	e_0	e_1	e_2	e_3	e_4	e_5	e_6	e_7	e_8	e_9	e_{10}
$\bar{E}_0(x)$	1										
$\bar{E}_1(x)$	$-\frac{1}{2}$	1									
$\bar{E}_2(x)$	0	-1	1								
$\bar{E}_3(x)$	$\frac{1}{4}$	0	$-\frac{3}{2}$	1							
$\bar{E}_4(x)$	0	1	0	-2	1						
$\bar{E}_5(x)$	$-\frac{1}{2}$	0	$\frac{5}{2}$	0	$-\frac{5}{2}$	1					
$\bar{E}_6(x)$	0	-3	0	5	0	-3	1				
$\bar{E}_7(x)$	$\frac{17}{8}$	0	$-\frac{21}{2}$	0	$\frac{35}{4}$	0	$-\frac{7}{2}$	1			
$\bar{E}_8(x)$	0	17	0	-28	0	14	0	-4	1		
$\bar{E}_9(x)$	$-\frac{31}{2}$	0	$\frac{153}{2}$	0	-65	0	21	0	$-\frac{9}{2}$	1	
$\bar{E}_{10}(x)$	0	-155	0	255	0	-126	0	30	0	-5	1

(2) Numbers　E_m **and**　$\bar{E}_m$　　(Sec. 8.06−1)

E_m	$\bar{E}_m$	Integers
	$\bar{E}_0$	1
	$\bar{E}_1$	0
E_1	$-\bar{E}_2$	1
E_2	$\bar{E}_4$	5
E_3	$-\bar{E}_6$	61
E_4	$\bar{E}_8$	1 385
E_5	$-\bar{E}_{10}$	50 521
E_6	$\bar{E}_{12}$	2 702 765
E_7	$-\bar{E}_{14}$	199 360 981
E_8	$\bar{E}_{16}$	19 391 512 145
E_9	$-\bar{E}_{18}$	2 404 879 675 441
E_{10}	$\bar{E}_{20}$	370 371 188 237 525

$$\bar{E}_3 = \bar{E}_5 = \bar{E}_7 = \cdots = 0$$

$\mathscr{S}_k^{(p)}$ = Stirling number

$X_h^{(p)}$ = factorial polynomial

k, p = positive integers

$\Gamma(x)$ = gamma function (Sec. 13.03 – 1)

$\binom{x}{p}$ = binomial coefficient (Sec. 1.04 – 2)

h, x = real numbers

(1) Relations

$$X_1^{(p)} = x(x-1)(x-2) \cdots (x-p+1) = \binom{x}{p} p! = \sum_{k=1}^{p} x^k \mathscr{S}_k^{(p)}$$

$$X_h^{(p)} = x(x-h)(x-2h) \cdots (x-ph+h) = \frac{h^p \Gamma\left(\frac{x}{h}+1\right)}{\Gamma\left(\frac{x}{h}-p+1\right)} = h^p \sum_{k=1}^{p} \left(\frac{x}{h}\right)^k \mathscr{S}_k^{(p)}$$

$$X_1^{(-p)} = \frac{1}{(x+1)(x+2)(x+3) \cdots (x+p)} = \left[\binom{x+p}{p} p!\right]^{-1} = \left[\sum_{k=1}^{p} (x+p)^k \mathscr{S}_k^{(p)}\right]^{-1}$$

$$X_h^{(-p)} = \frac{1}{(x+h)(x+2h)(x+3h) \cdots (x+ph)} = \frac{\Gamma\left(\frac{x}{h}+1\right)}{h^p \Gamma\left(\frac{x}{h}+p+1\right)} = \left[h^p \sum_{k=1}^{p} \left(\frac{x}{h}+p\right)^k \mathscr{S}_k^{(p)}\right]^{-1}$$

(2) Numerical Values

k	$\mathscr{S}_k^{(1)}$	$\mathscr{S}_k^{(2)}$	$\mathscr{S}_k^{(3)}$	$\mathscr{S}_k^{(4)}$	$\mathscr{S}_k^{(5)}$	$\mathscr{S}_k^{(6)}$	$\mathscr{S}_k^{(7)}$	$\mathscr{S}_k^{(8)}$	$\mathscr{S}_k^{(9)}$
1	1	−1	2	−6	24	−120	720	−5 040	40 320
2		1	−3	11	−50	274	−1 764	13 068	−109 584
3			1	−6	35	−225	1 624	−13 132	118 121
4				1	−10	85	−735	6 769	−67 284
5					1	−15	175	−1 960	22 449
6						1	−21	322	−4 536
7							1	−28	546
8								1	−36
9									1

k	$\mathscr{S}_k^{(10)}$	$\mathscr{S}_k^{(11)}$	$\mathscr{S}_k^{(12)}$	$\mathscr{S}_k^{(13)}$
1	−362 880	3 628 800	−39 916 800	479 001 600
2	1 026 576	−10 628 640	120 543 840	−1 486 442 880
3	−1 172 700	12 753 576	−150 917 976	1 931 559 552
4	723 680	−8 409 500	105 258 076	−1 414 014 888
5	−269 325	3 416 930	−45 995 730	657 206 836
6	63 273	−902 055	13 339 535	−206 070 150
7	−9 450	157 773	−2 637 558	44 990 231
8	870	−18 150	357 423	6 926 634
9	−45	1 320	−32 670	749 463
10	1	−55	1 925	−55 770
11		1	−66	2 717
12			1	−78
13				1

[1]For applications see Secs. 1.03 and 19.03.

$Z(m)$ = zeta function	$\bar{Z}(m)$ = complementary zeta function
$\bar{B}_m$ = Bernoulli number (Sec. A.03)	$\bar{E}_m$ = Euler number (Sec. A.04)
m, k = positive integers	$\beta = (-1)^{k+1}$

(1) Relations

$$\sum_{k=1}^{\infty} \frac{1}{k^m} = Z(m)$$

$$\sum_{k=1}^{\infty} \frac{1}{(2k)^m} = 2^{-m} Z(m)$$

$$\sum_{k=1}^{\infty} \frac{1}{(2k-1)^m} = (2^m - 1) 2^{-m} Z(m)$$

$$\sum_{k=1}^{\infty} \frac{1}{k^{2m}} = \frac{(2\pi)^{2m}}{(2m)!2} |\bar{B}_{2m}|$$

$$\sum_{k=1}^{\infty} \frac{1}{(2k)^{2m}} = \frac{(\pi)^{2m}}{(2m)!2} |\bar{B}_{2m}|$$

$$\sum_{k=1}^{\infty} \frac{1}{(2k-1)^{2m}} = \frac{(\pi)^{2m}(2^{2m}-1)}{(2m)!2} |\bar{B}_{2m}|$$

$$\sum_{k=1}^{\infty} \frac{\beta}{k^m} = (2^m - 2) 2^{-m} Z(m)$$

$$\sum_{k=1}^{\infty} \frac{\beta}{(2k)^m} = (2^m - 2) 4^{-m} Z(m)$$

$$\sum_{k=1}^{\infty} \frac{\beta}{(2k-1)^m} = \bar{Z}(m)$$

$$\sum_{k=1}^{\infty} \frac{\beta}{k^{2m}} = \frac{(\pi)^{2m}(2^{2m}-2)}{(2m)!2} |\bar{B}_{2m}|$$

$$\sum_{k=1}^{\infty} \frac{\beta}{(2k)^{2m}} = \frac{(\pi/2)^{2m}(2^{2m}-2)}{(2m)!2} |\bar{B}_{2m}|$$

$$\sum_{k=1}^{\infty} \frac{\beta}{(2k-1)^{2m-1}} = \frac{(\pi/2)^{2m-1}}{(2m)!2} |\bar{E}_{2m}|$$

(2) Numerical Values

m	$Z(m)$	$(2^m - 1) 2^{-m} Z(m)$	$(2^m - 2) 2^{-m} Z(m)$	$\bar{Z}(m)$
1	∞	∞	0.693 147 181	0.785 398 163
2	1.644 934 067	1.233 700 550	0.822 467 033	0.915 965 594
3	1.202 056 903	1.051 799 790	0.901 542 677	0.968 946 146
4	1.082 323 234	1.014 678 032	0.947 032 829	0.988 944 552
5	1.036 927 755	1.004 523 763	0.972 119 770	0.996 157 828
6	1.017 343 062	1.001 447 077	0.985 551 091	0.998 685 222
7	1.008 349 277	1.000 471 549	0.992 593 820	0.999 554 508
8	1.004 077 356	1.000 155 179	0.996 233 002	0.999 849 990
9	1.002 008 393	1.000 051 345	0.998 094 298	0.999 949 684
10	1.000 994 575	1.000 017 041	0.999 039 508	0.999 983 164
11	1.000 494 189	1.000 005 666	0.999 517 143	0.999 994 375
12	1.000 246 087	1.000 001 886	0.999 757 685	0.999 998 122
13	1.000 122 713	1.000 000 628	0.999 878 543	0.999 999 374
14	1.000 061 248	1.000 000 209	0.999 939 170	0.999 999 791
15	1.000 030 589	1.000 000 070	0.999 969 551	0.999 999 930
16	1.000 015 282	1.000 000 023	0.999 984 764	0.999 999 977
17	1.000 007 637	1.000 000 008	0.999 923 782	0.999 999 992
18	1.000 003 817	1.000 000 003	0.999 996 188	0.999 999 997
19	1.000 001 908	1.000 000 001	0.999 998 094	0.999 999 999
20	1.000 000 956	1.000 000 000	0.999 999 047	1.000 000 000
21	1.000 000 477	1.000 000 000	0.999 999 523	1.000 000 000
22	1.000 000 238	1.000 000 000	0.999 999 762	1.000 000 000
23	1.000 000 119	1.000 000 000	0.999 999 881	1.000 000 000
24	1.000 000 060	1.000 000 000	0.999 999 440	1.000 000 000
25	1.000 000 030	1.000 000 000	0.999 999 970	1.000 000 000

[1]For applications see Secs. 8.08, 13.04, 20.12, 20.14, 20.25, and 20.27.

$$x = 0.01 - 0.50$$

x, rad	$\sin x$	$\cos x$	$\tan x$	e^x	e^{-x}	$\sinh x$	$\cosh x$	$\tanh x$	x, deg
0.01	0.01000	0.99995	0.01000	1.01005	0.99005	0.01000	1.00005	0.01000	0.57
0.02	0.02000	0.99980	0.02000	1.02020	0.98020	0.02000	1.00020	0.02000	1.15
0.03	0.03000	0.99955	0.03001	1.03045	0.97045	0.03000	1.00045	0.02999	1.72
0.04	0.03999	0.99920	0.04002	1.04081	0.96079	0.04001	1.00080	0.03998	2.29
0.05	0.04998	0.99875	0.05004	1.05127	0.95123	0.05002	1.00125	0.04996	2.86
0.06	0.05996	0.99820	0.06007	1.06184	0.94176	0.06004	1.00180	0.05993	3.44
0.07	0.06994	0.99755	0.07011	1.07251	0.93239	0.07006	1.00245	0.06989	4.01
0.08	0.07991	0.99680	0.08017	1.08329	0.92312	0.08009	1.00320	0.07983	4.58
0.09	0.08988	0.99595	0.09024	1.09417	0.91393	0.09012	1.00405	0.08976	5.16
0.10	0.09983	0.99500	0.10033	1.10517	0.90484	0.10017	1.00500	0.09967	5.73
0.11	0.10978	0.99396	0.11045	1.11628	0.89583	0.11022	1.00606	0.10956	6.30
0.12	0.11971	0.99281	0.12058	1.12750	0.88692	0.12029	1.00721	0.11943	6.88
0.13	0.12963	0.99156	0.13074	1.13883	0.87810	0.13037	1.00846	0.12927	7.45
0.14	0.13954	0.99022	0.14092	1.15027	0.86936	0.14046	1.00982	0.13909	8.02
0.15	0.14944	0.98877	0.15114	1.16183	0.86071	0.15056	1.01127	0.14889	8.59
0.16	0.15932	0.98723	0.16138	1.17351	0.85214	0.16068	1.01283	0.15865	9.17
0.17	0.16918	0.98558	0.17166	1.18530	0.84366	0.17082	1.01448	0.16838	9.74
0.18	0.17903	0.98384	0.18197	1.19722	0.83527	0.18097	1.01624	0.17808	10.31
0.19	0.18886	0.98200	0.19232	1.20925	0.82696	0.19115	1.01810	0.18775	10.89
0.20	0.19867	0.98007	0.20271	1.22140	0.81873	0.20134	1.02007	0.19738	11.46
0.21	0.20846	0.97803	0.21314	1.23368	0.81058	0.21155	1.02213	0.20697	12.03
0.22	0.21823	0.97590	0.22362	1.24608	0.80252	0.22178	1.02430	0.21652	12.61
0.23	0.22798	0.97367	0.23414	1.25860	0.79453	0.23203	1.02657	0.22603	13.18
0.24	0.23770	0.97134	0.24472	1.27125	0.78663	0.24231	1.02894	0.23550	13.75
0.25	0.24740	0.96891	0.25534	1.28403	0.77880	0.25261	1.03141	0.24492	14.32
0.26	0.25708	0.96639	0.26602	1.29693	0.77105	0.26294	1.03399	0.25430	14.90
0.27	0.26673	0.96377	0.27676	1.30996	0.76338	0.27329	1.03667	0.26362	15.47
0.28	0.27636	0.96106	0.28755	1.32313	0.75578	0.28367	1.03946	0.27291	16.04
0.29	0.28595	0.95824	0.29841	1.33643	0.74826	0.29408	1.04235	0.28213	16.62
0.30	0.29552	0.95534	0.30934	1.34986	0.74082	0.30452	1.04534	0.29131	17.19
0.31	0.30506	0.95233	0.32033	1.36343	0.73345	0.31499	1.04844	0.30044	17.76
0.32	0.31457	0.94924	0.33139	1.37713	0.72615	0.32549	1.05164	0.30951	18.33
0.33	0.32404	0.94604	0.34252	1.39097	0.71892	0.33602	1.05495	0.31852	18.91
0.34	0.33349	0.94275	0.35374	1.40495	0.71177	0.34659	1.05836	0.32748	19.48
0.35	0.34290	0.93937	0.36503	1.41907	0.70469	0.35719	1.06188	0.33638	20.05
0.36	0.35227	0.93590	0.37640	1.43333	0.69768	0.36783	1.06550	0.34521	20.63
0.37	0.36162	0.93233	0.38786	1.44773	0.69073	0.37850	1.06923	0.35399	21.20
0.38	0.37092	0.92866	0.39941	1.46228	0.68386	0.38921	1.07307	0.36271	21.77
0.39	0.38019	0.92491	0.41105	1.47698	0.67706	0.39996	1.07702	0.37136	22.35
0.40	0.38942	0.92106	0.42279	1.49182	0.67032	0.41075	1.08107	0.37995	22.92
0.41	0.39861	0.91712	0.43463	1.50682	0.66365	0.42158	1.08523	0.38847	23.49
0.42	0.40776	0.91309	0.44657	1.52196	0.65705	0.43246	1.08950	0.39693	24.06
0.43	0.41687	0.90897	0.45862	1.53726	0.65051	0.44337	1.09388	0.40532	24.64
0.44	0.42594	0.90475	0.47078	1.55271	0.64404	0.45434	1.09837	0.41364	25.21
0.45	0.43497	0.90045	0.48306	1.56831	0.63763	0.46534	1.10297	0.42190	25.78
0.46	0.44395	0.89605	0.49545	1.58407	0.63128	0.47640	1.10768	0.43008	26.36
0.47	0.45289	0.89157	0.50797	1.59999	0.62500	0.48750	1.11250	0.43820	26.93
0.48	0.46178	0.88699	0.52061	1.61607	0.61878	0.49865	1.11743	0.44624	27.50
0.49	0.47063	0.88233	0.53339	1.63232	0.61263	0.50984	1.12247	0.45422	28.07
0.50	0.47943	0.87758	0.54630	1.64872	0.60653	0.52110	1.12763	0.46212	28.65

x, rad	$\sin x$	$\cos x$	$\tan x$	e^x	e^{-x}	$\sinh x$	$\cosh x$	$\tanh x$	x, deg
0.51	0.48818	0.87274	0.55936	1.66529	0.60050	0.53240	1.13289	0.46995	29.22
0.52	0.49688	0.86782	0.57256	1.68203	0.59452	0.54375	1.13827	0.47770	29.79
0.53	0.50553	0.86281	0.58592	1.69893	0.58860	0.55516	1.14377	0.48538	30.37
0.54	0.51414	0.85771	0.59943	1.71601	0.58275	0.56663	1.14938	0.49299	30.94
0.55	0.52269	0.85252	0.61311	1.73325	0.57695	0.57815	1.15510	0.50052	31.51
0.56	0.53119	0.84726	0.62695	1.75067	0.57121	0.58973	1.16094	0.50798	32.09
0.57	0.53963	0.84190	0.64097	1.76827	0.56553	0.60137	1.16690	0.51536	32.66
0.58	0.54802	0.83646	0.65517	1.78604	0.55990	0.61307	1.17297	0.52267	33.23
0.59	0.55636	0.83094	0.66956	1.80399	0.55433	0.62483	1.17916	0.52990	33.80
0.60	0.56464	0.82534	0.68414	1.82212	0.54881	0.63665	1.18547	0.53705	34.38
0.61	0.57287	0.81965	0.69892	1.84043	0.54335	0.64854	1.19189	0.54413	34.95
0.62	0.58104	0.81388	0.71391	1.85893	0.53794	0.66049	1.19844	0.55113	35.52
0.63	0.58914	0.80803	0.72911	1.87761	0.53259	0.67251	1.20510	0.55805	36.10
0.64	0.59720	0.80210	0.74454	1.89648	0.52729	0.68459	1.21189	0.56490	36.67
0.65	0.60519	0.79608	0.76020	1.91554	0.52205	0.69675	1.21879	0.57167	37.24
0.66	0.61312	0.78999	0.77610	1.93479	0.51685	0.70897	1.22582	0.57836	37.82
0.67	0.62099	0.78382	0.79225	1.95424	0.51171	0.72126	1.23297	0.58498	38.39
0.68	0.62879	0.77757	0.80866	1.97388	0.50662	0.73363	1.24025	0.59152	38.96
0.69	0.63654	0.77125	0.82534	1.99372	0.50158	0.74607	1.24765	0.59798	39.53
0.70	0.64422	0.76484	0.84229	2.01375	0.49659	0.75858	1.25517	0.60437	40.11
0.71	0.65183	0.75836	0.85953	2.03399	0.49164	0.77117	1.26282	0.61068	40.68
0.72	0.65938	0.75181	0.87707	2.05443	0.48675	0.78384	1.27059	0.61691	41.25
0.73	0.66687	0.74517	0.89492	2.07508	0.48191	0.79659	1.27849	0.62307	41.83
0.74	0.67429	0.73847	0.91309	2.09594	0.47711	0.80941	1.28652	0.62915	42.40
0.75	0.68164	0.73169	0.93160	2.11700	0.47237	0.82232	1.29468	0.63515	42.97
0.76	0.68892	0.72484	0.95045	2.13828	0.46767	0.83530	1.30297	0.64108	43.54
0.77	0.69614	0.71791	0.96967	2.15977	0.46301	0.84838	1.31139	0.64693	44.12
0.78	0.70328	0.71091	0.98926	2.18147	0.45841	0.86153	1.31994	0.65271	44.69
0.79	0.71035	0.70385	1.00925	2.20340	0.45384	0.87478	1.32862	0.65841	45.26
0.80	0.71736	0.69671	1.02964	2.22554	0.44933	0.88811	1.33743	0.66404	45.84
0.81	0.72429	0.68950	1.05046	2.24791	0.44486	0.90152	1.34638	0.66959	46.41
0.82	0.73115	0.68222	1.07171	2.27050	0.44043	0.91503	1.35547	0.67507	46.98
0.83	0.73793	0.67488	1.09343	2.29332	0.43605	0.92863	1.36468	0.68048	47.56
0.84	0.74464	0.66746	1.11563	2.31637	0.43171	0.94233	1.37404	0.68581	48.13
0.85	0.75128	0.65998	1.13833	2.33965	0.42741	0.95612	1.38353	0.69107	48.70
0.86	0.75784	0.65244	1.16156	2.36316	0.42316	0.97000	1.39316	0.69626	49.27
0.87	0.76433	0.64483	1.18532	2.38691	0.41895	0.98398	1.40293	0.70137	49.85
0.88	0.77074	0.63715	1.20966	2.41090	0.41478	0.99806	1.41284	0.70642	50.42
0.89	0.77707	0.62941	1.23460	2.43513	0.41066	1.01224	1.42289	0.71139	50.99
0.90	0.78333	0.62161	1.26016	2.45960	0.40657	1.02652	1.43309	0.71630	51.57
0.91	0.78950	0.61375	1.28637	2.48432	0.40252	1.04090	1.44342	0.72113	52.14
0.92	0.79560	0.60582	1.31326	2.50929	0.39852	1.05539	1.45390	0.72590	52.71
0.93	0.80162	0.59783	1.34087	2.53451	0.39455	1.06998	1.46453	0.73059	53.29
0.94	0.80756	0.58979	1.36923	2.55998	0.39063	1.08468	1.47530	0.73522	53.86
0.95	0.81342	0.58168	1.39838	2.58571	0.38674	1.09948	1.48623	0.73978	54.43
0.96	0.81919	0.57352	1.42836	2.61170	0.38289	1.11440	1.49729	0.74428	55.00
0.97	0.82489	0.56530	1.45920	2.63794	0.37908	1.12943	1.50851	0.74870	55.58
0.98	0.83050	0.55702	1.49096	2.66446	0.37531	1.14457	1.51988	0.75307	56.15
0.99	0.83603	0.54869	1.52368	2.69123	0.37158	1.15983	1.53141	0.75736	56.72
1.00	0.84147	0.54030	1.55741	2.71828	0.36788	1.17520	1.54308	0.76159	57.30

$$x = 1.01 - 1.50$$

x, rad	$\sin x$	$\cos x$	$\tan x$	e^x	e^{-x}	$\sinh x$	$\cosh x$	$\tanh x$	x, deg
1.01	0.84683	0.53186	1.59221	2.74560	0.36422	1.19069	1.55491	0.76576	57.87
1.02	0.85211	0.52337	1.62813	2.77319	0.36059	1.20630	1.56689	0.76987	58.44
1.03	0.85730	0.51482	1.66524	2.80107	0.35701	1.22203	1.57904	0.77391	59.01
1.04	0.86240	0.50622	1.70361	2.82922	0.35345	1.23788	1.59134	0.77789	59.59
1.05	0.86742	0.49757	1.74332	2.85765	0.34994	1.25386	1.60379	0.78181	60.16
1.06	0.87236	0.48887	1.78442	2.88637	0.34646	1.26996	1.61641	0.78566	60.73
1.07	0.87720	0.48012	1.82703	2.91538	0.34301	1.28619	1.62919	0.78946	61.31
1.08	0.88196	0.47133	1.87122	2.94468	0.33960	1.30254	1.64214	0.79320	61.88
1.09	0.88663	0.46249	1.91709	2.97427	0.33622	1.31903	1.65525	0.79688	62.45
1.10	0.89121	0.45360	1.96476	3.00417	0.33287	1.33565	1.66852	0.80050	63.03
1.11	0.89570	0.44466	2.01434	3.03436	0.32956	1.35240	1.68196	0.80406	63.60
1.12	0.90010	0.43568	2.06596	3.06485	0.32628	1.36929	1.69557	0.80757	64.17
1.13	0.90441	0.42666	2.11975	3.09566	0.32303	1.38631	1.70934	0.81102	64.74
1.14	0.90863	0.41759	2.17588	3.12677	0.31982	1.40347	1.72329	0.81441	65.32
1.15	0.91276	0.40849	2.23450	3.15819	0.31664	1.42078	1.73741	0.81775	65.89
1.16	0.91680	0.39934	2.29580	3.18993	0.31349	1.43822	1.75171	0.82104	66.46
1.17	0.92075	0.39015	2.35998	3.22199	0.31037	1.45581	1.76618	0.82427	67.04
1.18	0.92461	0.38092	2.42727	3.25437	0.30728	1.47355	1.78083	0.82745	67.61
1.19	0.92837	0.37166	2.49790	3.28708	0.30422	1.49143	1.79565	0.83058	68.18
1.20	0.93204	0.36236	2.57215	3.32012	0.30119	1.50946	1.81066	0.83365	68.75
1.21	0.93562	0.35302	2.65032	3.35348	0.29820	1.52764	1.82584	0.83668	69.33
1.22	0.93910	0.34365	2.73275	3.38719	0.29523	1.54598	1.84121	0.83965	69.90
1.23	0.94249	0.33424	2.81982	3.42123	0.29229	1.56447	1.85676	0.84258	70.47
1.24	0.94578	0.32480	2.91193	3.45561	0.28938	1.58311	1.87250	0.84546	71.05
1.25	0.94898	0.31532	3.00957	3.49034	0.28650	1.60192	1.88842	0.84828	71.62
1.26	0.95209	0.30582	3.11327	3.52542	0.28365	1.62088	1.90454	0.85106	72.19
1.27	0.95510	0.29628	3.22363	3.56085	0.28083	1.64001	1.92084	0.85380	72.77
1.28	0.95802	0.28672	3.34135	3.59664	0.27804	1.65930	1.93734	0.85648	73.34
1.29	0.96084	0.27712	3.46721	3.63279	0.27527	1.67876	1.95403	0.85913	73.91
1.30	0.96356	0.26750	3.60210	3.66930	0.27253	1.69838	1.97091	0.86172	74.48
1.31	0.96618	0.25785	3.74708	3.70617	0.26982	1.71818	1.98800	0.86428	75.06
1.32	0.96872	0.24818	3.90335	3.74342	0.26714	1.73814	2.00528	0.86678	75.63
1.33	0.97115	0.23848	4.07231	3.78104	0.26448	1.75828	2.02276	0.86925	76.20
1.34	0.97348	0.22875	4.25562	3.81904	0.26185	1.77860	2.04044	0.87167	76.78
1.35	0.97572	0.21901	4.45522	3.85743	0.25924	1.79909	2.05833	0.87405	77.35
1.36	0.97786	0.20924	4.67344	3.89619	0.25666	1.81977	2.07643	0.87639	77.92
1.37	0.97991	0.19945	4.91306	3.93535	0.25411	1.84062	2.09473	0.87869	78.50
1.38	0.98185	0.18964	5.17744	3.97490	0.25158	1.86166	2.11324	0.88095	79.07
1.39	0.98370	0.17981	5.47069	4.01485	0.24908	1.88289	2.13196	0.88317	79.64
1.40	0.98545	0.16997	5.79788	4.05520	0.24660	1.90430	2.15090	0.88535	80.21
1.41	0.98710	0.16010	6.16536	4.09596	0.24414	1.92591	2.17005	0.88749	80.79
1.42	0.98865	0.15023	6.58112	4.13712	0.24171	1.94770	2.18942	0.88960	81.36
1.43	0.99010	0.14033	7.05546	4.17870	0.23931	1.96970	2.20900	0.89167	81.93
1.44	0.99146	0.13042	7.60183	4.22070	0.23693	1.99188	2.22881	0.89370	82.51
1.45	0.99271	0.12050	8.23809	4.26311	0.23457	2.01427	2.24884	0.89569	83.08
1.46	0.99387	0.11057	8.98861	4.30596	0.23224	2.03686	2.26910	0.89765	83.65
1.47	0.99492	0.10063	9.88737	4.34924	0.22993	2.05965	2.28958	0.89958	84.22
1.48	0.99588	0.09067	10.98338	4.39295	0.22764	2.08265	2.31029	0.90147	84.80
1.49	0.99674	0.08071	12.34986	4.43710	0.22537	2.10586	2.33123	0.90332	85.37
1.50	0.99749	0.07074	14.10142	4.48169	0.22313	2.12928	2.35241	0.90515	85.94

$$x = 1.51 - 2.00$$

x, rad	$\sin x$	$\cos x$	$\tan x$	e^x	e^{-x}	$\sinh x$	$\cosh x$	$\tanh x$	x, deg
1.51	0.99815	0.06076	16.42809	4.52673	0.22091	2.15291	2.37382	0.90694	86.52
1.52	0.99871	0.05077	19.66953	4.57223	0.21871	2.17676	2.39547	0.90870	87.09
1.53	0.99917	0.04079	24.49841	4.61818	0.21654	2.20082	2.41736	0.91042	87.66
1.54	0.99953	0.03079	32.46114	4.66459	0.21438	2.22510	2.43949	0.91212	88.24
1.55	0.99978	0.02079	48.07848	4.71147	0.21225	2.24961	2.46186	0.91379	88.81
1.56	0.99994	0.01080	92.62050	4.75882	0.21014	2.27434	2.48448	0.91542	89.38
1.57	1.00000	0.00080	1255.76559	4.80665	0.20805	2.29930	2.50735	0.91703	89.95
1.58	0.99996	−0.00920	−108.64920	4.85496	0.20598	2.32449	2.53047	0.91860	90.53
1.59	0.99982	−0.01920	−52.06697	4.90375	0.20393	2.34991	2.55384	0.92015	91.10
1.60	0.99957	−0.02920	−34.23253	4.95303	0.20190	2.37557	2.57746	0.92167	91.67
1.61	0.99923	−0.03919	−25.49474	5.00281	0.19989	2.40146	2.60135	0.92316	92.25
1.62	0.99879	−0.04918	−20.30728	5.05309	0.19790	2.42760	2.62549	0.92462	92.82
1.63	0.99825	−0.05917	−16.87110	5.10387	0.19593	2.45397	2.64990	0.92606	93.39
1.64	0.99761	−0.06915	−14.42702	5.15517	0.19398	2.48059	2.67457	0.92747	93.97
1.65	0.99687	−0.07912	−12.59926	5.20698	0.19205	2.50746	2.69951	0.92886	94.54
1.66	0.99602	−0.08909	−11.18055	5.25931	0.19014	2.53459	2.72472	0.93022	95.11
1.67	0.99508	−0.09904	−10.04718	5.31217	0.18825	2.56196	2.75021	0.93155	95.68
1.68	0.99404	−0.10899	−9.12077	5.36556	0.18637	2.58959	2.77596	0.93286	96.26
1.69	0.99290	−0.11892	−8.34923	5.41948	0.18452	2.61748	2.80200	0.93415	96.83
1.70	0.99166	−0.12884	−7.69660	5.47395	0.18268	2.64563	2.82832	0.93541	97.40
1.71	0.99033	−0.13875	−7.13726	5.52896	0.18087	2.67405	2.85491	0.93665	97.98
1.72	0.98889	−0.14865	−6.65244	5.58453	0.17907	2.70273	2.88180	0.93786	98.55
1.73	0.98735	−0.15853	−6.22810	5.64065	0.17728	2.73168	2.90897	0.93906	99.12
1.74	0.98572	−0.16840	−5.85353	5.69734	0.17552	2.76091	2.93643	0.94023	99.69
1.75	0.98399	−0.17825	−5.52038	5.75460	0.17377	2.79041	2.96419	0.94138	100.27
1.76	0.98215	−0.18808	−5.22209	5.81244	0.17204	2.82020	2.99224	0.94250	100.84
1.77	0.98022	−0.19789	−4.95341	5.87085	0.17033	2.85026	3.02059	0.94361	101.41
1.78	0.97820	−0.20768	−4.71009	5.92986	0.16864	2.88061	3.04925	0.94470	101.99
1.79	0.97607	−0.21745	−4.48866	5.98945	0.16696	2.91125	3.07821	0.94576	102.56
1.80	0.97385	−0.22720	−4.28626	6.04965	0.16530	2.94217	3.10747	0.94681	103.13
1.81	0.97153	−0.23693	−4.10050	6.11045	0.16365	2.97340	3.13705	0.94783	103.71
1.82	0.96911	−0.24663	−3.92937	6.17186	0.16203	3.00492	3.16694	0.94884	104.28
1.83	0.96659	−0.25631	−3.77118	6.23389	0.16041	3.03674	3.19715	0.94983	104.85
1.84	0.96398	−0.26596	−3.62449	6.29654	0.15882	3.06886	3.22768	0.95080	105.42
1.85	0.96128	−0.27559	−3.48806	6.35982	0.15724	3.10129	3.25853	0.95175	106.00
1.86	0.95847	−0.28519	−3.36083	6.42374	0.15567	3.13403	3.28970	0.95268	106.57
1.87	0.95557	−0.29476	−3.24187	6.48830	0.15412	3.16709	3.32121	0.95359	107.14
1.88	0.95258	−0.30430	−3.13038	6.55350	0.15259	3.20046	3.35305	0.95449	107.72
1.89	0.94949	−0.31381	−3.02566	6.61937	0.15107	3.23415	3.38522	0.95537	108.29
1.90	0.94630	−0.32329	−2.92710	6.68589	0.14957	3.26816	3.41773	0.95624	108.86
1.91	0.94302	−0.33274	−2.83414	6.75309	0.14808	3.30250	3.45058	0.95709	109.43
1.92	0.93965	−0.34215	−2.74630	6.82096	0.14661	3.33718	3.48378	0.95792	110.01
1.93	0.93618	−0.35153	−2.66316	6.88951	0.14515	3.37218	3.51733	0.95873	110.58
1.94	0.93262	−0.36087	−2.58433	6.95875	0.14370	3.40752	3.55123	0.95953	111.15
1.95	0.92896	−0.37018	−2.50948	7.02869	0.14227	3.44321	3.58548	0.96032	111.73
1.96	0.92521	−0.37945	−2.43828	7.09933	0.14086	3.47923	3.62009	0.96109	112.30
1.97	0.92137	−0.38868	−2.37048	7.17068	0.13946	3.51561	3.65507	0.96185	112.87
1.98	0.91744	−0.39788	−2.30582	7.24274	0.13807	3.55234	3.69041	0.96259	113.45
1.99	0.91341	−0.40703	−2.24408	7.31553	0.13670	3.58942	3.72611	0.96331	114.02
2.00	0.90930	−0.41615	−2.18504	7.38906	0.13534	3.62686	3.76220	0.96403	114.59

$$x = 2.01 - 2.50$$

x, rad	$\sin x$	$\cos x$	$\tan x$	e^x	e^{-x}	$\sinh x$	$\cosh x$	$\tanh x$	x, deg
2.01	0.90509	−0.42522	−2.12853	7.46332	0.13399	3.66466	3.79865	0.96473	115.16
2.02	0.90079	−0.43425	−2.07437	7.53832	0.13266	3.70283	3.83549	0.96541	115.74
2.03	0.89641	−0.44323	−2.02242	7.61409	0.13134	3.74138	3.87271	0.96609	116.31
2.04	0.89193	−0.45218	−1.97252	7.69061	0.13003	3.78029	3.91032	0.96675	116.88
2.05	0.88736	−0.46107	−1.92456	7.76790	0.12873	3.81958	3.94832	0.96740	117.46
2.06	0.88271	−0.46992	−1.87841	7.84597	0.12745	3.85926	3.98671	0.96803	118.03
2.07	0.87796	−0.47873	−1.83396	7.92482	0.12619	3.89932	4.02550	0.96865	118.60
2.08	0.87313	−0.48748	−1.79111	8.00447	0.12493	3.93977	4.06470	0.96926	119.18
2.09	0.86821	−0.49619	−1.74977	8.08492	0.12369	3.98061	4.10430	0.96986	119.75
2.10	0.86321	−0.50485	−1.70985	8.16617	0.12246	4.02186	4.14431	0.97045	120.32
2.11	0.85812	−0.51345	−1.67127	8.24824	0.12124	4.06350	4.18474	0.97103	120.89
2.12	0.85294	−0.52201	−1.63396	8.33114	0.12003	4.10555	4.22558	0.97159	121.47
2.13	0.84768	−0.53051	−1.59785	8.41487	0.11884	4.14801	4.26685	0.97215	122.04
2.14	0.84233	−0.53896	−1.56288	8.49944	0.11765	4.19089	4.30855	0.97269	122.61
2.15	0.83690	−0.54736	−1.52898	8.58486	0.11648	4.23419	4.35067	0.97323	123.19
2.16	0.83138	−0.55570	−1.49610	8.67114	0.11533	4.27791	4.39323	0.97375	123.76
2.17	0.82578	−0.56399	−1.46420	8.75828	0.11418	4.32205	4.43623	0.97426	124.33
2.18	0.82010	−0.57221	−1.43321	8.84631	0.11304	4.36663	4.47967	0.97477	124.90
2.19	0.81434	−0.58039	−1.40310	8.93521	0.11192	4.41165	4.52356	0.97526	125.48
2.20	0.80850	−0.58850	−1.37382	9.02501	0.11080	4.45711	4.56791	0.97574	126.05
2.21	0.80257	−0.59656	−1.34534	9.11572	0.10970	4.50301	4.61271	0.97622	126.62
2.22	0.79657	−0.60455	−1.31761	9.20733	0.10861	4.54936	4.65797	0.97668	127.20
2.23	0.79048	−0.61249	−1.29061	9.29987	0.10753	4.59617	4.70370	0.97714	127.77
2.24	0.78432	−0.62036	−1.26429	9.39333	0.10646	4.64344	4.74989	0.97759	128.34
2.25	0.77807	−0.62817	−1.23863	9.48774	0.10540	4.69117	4.79657	0.97803	128.92
2.26	0.77175	−0.63592	−1.21359	9.58309	0.10435	4.73937	4.84372	0.97846	129.49
2.27	0.76535	−0.64361	−1.18916	9.67940	0.10331	4.78804	4.89136	0.97888	130.06
2.28	0.75888	−0.65123	−1.16530	9.77668	0.10228	4.83720	4.93948	0.97929	130.63
2.29	0.75233	−0.65879	−1.14200	9.87494	0.10127	4.88684	4.98810	0.97970	131.21
2.30	0.74571	−0.66628	−1.11921	9.97418	0.10026	4.93696	5.03722	0.98010	131.78
2.31	0.73901	−0.67370	−1.09694	10.07442	0.09926	4.98758	5.08684	0.98049	132.35
2.32	0.73223	−0.68106	−1.07514	10.17567	0.09827	5.03870	5.13697	0.98087	132.93
2.33	0.72538	−0.68834	−1.05381	10.27794	0.09730	5.09032	5.18762	0.98124	133.50
2.34	0.78146	−0.69556	−1.03293	10.38124	0.09633	5.14245	5.23878	0.98161	134.07
2.35	0.71147	−0.70271	−1.01247	10.48557	0.09537	5.19510	5.29047	0.98197	134.65
2.36	0.70441	−0.70979	−0.99242	10.59095	0.09442	5.24827	5.34269	0.98233	135.22
2.37	0.69728	−0.71680	−0.97276	10.69739	0.09348	5.30196	5.39544	0.98267	135.79
2.38	0.69007	−0.72374	−0.95349	10.80490	0.09255	5.35618	5.44873	0.98301	136.36
2.39	0.68280	−0.73060	−0.93458	10.91349	0.09163	5.41093	5.50256	0.98335	136.94
2.40	0.67546	−0.73739	−0.91601	11.02318	0.09072	5.46623	5.55695	0.98367	137.51
2.41	0.66806	−0.74411	−0.89779	11.13396	0.08982	5.52207	5.61189	0.98400	138.08
2.42	0.66058	−0.75075	−0.87989	11.24586	0.08892	5.57847	5.66739	0.98431	138.66
2.43	0.65304	−0.75732	−0.86230	11.35888	0.08804	5.63542	5.72346	0.98462	139.23
2.44	0.64543	−0.76382	−0.84501	11.47304	0.08716	5.69294	5.78010	0.98492	139.80
2.45	0.63776	−0.77023	−0.82802	11.58835	0.08629	5.75103	5.83732	0.98522	140.37
2.46	0.63003	−0.77657	−0.81130	11.70481	0.08543	5.80969	5.89512	0.98551	140.95
2.47	0.62223	−0.78283	−0.79485	11.82245	0.08458	5.86893	5.95352	0.98579	141.52
2.48	0.61437	−0.78901	−0.77866	11.94126	0.08374	5.92876	6.01250	0.98607	142.09
2.49	0.60645	−0.79512	−0.76272	12.06128	0.08291	5.98918	6.07209	0.98635	142.67
2.50	0.59847	−0.80114	−0.74702	12.18249	0.08208	6.05020	6.13229	0.98661	143.24

$$x = 2.51 - 3.00$$

x, rad	sin x	cos x	tan x	e^x	e^{-x}	sinh x	cosh x	tanh x	x, deg
2.51	0.59043	−0.80709	−0.73156	12.30493	0.08127	6.11183	6.19310	0.98688	143.81
2.52	0.58233	−0.81295	−0.71632	12.42860	0.08046	6.17407	6.25453	0.98714	144.39
2.53	0.57417	−0.81873	−0.70129	12.55351	0.07966	6.23692	6.31658	0.98739	144.96
2.54	0.56596	−0.82444	−0.68648	12.67967	0.07887	6.30040	6.37927	0.98764	145.53
2.55	0.55768	−0.83005	−0.67186	12.80710	0.07808	6.36451	6.44259	0.98788	146.10
2.56	0.54936	−0.83559	−0.65745	12.93582	0.07730	6.42926	6.50656	0.98812	146.68
2.57	0.54097	−0.84104	−0.64322	13.06582	0.07654	6.49464	6.57118	0.98835	147.25
2.58	0.53253	−0.84641	−0.62917	13.19714	0.07577	6.56068	6.63646	0.98858	147.82
2.59	0.52404	−0.85169	−0.61530	13.32977	0.07502	6.62738	6.70240	0.98881	148.40
2.60	0.51550	−0.85689	−0.60160	13.46374	0.07427	6.69473	6.76901	0.98903	148.97
2.61	0.50691	−0.86200	−0.58806	13.59905	0.07353	6.76276	6.83629	0.98924	149.54
2.62	0.49826	−0.86703	−0.57468	13.73572	0.07280	6.83146	6.90426	0.98946	150.11
2.63	0.48957	−0.87197	−0.56145	13.87377	0.07208	6.90085	6.97292	0.98966	150.69
2.64	0.48082	−0.87682	−0.54837	14.01320	0.07136	6.97092	7.04228	0.98987	151.26
2.65	0.47203	−0.88158	−0.53544	14.15404	0.07065	7.04169	7.11234	0.99007	151.83
2.66	0.46319	−0.88626	−0.52264	14.29629	0.06995	7.11317	7.18312	0.99026	152.41
2.67	0.45431	−0.89085	−0.50997	14.43997	0.06925	7.18536	7.25461	0.99045	152.98
2.68	0.44537	−0.89534	−0.49743	14.58509	0.06856	7.25827	7.32683	0.99064	153.55
2.69	0.43640	−0.89975	−0.48502	14.73168	0.06788	7.33190	7.39978	0.99083	154.13
2.70	0.42738	−0.90407	−0.47273	14.87973	0.06721	7.40626	7.47347	0.99101	154.70
2.71	0.41832	−0.90830	−0.46055	15.02928	0.06654	7.48137	7.54791	0.99118	155.27
2.72	0.40921	−0.91244	−0.44848	15.18032	0.06587	7.55722	7.62310	0.99136	155.84
2.73	0.40007	−0.91648	−0.43653	15.33289	0.06522	7.63383	7.69905	0.99153	156.42
2.74	0.39088	−0.92044	−0.42467	15.48699	0.06457	7.71121	7.77578	0.99170	156.99
2.75	0.38166	−0.92430	−0.41292	15.64263	0.06393	7.78935	7.85328	0.99186	157.56
2.76	0.37240	−0.92807	−0.40126	15.79984	0.06329	7.86828	7.93157	0.99202	158.14
2.77	0.36310	−0.93175	−0.38970	15.95863	0.06266	7.94799	8.01065	0.99218	158.71
2.78	0.35376	−0.93533	−0.37822	16.11902	0.06204	8.02849	8.09053	0.99233	159.28
2.79	0.34439	−0.93883	−0.36683	16.28102	0.06142	8.10980	8.17122	0.99248	159.86
2.80	0.33499	−0.94222	−0.35553	16.44465	0.06081	8.19192	8.25273	0.99263	160.43
2.81	0.32555	−0.94553	−0.34431	16.60992	0.06020	8.27486	8.33506	0.99278	161.00
2.82	0.31608	−0.94873	−0.33316	16.77685	0.05961	8.35862	8.41823	0.99292	161.57
2.83	0.30657	−0.95185	−0.32208	16.94546	0.05901	8.44322	8.50224	0.99306	162.15
2.84	0.29704	−0.95486	−0.31108	17.11577	0.05843	8.52867	8.58710	0.99320	162.72
2.85	0.28748	−0.95779	−0.30015	17.28778	0.05784	8.61497	8.67281	0.99333	163.29
2.86	0.27789	−0.96061	−0.28928	17.46153	0.05727	8.70213	8.75940	0.99346	163.87
2.87	0.26827	−0.96334	−0.27847	17.63702	0.05670	8.79016	8.84686	0.99359	164.44
2.88	0.25862	−0.96598	−0.26773	17.81427	0.05613	8.87907	8.93520	0.99372	165.01
2.89	0.24895	−0.96852	−0.25704	17.99331	0.05558	8.96887	9.02444	0.99384	165.58
2.90	0.23925	−0.97096	−0.24641	18.17415	0.05502	9.05956	9.11458	0.99396	166.16
2.91	0.22953	−0.97330	−0.23582	18.35680	0.05448	9.15116	9.20564	0.99408	166.73
2.92	0.21978	−0.97555	−0.22529	18.54129	0.05393	9.24368	9.29761	0.99420	167.30
2.93	0.21002	−0.97770	−0.21481	18.72763	0.05340	9.33712	9.39051	0.99431	167.88
2.94	0.20023	−0.97975	−0.20437	18.91585	0.05287	9.43149	9.48436	0.99443	168.45
2.95	0.19042	−0.98170	−0.19397	19.10595	0.05234	9.52681	9.57915	0.99494	169.02
2.96	0.18060	−0.98356	−0.18362	19.29797	0.05182	9.62308	9.67490	0.99464	169.60
2.97	0.17075	−0.98531	−0.17330	19.49192	0.05130	9.72031	9.77161	0.99475	170.17
2.98	0.16089	−0.98697	−0.16301	19.68782	0.05079	9.81851	9.86930	0.99485	170.74
2.99	0.15101	−0.98853	−0.15276	19.88568	0.05029	9.91770	9.96798	0.99496	171.31
3.00	0.14112	−0.98999	−0.14255	20.08554	0.04979	10.01787	10.06766	0.99505	171.89

x, rad	sin x	cos x	tan x	e^x	e^{-x}	sinh x	cosh x	tanh x	x, deg
3.05	0.09146	−0.99581	−0.09185	21.11534	0.04736	10.53399	10.58135	0.99552	174.75
3.10	0.04158	−0.99914	−0.04162	22.19795	0.04505	11.07645	11.12150	0.99595	177.62
3.15	−0.00841	−0.99996	0.00841	23.33606	0.04285	11.64661	11.68946	0.99633	180.48
3.20	−0.05837	−0.99829	0.05847	24.53253	0.04076	12.24588	12.28665	0.99668	183.35
3.25	−0.10820	−0.99413	0.10883	25.79034	0.03877	12.87578	12.91456	0.99700	186.21
3.30	−0.15775	−0.98748	0.15975	27.11264	0.03688	13.53788	13.57476	0.99728	189.08
3.35	−0.20690	−0.97836	0.21148	28.50273	0.03508	14.23382	14.26891	0.99754	191.94
3.40	−0.25554	−0.96680	0.26432	29.96410	0.03337	14.96536	14.99874	0.99777	194.81
3.45	−0.30354	−0.95282	0.31857	31.50039	0.03175	15.73432	15.76607	0.99799	197.67
3.50	−0.35078	−0.93646	0.37459	33.11545	0.03020	16.54263	16.57282	0.99818	200.54
3.55	−0.39715	−0.91775	0.43274	34.81332	0.02872	17.39230	17.42102	0.99835	203.40
3.60	−0.44252	−0.89676	0.49347	36.59823	0.02732	18.28546	18.31278	0.99851	206.26
3.65	−0.48679	−0.87352	0.55727	38.47467	0.02599	19.22434	19.25033	0.99865	209.13
3.70	−0.52984	−0.84810	0.62473	40.44730	0.02472	20.21129	20.23601	0.99878	211.99
3.75	−0.57156	−0.82056	0.69655	42.52108	0.02352	21.24878	21.27230	0.99889	214.86
3.80	−0.61186	−0.79097	0.77356	44.70118	0.02237	22.33941	22.36178	0.99900	217.72
3.85	−0.65063	−0.75940	0.85676	46.99306	0.02128	23.48589	23.50717	0.99909	220.59
3.90	−0.68777	−0.72593	0.94742	49.40245	0.02024	24.69110	24.71135	0.99918	223.45
3.95	−0.72319	−0.69065	1.04711	51.93537	0.01925	25.95806	25.97731	0.99926	226.32
4.00	−0.75680	−0.65364	1.15782	54.59815	0.01832	27.28992	27.30823	0.99933	229.18
4.05	−0.78853	−0.61500	1.28215	57.39746	0.01742	28.69002	28.70744	0.99939	232.05
4.10	−0.81828	−0.57482	1.42353	60.34029	0.01657	30.16186	30.17843	0.99945	234.91
4.15	−0.84598	−0.53321	1.58659	63.43400	0.01576	31.70912	31.72488	0.99950	237.78
4.20	−0.87158	−0.49026	1.77778	66.68633	0.01500	33.33567	33.35066	0.99955	240.64
4.25	−0.89499	−0.44609	2.00631	70.10541	0.01426	35.04557	35.05984	0.99959	243.51
4.30	−0.91617	−0.40080	2.28585	73.69979	0.01357	36.84311	36.85668	0.99963	246.37
4.35	−0.93505	−0.35451	2.63760	77.47846	0.01291	38.73278	38.74568	0.99967	249.24
4.40	−0.95160	−0.30733	3.09632	81.45087	0.01228	40.71930	40.73157	0.99970	252.10
4.45	−0.96577	−0.25939	3.72327	85.62694	0.01168	42.80763	42.81931	0.99973	254.97
4.50	−0.97753	−0.21080	4.63733	90.01713	0.01111	45.00301	45.01412	0.99975	257.83
4.55	−0.98684	−0.16168	6.10383	94.63241	0.01057	47.31092	47.32149	0.99978	260.70
4.60	−0.99369	−0.11215	8.86017	99.48432	0.01005	49.73713	49.74718	0.99980	263.56
4.65	−0.99805	−0.06235	16.00767	104.58499	0.00956	52.28771	52.29727	0.99982	266.43
4.70	−0.99992	−0.01239	80.71276	109.94717	0.00910	54.96904	54.97813	0.99983	269.29
4.75	−0.99929	0.03760	−26.57541	115.58428	0.00865	57.78782	57.79647	0.99985	272.15
4.80	−0.99616	0.08750	−11.38487	121.51042	0.00823	60.75109	60.75932	0.99986	275.02
4.85	−0.99055	0.13718	−7.22093	127.74039	0.00783	63.86628	63.87411	0.99988	277.88
4.90	−0.98245	0.18651	−5.26749	134.28978	0.00745	67.14117	67.14861	0.99989	280.75
4.95	−0.97190	0.23538	−4.12906	141.17496	0.00708	70.58394	70.59102	0.99990	283.61
*5.00	−0.95892	0.28366	−3.38052	148.41316	0.00674	74.20321	74.20995	0.99991	286.48

*For x > 5.00 and other numerical data, refer to:

Abramowitz, M., and I. A. Stegun: "Handbook of Mathematical Functions," National Bureau of Standards, Washington, D.C., 1964.

Dwight, H. B.: "Mathematical Tables," 3d ed., Dover, New York, 1961.

Flecher, A., J. C. P. Miller, L. Rosenhead, and L. J. Comrie: "An Index of Mathematical Tables," 2d ed., Addison-Wesley, Reading, Mass., 1962.

Lebedev, A. V., and R. M. Fedorova: "A Guide to Mathematical Tables," Pergamon, New York, 1960.

(1) Legendre Polynomials $P_n(x)$ (Sec. 14.18) $\boxed{P_0(x) = 1}$

$P_1(x) = x$

$P_2(x) = \dfrac{1}{2}(3x^2 - 1)$

$P_3(x) = \dfrac{1}{2}(5x^3 - 3x)$

$P_4(x) = \dfrac{1}{8}(35x^4 - 30x^2 + 3)$

$P_5(x) = \dfrac{1}{8}(63x^5 - 70x^3 + 15x)$

$$P_n(x) = \frac{2n-1}{n}(x)P_{n-1}(x) - \frac{n-1}{n}P_{n-2}(x)$$

(2) Legendre Polynomials $Q_n(x)$ (Sec. 14.18) $\boxed{Q_0(x) = \ln\sqrt{\dfrac{1+x}{1-x}}}$

$Q_1(x) = x\,Q_0(x) - 1$

$Q_2(x) = P_2(x)\,Q_0(x) - \dfrac{3}{2}x$

$Q_3(x) = P_3(x)\,Q_0(x) - \dfrac{5}{2}x^2 + \dfrac{2}{3}$

$Q_4(x) = P_4(x)\,Q_0(x) - \dfrac{35}{8}x^3 + \dfrac{55}{24}x$

$Q_5(x) = P_5(x)\,Q_0(x) - \dfrac{63}{8}x^4 + \dfrac{49}{8}x^2 - \dfrac{8}{15}$

$$Q_n(x) = \frac{2n-1}{n}(x)Q_{n-1}(x) - \frac{n-1}{n}Q_{n-2}(x)$$

(3) Chebyshev Polynomials $T_n(x)$ (Sec. 14.20) $\boxed{T_0(x) = 1}$

$T_1(x) = x$

$T_2(x) = 2x^2 - 1$

$T_3(x) = 4x^3 - 3x$

$T_4(x) = 8x^4 - 8x^2 + 1$

$T_5(x) = 16x^5 - 20x^3 + 5x$

$$T_n(x) = 2x\,T_{n-1}(x) - T_{n-2}(x)$$

(4) Chebyshev Polynomials $U_n(x)$ (Sec. 14.20) $\boxed{U_0(x) = \sin^{-1} x}$

$U_1(x) = \sqrt{1 - x^2}$

$U_2(x) = 2x\sqrt{1 - x^2}$

$U_3(x) = (4x^2 - 1)\sqrt{1 - x^2}$

$U_4(x) = (8x^3 - 4x)\sqrt{1 - x^2}$

$U_5(x) = (16x^4 - 12x^2 + 1)\sqrt{1 - x^2}$

$$U_n(x) = 2x\,U_{n-1}(x) - U_{n-2}(x)$$

(5) Laguerre Polynomials $L_n(x)$ (Sec. 14.22) $\boxed{L_0(x) = 1}$

$L_1(x) = 1 - x$

$L_2(x) = 2 - 4x + x^2$

$L_3(x) = 6 - 18x + 9x^2 - x^3$

$L_4(x) = 24 - 96x + 72x^2 - 16x^3 + x^4$

$L_5(x) = 120 - 600x + 600x^2 - 200x^3 + 25x^4 - x^5$

$$L_n(x) = (2n - 1 - x)L_{n-1}(x) - (n-1)^2 L_{n-2}(x)$$

(6) Hermite Polynomials $H_n(x)$ (Sec. 14.23) $\boxed{H_0(x) = 1}$

$H_1(x) = 2x$

$H_2(x) = 4x^2 - 2$

$H_3(x) = 8x^3 - 12x$

$H_4(x) = 16x^4 - 48x^2 + 12$

$H_5(x) = 32x^5 - 160x^3 + 120x$

$H_n(x) = 2x\,H_{n-1}(x) - 2(n-1)\,H_{n-2}(x)$

(1) Archimedes Number π

(a) Definition. The symbol π denotes the ratio of the circumference of a circle to its diameter,

$$\pi = 3.141\ 592\ 653\ 589\ 793\ 238\ 462\ 643\ldots$$

In 1882 C.L.P. Lindemann proved that π is both an irrational and a transcendental number and thus has shown that the problem of rectification and squaring of circle with ruler and compass alone is a mathematical impossibility.

(b) Approximation by fraction

$$\pi = \frac{22}{7} - \epsilon = 3.142\ 857\ 143 - \epsilon \qquad \epsilon < 1.3 \times 10^{-3}$$

$$\pi = \frac{355}{113} - \epsilon = 3.141\ 592\ 920 - \epsilon \qquad \epsilon < 2.7 \times 10^{-7}$$

(c) Evaluation by series

$$\pi = 4 \sum_{k=1}^{\infty} \frac{(-1)^{k-1}}{2k-1} \left(\frac{4}{5^{2k-1}} - \frac{1}{239^{2k-1}} \right) \qquad n = 7, \epsilon < 5 \times 10^{-10}$$

(2) Base of Natural Logarithms e

(a) Definition. The symbol e denotes the limit

$$e = \lim_{m \to \infty} \left(1 + \frac{1}{m} \right)^m = \lim_{n \to 0} (1+n)^{1/n} = 2.718\ 281\ 828\ 459\ 045\ 235\ 360\ldots$$

and is the base of the natural system of logarithms. In 1873, C. Hermite proved that e is both an irrational and a transcendental number.

(b) Approximation by fraction

$$e = \frac{19}{7} + \epsilon = 2.714\ 285\ 714 + \epsilon \qquad \epsilon < 4 \times 10^{-3}$$

$$e = \frac{1264}{465} + \epsilon = 2.718\ 279\ 570 + \epsilon \qquad \epsilon < 2.3 \times 10^{-6}$$

(c) Evaluation by series

$$e = 2\left(\frac{1}{1!} + \frac{2}{3!} + \frac{3}{5!} + \cdots \right) = 2 \sum_{k=1}^{\infty} \frac{k}{(2k-1)!} \qquad n = 7, \epsilon < 4.6 \times 10^{-10}$$

(3) Euler Constant C

(a) Definition. The symbol C denotes the limit

$$C = \lim_{n \to \infty} \left(1 + \frac{1}{2} + \frac{1}{3} + \cdots + \frac{1}{n} - \ln n \right) = 0.577\ 215\ 664\ 901\ 532\ 860\ 606\ \ldots$$

No proof is yet available whether C is an irrational number.

(b) Approximation by fraction

$$C = \sqrt{\frac{1}{3}} - \epsilon = 0.577\ 350\ 269 - \epsilon \qquad \epsilon < 1.3 \times 10^{-4}$$

$$C = \frac{228}{395} - \epsilon = 0.577\ 215\ 190 - \epsilon \qquad \epsilon < 4.8 \times 10^{-7}$$

(c) Evaluation by series

$$C = \sum_{k=1}^{n} \frac{1}{k} - \ln n - \epsilon$$

$$n = 10^3, \epsilon < 5 \times 10^{-4}$$
$$n = 10^6, \epsilon < 5 \times 10^{-7}$$
$$n = 10^9, \epsilon < 5 \times 10^{-10}$$

Appendix B
GLOSSARY
OF SYMBOLS

$=$ or $::$	Equals	$\pm$ or $\neq$	Does not equal
$>$	Greater than	$<$	Less than
$\geq$	Greater than or equal	$\leq$	Less than or equal
$\equiv$	Identical	$\approx$	Approximately equal

B.2 ALGEBRA

$+$	Plus or positive	$-$	Minus or negative
$\pm\}$	Plus or minus / Positive or negative	$\mp\}$	Minus or plus / Negative or positive
$\times$	Multiplied by	$\div$ or $:$	Divided by
a^n	nth power of a	$\sqrt[n]{a}$	nth root of a
$\log\}$ $\log_{10}\}$	Common logarithm or **Briggs' logarithm**	$\ln\}$ $\log_e\}$	Natural logarithm or Napier's logarithm
$(\ )$	Parentheses	$[\ ]$ Brackets	$\{\ \}$ Braces

$$\begin{vmatrix} a_1 & a_2 & \cdots \\ b_1 & b_2 & \cdots \\ \cdots\cdots\cdots \end{vmatrix} \quad \text{Determinant} \qquad \begin{bmatrix} a_1 & a_2 & \cdots \\ b_1 & b_2 & \cdots \\ \cdots\cdots\cdots \end{bmatrix} \quad \text{Matrix}$$

I	Unit matrix	Adj	Adjoint matrix
A^{-1}	Inverse of the A matrix	A^T	Transpose of the A matrix
$n!$	n factorial	$\binom{n}{k}$	Binomial coefficient
$n!!$	n double factorial	$X_h^{(p)}$	Factorial polynomial

$k^P n$ The number of all possible permutations of n elements, among which there are k elements of equal value.

$k^V n$ The number of all possible permutations of n elements taken k at a time.

$k^C n$ The number of all possible permutations (without repetition) of n elements taken k at a time.

B.3 COMPLEX NUMBERS

$i = \sqrt{-1}$	Unit imaginary number	$z = x + iy$	Complex variable
$|z|$	Absolute value of z	$\bar{z}$	Conjugate of z
$\operatorname{Re}(z), \mathscr{R}(z)$	Real part of z	$\operatorname{Im}(z), \mathscr{I}(z)$	Imaginary part of z

$\parallel$	Parallel to		$\alpha°$	α in degrees
$\perp$	Perpendicular to		α'	α in minutes
$\angle$	Angle		α''	α in seconds
$\cong$	Congruent to		$\sim$	Similar to

$\triangle$	Triangle		$\bigcirc$	Circle
$\square$	Parallelogram		$\square$	Square

$\overline{AB}$	The line segment between A and B
$P(x, y, z)$	Point P given by the cartesian coordinates x, y, z
$P(r, \theta, z)$	Point P given by the cylindrical coordinates r, θ, z
$P(\rho, \theta, \phi)$	Point P given by the spherical coordinates ρ, θ, ϕ

$\widehat{AB}$	The arc segment between A and B

B.5   CIRCULAR AND HYPERBOLIC FUNCTIONS

sin	Sine		sinh	Hyperbolic sine
cos	Cosine		cosh	Hyperbolic cosine
tan	Tangent		tanh	Hyperbolic tangent
cot	Cotangent		coth	Hyperbolic cotangent
sec	Secant		sech	Hyperbolic secant
csc	Cosecant		csch	Hyperbolic cosecant
vers	Versine		covers	Coversine
$\sin^{-1}$	Inverse sine		$\sinh^{-1}$	Inverse hyperbolic sine
$\cos^{-1}$	Inverse cosine		$\cosh^{-1}$	Inverse hyperbolic cosine
$\tan^{-1}$	Inverse tangent		$\tanh^{-1}$	Inverse hyperbolic tangent
$\cot^{-1}$	Inverse cotangent		$\coth^{-1}$	Inverse hyperbolic cotangent
$\sec^{-1}$	Inverse secant		sech^{-1}	Inverse hyperbolic secant
$\csc^{-1}$	Inverse cosecant		csch^{-1}	Inverse hyperbolic cosecant

B.6   VECTOR ANALYSIS

$\mathbf{i}, \mathbf{j}, \mathbf{k}$	Unit vectors, cartesian system of coordinates		$\mathbf{e}_s$	Unit vector in s direction

$\mathbf{r} = \mathbf{i}x + \mathbf{j}y + \mathbf{k}z$	Position vector, cartesian coordinates
$\mathbf{r} = r_a\mathbf{e}_a + r_\theta\mathbf{e}_\theta + r_z\mathbf{e}_z$	Position vector, cylindrical coordinates
$\mathbf{r} = r_b\mathbf{e}_b + r_\theta\mathbf{e}_\theta + r_\phi\mathbf{e}_\phi$	Position vector, spherical coordinates

$\mathbf{r}_1 \bullet \mathbf{r}_2$	Scalar product		$\mathbf{r}_1 \times \mathbf{r}_2$	Vector product
∇	Vector differential operator		∇^2	Laplacian operator

(a, b)	The bounded open interval	$[a, b]$	The bounded closed interval
$f(x), F(x)$	The function of x	$f^{-1}(x), F^{-1}(x)$	The inverse function of x
$\displaystyle\sum_{i=1}^{n} u_i$	The sum of n terms	$\displaystyle\prod_{i=1}^{n} u_i$	The product of n terms

$$\bigwedge_{k=1}^{n} [1 \pm a_k s]$$ **The nested sum of $n+1$ terms**

Δu	The increment of u	du	The differential of u

$\dfrac{dy}{dx}, y', D_x y$

$\dfrac{df(x)}{dx}, f'(x), D_x f(x)$ $\Bigg\}$ The first-order derivative of $y = f(x)$ with respect to x

$\dfrac{d^n y}{dx^n}, y^{(n)}, D_x^{(n)} y$

$\dfrac{d^n f(x)}{dx}, f^{(n)}(x), D_x^{(n)} f(x)$ $\Bigg\}$ The nth-order derivative of $y = f(x)$ with respect to x

$\dfrac{\partial w}{\partial x}, w_x, D_x w$

$\dfrac{\partial f}{\partial x}, f_x, F_x$ $\Bigg\}$ The first-order partial derivative of $w = f(x, y, \ldots)$ with respect to x

$\dfrac{\partial^2 w}{\partial x \partial y}, w_{xy}, D_{xy} w$

$\dfrac{\partial^2 f}{\partial x \partial y}, f_{xy}, F_{xy}$ $\Bigg\}$ The second-order partial derivative of $w = f(x, y, \ldots)$ with respect to x and then with respect to y

$f'(a+)$ The derivative on the right of $x = a$ $f'(a-)$ The derivative on the left of $x = a$

$\dfrac{\partial(f_1, f_2, \ldots, f_n)}{\partial(x_1, x_2, \ldots, x_n)}, J \dfrac{(f_1, f_2, \ldots, f_n)}{(x_1, x_2, \ldots, x_n)}$ Jacobian determinant (Sec. 7.08)

$\displaystyle\int f(x)\,dx$	The indefinite integral of $y = f(x)$	$\displaystyle\int_a^b f(x)\,dx$	The definite integral of $y = f(x)$ between limits a and b
$\displaystyle\iint$	The double integral	$\displaystyle\iiint$	The triple integral

$\displaystyle\int_C$ The line integral $\displaystyle\int_S$ The surface integral $\displaystyle\int_V$ The volume integral

A	Area	M	Static moment
V	Volume	I	Moment of inertia
k	Radius of gyration	J	Polar moment of inertia
ρ	Radius of curvature	κ	Curvature
L	Length of curve	x_C, y_C, z_C	Coordinates of centroid

π	Archimedes number	$\bar{B}_m$	Bernoulli number
e	Euler's number	B_m	Auxiliary Bernoulli number
C	Euler's constant	$\bar{E}_m$	Euler number
$\mathscr{S}_h^{(p)}$	Stirling number	E_m	Auxiliary Euler number

B.9   SPECIAL FUNCTIONS

$\Gamma(x)$	Gamma function	$\Pi(x)$	Pi function
$B(x)$	Beta function	$\mathrm{erf}(x)$	Error function
$\Psi(x)$	Digamma function	$\Psi^{(m)}(x)$	Polygamma function
$Z(x)$	Zeta function	$\bar{Z}(x)$	Complementary zeta function

$\mathrm{Si}(x), \mathrm{Ci}(x), \mathrm{Ei}(x), \bar{\mathrm{Ei}}(x)$	Integral functions
$S(x), C(x)$	Fresnel integrals

$$\left.\begin{array}{l} E(k, x), F(k, x), \Pi(n, k, x) \\ E(k, \phi), F(k, \phi), \Pi(n, k, \phi) \end{array}\right\} \quad \text{Elliptic integrals}$$

$\mathrm{sn}\, u, \mathrm{cn}\, u, \mathrm{dn}\, u$	Jacobi's elliptic functions
$\mathscr{L} f(t) = f(s)$	Laplace transform
$\mathscr{D}(x - t)$	**Dirac delta functions**

B.10   BESSEL FUNCTIONS

J_n	Bessel function of the first kind of order n
Y_n	Bessel function of the second kind of order n
I_n	Modified Bessel function of the first kind of order n
K_n	Modified Bessel function of the second kind of order n

$H_n^{(1)}$	Hankel function of the first kind of order n
$H_n^{(2)}$	Hankel function of the second kind of order n

Ber_n	Ber function of order n	Ker_n	Ker function of order n
Bei_n	Bei function of order n	Kei_n	Kei function of order n

B.11   ORTHOGONAL POLYNOMIALS

$p_n(x)$	Orthogonal polynomial in n	$w(x)$	Weight function
$F(\alpha, \beta, \gamma, x)$	Hypergeometric series		
$P_n(x)$	Legendre polynomial in n		
$T_n(x)$	Chebyshev polynomial in n		
$L_n(x)$	Laguerre polynomial in n		
$H_n(x)$	Hermite polynomial in n		

ϵ	Absolute error	$\bar{\epsilon}$	Relative error
ϵ_T	Truncation error	$\bar{y}$	Substitute function
r_{ij}	Carryover value	m_j	Starting value
Δy_n	Forward difference	∇y_n	Backward difference
δy_n	Central difference	Δx_n	Divided difference

B.13 PROBABILITY AND STATISTICS

$\cup$	Union	$\cap$	Intersection
$P(E)$	Probability of occurrence	$P(\bar{E})$	Probability of nonoccurrence
$\phi(x)$	Probability density	$\phi(x, y)$	Joint probability density
$\bar{X}$	Arithmetic means	$\bar{G}$	Geometric means
$\bar{H}$	Harmonic means	$\bar{Q}$	Quadratic means
D	Deviation	$\bar{D}$	Mean deviation
σ	Standard deviation	σ^2	Variance
μ_k	Moment of degree k	β_2	Kurtosis
γ_1	Coefficient of skewness	γ_2	Coefficient of excess
$\phi_N(t)$	Ordinate of the standard normal curve	$F_N(t)$	Area under the standard normal curve

B.14 GREEK ALPHABET

A	α	Alpha	H	η	Eta	N	ν	Nu	T	τ	Tau
B	β	Beta	Θ	θ	Theta	Ξ	ξ	Xi	Y	υ	Upsilon
Γ	γ	Gamma	I	ι	Iota	O	o	Omicron	Φ	ϕ	Phi
Δ	δ	Delta	K	κ	Kappa	Π	π	Pi	X	χ	Chi
E	ϵ	Epsilon	Λ	λ	Lambda	P	ρ	Rho	Ψ	ψ	Psi
Z	ζ	Zeta	M	μ	Mu	Σ	σ	Sigma	Ω	ω	Omega

B.15 GERMAN ALPHABET

Aa	Bb	Cc	Dd	Ee	Ff	Gg	Hh	Ii	Jj	Kk	Ll	Mm

Nn	Oo	Pp	Qq	Rr	Ss	Tt	Uu	Vv	Ww	Xx	Yy	Zz

B.16 RUSSIAN ALPHABET

Upright	Cursive	Transliteration (pronunciation)	Upright	Cursive	Transliteration (pronunciation)
А а	*А а*	a	Р р	*Р р*	r
Б б	*Б б*	b	С с	*С с*	s
В в	*В в*	v	Т т	*Т т*	t
Г г	*Г г*	g	У у	*У у*	u (as in moon)
Д д	*Д д*	d	Ф ф	*Ф ф*	f
Е е	*Е е*	ye (as in yell)	Х х	*Х х*	kh
Ё ё	*Ё ё*	yë (yo as in yore)	Ц ц	*Ц ц*	ts
Ж ж	*Ж ж*	zh	Ч ч	*Ч ч*	ch (church)
З з	*З з*	z	Ш ш	*Ш ш*	sh
И и	*И и*	i	Щ щ	*Щ щ*	shch
Й й	*Й й*	y	Ъ ъ	*Ъ ъ*	[double apostrophe, no sound]
К к	*К к*	k	Ы ы	*Ы ы*	y (rhythm)
Л л	*Л л*	l	Ь ь	*Ь ь*	[apostrophe; palatalizes preceding consonant]
М м	*М м*	m			
Н н	*Н н*	n	Э э	*Э э*	e (elder)
О о	*О о*	o	Ю ю	*Ю ю*	yu (union)
П п	*П п*	p	Я я	*Я я*	ya (yard)

Symbol	Name	Quantity	Relations
cd	candela	luminous intensity	basic unit
kg	kilogram	mass	basic unit
lm	lumen	luminous flux	$cd \cdot sr$
lx	lux	illumination	$lm \cdot m^{-2}$
m	meter	length	basic unit
mol	mole	amount of substance	basic unit
rad	radian	plane angle	supplementary unit
s	second	time	basic unit
sr	steradian	solid angle	supplementary unit
A	ampere	electric current	basic unit
C	coulomb	electric charge	$A \cdot s$
F	farad	electric capacitance	$A \cdot s \cdot V^{-1}$
H	henry	electric inductance	$V \cdot s \cdot A^{-1}$
Hz	hertz	frequency	s^{-1}
J	joule	work, energy	$N \cdot m$
K	kelvin	temperature degree	basic unit
N	newton	force	$kg \cdot m \cdot s^{-2}$
Pa	pascal	pressure, stress	$N \cdot m^{-2}$
S	siemens	electric conductance	Ω^{-1}
T	tesla	magnetic flux density	$Wb \cdot m^{-2}$
V	volt	voltage, electromotive force	$W \cdot A^{-1}$
W	watt	power	$J \cdot s^{-1}$
Wb	weber	magnetic flux	$V \cdot s$
Ω	ohm	electric resistance	$V \cdot A^{-1}$

B.18   DECIMAL MULTIPLES AND FRACTIONS OF UNITS

Factor	Prefix	Symbol	Factor	Prefix	Symbol
10^1	deka	D*	10^{-1}	deci	d
10^2	hecto	h	10^{-2}	centi	c
10^3	kilo	k	10^{-3}	milli	m
10^6	mega	M	10^{-6}	micro	μ
10^9	giga	G	10^{-9}	nano	n
10^{12}	tera	T	10^{-12}	pico	p
10^{15}	femta	F	10^{-15}	femto	f
10^{18}	atta	A	10^{-18}	atto	a

*In some literature, the symbol "da" is used for deka

Symbol*	Name	Quantity	Relations
deg	degree	plane angle	supplementary unit
ft	foot	length	basic unit
g	standard gravity	acceleration	$32.174 \text{ ft} \cdot \text{sec}^{-2}$
hp	horsepower	power	$550 \text{ lbf} \cdot \text{ft} \cdot \text{sec}^{-1}$
lb	pound-mass	mass	$\text{lbf} \cdot \text{g}^{-1}$
lbf	pound-force	force	basic unit
pd	poundal	force	$\text{lb} \cdot \text{ft} \cdot \text{s}^{-2}$
sec	second	time	basic unit
sl	slug	mass	$\text{lbf} \cdot \text{ft}^{-1} \cdot \text{sec}^2$
Btu	British thermal unit	work, energy	$778.128 \text{ ft} \cdot \text{lbf}$
°F	degree Fahrenheit	temperature	basic unit

*For A, C, F, ... refer to Sec. B.17.

B.20 METRIC SYSTEM OF UNITS (MKS SYSTEM)

Symbol*	Name	Quantity	Relations
cal	calorie	work, energy	$0.42665 \text{ m} \cdot \text{kgf}$
deg	degree	plane angle	supplementary unit
g	standard gravity	acceleration	$9.80665 \text{ m} \cdot \text{sec}^{-2}$
hp	horsepower	power	$75 \text{ kgf} \cdot \text{m} \cdot \text{sec}^{-1}$
kg	kilogram-mass	mass	$\text{kgf} \cdot \text{g}^{-1}$
kgf	kilogram-force	force	basic unit
m	meter	length	basic unit
sec	second	time	basic unit
°C	degree Celsius	temperature	basic unit

*For A, C, F, ... refer to Sec. B.17.

Appendix C
REFERENCES AND BIBLIOGRAPHY

C.1 Algebra

1.01   Aitken, A. C.: "Determinants and Matrices," 8th ed., Interscience, New York, 1956.
1.02   Ayres, F., Jr.: "Matrices," McGraw-Hill, New York, 1962.
1.03   Birkhoff, G., and S. MacLane: "A Survey of Modern Algebra," Macmillan, New York, 1941.
1.04   Ferrar, W. L.: "Algebra," Oxford, London, 1941.
1.05   Frazer, R. A., W. J. Duncan, and A. R. Collar: "Elementary Matrices," Cambridge, London, 1938.
1.06   Lipschutz, S.: "Linear Algebra," McGraw-Hill, New York, 1968.
1.07   Middlemiss, R. R.: "College Algebra," McGraw-Hill, New York, 1952.
1.08   Smirnov, V. I.: "Linear Algebra and Group Theory," McGraw-Hill, New York, 1961.
1.09   Upensky, J. V.: "Theory of Equations," McGraw-Hill, New York, 1948.
1.10   Zurmühl, R.: "Matrizen," 3d ed., Springer, Berlin, 1961.

C.2 Geometry

2.01   Bronstein, I. N., and K. A. Semendjajew: "Taschenbuch der Mathematik," 7th ed., Deutsch, Zurich, 1967
2.02   Court, N. A.: "College Geometry," 2d ed., Barnes & Noble, New York, 1952.
2.03   Coxeter, H. S. M.: "Introduction to Geometry," Wiley, New York, 1961.
2.04   Klein, F.: "Famous Problems of Elementary Geometry," 2d ed., Dover, New York, 1956.

C.3 Trigonometry

3.01   Kells, L. M., W. F. Kern, and J. R. Bland: "Plane and Spherical Trigonometry," 3d ed., McGraw-Hill, New York, 1951.
3.02   Loney, S. L.: "Plane Trigonometry," Cambridge, London, 1900.

C.4 Plane Analytic Geometry

4.01   Carmichael, R. D., and E. R. Smith: "Mathematical Tables and Formulas," Dover, New York, 1962.
4.02   Cell, J. W.: "Analytic Geometry," 3d ed., Wiley, New York, 1954.
4.03   Middlemiss, R. R.: "Analytic Geometry," 2d ed., McGraw-Hill, New York, 1955.
4.04   Oakley, C. O.: "Analytic Geometry," Barnes & Noble, New York, 1961.

C.5 Space Analytic Geometry

5.01   Albert, A.: "Solid Analytic Geometry," McGraw-Hill, New York, 1949.
5.02   Bartsch, H. J.: "Mathematische Formeln," 5th ed., Veb Fachbuchverlag, Leipzig, 1966.
5.03   Láska, W.: "Sammlung von Formeln der Mathematik," Vieweg, Braunschweig, 1894.
5.04   Tideström, S. M.: "Manuel de Base de L'ingénieur," Dunod, Paris, 1959.

C.6 Elementary Functions

6.01   Abramowitz, M., and I. A. Stegun: "Handbook of Mathematical Functions," National Bureau of Standards, Washington, D.C., 1964.
6.02   Hayashi, K.: "Fünfstellige Tafeln der Kreis- und Hyperbelfunktionen," De Gruyter, Berlin, 1955.

C.7 Differential Calculus

7.01   Courant, R.: "Differential and Integral Calculus," vol. 1., Interscience, New York, 1936.
7.02   Goursat, E.: "A Course in Mathematical Analysis," vol. 1., Dover, New York, 1959.
7.03   Guggenheim, H. W.: "Differential Geometry," McGraw-Hill, New York, 1963.
7.04   Vojtěch, J.: "Základy Matematiky," vol. 1., Jednota ČMF, Prague, 1946.
7.05   Yakolev, K. P.: "Handbook for Engineers," vol. 1., Pergamon, New York, 1965.

C.8 Infinite Series

8.01   Fort, T.: "Infinite Series," Oxford, New York, 1930.
8.02   Jolley, L. B. W.: "Summation of Series," 2d ed., Dover, New York, 1961.
8.03   Knopp, K.: "Theory and Application of Infinite Series," Hafner, New York, 1948.

C.9 Integral Calculus

9.01 Sokolnikoff, I. S.: "Advanced Calculus," McGraw-Hill, New York, 1939.
9.02 Taylor, A.: "Advanced Calculus," Ginn., Boston, 1955.
9.03 Thomas, G. B.: "Calculus and Analytic Geometry," Addison-Wesley, Reading, Mass., 1953.
9.04 Vojtěch, J.: "Základy Matematiky," vol. 2., Jednota ČMF, Prague, 1946.

C.10 Vector Analysis

10.01 Coffin, J. G.: "Vector Analysis," Wiley, New York, 1938.
10.02 Hawkins, G. A.: "Multilinear Analysis. . . ," Wiley, New York, 1963.
10.03 Margenau, H., and G. M. Murphy: "Mathematics of Physics and Chemistry," Van Nostrand, Princeton, N.J., 1943.
10.04 Whittaker, E. T., and G. N. Watson: "A Course of Modern Analysis," Macmillan, New York, 1944.

C.11 Functions of Complex Variable

11.01 Churchill, R. V.: "Complex Variables and Applications," 2d ed., McGraw-Hill, New York, 1960.
11.02 Goursat, E.: "A Course in Mathematical Analysis," vol. 2., part 1, Dover, New York, 1959.
11.03 Souders, M.: "Engineer's Companion," Wiley, New York, 1966.
11.04 Spiegel, M. R.: "Complex Variables," McGraw-Hill, New York, 1964.

C.12 Fourier Series

12.01 Churchill, R. V.: "Fourier Series and Boundary Value Problems," McGraw-Hill, New York, 1941.
12.02 Miller, K. S.: "Partial Differential Equations in Engineering Problems," Prentice-Hall, Englewood Cliffs, N.J., 1953.
12.03 Salvadori M. G., and R. J. Schwartz: "Differential Equations in Engineering Problems," Prentice-Hall, Englewood Cliffs, N.J., 1954.
12.04 Sokolnikoff, I. S., and R. M. Redheffer: "Mathematics of Physics and Modern Engineering," 2d ed., McGraw-Hill, New York, 1966.

C.13 Higher Transcendent Functions

13.01 Davis, H.: "Tables of Higher Mathematical Functions," Principia, Bloomington, Ind., 1935.
13.02 Erdelyi, A.: "Higher Transcendental Functions," 3 vols., McGraw-Hill, New York, 1953–1954.
13.03 Janke, E., and F. Emde: "Tables of Functions," Dover, New York, 1945.

C.14 Ordinary Differential Equations

14.01 Abramowitz, M., and I. A. Stegun: "Handbook of Mathematical Functions," National Bureau of Standards, Washington, D.C., 1964.
14.02 Dwight, H. B.: "Tables of Integrals and Other Mathematical Data," 4th ed., Macmillan, New York, 1967.
14.03 Frank, R., and R. V. Mises: "Die Differential und Integral Gleichungen Der Mechanik und Physik," 2 vols., Dover, New York, 1961.
14.04 Ince, E. L.: "Ordinary Differential Equations," Dover, New York, 1956.
14.05 Kamke, E.: "Differential Gleichungen, Lösungsmethoden und Lösungen," 3d ed., Chelsea, New York, 1948.
14.06 MacLachlan, N. W.: "Bessel Functions for Engineers," Oxford, London, 1946.
14.07 Madelung, E.: "Die Mathematischen Hilfsmittel des Physikers," 7th ed., Springer, Berlin, 1964.
14.08 Sauer, R., and I. Szabo: "Mathematische Hilfsmittel des Ingenieurs," vol. I., Springer, Berlin, 1967.
14.09 Szabo, I.: "Hütte Mathematische Formeln und Tafeln," Ernst, Berlin, 1959.
14.10 Szegö, G.: "Orthogonal Polynomials," American Mathematics Society, New York, 1939.

C.15 Partial Differential Equations

15.01 Korn, G. A., and T. M. Korn: "Mathematical Handbook for Scientists and Engineers," 2d ed., McGraw-Hill, New York, 1968.

15.02 Sommerfeld, A.: "Partial Differential Equations of Physics," Academic, New York, 1949.

C.16 Laplace Transforms

16.01 Churchill, R. V.: "Operational Mathematics," 3d ed., McGraw-Hill, New York, 1971.

16.02 Doetsch, G.: "Handbuch der Laplace-Transformation," 3 vols., Birkhäuser, Basel, 1950–1956.

16.03 Nixon, F.: "Handbook of Laplace Transforms," Prentice-Hall, Englewood Cliffs, N.J., 1960.

C.17 Numerical Methods

17.01 Hamming, R.: "Numerical Methods for Scientists and Engineers," 2d ed., McGraw-Hill, New York, 1973.

17.02 Hildebrand, F. B.: "Introduction to Numerical Analysis," McGraw-Hill, New York, 1956.

17.03 Salvadori, M. G., and M. L. Baron: "Numerical Methods in Engineering," 2d ed., Prentice-Hall, Englewood Cliffs, N.J., 1961.

17.04 Sanden, V.: "Praktische Mathematik," Teubner, Leipzig, 1953.

17.05 Scarborough, J. B.: "Numerical Mathematical Analysis," 3d ed., Johns Hopkins Press, Baltimore, 1955.

17.06 Yakolev, K. P.: "Handbook for Engineers," vol. 1., Pergamon, New York, 1965.

C.18 Probability and Statistics

18.01 Burington, R. S., and D. C. May: "Handbook of Probability and Statistics with Tables," Handbook Publishers, Sandusky, Ohio, 1953.

18.02 Cramer, H.: "Mathematical Methods of Statistics," Princeton University Press, Princeton, N.J., 1951.

18.03 Hald, A.: "Statistical Theory and Engineering Applications," Wiley, New York, 1952.

18.04 Korn, G. A., and T. M. Korn: "Mathematical Handbook for Scientists and Engineers," 2d ed., McGraw-Hill, New York, 1968.

18.05 Spiegel, M. R.: "Statistics," McGraw-Hill, New York, 1961.

C.19 Tables of Indefinite Integrals

19.01 Bois, G. P.: "Tables of Indefinite Integrals," Dover, New York, 1964.

19.02 Bronstein, I., and K. Semendjajev: "Pocketbook of Mathematics," 6th ed., Soviet Government Press, Moscow, 1956.

19.03 Burington, R. S.: "Handbook of Mathematical Tables and Formulas," 4th ed., McGraw-Hill, New York, 1965.

19.04 Dwight, H. B.: "Tables of Integrals and Other Mathematical Data," 4th ed., Macmillan, New York, 1961.

19.05 Gröbner, W. and N. Hofreiter: "Integraltafeln," vol. 1, 4th ed., Springer, Vienna, 1965.

19.06 Hirsch, M.: "Integral Tables," Baynes, London, 1823.

19.07 Meyer Zur Capellen, W.: "Integraltafeln," Springer, Berlin, 1950.

19.08 Spiegel, M. R.: "Mathematical Handbook of Formulas and Tables," McGraw-Hill, New York, 1968.

C.20 Tables of Definite Integrals

20.01 de Haan, B.: "Nouvelles Tables d'Intégrales Définies," Hafner, New York, 1957.

20.02 Gradshteyn, I. S., and I. M. Ryzhik: "Table of Integrals, Series and Products," 4th ed., Academic, New York, 1965.

20:03 Gröbner, W., and N. Hofreiter: "Integraltafeln," vol. 2, 4th ed., Springer, Vienna, 1965.

INDEX

INDEX

References are made to section numbers for the text material. Numbers preceded by the letters A and B refer to the Appendixes. In the designation of units and systems of units the following abbreviations are used:

avd = avoirdupois   thm = tnermochemical
liq = liquid   try = troy(apothecary)
nat = nautical   FPS = English system
phs = physical   IST = international steam tables
stu = statute   MKS = metric system
tec = technical   SI = international system

Conversion Factors

Selected Unit	Conversion Factor SI Units	Conversion Factor FPS Units
Acre	4.046 856*(+03) m²	4.356 000*(+04) ft²
Acre-foot	1.233 482 (+03) m³	4.356 000*(+04) ft³
Angstrom	1.000 000 (−10) m	3.937 008 (−09) in
Astronomical unit	1.495 980 (+11) m	4.908 807 (+11) ft
Atmosphere (phs)	1.013 250 (+05) Pa	1.469 393 (+01) lbf · in⁻²
Atmosphere (tec)	9.806 650*(+04) Pa	1.422 334 (+01) lbf · in⁻²
Bar	1.000 000*(+05) Pa	1.450 377 (+01) lbf · in⁻²
British thermal unit (IST)	1.055 056 (+03) J	7.781 693 (+02) ft · lbf
British thermal unit (mean)	1.055 870 (+03) J	7.787 697 (+02) ft · lbf
British thermal unit (thm)	1.054 350 (+03) J	7.776 488 (+02) ft · lbf
Calorie (IST)	4.186 800*(+00) J	3.088 026 (+00) ft · lbf
Calorie (mean)	4.190 020 (+00) J	3.090 400 (+00) ft · lbf
Calorie (thm)	4.184 000*(+00) J	3.085 959 (+00) ft · lbf
Chain (gunter)	2.011 680*(+01) m	6.600 000*(+01) ft
Chain (ramsden)	3.048 000*(+01) m	1.000 000*(+02) ft
Circular inch	5.067 075 (+00) cm²	7.854 000 (−01) in²
Circular mil	5.067 075 (−06) cm²	7.854 000 (−07) in²
Degree (angle)	1.745 329 (−02) rad	1.745 329 (−02) rad
Degree Celsius (°C)	1.000 000*(+00) K	1.800 000*(+00) °F
Degree Fahrenheit (°F)	5.555 556 (−01) K	1.000 000*(+00) °F
Kelvin (K)	1.000 000*(+00) K	1.800 000*(+00) °F
Dyne	1.000 000*(−05) N	2.248 089 (−06) lbf
Electronvolt	1.602 189 (−19) J	1.181 714 (−19) ft · lbf
Erg	1.000 000*(−07) J	7.375 621 (−08) ft · lbf
Fathom	1.828 800*(+00) m	6.000 000*(+00) ft
Foot	3.048 000*(−01) m	1.000 000*(+00) ft
Foot (U.S. survey)	3.048 006 (−01) m	1.000 002*(+00) ft
Foot²	9.290 304*(−02) m²	1.000 000*(+00) ft²
Foot³	2.831 685 (−02) m³	1.000 000*(+00) ft³
Foot · pound-force	1.355 818 (+00) J	1.000 000*(+00) ft · lbf
Foot · pound-force/hour	3.766 161 (−04) W	5.050 505 (−07) hp
Furlong	2.011 680*(+02) m	6.600 000*(+02) ft
Gallon (U.S. liq)	3.785 412 (−03) m³	1.336 806 (−01) ft³
Gallon (U.S. liq)/hour	1.051 503 (−06) m³ · s⁻¹	3.713 350 (−05) ft³ · s⁻¹
Grain	6.479 891*(−05) kg	1.428 571 (−04) lb
Horsepower (FPS)	7.456 999 (+02) W	5.500 000*(+02) ft · lbf · s⁻¹
Horsepower (MKS)	7.354 990 (+02) W	9.863 204 (−01) hp
Inch	2.540 000*(−02) m	8.333 333 (−02) ft
Inch²	6.451 600*(−04) m²	6.944 444 (−03) ft²
Inch³	1.638 706 (−05) m³	5.787 037 (−04) ft³
Joule	1.000 000*(+00) J	7.375 622 (−01) ft · lbf
Kilogram	1.000 000*(+00) kg	2.204 623 (+00) lb
Kilogram/meter³	1.000 000*(+00) kg · m⁻³	6.242 794 (−02) lb · ft⁻³
Kilogram-force	9.806 650*(+00) N	2.204 623 (+00) lbf
Kilogram-force/centimeter²	9.806 650*(+04) Pa	1.422 334 (+01) lbf · in⁻²

*Asterisk after the last digit indicates that the conversion factor is exact by the definition of National Bureau of Standards. For symbols of units see Appendix B, for abbreviations see p. 381.